DEVELOPMENTS IN **SEDIMENTOLOGY** 57

CYCLIC DEVELOPMENT OF SEDIMENTARY BASINS

DEVELOPMENTS IN **SEDIMENTOLOGY** 57

CYCLIC DEVELOPMENT OF SEDIMENTARY BASINS

JANNES M. MABESOONE AND VIRGÍNIO H. NEUMANN
C/O LABORATÓRIO GEOLOGIA SEDIMENTAR
DEPARTAMENTO GEOLOGIA
UNIVERSIDADE FEDERAL DE PERNAMBUCO
AV. ACADÊMICO HÉLIO RAMOS, S/N
CIDADE UNIVERSITÃRIA
50741-530 RECIFE (PE)
BRAZIL

SERIES EDITOR: A.J. VAN LOON

2005

ELSEVIER

Amsterdam – Boston – Heidelberg – London – New York – Oxford
Paris – San Diego – San Francisco – Singapore – Sydney – Tokyo

ELSEVIER B.V.
Radarweg 29
P.O. Box 211, 1000 AE
Amsterdam
The Netherlands

ELSEVIER Inc.
525 B Street, Suite 1900
San Diego, CA 92101-4495
USA

ELSEVIER Ltd
The Boulevard,
Langford Lane
Kidlington, Oxford OX5
1GB UK

ELSEVIER Ltd.
84 Theobalds Road
London WC1X 8RR
UK

First edition 2005

Library of Congress Cataloging in Publication Data
A catalog record is available from the Library of Congress.

British Library Cataloguing in Publication Data
A catalogue record is available from the British Library.

ISBN: 0-444-52070-8

Series ISSN: 0070-4571

∞ The paper used in this publication meets the requirements of ANSI/NISO Z39.48-1992 (Permanence of Paper).

Transferred to digital print 2007
Printed and bound by CPI Antony Rowe, Eastbourne

PREFACE

Certain topics within geology have been subjects of fundamental controversy, and they have raised heated discussions for a long time – and some of them still do so. A well-known example is continental drift, phenomenon which is now no longer subject to debates, and which is currently even considered to be one of the most obvious expressions of a process that is fundamental for the geological evolution of the Earth: sea-floor spreading. The "father" of the theory of continental drift, Alfred Wegener, was – through his father-in-law, the well-known climatologist Wladimir Köppen – also involved in the acceptance by the earth-scientific community of another controversial item: astronomical parameters as the main factor determining the alternation of ice ages and interglacials during the Pleistocene.

This alternation of glacials and interglacials was the first earth-scientific phenomenon that has been recognized as a rhythmic alternation due to astronomical factors at a large scale. Rhythmic sedimentation on a small scale had, obviously, been recognized much earlier already, the simplest form being the herringbone structures in deposits formed under the influence of the tides, which are due to the interaction between the gravity fields of Earth, Sun, and Moon. In the meantime many more types of rhythmic or cyclic sedimentation have been found, and in the past decade it has become clear that astronomical factors determine many more sedimentary cycles than has been assumed in the past. The phenomenon of "orbital forcing" is now a hot topic among sedimentologists.

The increased understanding of the great influence of astronomical parameters on the geological record makes one wonder whether these parameters have also affected the earth's history in a way not known thus far. There have been several hypotheses in the past in this context. Stratigraphic thinking in North America, for instance, has been dominated in the early part of the twentieth century by theories of large-scale cycles of sedimentation worldwide. Other geologists, however, doubted about this synchroneity, emphasizing that the greater stratigraphical units, such as systems, should be defined through type sections and faunal successions rather than by major unconformities that may be due to cyclic processes. This line of thinking prevailed until about the 1970s. The concept of large-scale sedimentation cycles – triggered by tectonic pulses – has, however, never been abandoned by Russian earth scientists (Ruchin, 1958, and later authors). The cyclic repetition of geologic phenomena on a large scale, with cycles of about 200 Ma duration as the most important, due to tectonic processes finding their origin in the earth's interior, and probably triggered by orbital forcing, is too obvious to be a mere coincidence.

It is the objective of the present book to assemble and integrate the facts, principles, and hypotheses upon the cyclicity of the earth's processes and phenomena as far as they are concerned with the development of sedimentary basins and their lithic infillings. And we hope to convince the reader that the cyclicity found in basin formation is closely related to cyclicity in the earth's interior, which has also numerous other consequences: continental assembly and breaking up, opening and closing of oceans, as well as features, such as accentuated and attenuated climate zones, glaciations, and volcanism. Their bearing upon sediment accumulation and basin development is outstanding; it was so in the Mesoproterozoic and maybe even earlier.

It could not be avoided that the style of presentation varies from chapter to chapter due to the provenance of co-authors from different parts of the world. Furthermore, the use of specific technical terms may differ from country to country, but it has been tried to avoid as much as possible such differences. With respect to definitions of not much-used terms, those given by Bates and Jackson (1987) have been followed.

The editors are indebted to many individuals without whose cooperation the presentation of this volume would have been impossible. They gratefully express their thanks in the first place to the editor-in-chief of the book series "Developments in Sedimentology" who encouraged the publication of a volume about a controversial subject that never was considered in this way before. Secondly, they are also very much indebted to the authors of several chapters that deal with long-term cyclic sequences in different sedimentary basins of the world.

Furthermore, the editors like to acknowledge the technical help given to them by Mr. Jonas Melo who improved the preparation of the compuscript of this book, and also to Mr. José Antônio Barbosa who did part of the drafting for many chapters.

J.M. Mabesoone
V.H. Neumann

TABLE OF CONTENTS

LIST OF CONTRIBUTORS

Abdulkader M. Abed: Department of Geology and Mineralogy, University of Jordan, AMMAN, Jordan

Ananya Biswas: Department of Mining and Geology, Bengal Engineering and Science University, HOWRAH, India

Pradip K. Dasgupta: Department of Geology, University of Calcutta, CALCUTTA, India

Frank R. Ettensohn: University of Kentucky, Department of Geological Sciences, LEXINGTON (Kentucky), U.S.A.

V.V. Golozubov: Far East Geological Institute, Far Eastern Branch Russian Academy of Sciences, VLADIVOSTOK, Russia

Vadim M. Golubev: VNIIOkeangeologiya, ST. PETERSBURG, Russia

Mikhail V. Goroshko: Institute of Tectonics and Geophysics, Far Eastern Institute of Geology and Geophysics, KHABAROVSK, Russia

I.V. Kemkin: Far East Geological Institute, Far Eastern Branch Russian Academy of Sciences, VLADIVOSTOK, Russia

A.I. Khanchuk: Far East Geological Institute, Far Eastern Branch Russian Academy of Sciences, VLADIVOSTOK, Russia

M.K. Kos'ko: VNIIOkeangeologiya, ST. PETERSBURG, Russia

A.J. van Loon: Faculty of Earth Sciences, Univeristy of Silesia, SOSNOWIEC, Poland

Jannes M. Mabesoone: Laboratory of Sedimentary Geology, Department of Geology, Federal University of Pernambuco, RECIFE (Pernambuco), Brazil

Pavel V. Markevich: Far East Geological Institute, Far Eastern Branch Russian Academy of Sciences, VLADIVOSTOK, Russia

Rajesh Mukherjee: Department of Geology, University of Calcutta, CALCUTTA, India

Virgínio H. Neumann: Laboratory of Sedimentary Geology, Department of Geology, Federal University of Pernambuco, RECIFE (Pernambuco), Brazil

Davor Pavelić : Faculty of Mining, Geology and Petroleum Exploration, ZAGREB, Croatia

A.N. Philippov: Far East Geological Institute, Far Eastern Branch Russian Academy of Sciences, VLADIVOSTOK, Russia

S.A. Shorokova: Far East State Technical University, VLADIVOSTOK, Russia
Oleg I. Suprunenko: VNIIOkeangeologiya, ST. PETERSBURG, Russia
Y.D. Zakharov: Far East Geological Institute, Far Eastern Branch Russian Academy of Sciences, VLADIVOSTOK, Russia

1. LONG-TERM CYCLICITIES: INTRODUCTION

J.M. MABESOONE, F.R. ETTENSOHN, AND A.J. VAN LOON,
AND V.H. NEUMANN

One of the most intriguing characteristics of the geologic record is the presence of cycles with different periodicities. Cyclic sediments show a repetition of rock units in a particular order within a sedimentary succession. Cycles of a specific nature may occur within other cycles, thus reflecting repetition processes of various nature, duration and causes. The simplest form of repetition involves only two components. At the lower end of the scale, these are alternating laminae of, for instance, fine sand and clay. At the other end of the scale, repetitions can span systems or even longer intervals, involving the development of sedimentary basins. Von Bubnoff (1948) and Sloss (1964) referred already to such cycles of the order of geological systems (or more), which are mutually separated by regional unconformities.

Cycles with external causes (i.e. conditions or processes outside the sedimentary basin) have been designated as "allocyclic", whereas those with causes internal to the system are called "autocyclic" (see Beerbower, 1964). In the last half century, an increasing number of long-term (0.5–1.0 Ga) allocyclic cycles have been recognized. They have largely been ascribed to tectonic causes outside the systems under study. Tectonic-sedimentary cycles, with periodicities of about 0.1–0.2 Ga, as suggested by Mabesoone (1988) – inspired by the work of Sloss and Speed (1974) – seem to be extremely important for the origin and development of sedimentary basins, but have – remarkably enough – raised relatively little interest until recently. However, their recognition and precise reconstruction become increasingly difficult with age. Not only becomes the geologic record scarcer with time, but the older the rocks are, the more they have been affected by all kinds of geological processes. Their original characteristics thus become increasingly difficult to unravel when it comes to more and more remote periods.

Some aspects of this longer term cyclicity have been dealt with by several investigators, indeed, but their considerations have never been brought together in special volumes. The available data are thus scattered over works on other themes, such as basin analysis (see Miall, 1984) and cyclic stratigraphy (see Einsele et al., 1991; Miall, 1997). It is the intention of the present book to give an overview of these long-term cycles, and to reconstruct their origin, in order to find out in how far large-scale and long-term cyclicity has determined basin development.

Cyclic Development of Sedimentary Basins
J.M. Mabesoone and V.H. Neumann, editors.

Chiefly large-scale and long-term cycles are taken into account in the present book. Cycles and rhythms of smaller scales, especially those occurring in specific sedimentary depositional environments have been dealt with extensively by Einsele et al. (1991) and Miall (1997). However, in many of the following chapters these smaller scale cycles have also been considered in the description of various sedimentary basins.

1.1 TYPES OF LONG-TERM CYCLES

Because the various types of large-scale and long-term cyclicity were recognized only one by one in the course of geological investigations, there is no logical or consistent terminology for these cycles. We will adhere in the present book to the terms that are most commonly used in this context, even though this terminology may confuse the reader a bit in the beginning. We consider this the best approach, however, because the introduction of new terms might result in more confusion if previous literature is to be compared with the present book.

The most commonly used terms are the following (in order of duration):

- The *chelogenous* (also called chelogenic) *cycles* (Sutton, 1963) that can be distinguished in continental evolution and craton formation (>1 Ga cycles).
- The so-called *first-order cycles* (also called global supercontinent cycles) that have a regional to global distribution in space and time, and that represent the stratigraphic signatures resulting from the interaction of tectonic, eustatic, sedimentation, and climatic processes (Vail et al., 1991), and that are caused by crustal extension, thermal cooling, flexural loading, and changes in ocean basin volume; these cycles are in the 200–500 Ma range.
- The well-known and generally accepted *Wilson cycles* (Dewey and Burke, 1974) that present the successive recurrence of continental spreading and convergence induced by plate tectonics, with periods that are generally in the 100 Ma range.
- The so-called *second-order cycles* (also called supercycles) that, like the first-order cycles, have a regional to global distribution in space and time, and that represent the stratigraphic signatures resulting from the interaction of tectonic, eustatic, sedimentation, and climatic processes (Vail et al., 1991). They are, however, generated by continental-scale thermal processes in the earth's mantle and by plate kinematics. The cycles, which are in the 10–100 Ma range, include the tectonic-sedimentary episodes of Sloss and Speed (1974) and Mabesoone (1988) as well as the so-called basinal cycles of Ruchin (1958).
- In certain cases, *third-order cycles*, of regional to local significance, are distinguished, chiefly when the concerning basin history spans only one

first-order episodic event. These third-order cycles, which are in the 2–10 Ma range, are characterized by basement movement caused by regional plate kinematics. They include the paralic cycles of Ruchin (1958).

– *Milankovitch cycles* (global cycles generated by orbital forcing) are typically in the 0.01–2 Ma range. They include the lacustrine cycles of Ruchin (1958).

It is obvious that most of the above cycles or rhythms correspond roughly to those distinguished by Ruchin (1958), although the latter are chiefly concerned with sedimentary infillings, and their depositional environments. He distinguished:

(1) basinal rhythms, dealing with the origin and further development of whole basins, generally with marine and continental sediment systems, in continuously subsiding basins;
(2) paralic rhythms, i.e. cycles in the transitional zone between continents and seas or oceans, in the same basin (coal cyclothems are a classical example);
(3) lacustrine rhythms, occurring preferentially in continental successions (molasse deposits are common representatives).

Another subdivision of cycles was presented by Haq et al. (1988) on the basis of relative sea-level movements. Haq and his co-workers distinguished the Mesozoic and the Cenozoic:

(a) megacycles [T(ejas), Z(uni), A(bsaroka)], which may be subdivided (TB, TA – Tejas B and A; UZA – Upper Zuni A; LZB, LZA – Lower Zuni B and A; UAB, UAA – Upper Absaroka B and A)], with names after Sloss (1963; Fig. 3.1, Table 3.1). They represent tectonic-sedimentary intervals of some 100 Ma, and may be compared with the above-mentioned second-order cycles.
(b) Supercycles with a duration of the order of 10 Ma, which can be compared with the above-mentioned third-order cycles. They are subdivided into units (TB 1–3, TA 1–4, UZA 1–4, LZB 1–4, LZA 1–4, UAB 1–4, UAA 1–4) that may be combined into sets.

1.2 EARLY IDEAS

Earlier geologists clearly felt that the forces that had formed the earth's major mountain systems "slumbered" for long time, only to "awaken" suddenly at specific moments to start orogenesis with great intensity across widespread, though narrow, segments of the earth's crust. They found that many of these mountain-building events occurred simultaneously across very distant parts of the earth's crust, and that these events consisted of short, well-defined pulses of tectonic activity. In addition, it became clear that the intrusion of granitic bodies, which commonly follows phases of thrusting and folding, was also related with these pulses, but typically occurred in their later stages. Such observations mainly concerned Phanerozoic orogenes, but the question was already raised early whether Precambrian orogenetic phases would be characterized by similar features. Or would the pulses known from Phanerozoic orogenes be different from the Precambrian ones because of different conditions during the (early) Precambrian related to the cooling of the earth and the consolidation of its crust? Nowadays, it can be seen – thanks to radiometric dating – that, when periods of major orogenetic activity are plotted on a linear time scale, a general periodicity exists in the cyclicity of orogenetic events, even for different parts of the world: the phases of major tectonic activity typically lasted 50–80 Ma, and alternated with quiet periods that persisted for about 120 Ma (Fig. 1.1).

Remarkably, all of this had already been suggested before 1940, based on – then still highly imprecise – Ra/Pb-method datings. An example of such cyclic orogenetic activity has been described by Van der Vlerk and Kuenen (1948) from the Precambrian of Finland. They demonstrated in those rocks the presence of successive cycles, each one generally characterized by a "quiet" period of sedimentation, followed by folding and mountain building, and ending with a period of granitic intrusion. Their largest difficulty was how to correlate the cycles in one area with those from other parts of the country.

1.3 FURTHER DEVELOPMENT

The principle of cyclicity has been implicit in the plate-tectonic paradigm almost since its conception in the 1960s. Although the idea of cyclic tectonic states in the earth's crust dates back to at least the work by Stille (1924), attempts to explain this cyclicity clearly have come about only since the idea of crustal mobility had been accepted (e.g. Runcorn, 1962; Sutton, 1963; Fitch and Miller, 1965). Most of this work carried out approximately half a century ago focused on igneous, metamorphic, and deformational manifestations in predominantly Precambrian settings, but Wilson (1966) generated a more comprehensive explanation for this cyclicity, based on the recurrence of Phanerozoic oceanic basins. In contrast to earlier work, Wilson's explanation relied on evidence from fauna, sedimentation and paleogeography. It was so successful that Dewey and Burke (1974) later coined the term "Wilson cycle" for his explanation. Although subsequently refined and modified (Wilson, 1968, 1970, 1974), concepts behind the Wilson cycle still provide the basic foundation and framework for the analysis of foreland basins and related orogenic belts.

Wilson (1966, 1968) also clearly realized that distinct sedimentary regimes could be generated by the proposed cycle, but indicated that the resulting

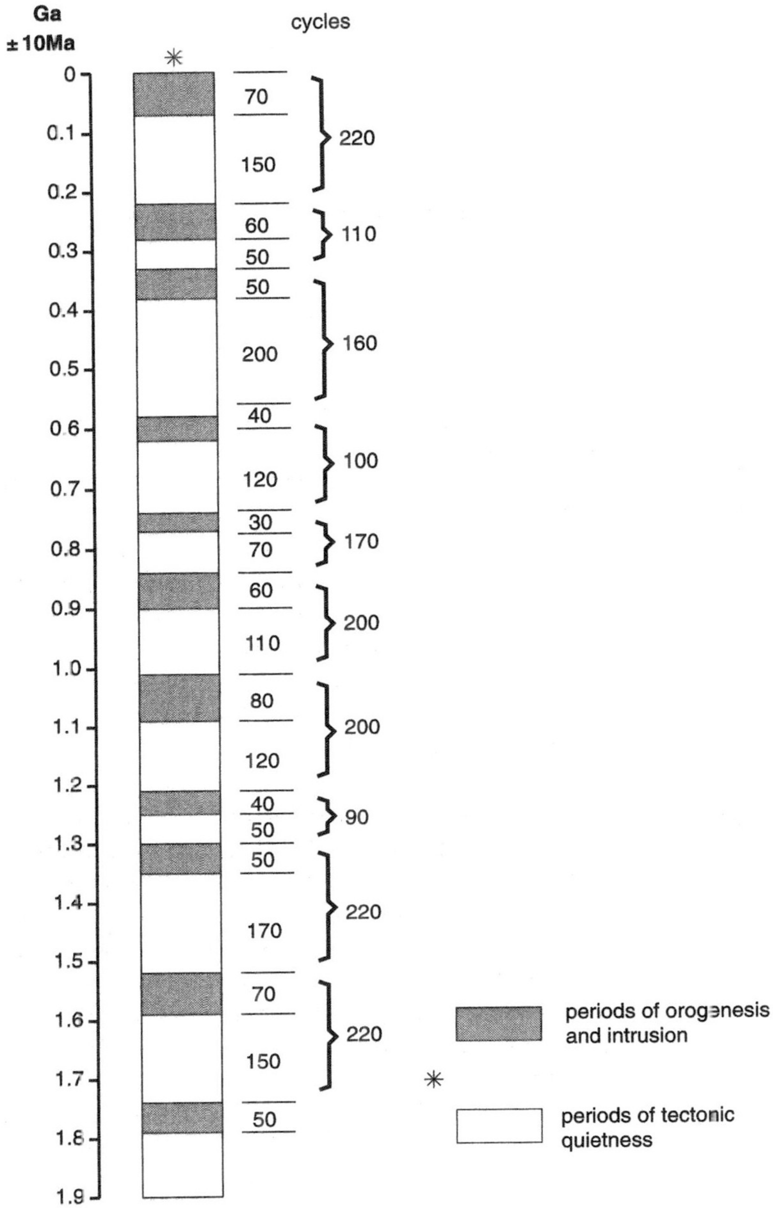

Fig. 1.1: Cycles in the development of the earth's crust, as suggested in 1940 (after Van der Vlerk and Kuenen, 1948). Right: periods in Ma and cycles.

sedimentary manifestation might change, depending upon the location of, and the stage within the cycle; his sedimentary characterizations were, however, very general.

In spite of much evidence for cyclicity in many aspects of the earth's geological history and a symposium on cyclic sedimentation organized by the Kansas Geological Survey (Merriam, 1966), the subject received little attention from sedimentologists and stratigraphers between the 1960s and 1970s. Much more attention was devoted to process–response models, authigenic processes, and the implications of plate tectonics for basin development, even though not all the necessary kinds of data were yet available at the time.

In Russia, however, the idea of cyclicity – which had been recognized there in the rhythmic development of sedimentary successions – had never been abandoned. The ideas had been published there (in Russian) by Ruchin as early as 1953. Ruchin (1958) emphasized that epeirogenic movements of the earth's crust were the underlying causes of such sedimentary rhythms, and that the determining forces that generated these "rhythms" were not the largest, undulating, epeirogenic movements, but rather the smaller pulses or phases that comprised them. He indicated, moreover, that large, long-term rhythms could only be recognized in the thickest sedimentary successions that affected entire basins, and he called them "first-order" rhythms. Although Ruchin also recognized the importance of shorter, smaller scale rhythms, his classification of them is too complex to detail here.

Previously to, and contemporaneously with the work of Wilson, stratigraphers like Sloss (1963) and Ham and Wilson (1967) had recognized large, craton-wide, stratigraphical successions separated by interregional unconformities. They suggested that some of these successions probably reflected changing tectonic behavior (Sloss, 1964; Ham and Wilson, 1967). Although some of these successions were interpreted to exhibit a distinct series of sedimentary and stratigraphic responses to apparent tectonic states, Sloss (1964) found no systematic repetition indicative of cyclicity. Similar "Sloss sequences" were subsequently recognized in Russia (Ronov et al., 1969), parts of Eastern Europe (Sloss, 1972, 1976), and Brazil (Soares et al., 1974). By 1974, Sloss and Speed ascribed these sequences to three types of global tectonic episodes (emergent, submergent, and oscillatory) with some systematic repetition. During parts of the Phanerozoic, several of the episodes appear to have been cyclic, although work by Mabesoone (1988) in Brazil suggests that a more perfect cyclicity may have existed, even as early as during the Proterozoic. Other workers, however, who have been able to acquire more detail from the geological record, have cast doubt on such cyclicity, suggesting a more irregular, even chaotic, development.

1.3.1 RELATIONSHIP WITH SEA-LEVEL FLUCTUATIONS

With Sloss's (1963) recognition of six large-scale sedimentary successions and their bounding regional unconformities on the North American platform, the first ideas of what is today called "sequence stratigraphy" were conceived. Although Sloss and his collaborators continued to work on this theme, the general acceptance of these ideas took longer, and the interest generated since then is still ongoing, in part reflected by this volume.

Only after the development of modern seismic stratigraphy and the publication and interpretation of its results by Vail et al. (1977), did the subject really raise the interest of numerous researchers. As a result, the number of studies on this topic increased greatly, though often without a truly critical analysis of the applied methods and their results. The study by Vail and his collaborators re-introduced ideas of cyclicity and the subdivision of larger cycles into smaller, nested cycles of different orders. Although they have suffered criticism, these very ideas still form the basis of many recent and current studies on the subject.

By the late 1970s, many of the interregional unconformities used by Sloss (1963) to define his sequences had been related to cycles of relative, global highstand and lowstand of sea level (Vail et al., 1977), the causes of which are still being debated. On the largest scale, however, there has been general agreement that the first-order cycles and many of the second-order cycles are related to global tectonics. Moreover, once it became apparent that the eustatic curves could be used as indirect indices of plate activity at a global scale (e.g. Fischer, 1981, 1984), the idea of the 500 Ma supercontinent cycle was born (Worsley et al., 1984, 1985, 1986; Nance et al., 1986). It is now understood that the occurrences of first-order, eustatic, sea-level highstands generally correspond to stages of continental dispersal, whereas first-order periods of sea-level lowstand reflect stages of continent re-aggregation into supercontinents like Rodinia and Pangea (Fischer, 1981, 1984).

The origin of the second-order cycles, the manifestations of which are particularly apparent in foreland basins and adjacent areas, was not so clear, however, although Vail et al. (1977) thought that they were probably related to specific orogenic movements. During such movements, higher rates of seafloor spreading and increased subduction were thought to have increased sea-level highstands (characterizing these second-order cycles), flooding continental and basin margins (Hays and Pitman, 1973; Vail et al., 1977). Later work by Pitman (1978) showed, however, that the volumetric effects of seafloor spreading are too gradual to be the principal cause of such eustatic changes, and Sloss (1984) showed that other eustatic causes do not last sufficiently long, and that they lack the amplitude necessary to form unconformities.

By the 1970s, the idea of orogenic influence via lithospheric flexuring from craton-margin orogenes provided another mechanism for understanding relative sea-level fluctuations in foreland basins and across large areas of the adjacent craton. Work carried out on the problem during the 1970s has actually shown that, the proximal and distal epeirogenic movements responsible for such fluctuations were largely controlled during non-glacial periods by subcrustal and supracrustal loading at and near active orogenic belts that may generate long-wavelength subsidence and tilting of the crust up to 2000 km away (Walcoot, 1970; Price, 1973; Cross and Pilger, 1978; Mitrovica et al., 1989; Gurnis, 1991, 1992, 1993; Kominz and Bond, 1991; Coakley and Gurnis, 1995; Moresi and Gurnis, 1996). This means that the effects of tectonic loading can influence relative sea-level fluctuations, as well as cratonic and foreland-basin sedimentary sequences across large parts of continents with converging margins, and that the nature and rate of resulting changes will increase substantially toward the convergent margins (Gurnis, 1992, 1993).

The extent of relative sea-level changes and the creation of new accommodation space associated with convergent margins and related cratonic flexuring are, obviously, best considered to be regional phenomena. If dispersing Paleozoic continents were largely surrounded by subduction zones as suggested by Scotese (1998), plate reorganization initiated on one margin may have conditioned nearly coeval tectonism and associated flexuring on other margins as well. Moreover, given the apparent interconnectedness of subduction zones during continental dispersion (e.g. Scotese, 1998), a series of more or less synchronous regionally tectonic events, as apparently existed during parts of the Silurian (Ettensohn and Brett, 1998) and Devonian (Johnson, 1971), may have combined to generate continental-to-global-scale epeirogenic movements that contributed to sedimentary signals of wide extent such as the unconformity-bound sequences of Sloss (1964, 1972, 1976).

1.3.2 RELATIONSHIP WITH TECTONIC CYCLES

"The major control of all sedimentation is tectonics. If there were no uplift, there would be soon no land to erode and hence no sediment to deposit. The question, therefore, is not whether or not tectonics controls sedimentation but how the control operates." Blatt et al. (1972, p. 591) thus expressed the fact that tectonics and sedimentation are closely linked, so that both subjects must be considered together.

As early stratigraphers classified sedimentary rocks into major units or systems separated by major unconformities, they regarded each unit as the record of a major cycle of transgression and regression of the sea (e.g. Suess, 1883–1908).

Because many other geologists doubted these ideas and ignored the evidence of major unconformities, defining major stratigraphic units instead by type section and faunal succession, the above idea was almost discarded.

Dana (1873, 1880), however, had already suggested that mountain chains pass through predictable stages in a sequence of events that appear to be cyclic. Much later, it was Ruchin (1958) who elaborated in detail about how the rhythmic epeirogenic movements determine not only the thickness and composition of the sedimentary sequences, but also their mode of accumulation, in which the smaller, pulsatory, crustal movements become very important. Hence, the origin of these cycles or rhythms was interpreted as depending on periodic uplift and downwarp of the earth's crust. As a consequence of these alternating movements, a repetition of erosion and sediment accumulation was suggested to have taken place. Such alternating movements of the crust, moreover, favor transgressions and regressions of the sea, which are also recorded in the character of deposited sediments.

Each of these longest, most robust, crustal cycles with durations of more than one geological period (first-order cycles) also seems to coincide with one succession of submergent and oscillatory-emergent tectonic-sedimentary episodes, first described by Sloss and Speed (1974). In a similar way, Bucher (1933) derived his own expression of a geological "law", namely that a typical orogenic cycle begins with a geosynclinal depression and ends with major uplift, and that the interval between these limiting events comprises two phases. The first phase is one of quiet sinking, only occasionally interrupted by uplifts; the second phase, in contrast, is one of crustal-folding events separated by ever diminishing periods of geosynclinal subsidence.

Hence, if any of the above ideas is valid, sedimentation must largely be controlled by tectonics at nearly every stage in a cycle through the rate of uplift and erosion, through the gradient in the transport area, through the subsidence rate in the receiving basin, and through changing pressures and temperatures in the sediment after deposition due to burial, folding and faulting.

1.4 CURRENT RESEARCH

More recent ideas about cyclic development during the formation of sedimentary basins have arisen from detailed studies of the geological history of the northeast Brazilian Borborema tectonic province (Mabesoone, 2001; Mabesoone and Neumann, 2002). The tectonic-sedimentary episodes of Sloss and Speed (1974), later revised by Mabesoone (1988), provided the framework for these studies, and it appeared possible, by using this framework, to trace cyclic development back to the beginning of the Mesoproterozoic.

In the Borborema Province (NE Brazil), two types of sedimentary-basin series can be recognized: (1) marginal-basin successions developed along boundaries where one province was amalgamated to neighboring cratons, or was separated from them; (2) intracontinental-basin successions developed in tectonically weak zones within the province proper (Fig. 1.2). Marginal basins formed at the end of the Paleoproterozoic during the amalgamation processes in a convergent setting; they were often reactivated later during periods of crustal extension. During the Cretaceous, marginal basins once again formed along the opening South Atlantic Ocean rift during a major episode of submergence. Several intervals of reactivation also occurred cyclically during the same episode, but not all basins were equally affected (Table 1.1). The intracontinental basins formed through intraplate stresses that caused pull-apart movements in weak crustal zones, commonly as transtensional basins in hybrid settings. These latter basins also formed cyclically from the beginning of the Mesoproterozoic onward, as rift basins or centroclines (as defined by Bates and Jackson, 1987), or as rift basins or syneclises during the Neoproterozoic and Phanerozoic (Table 1.2), depending upon the intensity of

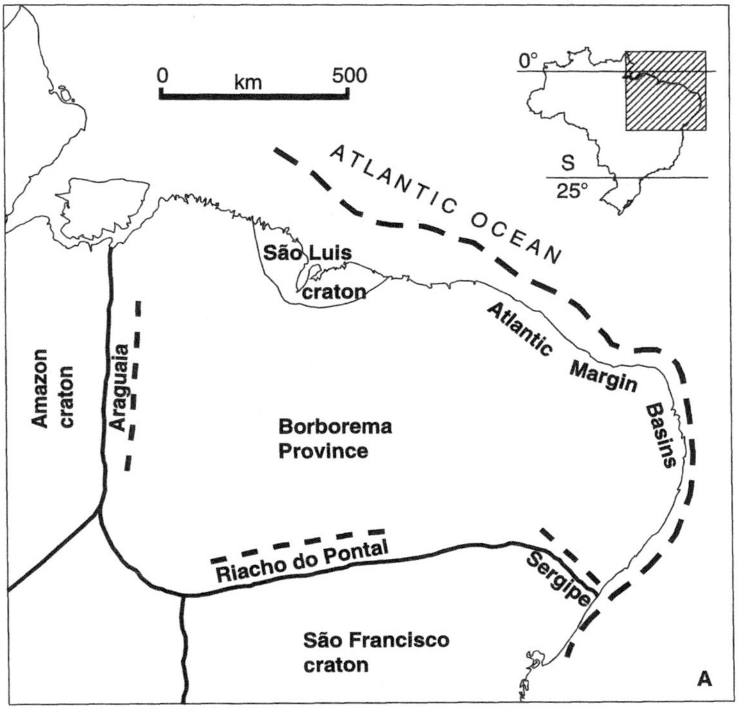

Fig. 1.2: The NE Brazilian Borborema tectonic province. (A) marginal-belt basins;

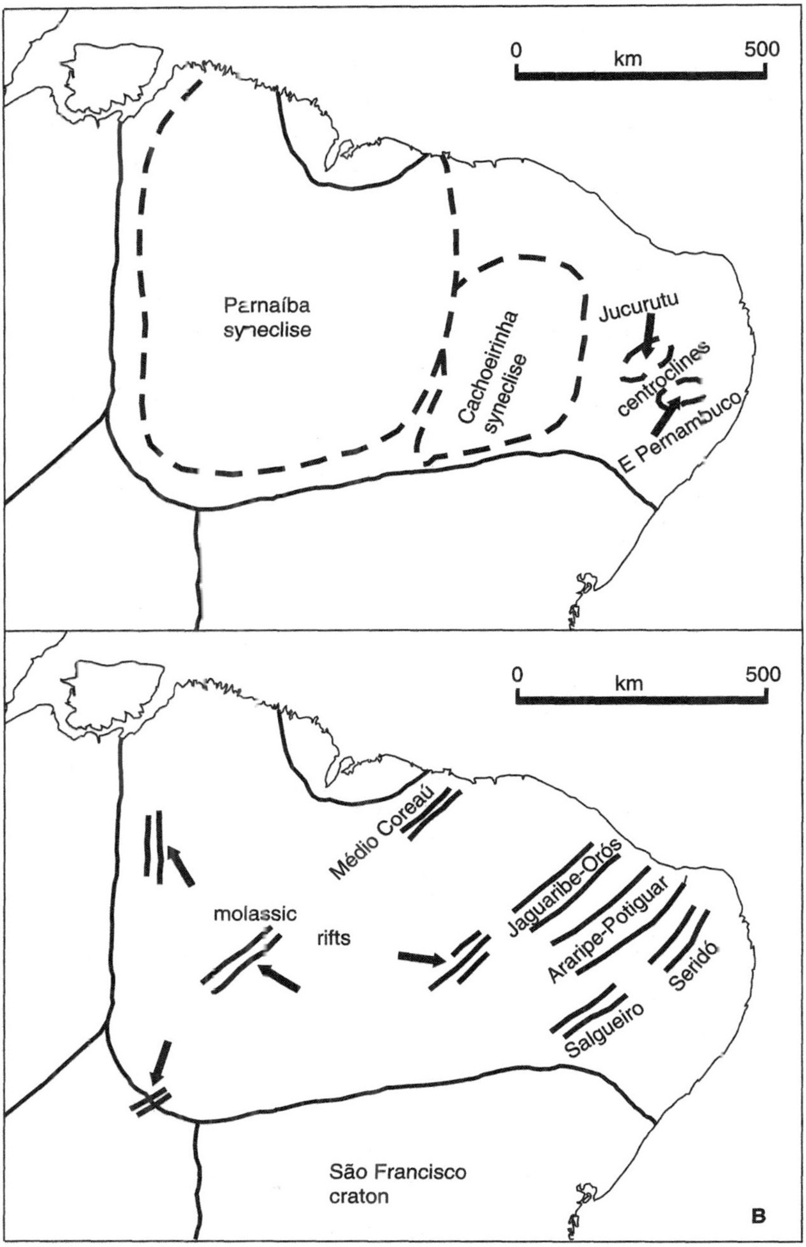

Fig. 1.2: Cont'd (B) intracontinental basins (syneclises and centroclines, after the definition by Bates and Jackson, 1987).

Table 1.1: NE Brazilian Borborema Province: marginal basin formation and reactivation.

Ga ±10 Ma	age: era	age: period	tectonic-sedimentary episode	sediment accumulation in marginal basins
0	phanerozoic	Quaternary-Tertiary	oscillatory	
0.08		Cretaceous-Jurassic	submergent	Atlantic margin
0.18		Triassic-Late Carboniferous	oscillatory	
0.28		Early Carboniferous-Devonian	submergent	–
0.38		Silurian-Late Ordovician	oscillatory	
0.48		Early Ordovician-Cambrian	submergent	Médio Coreaú, Gurupí, Araguaia, Sergipe, N-C-S Brasília
0.58	neoproterozoic	Vendian	oscillatory	
0.68		Late Cryogenian	submergent	Médio Coreaú, Sergipe, Riacho do Pontal, N-C-S Brasília
0.78		Early Cryogenian	oscillatory	
0.88		Tonian	submergent	Médio Coreaú, Gurupí, Araguaia, Sergipe, Riacho do Pontal, N-C-S Brasília
0.98	mesoproterozoic	Late Stenian	oscillatory	
1.08		Early Stenian	submergent	Araguaia, Sergipe, Riacho do Pontal
1.18		Late Ectasian	oscillatory	
1.28		Early Ectasian	submergent	N-S Brasília
1.38		Late Calymmian	oscillatory	
1.48		Early Calymmian	submergent	Araguaia
1.58		Late Statherian	oscillatory	
1.68		Early Statherian	submergent	Médio Coreaú, Araguaia, Sergipe, N-C-S Brasília
1.78 1.88		Orosirian	?	

the tectonic activity. Occasionally basins were reactivated, although renewed subsidence usually did not persist long enough to start new sedimentation phases.

In 1994, Petrobrás (Feijó, 1994a) integrated geological and biostratigraphical data from the Phanerozoic sedimentary basins of Brazil, obtained during more than 40 years of uninterrupted activity in the area. The resulting formal stratigraphic units followed, as much as possible, rules established by the Brazilian Code of Stratigraphic Nomenclature (Petri et al., 1986) and the Stratigraphic Lexicon of Brazil (Baptista et al., 1984). Depositional successions were distinguished on the basis of regional unconformities interpreted through seismic

Table 1.2: NE Brazilian Borborema tectonic province: intracontinental basin type formation and epochs of glaciations.

Ga ± 10 Ma	age		tectonic-sedimentary episode	orogenetic phase	basin type	other phenomena
	era	period				
0						
	phanerozoic	Quaternary-Tertiary	oscillatory-emergent		inversion	glaciation
0.08				alpine		
		Cretaceous-Jurassic	submergent		rift	
0.18						
		Triassic-Late Carboniferous	oscillatory-emergent		inversion	strong glaciation
0.28				variscan		
		Early Carboniferous-Devonian	submergent		syneclise	
0.38						
		Silurian-Late Ordovician	oscillatory-emergent		inversion	glaciation
0.48				caledonian		
		Early Ordovician-Cambrian	submergent		rift	
0.58						
	neoproterozoic	Vendian	oscillatory-emergent		inversion	strong glaciation
0.68				baikalian-avalonian		
		Late Cryogenian	submergent		syneclise	
0.78						
		Early Cryogenian	oscillatory-emergent		inversion	glaciation
0.88				?		
		Tonian	submergent		rift	
0.98						
	mesoproterozoic	Late Stenian	oscillatory-emergent		inversion	strong? glaciation
1.08				grenvillian		
		Early Stenian	submergent		centrocline	
1.18						
		Late Ectasian	oscillatory-emergent		inversion	
1.28				elzevitian		
		Early Ectasian	submergent		rift	
1.38						
		Late Calymmian	oscillatory-emergent		inversion	
1.48				kilarnean		
		Early Calymmian	submergent		centrocline	
1.58						
		Late Statherian	oscillatory-emergent		inversion	
1.68				hudsonian		
		Early Statherian	submergent		rift	
1.78						
	paleoproterozoic					

obs.: syneclise and centrocline, after definition by Bates and Jackson (1987).

records, biostratigraphic data, or visible breaks in the sedimentary record. Further definition of the successions, based on unconformities and their related conformities, followed models suggested by Vail et al. (1977, 1991) and Vail (1984) and detailed by Posamentier et al. (1988). The completed work shows successions that correspond more or less to the tectonic-sedimentary episodes proposed by Sloss and Speed (1974) and suggest, although not explicitly mentioned, a succession of cycles with durations of about 100 Ma. Six of such successions, called Silurian, Devonian, Carboniferous-Triassic, Jurassic, Cretaceous, and Cenozoic, were distinguished; these were confirmed by Milani

(2002) throughout Brazilian intracontinental sedimentary basins. Their presence and approximate coincidence with the episodes of Sloss and Speed (1974) on the Amazonian craton is part of the growing evidence for a succession of first- and second-order cycles that may have global significance.

2. CAUSES OF CYCLICITY

V.M. GOLUBEV

2.1 INTRODUCTION

Cyclic processes in the formation of sedimentary basins and the sedimentary cover of the Earth are the clue for the development of the crust. Such study is only possible through the understanding of the driving forces of geotectonic mechanisms. Neither the geosynclinal-platform theory that addresses cyclicity, nor the lithosphere plates theory, offer a complete and noncontradictory explanation of the planetary evolution. It is impossible to understand this evolution without considering the fact that the earth, being a celestial body, is an open system, and exists in accordance with laws governing the Solar System. Herein, an attempt is made to formulate the principles of geodynamics and geotectonics, acting forces and evolutionary factors in the universe as well as within the interior of the earth (Golubev, 1996a, b, c, 2000a, b, c, d).

2.2 THESES OF SYSTEM GEOTECTONICS

The basic theses of system geotectonics that fundamentally solve key problems of geology, generates protean concepts concerning cyclic processes involved in the evolution of the earth.

2.2.1 ABYSSAL HETEROGENEITY AND ECCENTRICITY OF THE EARTH

The division of the crust into lithospheric plates reveals the heterogeneity of the mantle (Golubev, 1992a, b) with respect to convection movements. The heterogeneity has resulted from resonance lunar–solar gravitational and magnetic influences that result from eclipses. Antipodal Pacific and African heterogeneities mark the displacement of the core of the earth towards the Pacific hemisphere (the possible origin of the Moon), due to cyclical influences of the lunar perigee. Antipodal Antarctic and Arctic heterogeneities have reflected the displacement of the core of our planet into the southern hemisphere due to cyclic

Cyclic Development of Sedimentary Basins
J.M. Mabesoone and V.H. Neumann, editors.

influences of the Sun (origin of the earth) in perihelion. The sphere of influence of these two pairs of heterogeneities has caused the segregation of seven main plates: Pacific, Australian and African, Antarctic and South American, Eurasian and North American. As a result of structural differentiation, the oceanic South Pacific and continental African-Eurasian hemispheres became separated, having resulted in displacement of the planet core.

2.2.2 SEGREGATION AND KINEMATICS OF LITHOSPHERIC PLATES

Lithospheric plates have no continuous boundaries, rotating strictly (following from Euler's theorem) under influence of the shear stresses as the earth rotates. Small (up to 10–15°) alternate turns of the plates in combination with a fragile plastic pattern of deformations between the plates provide integrity of the lithosphere, which undergoes stress only in mobile belts and rift zones. The direction of plate rotation changes as a result of inversion of planetary rotational stresses caused by the acceleration of rotation of the earth in the beginning of the geodynamic (geotectonic) cycle and deceleration of rotation in the middle of the cycle. Segregation of the plates has been determined by consolidation of the lithosphere and accompanied by breaking up of thinned plate margins into a number of small plates and platforms; the latter eventually forming oceanic plates. Central parts of the plates have also broken apart and formed platforms, which have eventually become continental in nature (except for the Pacific plate). Plate collisions are relieved by collisions of their platforms, resulting in the formation of fold belts. Collisions of smaller blocks of crust result in various kinds of intraplate tectonics, different in continents than in the ocean.

2.2.3 GEOTECTONIC BELTS AND MANTLE CONVECTION

Lithospheric plates are separated by transoceanic and transcontinental geotectonic belts, in general antipodal (Fig. 2. 1). Global sutures of the lithosphere are marked by the greatest seismic and volcanic activity. The Pacific belt consists of mid-oceanic ridges, while the transcontinental belt is expressed in an enormous orogenic folded belt, which includes the southwestern and southeastern margins of Eurasia and North America, along with a branch on the western margin of South America. Plate boundaries are primarily shift related: generally shift-tension, divergent – in the transoceanic belt, and shift-compression, convergent – in the transcontinental belt. Differences between the two types of boundaries are caused by their position: the first positioned above mantle upwelling zones and

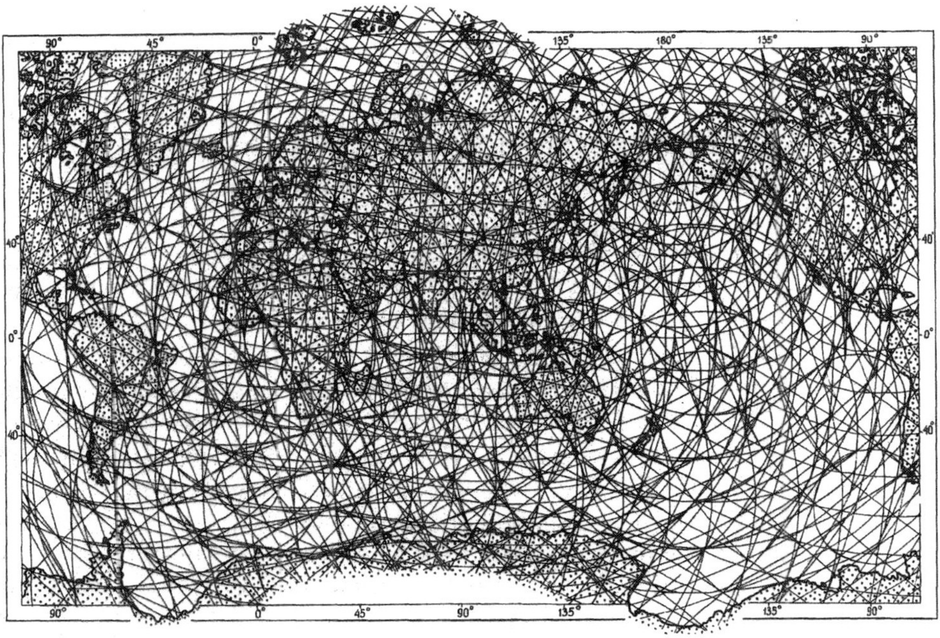

Fig. 2.1: Scheme of planetary jointing based on the map of bottom relief of the World Ocean, scale 1:120,000,000 (Golubev, 1994a).

the second positioned above mantle downwelling zones. Two-tiered thermochemical convection is stimulated by pulsation of the planetary core, increasing during cyclical lunar–solar influences, displacements of the core, and changes in the rotation velocity and vibrations of the earth.

2.2.4 MOBILE FOLDED BELTS

Mobile belt reforming of active plate margins by direct and reverse collisions of plates during the geodynamic cycle is peculiar to convergent plate boundaries (Golubev, 1994b, c, 1996a, b, c). Large-block fragmentation of the lithosphere is reflected in the segmental structure of folded belts. Phases and stages of mobile belt evolution are determined by the undulating discharges of shear stresses between platforms and massifs, which turn in response to collisions of adjacent plates. Initial shift-tension faults create chains of mobile belt sedimentary basins, and return shift-compression faults cause folding and epiplatform orogeny. Metamorphism and magmatism of folded systems result from thermodynamic discharge of shear stresses. Adjacent volcanic belts are formed by astenospheric diapirism in mobile zones of the lithosphere. Ancient crystalline massifs in front of folded systems undergo diffuse spreading, and are covered by plateau basalts. They subside due to deserpentinization of their basements, forming deep seas.

2.2.5 DEEP OCEANIC RIDGES AND ASTENOSPHERIC TECTONICS

Divergent plate boundaries are characterized by tectonic-volcanic reworking of the lithosphere through astenospheric flows eating away at both sides from zones of mantle upwelling in mid-oceanic ridges (Golubev, 1996a, b, c, 2000b, c, d). These flows initially augmented the primary subcontinental lithosphere, forming its global heterogeneities, at the same time producing their petrochemical differentiation. Powerful flows propped up relatively thin proto-oceanic platforms, causing their uplift, along with erosion of the "granite" layer, their partial melting, and degranitization. Enriched with silica and alkalis, flows were pushed down by convection into zones of astenospheric subduction below protocontinents, growing and granitizing the lithosphere. Partial melting of the lithosphere forms archlike domes, with thinning from 70–80 km at oceanic margins to 10–30 km below mid-oceanic ridges. The transfer of sialic components below continents caused triple thickening of their lithosphere in comparison with oceans.

2.2.6 OCEANIC TRANSITION ZONES AND ASTENOSPHERIC SUBDUCTION

Astenospheric subduction in mantle downwelling zones at active continental margins is marked by zones of seismic focuses, which are inclined only at depths of 50–100 km, that is to say, below the bottom of the oceanic lithosphere. Chains of zones of seismic focuses and island arcs suggest the division of astenospheric flows into separate jets. Intense astenospheric tectonic surroundings of the Pacific Ocean is emphasized by the highest seismic and volcanic activity, high heat flow, and significant gravity anomalies. The rising-up and melting-through of continental plate margins because of astenospheric flows is expressed in island arcs and an enormous ring of andesite volcanism. Astenospheric diapirs, accompanying subduction, appear in deep seas lined with basalts. There are also zones of less intense astenospheric subduction below passive continental margins, expressed in peri-oceanic subsidence, gravity and magnetic anomalies, whereas astenospheric diapirism is expressed in trapped fields and "basaltic windows" below sedimentary basins of the Arctic shelf.

2.2.7 OCEANIC MAGNETIC ANOMALIES AND DIFFUSE-LINEAR SPREADING

Linear magnetic anomalies are superimposed upon a mosaic magnetic field of pre-oceanic lithosphere, characterizing zones of diffuse-linear spreading and outpouring of basalts. Lava flows, widespread over half of the earth-induced trapped volcanism, caused by intensified pulsations of the planet core and thinning of the lithosphere as a result of astenospheric flows. The diffuse-linear spreading originates from the impetus of the flows, which separate ridges of thinned lithosphere, splashing out in the form of fissure volcanism. When the spreading zones become closed, basification of the lithosphere occurs. The rate of formation of oceanic lithosphere is proportional to the flow velocity and inversely proportional to the thickness of the lithosphere. Intense astenospheric flows above the displaced planetary core have predetermined a complete oceanization of the Pacific plate. Overheated and liquid flows have manifested in gentle South- and East-Pacific rises. Powerful hydrothermal inflow has produced enormous fields of ferromanganese nodules.

2.2.8 INVERSIONS AND POLE DISPLACEMENTS OF THE PLANETARY MAGNETIC FIELD

Spreading-related anomalies have countered inversions of the magnetic field, resulting from changes in the rotational velocity and inclination of the earth on its orbital plane axis. Telluric currents, abruptly increasing as a result of inversions, cause melting of the crust and formation of magmatic centers in the geodynamic nodes and mobile zones. Changes in axis inclination are accompanied by displacement of the axis (poles) of the magnetic dipole, which is oriented relative to the interplanetary magnetic field, resulting in collisions of lithospheric plates and crust blocks, generating paleomagnetic data. Only the Cenozoic series of spreading-related anomalies has exactly recorded the narrowing of plateau-basaltic volcanism areas towards mid-oceanic ridges, occurring contemporaneously with the closure of fissure zones and the "erasing" of earlier anomalies. Mesozoic anomalies have reflected an initial expansion of volcanic areas from zones of mantle upwelling, assymetrical due to a more intense erosion of the sublithospheric dome on the western oceanic margins.

2.2.9 DYNAMICS OF SPREADING AND TRANSFORM FAULTS

Late Cretaceous-Cenozoic narrowing of areas of spreading-related volcanism has followed the displacement of epicenters of mantle upwelling during deceleration of the rotation, and displacement of the planetary core, starting at the orogenic stage of the Alpine geodynamic cycle. Inversion of rotational stresses has resulted in compression of submeridional compression zones in the earth, and the opening of transform faults. Transform faults are usually degenerative strike-slip faults possessing tension and compression components. They compensate tensional strike-slip faults in mid-oceanic ridges and the expansion of oceanic platforms by solidifying of magma-conducting fissures. The sequential closing of these fissures is demonstrated by the increased age of volcanic mountains at outer edges of the ridges positioned at the intersections of the spreading-related anomalies and transform faults. Transform faults continue on the continent, but they lose their significance. They primarily represent oceanic elements of the planetary structure.

2.2.10 PLANETARY JOINTING AND GEODYNAMIC MATRIX

A regular planetary jointing network evidences stability of the rotational axis of the earth, and of the lithospheric plates (Fig. 2.2). The ancient network has recorded patterns of rotational stresses, and vector paths of lunar–solar influences, in repeated multi-component cycles (Golubev, 1993a, b, 1994a). The network frame is formed by orthogonal and diagonal systems of jointing zones of various types, forming an elementary rhomboidal jointing of the crust. Intermediate and feathering fault systems complicate the rhomboids, forming concentric morph structures. Spacing between parallel jointing zones of different rank is 7740, 6300, 5160, 3780, 2580, 1740, 1200, 540, 225, 85, 63, 42, 21, and 10 km or less. The spacing of zones repeats in the cross section of the earth, exhibiting a three-dimensional matrix of stresses. The discharge of planetary rotational shear stresses is realized by the collisions and vertical motions of crustal blocks. In this manner, different fragments of the regular network constitute a variety of normal faults of diverse patterns and types. Full control of the network on geological and geophysical structures of the crust, along with the dynamics of the astenosphere and hydrosphere, acquires the importance of a geodynamic matrix force.

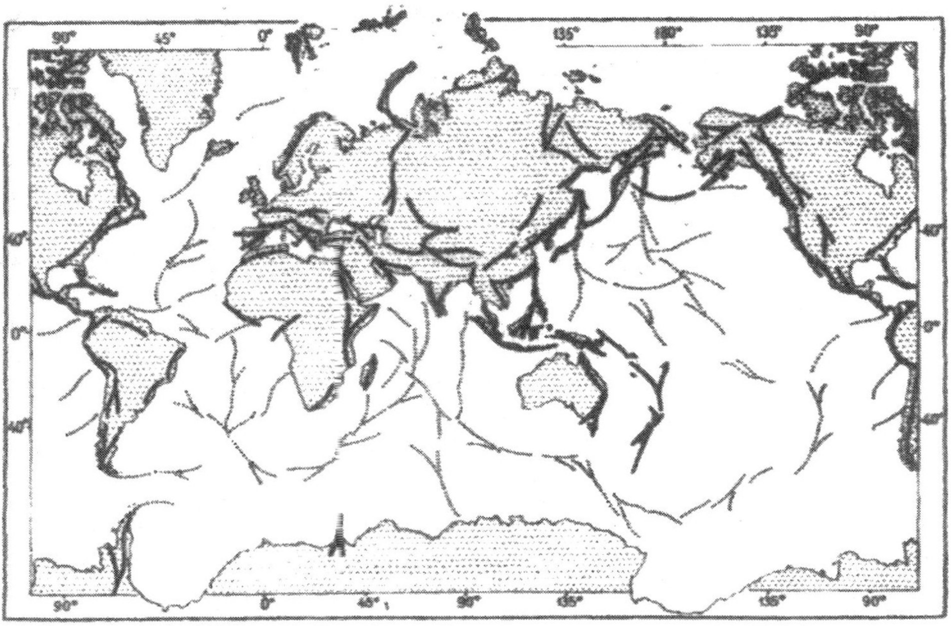

Fig. 2.2: Scheme of geotectonic belts. Heavy lines are the structural elements of the transcontinental belt; dotted lines are the structural elements of the transoceanic belt.

2.2.11 EARTH RHYTHMS AND THE GEODYNAMIC FIELD

Cyclic excitation of the geodynamic matrix reflects resonance amplification of inherent oscillations of the core of the planet, originated during the formation of the earth. Ultrashort oscillations of the core produce an energy-charged geodynamic field (a thin gravimagnetic field), transforming into gravitational and electromagnetic fields, directly responsible for endogenous tectonic-magmatic activity. Oscillations of the core, and of the earth itself, are stimulated and adjusted in amplitude, frequency and phase by basic cosmic rhythms, synchronizing within a multi-component geodynamic rhythm. Ultrashort, short, daily, monthly, annual, multi-annual, centurial and millennial earthly rhythms are mainly reflections of lunar tides Sun (the mediator of galactic dynamics) activity. Precisely oriented lunar–solar eclipse influences, cyclically passing through nodes of the geomatrix, excite the planet core and provoke a discharge of georotational elastic stresses in the lithosphere. Seismic and volcanic activities reflect all of the components of geodynamics and system geotectonics (Golubev, 1993a, b, 1994e, f).

2.2.12 GEOCHRONOLOGICAL SCALE AND GEOTECTONIC CYCLICITY

The motion of the planet and relative interactions within the Solar System determines the earth dynamics. The highest cycle is represented by circulation around the galaxy core (Golubev, 1992a, 1994e). The phases of anomalistic and sidereal periods of circulation (further referred to as *galaperiods*), 190 and 215 Ma, correspondingly mark geological periods and epochs, whereas coincidence of the beginning points of the galaperiods 4.5 Ga ago (*galactic resonance*) initiated the formation of the Solar System. Galaperiods, mere components of the actual circulation cycle (further referred to as *galacycle*), designate a turn circuit of solar-galactic barycenter and is expressed in irregular (155–195 Ma) geodynamic cycles. Phanerozoic geocycles in the middle of Cambrian, in the end of Devonian, Triassic and Neogene periods, correspond to the Caledonian, Hercynian, and Alpine geotectonic cycles and, neotectonics. Geocycles and their hemiperiods correspond to the Paleozoic, Mesozoic and Cenozoic eras. Undulatory amplitude changes in the geodynamic activity of geocycles, with periods of 1550–1700 Ma, mark the Priscoan, Archean, Proterozoic, and Phanerozoic boundaries (Fig. 2.3). The extreme points of cycles and megacycles are epochs of successive geosphere evolution, and contemporaneous complication of the biosphere (Golubev, 1995).

2.2.13 EVOLUTION OF THE LITHOSPHERE AND ORIGIN OF PLATES

Successive transformation of the lithosphere is caused by pulsating mantle convection, presenting differentiation due to stratification and redistribution relative to the age of the planetary composition (Golubev, 1992a, 1994d). At the beginning of the Archean, a subcontinental lithosphere had formed in the shape of a boundless Pangea, in part covered by shallow seas. Due to asthenospheric flows, the lithosphere grew downward, becoming stratified and serpentinized, and involving juvenile water. Plate heterogeneities of the lithosphere formed in a way that their marginal platforms were thinned and degranitized, while the central platforms grew downward and became granitized. Only the Pacific plate heterogeneity was thinned and basified as a whole. Splitting up of the already quite heterogeneous and rigid lithosphere into plates was marked by global Pan-African diastrophism, resulting from the increase of pulsations of the earth in the Phanerozoic. Since the Caledonian geodynamic cycle, the Pacific and African plates became separated. Since the Hercynian geocycle, the rest of

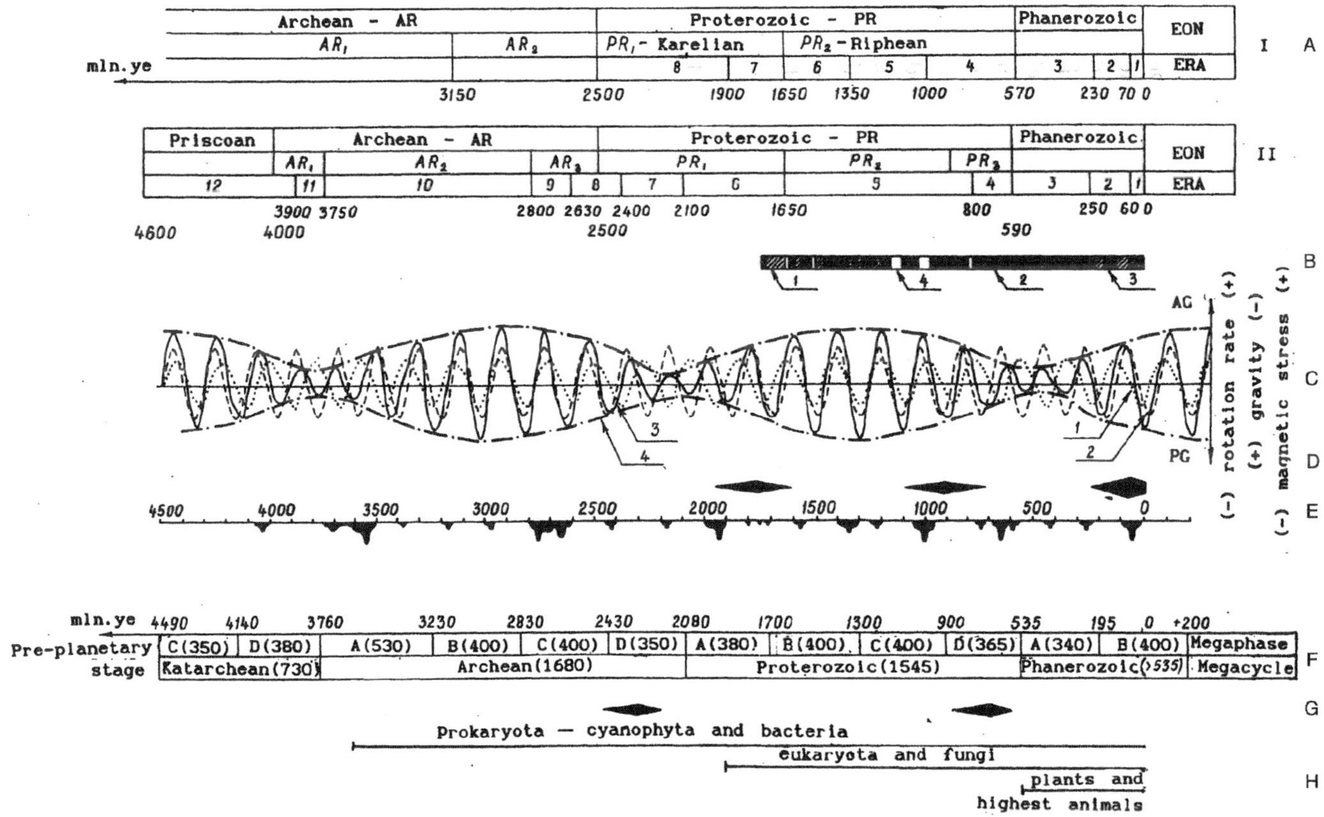

Fig. 2.3: Galactic cyclicity and megacyclicity of the geodynamic, geological, and biological processes since the time of origin of the earth. (A) General geochronological scale – era I: 1 – Cenozoic, 2 – Mesozoic, 3 – Paleozoic, 4 – Late Riphean, 5 – Middle Riphean, 6-Early Riphean, 7 – Late Karelian, 8 – Early Karelian (after Semikhatov et al., 1991); era II: 1 – Cenozoic, 2 – Mesozoic, 3 – Paleozoic, 4 – Sinian, 5 – Riphean, 6 – nameless, 7 – Huronian, 8 – nameless, 9 – Randian, 10 – Swazian, 11 – Isuan, 12 – Hadian (after Harland et al., 1982); (B) polarity of the geomagnetic field: 1 – direct, 2 – reverse, 3 – variable, 4 – not studied; (C) galactic geodynamics: (+) – increase, (–) – decrease of parameters, AG – apogalaction, PG – perigalaction; periods: 1 – siderial, 2 – anomalistic, 3 – geodynamic cycles, 4 – megacycles; (D) epochs of plateau-basalt volcanism (after Makarenko, 1983); (E) epochs of tectonic-magmatic diastrophisms (after Salop, 1982); (F) general geodynamic scale in Ma; (G) epochs of glaciation; (H) evolution of life (after Sokolov and Fedonkin, 1989).

the plates followed, with the separation of Gondwana and Laurasia. Since the Alpine geocycle the plates experienced oceanization and continentalization.

2.2.14 FORMATION OF OCEANS AND CONTINENTS

The formation of the oceans was the consequence of diffuse-linear spreading and fissure basaltic volcanism (Golubev, 1992a, 1996a, b, c). Final degranitization of the proto-crust occurred under conditions of astenospheric heating, under a heat-isolating basaltic cover. The residual upper layer of the crust became converted into strips in a gabbroid layer, whereas the lower layer underwent deserpentinization, accompanied by a Moho discontinuity uplift. The entire lithosphere experienced dehydration of disseminated serpentinites and subsequent consolidation. As volcanism receded and ocean formation boundaries advanced towards future mid-oceanic ridges, the lithosphere subsided into the astenosphere, as deep as the thickness of water layer, more than half of which had been supplied owing to dehydration of the lithosphere. Approximately 20% content of serpentinites in the lithosphere is enough to produce this amount of water. At the same time continents, subject to granitization, emerged as a result of isostasy. Supply and redistribution of water is recorded in the seismostratigraphic curve. The formation of oceans is characterized by the steps on continental slopes, by oceanic sedimentary complexes, by the presence of microcontinents, and metamorphic rocks with ages up to 1.8 Ga as well as indications of a crustal thickness up to 20–30 km beneath the oceans.

2.2.15 STAGES OF ORIGIN AND NEOTECTONICS OF THE OCEAN

Oceans formed as a process of zoning expansion and deepening, from the periphery towards mid-oceanic ridges, according to stages of the Alpine geodynamic cycle. Beginning from its second stage in the Early Cretaceous, the Pacific Ocean was formed. At the third stage in the Late Cretaceous, the Atlantic, Indian, and Arctic Oceans appeared. During the fourth stage in the Eocene, the oceans expanded, became deeper. Deep seas appeared at this point. Ocean formation occupied more and more the north continental hemisphere. Transition from the Alpine geodynamic cycle to the neotectonic cycle, occurred in the Pliocene, causing mass ocean formation with submergence of mid-oceanic ridges and formation of the World Ocean and continents. Lithospheric plates, along with their oceanic and continental platforms, interplate geotectonic belts and intraplate mobile belts of continental and oceanic types became the principal tectonic structures. Change of the geotectonic regime marked the beginning of the oceanic phase the evolution of the earth.

2.3 STRATIGRAPHY AND THE GALACTIC YEAR

The stages of ocean formation suggest we focus attention on cyclicity as an index of the driving forces in planetary evolution. The acceptance of cyclicity of geotectonic, magmatic, sedimentary, metamorphic, and bioevolutionary processes, reflected in the biostratigraphic time scale, has made geology a historical science. Through magnetostratigraphy, rhythmic similarities particular to our planet in regards magnetic field inversions have been identified. Although the cyclicity has been primarily attributed to endogenous activity, this only conceals the problem of rhythm complexity. The relation of earth's rhythms (e.g. seismic activity), ranging from hours and days to centuries and millennia, along with the rhythms of earth and the moon movements, together with solar activity, suggest the circulation of the earth around the galaxy core to be the highest order cycle of geologic time.

The galactic motion of the earth has attracted the attention of geologists since the beginning of the era of space exploration; however, spatial causes of tectogenesis are still more commonly considered as being exotic. The atavistic geocentric model disguised by the concept of self-development of the earth, along with lack of mechanisms explaining the galactic influence in cosmo-geological models can be held responsible for this view. This underestimation results primarily from uncertain general acceptance of global tectonic cycles, which are seldom recognized within the regional cyclicity. In addition, there exists uncertainty in the astronomic motion models of the Solar System, first of all, in estimating the duration and the beginning point of the so-called "galactic year". The concept of years, counted in different cosmo-geological models, from different subdivisions of the geochronological scale, as a set of periods and epochs, receives different tectonic significance. Moreover, the galactic year is correlated either with the anomalisitc, or with full circulation period of the Sun, suggesting imperfection in both models.

The motion of the Solar System around the galaxy core (black hole) has been described by Parenago (1952). This old model of the prominent astronomer provides the best correlation with geochronology. An anomalistic period, which means passage through the perigalaction, lasts 176 Ma. The next passage through this perigalaction will occur in 12 Ma. The apogalaction has passed 76 Ma ago. A draconic period, meaning passage through opposite nodes of the orbit, lasts 88 Ma. A complete circulation period lasts 212 Ma, ending when the Sun occupies its previous position relative to the other galaxies. This requires compensation in the galaxy orbit during an anomalistic period by backward displacement of orbital nodes. According to other models, a full period lasts on average from 165 to 260 Ma (precisely 200–220 Ma).

The anomalistic circulation period of the Solar System, according to Kepler's second law, is characterized by a gravity increase (1.5 times) in the galaxy core from apogalaction to perigalaction. The complete (sidereal) period should be distinguished according to variations of the galactic gravity field during resonance interactions between the mass centers of the galaxies. Stimulation of the galaxy core should in both cases stir up flashes of galactic magnetic intensity up to 10^{-6} oersted. However, the real orbital revolution of the Solar System, summing up its anomalistic and sidereal components and displaying circulation of the solar-galactic barycenter, should be more significant. The physical extremity of the turning points of the total galactic cycle (called *galacycle*) is amplified, coinciding with the extreme points of *galaperiods*. All these extreme points result in flashes of the gravimagnetic dynamic field of the galaxy, in this way controlling the dynamics of the Sun.

Stimulation of the Sun at extreme points of the galacycle is expressed as velocity changes in regard to circulation, rotation, amplifying of core oscillations, and an increase in solar activity. Pulsations of the gravimagnetic fields of the Sun are reflected in the dynamics of the Solar System planets through excitation of their gravimagnetic fields, initiating endogenous tectonic-magmatic activity and age transformations. A total orbital revolution cycle of the Solar System indicates an actual galactic year, whose duration fluctuates due to mutual displacement of the anomalistic and sidereal galaperiods. These changing periods of planetary dynamics can be plotted as sinusoidal curves (Fig. 2.3). The inflection points of the curves and the intersection points with the x-axis denote inversions of the kinematics, and physical fields of the earth. They are extreme points in geodynamic and geologic senses. These extreme points are recognized among the boundaries of the systems and series of the Phanerozoic stratigraphic scale, as well as on the graphs of the sediment cover that forms the earth, designed by Ronov (1961) as the boundaries of sedimentation cycles lasting from 215 to 190 Ma.

Three sedimentary cycles are most expressive: Late Vendian-Silurian, Devonian-Triassic, and uncompleted Jurassic-Quaternary, which began respectively at 620, 405, and 195 Ma. The mean duration of a cycle of about 215 Ma is equal to a sidereal galaperiod. Each such period corresponds to four geological periods with mean duration of 55, 70, 55, and 35 Ma, considered to be phases of the galasidereal sedimentary cycle. The boundary between the last two phases of the current cycle corresponds to the division of the Cenozoic into Tertiary and Quaternary, which are equal to periods of the Mesozoic and Paleozoic. Paleogene and Neogene appear to be subdivisions of the Tertiary period.

The galasidereal sedimentation cycles are controlled by global sea level oscillations, expressed in transgressions and regressions, which are described by the seismostratigraphic curve (i.e. the Vail curve). Water advances rather gradually over the continent from the beginning of the sidereal galaperiod onwards, and retreats suddenly from its middle. Less considerable sea-level falls mark the boundaries of all phases of the galaperiod. All this can be explained by galasidereal changes of rotation velocity, and displacements of the planet core, affecting the intensity of mantle convection and the supply of juvenile water, along with periodical redistribution of water in the oceans during isostatic movements of the lithosphere. Oscillatory epeirogenic movements are subordinate to the multi-component cyclicity of the pulsations of the planet.

The seismostratigraphic curve shows that ocean levels have risen about 1000 m above modern levels from the Jurassic until the end of the Cretaceous period, with two abrupt falls of 200–300 m in the beginning of the Early and the Late Cretaceous. From the beginning of the Cenozoic sea levels are decreasing, with minor rises and falls within each geological age. Considerable falls occurred in the beginning of the Paleocene (500 m), in the end of the Oligocene (900 m, i.e. 500 m below the modern level), and at the end of the Miocene, when the newly risen sea level fell 500 m. A similar fall occurred again in the Quaternary. These processes are caused by endogenous water influx and seafloor subsidence. The gradual influx is accompanied by prolonged transgressions, whereas rapid deepening of oceanic basins and the drainage of shelves result in abrupt regressions. The seismostratigraphic curve acquires the importance of an index of ocean forming processes beginning from the shelves to oceans. The oceans have periodically become shallower and drained at places because of basin subsidence. This model is in accordance with the small hiatuses in the sedimentary cover of the oceans, occurring most frequently during the Cenozoic. The drainage involved both shelves and submarine ridges, which became bridges between continents allowing the migration of flora and fauna.

Global transgressions are complicated by regional subsidence involving blocks of the planet crust, whereas contemporaneous rising of fold mountain systems and isostatic emergence of continents complicate regressions. Geotectonics appears to be the primary factor of cyclicity in sedimentation, which controls the formation of sedimentary basins and the supply of clastic material, directly related to the degree of relief contrasts. This relationship is evidenced by sedimentation rates, encountered in folded systems 3–4 times the height of platforms. Volcanogenic, chemogenic, and biogenic sediments are deposited with the same rhythm as clastics, emphasizing subordination of sedimentation to tectonics, and primarily to geodynamics.

Cyclic transformation of the planet crust and increasing contrast of relief are the immediate causes of increased sedimentation rates, thus increasing the thickness of stratigraphic systems of the Phanerozoic, especially the Tertiary. Additional effect provides gravity increase due to deceleration of the earth's rotation because of lunar tides. This is demonstrated by the natural slip angle, which has increased 5–7° since the Proterozoic, and preserved in cross bedding. Gravity increase leads to intensification of physical and chemical weathering, and denudation due to accelerated transport and differentiation of the clastic material resulting from atmospheric and surface water flows.

The anomalistic galaperiod is characterized by changes in geotectonic activity, with a less amplitude and larger gradient. It corresponds to four sedimentary cycles: Pre-Vendian-Cambrian, Ordovician-Middle Carboniferous, Late Carboniferous-Early Cretaceous, and incomplete Late Cretaceous-Quaternary, which began 680, 500, 310, and 110 Ma ago, respectively. The galactic anomalistic cycle lasts from about 190 Ma, and consists of four phases with durations of 45, 45, 55, and 45 Ma in average. Each phase contains deposits of several geological epochs. The duration of galactic anomalistic cycles is almost equal to an astronomical one: 190 Ma instead of 176 Ma. The boundary between the first and the second phases of the last cycle is close to the time of passage of the Solar System through apogalaction: 67 Ma instead of 76 Ma ago.

Most geological epochs are distinguished by periodical coincidences between the boundaries of the phases of sidereal and anomalistic galaperiods. The rest of the epochs are distinguished by coincidences of the boundaries of each of the four phases, lasting 20–30 Ma, of the rigorous galaperiod, which characterizes the oscillatory movement of the Solar System relative to the galaxy plane. Intensification of the extreme points of galaperiods by boundaries of rigorous phases is exhibited in the mean duration of the epochs of 20–39 Ma, while geological periods are mainly divided into two of such epochs.

The duration of geological epochs varies between 7 and 43 Ma, increasing towards the beginning of the Phanerozoic. An epoch is, as it were, expanded in duration to a whole period in the Precambrian, and reduced to an age in the Cenozoic. An age is also expanded from 1.5–2 Ma in the Neogene, to 7–16 Ma in the Cambrian. This is due to the incompleteness of the stratigraphic division of ancient units, monotonous because of multiple metamorphism and poor fossil diversity. Gradual complication of the earth's crust and organic world, leading to the enrichment of the structural and substantial content of small geocycles resulting as an effect of the increase in rank of the stratigraphic units.

2.4 THE GEODYNAMIC CYCLE AND OROGENIC EPOCHS

A complication of facies and formational composition is characteristic of sedimentary complexes, which conceal in this way galactic anomalistic and galasidereal sedimentary cycles. The complexes are related to geotectonic cycles, as established by Suess (1883–1909) and Bertrand (1886–1887): Baikalian, Caledonian, Hercynian, and Alpine. Unequal and differing cycles caused by tectonic-magmatic activity occurred during the evolution of mobile belts and platforms, and are explained by endogenous activity. Geotectonic cycles correspond to geodynamic cycles, which include galaperiods, and oscillate in duration and amplitude (Fig. 2.3).

Geodynamic cycles have been established as the result of graphical addition of sinusoidal curves of the anomalistic and sidereal galaperiods, constructed with an amplitude ratio of 3:2. The beginning of the geocycle has been established at the minimum point of the resulting geodynamic curve, which marks the opposition of galaperiods. The extreme point reflects the position of the Solar System near the perigalaction, where it originated and repeatedly undergoes strong excitation due to activation of the Sun together with the movement resonance of the planets. Displacement of the core and change of inclination of the earth, the intensification of core oscillations, mantle convection, and acceleration of rotation and equatorial extension accompany inversion of the geophysical fields, together with radical tectonic-magmatic reconstruction. Shocks celebrate the "birthday" of the earth and the beginning of a new geodynamic cycle.

Geodynamic cycles (*geocycles*) last between 155 and 195 Ma, and vary in geodynamic curve amplitude, reflecting the general level of geotectonic activity. The geocycles correspond by both of these parameters to geotectonic cycles, except for the rejuvenation of their boundaries during 10–15 Ma, usually dated as Middle Cambrian, Late Devonian, Late Triassic and Neogene. The sense of such rejuvenation is that the geocycle finishes with denudation of the planetary structures that it created. However, this epoch, which in fact erases the sedimentary record of a folded system, is more often attributed to the beginning of the next geotectonic cycle. The Late Baikalian geocycle began during Pre-Vendian times (700 Ma ago), the Caledonian began during the middle of the Cambrian (535 Ma ago), the Hercynian at the end of the Devonian (350 Ma ago), and the Alpine at the end of the Triassic (195 Ma ago). During the Pliocene (5 Ma ago), the neotectonic cycle started. Boundaries between the geocycles are conventional because they are transitional epochs in the evolution of the earth, lasting 5–10 Ma.

The maximum point of the geodynamic curve, in the middle of the geodynamic cycle, is the point of maximal coincidence of galactic periods in the same phase. This coincidence signifies the position of the Solar System near the apogalaction, the point opposite the place of its origin. However, from the physical point of view, the proximity of the galaxy core is not as important as the position of the Sun on the line of junction with the mass centers of the many galaxies. Due to the location at the boundary of the geodynamic cycle, the earth is strongly affected and assumes other characteristics resulting from inversion of the galactic fields. The resonance amplification of the pulsations of the earth result in more expressive changes during the second half of the geocycle.

The geodynamic hemicycle can be divided into two unequal parts at the intersecting points of the geodynamic curve with the x-axis, galactic field influence begins to change their character. As a result, four unequal (25–65 Ma) phases of the geocycle are distingushed: (a) preparatory, (b) early, (c) late, (d) final, which are similar to stages of the mobile belt cycle. Stages of the geocycle (*geostages*) refer to predominant stresses in the lithosphere, corresponding to early and late mobile belt forming and early and late orogenic stages, that is, they reflect oscillatory increases and decreases of tension and compression stresses, peculiar to the mobile belt formation itself, and to orogenic phases. Boundaries between the geostages are generally marked by orogenic epochs.

Orogenic epochs have been established by Stille (1924) for Western Europe as consisting of 19 folding phases. In the course of time these have been supplemented by numerous other epochs (over 50). This fact has cast doubt on the existence of geocycles, though they merely show the presence of internal divisions. Thirteen phases of folding are the most widespread. They are epochs of global tectonic-magmatic activation involving fold belts and platforms, and expressed in the oceans as impulses of oceanic crust formation. Despite the complexity of establishing, and the uncertainty of precise timing, orogenic epochs are timed to extreme points of the geodynamic cycles and last 10–38 Ma, proportionate to the scale of geostress. The Late Baikalian orogeny completes the Baikalian geocycle. The Salairic, Taconic, Late Caledonian and Bretonian orogenic epochs mark the stage boundaries of the Caledonian geocycle. The Sudetic, Uralian, Late Hercynian, and Early Kimmerian epochs separate the stages of the Hercynian geocycle. The Late Kimmerian, Austrian, Laramian, and Late Alpine epochs divide the Alpine geocycle. Within the Late Alpine epoch is concealed the initiation of the neotectonic geocycle.

Fundamental reconstructions of the lithosphere are connected exclusively with the extreme points of the geodynamic cycles. Their relative dimensions become fixed when approaching the extreme points of the galaperiods. The latter are expressed in less significant orogenic events lasting 5–10 Ma at the boundaries of geological periods and epochs. However, prolonged periods of tectonic calm between the orogenic epochs can scarcely be regarded as quiet. The visible calm hides a continuous succession of micro-evolutionary reconstructions of the crust, associated with the extreme points of geodynamic cycles of lesser rank.

The expected course of the geodynamic curve shows that the next orogenic period will begin after 25 Ma, marking the end of the Quaternary. This epoch will last for about 20 Ma, until the end of the first stage of the neotectonic geocycle. During this stage, mobile belt rifting will increase, and ocean development will continue. Afterwards, the present neotectonic epoch will obtain the undisputable status of initiating a new geotectonic cycle, characterized once again by continent and ocean formation. In this case, the beginning of the neotectonic epoch, in spite of its name and original application to the Pliocene-Quaternary, has been lowered to Eocene (i.e. the beginning of the last stage of the Alpine geocycle), tectonically significant due to late orogeny and ocean formation.

Uncertainty in regard to the neotectonic cycle is caused not only by its extreme youth, but also by the complexity of establishing geotectonic cycles resulting from the heterogeneity of the lithosphere and distribution of planetary rotational stresses. A geocycle is generally characterized by a succession of dominant stresses in the lithosphere, predetermined by rotational acceleration and equatorial extension of the earth during the first stage of the geocycle, along with deceleration of rotation, and equatorial compression, during the second stage. A reverse type of stresses in mutually orthogonal folded systems causes the establishment of regional tectonic cycles, moved or reduced to a single phase or stage relative to the geodynamic cycle.

The stratigraphic volume of geotectonic cycles is generally changeable because of the migration of mobile belts and the incompleteness of the stratigraphic record in folded systems. It is due also to the conventional character of stage determination of the mobile belt formation cycle caused by regional peculiarities. Regional cycles often are named originally by their geographical position, and in this way mask classical cycles. Conversely, they sometimes receive classical names, and in this manner the classical cycles become discredited because of their different stratigraphic volume. Both situations lead to a chaos in terminology. This is especially peculiar for younger structural-formational complexes exhibiting stages and substages of a most expressive geocycle. As a result, phases of a geocycle are regarded as separate cycles. The Caledonian geotectonic cycle is more or less uniform in different regions and equal to the Caledonian geocycle. The proper Hercynian and Kimmerian geotectonic cycles could be called incomplete and may have occurred earlier. They correspond only to the first and second phases of the Hercynian geocycle. The Late Mesozoic (Laramian) and Cenozoic (Alpine in its strict sense) geotectonic cycles are of the same category. They correspond only to phases of the Alpine geocycle. In the case of a polycyclic development of folded systems, incomplete geotectonic cycles achieve all the completeness of a geocycle.

The vagueness of geotectonic cycle boundaries is aggravated by the fact that even those folded areas that are considered to have completed their development, have in fact not done so. These areas have undergone activation and repeated orogeny in all subsequent geodynamic cycles, most recently during the orogenic phase of the Alpine geocycle. Epiplatform orogeny during the Cenozoic has covered the majority of intraplate folded areas, forming peri-oceanic orogens on passive continental margins. Epiplatform orogeny occurred most powerfully during the formation of the perigeosynclinal Central Asian orogenic belt, which joined with the Alpine-Himalayan and Pacific mobile belts.

In spite of all diversity found in tectonic conditions, peneplains are common, always marking epochs of relief formation. In this way they confirm the global occurrence of geodynamic cycles. Physical and chemical weathering immediately begins with flattening of the tectonic-magmatic buildups of each stage of the geocycle. The Triassic phase of relief formation corresponds to completion of the Hercynic geocycle. The Jurassic-Cretaceous, Late Cretaceous-Eocene and Oligocene-Miocene phases correspond to the combined first and second stages, then third, and fourth stages of the Alpine geocycle. The Pliocene-Quaternary phase corresponds to the beginning of the neotectonic geocycle. The contrast of the neotectonic relief, and the new formation of oceans, suggest the initiation of a revolutionary epoch in the development of the earth.

2.5 GEODYNAMIC MEGACYCLES OF EVOLUTION

The general process of evolution is caused by undulatory changes in amplitude of geodynamic cycles according to convergences and divergences of the beginning points of anomalistic and sidereal galaperiods. Four geodynamic megacycles, corresponding to the principal geological eras, are perfectly distinguished by inflection points of the geodynamic megacurve drawn through the extreme points of the geocycles. The points of maximum divergence of the galaperiods 3760, 2080, and 535 Ma ago, mark the beginnings of the Archean, Proterozoic, and Phanerozoic megacycles, lasting 1680, 1545, and more than 535 Ma, respectively. They are preceded by the Priscoan semi-megacycle lasting 730 Ma, beginning from the time of origin of the earth, this being the pre-geological era (Fig. 2.3).

Pulses and megapulses of the geodynamic curve reflect the double level pulses of the earth, expressed as changes of rotation velocity and pole oblateness, displacements of the planet core, and changes in planet inclination, together with alterations of the direction and intensity of mantle convection. These processes are caused by excitation of the entire Solar System and specifically by outbursts of solar activity combined with tidal influences of the moon and planets during movement resonances, stimulating progressive transformations of the lithosphere and entire earth. Geocycles of the least amplitude, denoting the maximum divergence of the beginning points of galaperiods, are considered to be boundaries of megacycles. Frequent variations in the galactic fields and solar activity result in extreme excitation of the entire spectrum of georhythms, together with radical transformations of the earth. In turn, middle geocycles of the megacycles are epochs of maximum coincidence of galaperiods and resonance stimulation of all georhythms, resulting in fundamental tectonic-magmatic transformations.

Inflections of the geodynamic megacurve (its extreme points), coincide approximately with important boundaries of the geochronological scale, dividing megacycles in parts of similar type. According to the degree of convergence of galaperiods, the megacycles can be divided into four megaphases lasting 350–530 Ma each: (A) preparatory, (B) early, (C) late, and (D) final. The tectonic-magmatic epochs occur at the same time at the boundaries of the megaphases, however, with a delay of 50–100 Ma, suggesting a long-period adaptation of the earth. Diastrophisms are generally related to stages of a geocycle, but at the boundaries of semi-megacycles, they totally involve the margins of geocycles.

Revolutionary epochs of the earth at the boundaries of megaphases determine the type of evolution, with the middle boundaries of the megacycles qualitatively developing revolutionary transformations, because the megacycles, like the geocycles, consist of hemiperiods of expansion and contraction of the earth. At the boundaries of the hemi-megacycles, approximately 3.8, 2.8, 2.1, 1.3, and 0.5 Ga ago, the character of mobile belts underwent changes. Granulitic and greenstone belts of Archean (early and late ones, aged 3.3–3.8 and 2.6–2.8 Ga, respectively) have been replaced by proto-mobile belts in the Proterozoic (early and late ones, aged 1.5–2.0 and 0.6–1.2 Ga), followed by mobile fold belts and the epiplatform orogenes of the Phanerozoic.

The completion of the Priscoan hemi-megacycle of contraction has led to isolation of the relatively rigid crust from the primary plastic lithosphere, whereas the end of the Archean hemi-megacycle of contraction has been marked by cratonization of the crust, which became subcontinental due to granitization and thickening. During the course of the Proterozoic hemi-megacycle of extension, marked by the development of proto-mobile belts and plateau-basalt volcanism, the crust has been thickened in central (proto-continental) parts of plate heterogeneities of the lithosphere, and thinned at their proto-oceanic margins. Thinning has completely covered the Pacific lithosphere heterogeneity.

The beginning of the Proterozoic hemi-megacycle of contraction 1.3 Ga ago has been accompanied by the increase of dynamothermal reworking of the planet crust and formation of intracratonic mobile belts with a zonal structure. In the platform areas, as the result of remobilization of the crust, granitoid volcanic-plutonic rock associations have been formed. This time is often considered to be the era of small plate tectonics, the time of repeated disintegration and reunification of the embryonic Pangea, although it can only be asserted that there occurred plastic collisions of blocks of the all-embracing subcontinental Pangea supercontinent, of which only one third remains now as continents.

Proterozoic contraction and cratonization reached its maximum during the Late Baikalian geocycle, completing this megacycle. The geodynamic cycle has been expressed in the Pan-African diastrophism 650–500 Ma ago, recorded by granitoid magmatism. This magmatism is unique among continents of the southern hemisphere, which became essentially an oceanic one later. A global orogeny, named Brazilian in South America, Cadomian in North America and Europe, and Baikalian in Asia, marked the division of the lithosphere in main plates, whose collisions caused the formation of folded belts in the Phanerozoic.

Pan-African diastrophism was the precursor of the violent era of the Phanerozoic megacycle. A new hemi-megacycle of expansion of the earth started, but the expansion was so powerful that it prevailed over the evolutionary contraction. This extreme expansion resulted in cracking of a relatively consolidated lithosphere into plates, along with their subsequent formation, in stages, into oceans and continents. High-amplitude pulsations of the earth's core resulted in intense mantle convection and tectonic-magmatic activity. The Phanerozoic style of evolution differed from the monotonous and dull Precambrian style so much, that during 0.5 Ga the lithosphere has changed beyond recognition. The earth has entered the megacycle of transitory age, marking the first extreme divergence of galaperiods occurring in the Paleozoic.

The Phanerozoic megacycle has already passed the preparatory Paleozoic megaphase and the Mesozoic-Cenozoic half of the early megaphase. Unlike the Precambrian eras, which are equal to megacycles, the Paleozoic era corresponds only to one megaphase, and the Mesozoic and Cenozoic eras, only to hemiperiods of the geocycle. This also demonstrates evidence of the acceleration of evolution. The Paleozoic megaphase is characterized by rather low geotectonic activity and consists of the Caledonian and Hercynian geocycles lasting 185 and 155 Ma, respectively. The much more tectonically active Mesozoic-Cenozoic megaphase includes the Alpine geocycle lasting 190 Ma, and the neotectonic geocycle, that started 5 Ma ago. Both megaphases are exhibited in the seismostratigraphic curve. The Alpine geocycle is analogous with respect to amplitude to the Late Baikalian geocycle, terminating the Proterozoic era, and is positioned at a remarkable, but middle point of the Phanerozoic megacycle.

2.6 FINAL REMARKS

The earth was formed 4490 Ma ago, at the mid-point of the Priscoan megacycle, due to a full coincidence of the beginning points of galaperiods. The time of the galaxy resonance, which gave birth to the Solar System, is proven by radiological dating evidence of the most ancient minerals of the planet crust (4.50–4.58 Ga), of the lunar crust (3.3–4.2 Ga), and meteorites (4.50–4.55 Ga). The age of the earth is estimated, and widely accepted according to this data, to be from 4.5 to 4.6 Ga. The data also demonstrate the perfection of geochronological dating of the anomalistic and sidereal galaperiods. However small, their displacement abruptly changes the time of galaxy resonance and origin of the earth.

The constancy of galaperiods has been proven. Their number is not great: 21 sidereal and 24 anomalistic periods. These periods are not short. The earth has entered the 25th galactic year, and lived more than one third of the measured years. The life of the earth will apparently end at the next galaxy resonance in 3.7 Ga, at the age of $190 \times 43 = 215 \times 38 = 8170$ Ma. Extreme excitation of the exhausted earth will lead to its thermal death. The middle turning point of its life is marked by the galaxy anti-resonance expressed in extreme divergence of the beginning points of the galaperiods, which occurred in 4.1 Ga after the birth of the earth, causing ocean formation. The next galaxy anti-resonance at the earth's age of 12.3 Ga (about 70 galactic years) will likely initiate the disintegration of the Solar System.

The galactic-solar standard of geological time emphasizes the cosmogenous-endogenous character of the geodynamics and system geotectonics, providing a new vision of the moving forces and mechanisms of tectogenesis, converting geology from a historical science into an evolutionary one. A perspective is opening which breaks away from the stagnation in geotectonic science, the science that is not only a structural basis, but also a philosophical one (being employed as a cognition method). Transformation and reconciliation of the fixistic geosynclinal-platform theory together with the mobilistic theory of plate tectonics on the basis of mobile fixism, or even ultramobilism, appears to be the solution for success.

In conclusion, the galaxy, the Sun and planets are rotating and pulsating, and the vectors of lunar–solar influences are moving over the earth cyclically and with the intervals of the celestial mechanics. Lithospheric plates and platforms are reciprocally colliding, rifts are opening, mobile belts are migrating and ridges of transcontinental geotectonic belts are rising. Convective plumes are continually emerging from the pulsating mantle below the transoceanic geotectonic belt, and astenospheric flows are moving below the lithosphere.

At certain times, areas of diffuse-linear spreading and volcanism have appeared, expanded, and then later contracted. Following this contraction, fronts of ocean formation moved and created oceans. Contemporaneously with submergence of the dehydrated oceanic lithosphere, granitized continents incremented by astenospheric subduction, are emerging.

A cyclically transformed lithosphere oscillates, but remains integral. It reflects deep changes of the increasingly eccentric heterogeneous earth, living solar and galactic years during its life, which are imprinted in the sedimentary record and biostratigraphic scale. The evidence of a logical unity of immutability and mutability, rest and motion, embodied in the inherited evolution of the planetary crust, affirms the replacement of the geocentric geotectonics by another that is cosmogenous-endogenous.

3. CYCLICITY IN EARTH'S GEOLOGICAL EVOLUTION

J.M. MABESOONE AND V.H. NEUMANN

3.1 EARTH'S HISTORY

3.1.1 PRELUDE

After the cessation of the bombarding of Earth and Moon by meteorites at about 3.9 Ga ago, a more definitive crust started to develop. All continents are embryonic from coalescing sheets or "seed" nuclei during the Archean. These nuclei record very special conditions (permobile stage), of great efficiency for the production of continental crust. The growth of the continental masses towards greater sheets, with several hundred kilometers of size, happened in chelogenic form from the primordial nuclei, which constitute stable elements in the cooling earth, a process that continued in the Paleoproterozoic. In spite of reworking during all following cycles, some Archean nuclei have still been preserved until today. The sediments formed during this era are very immature; they became available through weathering and erosion, were transported over very short distances and deposited immediately after. They always contain great quantities of volcanic detritus, what points to a high mobility of the Archean sedimentary basins; limestones are almost absent.

The Paleoproterozoic was an important stage of orogenesis in the whole world. Various cycles may be distinguished in different parts of the lithosphere, between 2.45 and 1.75 Ga, consubstantiating important continental landmasses. The pre-existing Archean nuclei became amalgamated during these collations, and new "seed" nuclei of chelogenic growth appeared. The proto-continents grew together into lithospheric plates. The sediments of this epoch record the existence of tectonically more quiet basins in which the supplied material became differentiated after their respective grain sizes and petrographical composition, e.g. in sands and clays. Where few of this type of material were supplied, even limestones could form. Thus, the deposited sediments assume a more modern character. One type of characteristic sediment has still to be mentioned: the banded iron formations (BIF) that absorbed the then appearing free oxygen of the atmosphere, and precipitated chemically or biochemically from sea water. Their maximum formation took place between 2.3 and 2.0 Ga. After that they became substituted with

Cyclic Development of Sedimentary Basins
J.M. Mabesoone and V.H. Neumann, editors.

red-colored continental sandstones, the first appearance of red beds. The Paleoproterozoic may be called in earth's history, its stabilization stage.

3.1.2 FURTHER DEVELOPMENT

The Mesoproterozoic seems to have been a period of transition in which the cyclic tectonic-sedimentary episodes became more evident. During this period predominate intraplate movements, folding of mobile belts, strong magmatism, and sedimentation, showing a good preservation of the Paleoproterozoic collated terrains. The orogenetic events show only a secondary character, or are of more local importance, becoming stronger from the Ectasian to the Stenian. The Late Mesoproterozoic (Stenian) Grenvillian orogeny saw the formation of an extensive junction of continental fragments, possibly into a supercontinent.

From the beginning of the Neoproterozoic onwards, the history of the earth acquires a more "modern" character, almost equal to the present time. The submergent and oscillatory-emergent episodes, with their accompanying orogenic phases become ever more evident and their record can be recognized. This situation continues during the Phanerozoic. The only difference between the two eras and their distinction is caused by the fossil assemblages, which determine the chronostratigraphical scale of this latter era.

In the Phanerozoic important mobile belts developed circumscribing the Precambrian platform areas. Some of these show an accretionary character, including subduction, occurring peripheral to the continents. Others presented a collisional character, and today occur within the continental interiors. The sedimentary sequences deposited since the beginning of the Neoproterozoic have completely modern aspects.

In Fig. 3.1 (inspired in Brouwer, 1984) a scheme of the above-sketched history is summarized.

3.2 CYCLICITY IN EARTH'S PROCESSES

The first clues for a cyclicity of some processes that affect the earth, were actually provided by the glaciations. Three of these glaciations in the Phanerozoic call the attention because of their irrefutable record (from recent to ancient): one in the Neogene, including the present; another in the Permo-Carboniferous; a third one in the beginning of the Silurian. There exists still an undeniable record of a fourth glaciation at the end of the Precambrian (Varangian, Late Neoproterozoic), and almost certainly a fifth at the beginning of the Neoproterozoic (Kaufman, 1998).

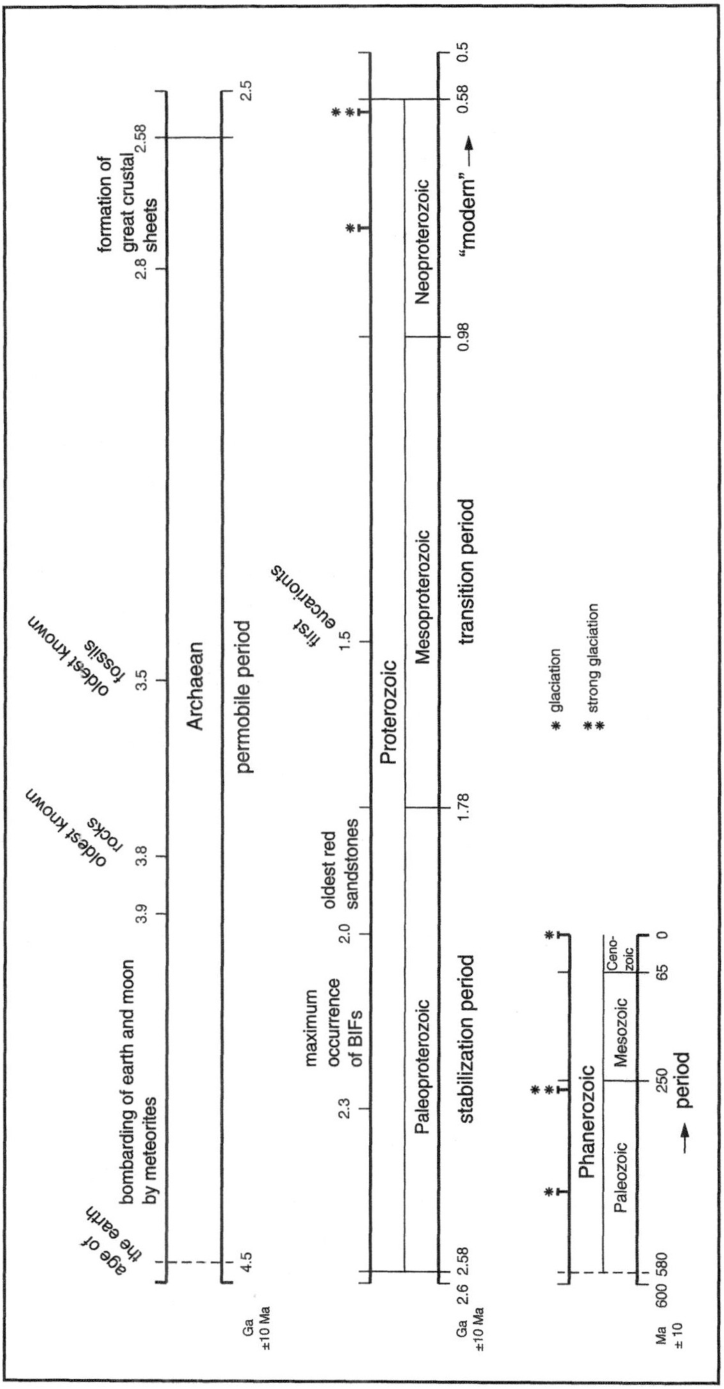

Fig. 3.1: Linear presentation of earth's geologic development and most important occurrences, inspired after Brouwer (1984).

The Varangian ice age is known from many areas which are actually found around the northern Atlantic Ocean and in the former Gondwana supercontinent; this glaciation must have been very extensive and intense as to have given rise to the *snowball earth* theory. The Silurian glaciation shows its record chiefly in North Africa and Northeast Brazil, at that time contiguous areas. Records of the Permo-Carboniferous glaciation are known since long from the southern hemisphere continents, including the Indian peninsula. And the fourth, Neogene ice age displays itself neatly in Europe, North America, and the Antarctic.

Between these intervals with extensive glaciations, which lasted for 10–20 Ma and sometimes longer, other intervals, well of longer duration, occurred with an almost absence of ice on the earth's surface.

After the recognition of this type of periodicity or even cyclicity, other cyclic features have been suggested, chiefly recognized in the Phanerozoic. In this case it is concerned with the mobility of the earth's crust and the big phases of global transgressions and regressions of the sea. What arose the interest of the researchers, was how to reach to know the causes of these cyclic phenomena.

Apparently, most available data indicate more that the earth's geological phenomena repeat themselves cyclically, as supposed already by the ancient geologists, rather than succeed themselves in nonlinear dynamics. Details hereabout have been presented by De Boer and Smith (1994).

The long and agitated 4.5 Ga history of the earth shows that there occurred, besides the cyclic processes also one-way processes. And these are the one-way processes that led many geologists to doubt about cyclicity.

The basic law of geology is still the principle of uniformitarianism, the basic approach to geology of Charles Lyell (1830–1833), "being an attempt to explain the former changes of the earth's surface by reference to causes now in operation". However, although these "causes now in operation" were the same all along the whole history of the earth, their intensity was not always the same. A young earth, with a major radioactive heat development, with a much greater mobility, with a thinner crust composed of small fragments, reacted otherwise than later under other circumstances and with other material.

A second example of a one-way development, generally called "evolution", is that of life. Although there exists a cyclicity in the explosion and extinction phases of life forms, accompanying the long-term geotectonic cycles, the direction is certainly one-way, an evolution from lower life forms that changes through time to yield other ever more advanced forms.

It is logical that there exist differences in the representation of cyclic phenomena on the diverse continents. However, these are only due to the behavior of the individual continents, as details that do not modify the general tendency. Thus, it remains evident that there exists a cyclicity in most of the processes which affect the earth, and that most phenomena are closely related between them.

3.3 TECTONIC-SEDIMENTARY EPISODES

When in 1963, Sloss revived the concept of large-scale unconformity-bounded sediment successions, calling them *sequences*, and six of them established in the Phanerozoic of the North American craton, similar units were also recognized in the platform areas of Russia (Ronov et al., 1969) and Brazil (Soares et al., 1974). This lead Sloss and Speed (1974) to propose their tectonic-sedimentary episodes, represented alternately by three global types in the Phanerozoic record of cratonic areas and their margins: oscillatory, emergent and submergent (Fig. 3.2). However, the authors failed to discern a major cyclicity in the six recognized sequences composed of variable lithologic associations.

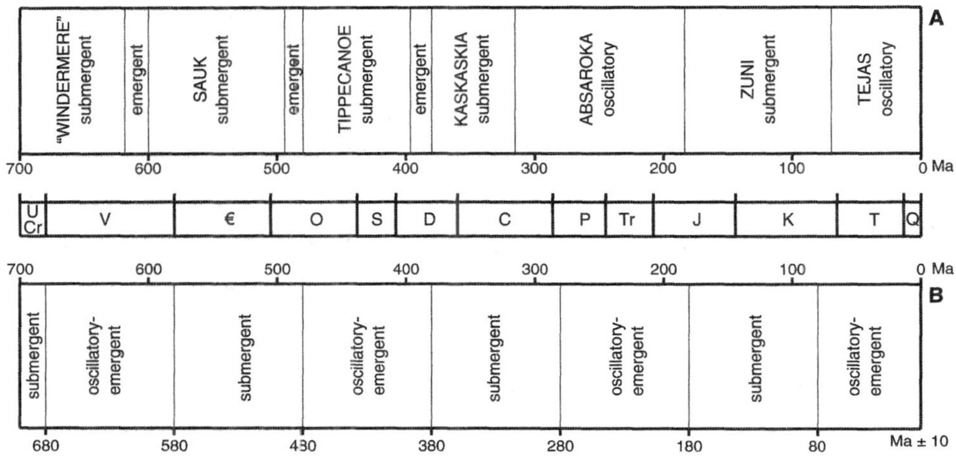

Fig. 3.2: Submergent and oscillatory-emergent episodes since Late Neoproterozoic and tectonic-sedimentary episodes proposed by Sloss and Speed (1974).

During the oscillatory mode pulsating vertical movements cause the gradual uplift of cratonic areas, resulting in many local and regional unconformities and at the end in an overall uplift of the craton. The emergent episode is defined by a slow rising of the craton, resulting in an insignificant relief and almost no sedimentation. During the submergent mode the craton is subject to a progressive subsidence, permitting the invasion of the sea. In the original paper of Sloss and Speed (1974) the submergent episodes are separated between them by either emergent or oscillatory periods. This led Mabesoone (1988) to reconsider the concept, finding out that only two emergent episodes were determined during the Paleozoic. A revision leads to the conclusion that these two emergent episodes separated by the Ordovician-Silurian submergent episode seem together to be nothing more than an oscillatory episode. And interpreting in this way, there surges a perfect repetition pointing even to a cyclicity. Probably, the recognition of the Ordovician-Silurian period as submergent and not as oscillatory is due to the precarious preservation of the chiefly continental sedimentary record in the North American craton, as the proper Sloss admitted (1993, in personal communication to the first author). Then, one may distinguish a couple of episodes with a submergent and oscillatory-emergent character, as a cycle of an approximately 200 Ma duration, that is successively repeated almost certainly since the Mesoproterozoic and maybe even earlier.

The now distinguished two types of tectonic-sedimentary cyclic episodes may be characterized as follows.

During a submergent episode that seems to start a cycle, a cratonic area undergoes progressively a subsidence that enables the sea to invade the area (*thalassocratic phase*). Along convergent margins may exist a considerable relief of ancient, partly denudated mountain chains, fact that does not occur at passive margins. During this episode the tendency is of continental separation and drift, plate collision with subsurface folding, increase in the number of epicontinental seas and littoral zones, explosion of life in these environments, and an attenuation of the differences between the earth's climate belts.

During the oscillatory-emergent episode the craton becomes progressively elevated in pulsating vertical movements that form island arcs near convergent plate boundaries. Along passive margins strike-slip movements seem to prevail. In the craton interior a considerable relief is developed due to epeirogenetic uplift and accompanying taphrogenesis. Erosion produces immature coarse clastic sediment that accumulates in intracratonic rift basins and alongside the cratonic margins (*geocratic phase*). The oscillating movements provoke numerous interruptions in the sedimentation, forming many local and regional unconformities. The general tendency is that of continental collision and the formation of one or two supercontinents, generally one in the northern hemisphere and the other in the southern one. Accompanying features are a decrease in the number of epicontinental seas and littoral belts, extinction of life, accentuated differences in relief and climate zones, and also glaciations. A maximum emergency results in a strong denudation of the mountainous relief and the formation of an extensive denudation surface that represents a great regional to worldwide unconformity.

A couple of submergent and oscillatory-emergent episodes represents also a so-called orogenetic phase, as is presented in Table 3.1, since the beginning of the Proterozoic.

Table 3.1: Couples of submergent and oscillatory-emergent tectonic-sedimentary episodes and orogenetic phases since Paleoproterozoic.

Ga ± 10 Ma	age: era	age: periods	tectonic-sedimentary episodes	orogenetic phase	period
0	Phanerozoic				"modern" period
		Quaternary-Tertiary	oscillatory-emergent		
0.08				alpine	
		Cretaceous-Jurassic	submergent		
0.18					
		Triassic-Late Carboniferous	oscillatory-emergent		
0.28				variscan (hercynian)	
		Early Carboniferous-Devonian	submergent		
0.38					
		Silurian-Late Ordovician	oscillatory-emergent		
0.48				caledonian	
		Early Ordovician-Cambrian	submergent		
0.58	Neoproterozoic				
		Vencian	oscillatory-emergent		
0.68				baikalian-avalonian*	
		Late Cryogenian	submergent		
0.78					
		Early Cryogenian	oscillatory-emergent		
0.88				?	
		Tonian	submergent		
0.98	Mesoproterozoic				transition period
		Late Stenian	oscillatory-emergent		
1.08				grenvillian*	
		Early Stenian	submergent		
1.18					
		Late Ectasian	oscillatory-emergent		
1.28				elzevitian*	
		Early Ectasian	submergent		
1.38					
		Late Calymmian	oscillatory-emergent		
1.48				kilarnean*	
		Early Calymmian	submergent		
1.58					
		Late Statherian	oscillatory-emergent		
1.68				hudsonian*	
		Early Statherian	submergent		
1.78	Paleoproterozoic				stabilization period
1.88			?		
1.98		Orosirian		moranian*	
2.08					
2.18		Rhyacian		blezardian*	
2.28					
2.38					
		Siderian		kenoran*	
2.48					
2.58					

* - after Canadian Shield orogenetic events (Harland et al., 1982).

— — - limit Paleoproterozoic-Mesoproterozoic after Brito Neves et al. (1990).

Subdivision Precambrian, partly after Cowie et al. (1989).

3.4 CYCLIC BASIN FORMATION

Sedimentary basins are formed by subsidence of the upper surface of the earth's crust by a number of processes, such as crustal thinning by extensional stretching; mantle to lithospheric thickening due to cooling of the lithosphere; sedimentary, volcanic, tectonic, and subcrustal loading; astenospheric flow; and crustal densification (Costa et al., 1992; Ingersoll and Busby, 1995). Essentially all basin formation settings involve a complex combination of processes, which may suffer modifying influences by ancillary effects.

The cyclicity of the tectonic events and their consequent sedimentary basin formation becomes clear from Table 1.2. During every pair of tectonic-sedimentary episodes (submergent and oscillatory-emergent), a number of basins is generated. The subsidence of the area prevailing in submergent episodes triggers the formation of a depression that becomes a sedimentary basin when enough relief difference is reached (Magnavita, 1996). When the tectonic activity during submergence is quite strong, rift formation prevails. But when the tectonic activity is less intense, the basins are generally centroclines or syneclises, implanted often upon an earlier rifted basement. More rifting and accompanying basin formation occurs at the end of the submergent episode and continues during the beginning of the following oscillatory-emergent episode, although on a smaller scale and as intracontinental small-sized rift basins. In the case of reactivation of already existing rifts and centroclines or syneclises, this happens also during submergent episodes but not always at every time, remaining restricted only to those episodes in areas with a rather strong tectonic activity.

The character of the lithic infillings of these sedimentary basins reflects their supposed origin. Record of oceanic crust has only been reported from the marginal basins hence their origin. From other periods the sedimentary sequences point only to basins implanted on continental crust, as is the case with the intracontinental basins.

3.5 SEDIMENTARY FILL

It is known that certain sediment types are preferably deposited during definite periods of the earth's history. Already in 1940, Pustowalow (after Ruchin, 1958) wrote – Certain geological periods are characterized by a dominant deposition of definite sediment types, in which the more intense formation of these types is periodically repeated during the course of earth's history. The succession in the formation of these dominant sediment types coincides with the scheme for the separation of sedimentary material, forming the big periods in sediment accumulation. In the background of these big periods, smaller sedimentation periods may occur, but these are only of local importance and due to local tectonic manifestations.

The periodicity in the formation of sedimentary rock types is determined by the repetition of geological phenomena of various types during the history of the earth. In this sense, the cyclicity of these phenomena is the obvious cause of the sediment type accumulation periodicity (cf. Mabesoone, 2003a).

Every tectonic cycle consists thus of two episodes, one submergent during which plate action causes folding in the crust, and the other oscillatory-emergent during which epeirogenetic uplift of these folded belts generates a significant relief of mountains and cordillera which become eroded near the end of the episode. Quite naturally, the differences between these episodes give rise to the deposition of different sediment types.

When basins start to form, the first accumulations of sediment occur near the borders when sufficient relief differences are attained. Generally such sediments are clastics, and their textural character depends on the relief type of the basin surroundings and provenance areas. In case these reliefs are young and vigorous as is the case of mobile belts, the sediments produced range between conglomerates and coarse sandstones, commonly of immature texture and composition, deposited more as a consequence of tectonic movements and subsequent erosion than of weathering. Rift basins produce coarser sediments because of the fault scarps, than centroclines and syneclises that have less abrupt surrounding reliefs. When these surrounding reliefs are platform areas, the produced sediments are medium- to fine-grained clastics, of a more mature character, deposited as a consequence of a rather accentuated weathering and removal more due to climatic changes than to tectonic activity (Mabesoone, 2003b). Towards the basin centers the coarse- to medium-sized clastic deposits become finer in a distal position, assuming a silty to clayey character. When the sea encroaches upon these basins, the terrigenous clastic deposits become substituted by carbonate sediments and, under dry climatic circumstances even by evaporites (Cecil et al., 2003). These types of sedimentary infillings occur chiefly during submergent episodes. The sedimentary response during the submergent episode mode is thus dominated by three factors: clastic sediment derived from erosion within the craton, non-clastic sediment deposited in epicontinental seas, which invaded the area, and some clastic wedges of sediment coming from remaining mountain ranges, which may exist near the cratonic margin.

A subsequent change in tectonic regime from extensional to contractional, as takes place during oscillatory-emergent episodes, generates basin inversion. Originally only applied in basins implanted on intraplate areas, the term may be used also for marginal basins, and even extensive to active mobile belts. The relief of such basins becomes elevated, the sea withdraws from the area, and the environment turns continental again. The then produced sediments are clastics, the coarser when the relief is accentuated, the finer when this relief gets a more platform character (Mabesoone, 2000a, b). Due to the oscillating uplift movements, the sedimentation is not continuous and a great number of regional and local unconformities develop. It is this phenomenon that enables the distinction between clastic relief-correlated sediments of oscillatory-emergent episodes and those of submergent episodes. Such differences could be identified, for instance, in the NE Brazilian Borborema tectonic province (Mabesoone, 2003b), since the uppermost Neoproterozoic. A complicating factor occurs when coarse deposits are reworked during later cycles, and accumulated in environments with flatter reliefs. During an oscillatory-emergent episode the craton is elevated with respect to sea level, and a considerable relief is developed in an interior shield area. Rather immature clastic sediments of psephitic and psammitic character are transported towards the cratonic margin or become trapped in intracratonic fault-bounded basins that occur frequently in this type of period. The episode starts with a regional unconformity that causes the substitution of carbonates and fine mature clastics at the cratonic margins by immature coarser clastics of internal derivation and by thick wedges of arkoses and lithic graywackes in craton-interior positions. The periodic or long-continued exclusion of the sea produces thick successions of non-marine sediments or of cyclically alternating marine and non-marine deposits. An eventual presence of carbonates and other non-clastic deposits is favored by the local climate of the area in question.

And because the tectonic-sedimentary episodes follow each other in a cyclic succession, the very character of the deposited sediments shows a same type of cyclicity. Nevertheless, in all these reconstructions of cyclic sedimentary basin development and their infillings, their preservation potential and consequent plate tectonic reconstruction should be seriously taken into account, as emphasized Ingersoll and Busby (1995), a subject that has rarely been discussed.

Thus, the abundance of different types and groups of sedimentary rocks changes regularly in the course of time, forming periodically repeated maxima and minima, which are strictly limited to definite stages of the tectonic-sedimentary cycles. The rather unimportant differences that may appear, are due to local circumstances, such as availability of rocks to be weathered and climate, as well as the rate of uplift or subsidence.

Figure 3.3 summarizes the features presented above.

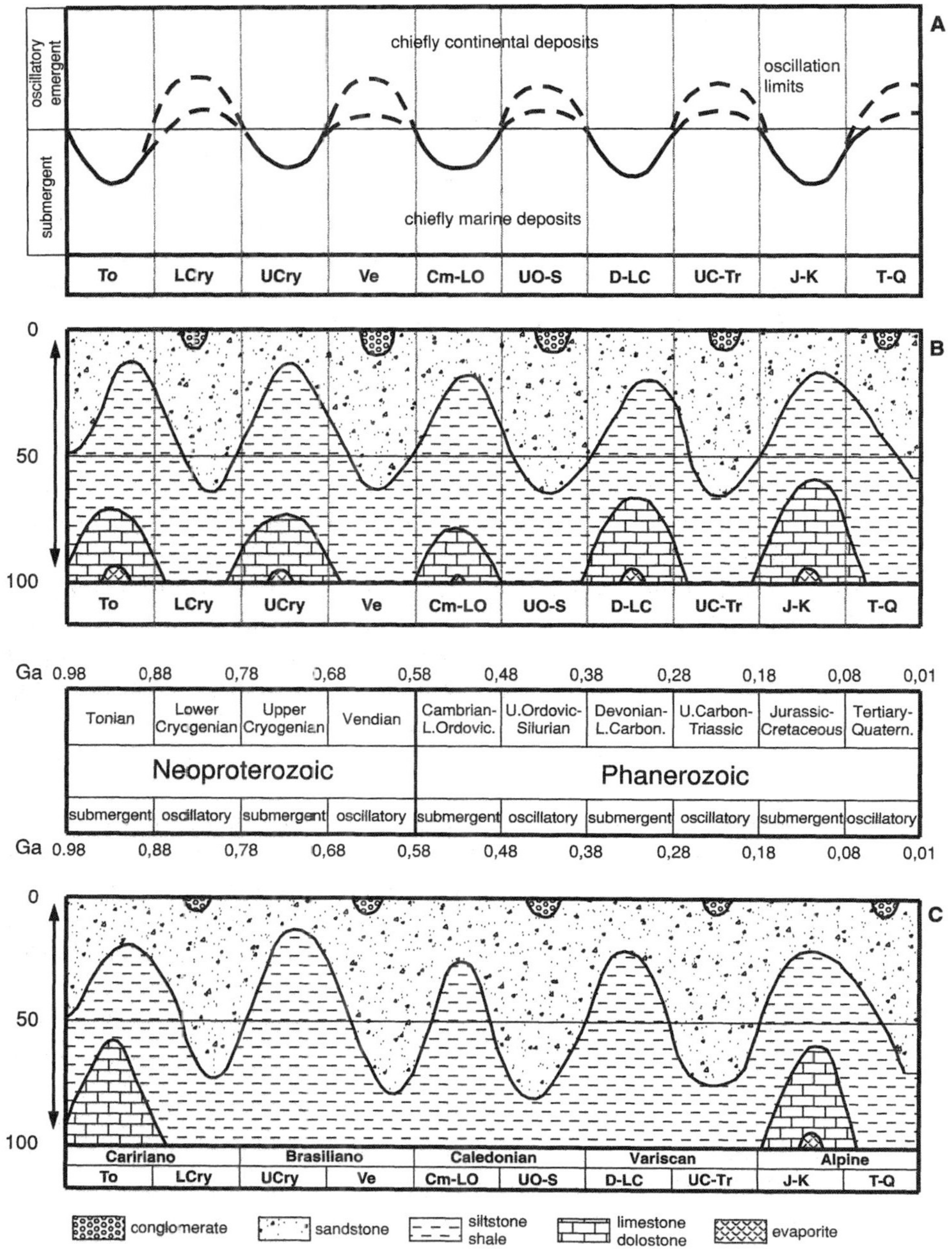

Fig. 3.3: Periodicity in sediment type accumulation since Neoproterozoic, after Mabesoone (2003a): (A) oscillation curve, (B) general sediment type accumulation, (C) examples from NE Brazilian Borborema tectonic province.

3.6 METALLOGENIC EPOCHS

Theoretically, mineral prediction involves basically two aspects: (1) laws of metallogenic epochs and metallogenic provinces, and the mutual function between them; (2) origin of ore deposits, or the procedure of their formation. In short, the difficult point of the problem lies in deciphering the law of concentration of ore-genetic elements in the span of crustal evolution. Sedimentary mineral deposits are nothing but a special type of sedimentary facies, a kind of mineralized lithosome related to its environment. They are therefore found in definite episodes of crustal evolution (Yeh Lien-tsun, 1977).

Thus, in this case a cyclicity becomes also apparent, dependent directly on the tectonic behavior of cratons and their margins. The cited author tried to decipher a law about the concentration of elements, which generate sedimentary ores within the crustal evolution. As this evolution is cyclic, the sedimentary mineral deposits must develop during certain determined epochs. And also in this case, the Phanerozoic is the best-known era, although the author (Yeh Lien-tsun) has considered in his conclusions the uppermost Proterozoic as well.

The existence of metallogenic periods and the completeness of a metallogenic sequence therein, relieves not simply on the evolution of the crustal movements, but also on the variations of the superficial geological environment, which play also a component part in the general process of the earth's evolution. The exogenic variations and the endogenic evolution for a pair of opposites and the unity of these opposites result in the formation of superficial ore deposits. Paleoclimatic and paleo-atmospheric changes, sea-level fluctuations and the evolution of ancient life may be regarded as the leading factors which control the formation of these superficial ore deposits (Fig. 3.4). Several metallogenic epochs of sedimentary mineral deposits may be recognized and each of them is marked roughly by a similar metallogenic sequence like this (from old to new): iron, manganese, phosphate, bauxite, coal, copper, and salts. This implies neatly in climates, since the deposits of iron represent commonly warm and humid climates and the salts arid climates. There exists also a relation with the transgressional and regressional cycles; the marine deposits: oolitic iron, manganese, and phosphorite, form during the transgressions of a submergent episode, and the continental deposits: bauxite, coal, copper in sandstones, and evaporites, develop mainly during the regression of an oscillatory-emergent episode.

The role of organisms and organic matter is also evident in the metallogenic periods and sequences. The Precambrian banded iron deposits occurred almost at the same time as the procaryotes. The well-known phosphorites accumulated in the beginning of the Phanerozoic appeared at about the same time when stromatolites flourished. Bauxite and coal came into being only when the

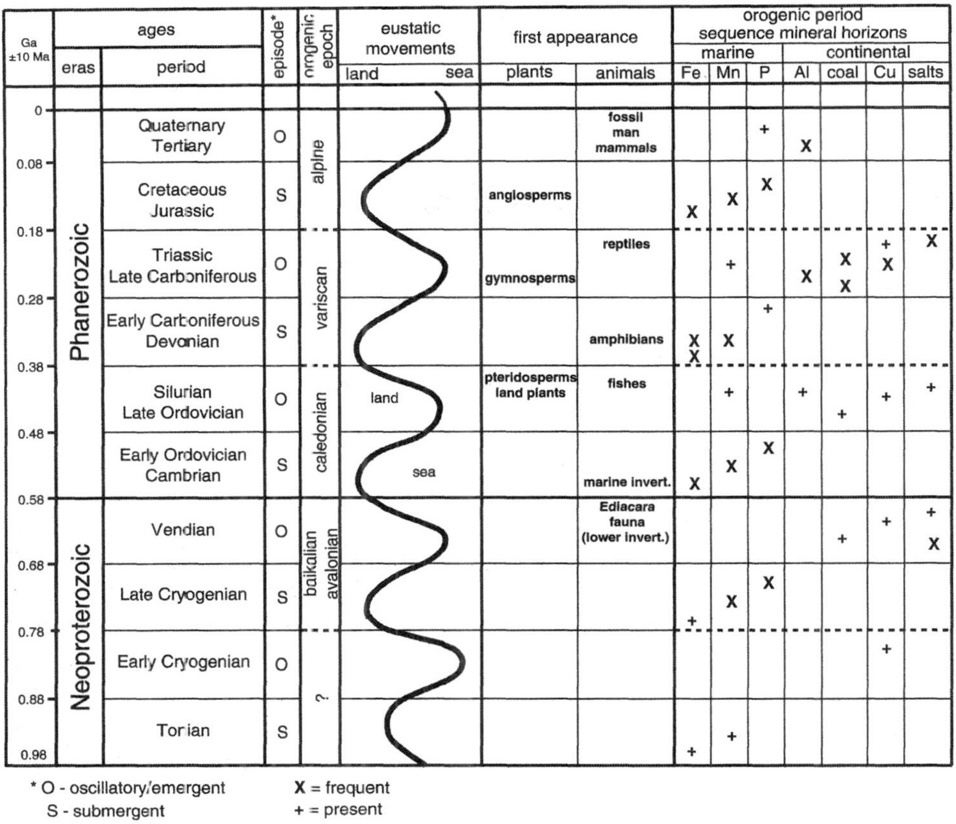

Fig. 3.4: Cyclic sedimentary mineral metallogenetic epochs, partly after data from Yeh Lien-tsun (1977).

terrestrial vegetation became well developed and abundant during the Variscan orogeny. In short, metallogenic periods and metallogenic sequences of sedimentary deposits are closely related to the development and evolution of organisms and vegetation.

To conclude, the whole history of the origin of sedimentary ore deposits is by no means concerned simply with a process of sedimentary differentiation. Provenance ore-forming materials of different composition and different mode of existence were formed at different time spans. The complete procedure of sedimentary ore genesis includes first weathering differentiation, then sedimentary differentiation, and finally diagenetic differentiation, each process having taken place during certain, well-defined tectonic-sedimentary episodes (see

Fig. 3.4). This also permits the conclusion about the existence of a cyclicity in the development of sedimentary ores.

3.7 FINAL REMARKS

It became clear that most of the processes that occur in the earth's history follow a cyclic path, probably caused by orbital forces. These cyclic sequences of events in tectonics, sedimentation, climate and sea level, are neatly reflected in the stratigraphical record. A 200 Ma periodicity results essentially from the assembly and dismemberment of supercontinents. The 100 Ma tectonic-sedimentary episodes result in the accumulation of different sediment types. This can be traced in many cratonic and marginal areas of the continents. The shorter term cyclic sequences are the result of regional to local geological phenomena such that they are generally not correlatable on an intercontinental scale.

The long-term cycles in earth's geological processes are still subject of controversy. However, the foregoing subjects seem to confirm that the existence of these long-term cycles can not be longer denied, although more data, especially for the Proterozoic, should be obtained.

4. EXAMPLES OF SEDIMENTARY BASINS OF WORLD

J.M. MABESOONE

4.1 INTRODUCTION

In this chapter a number of areas with sedimentary basins and their respective cyclic developments have been considered, based on interpretations and sometimes reinterpretations of data taken from literature and field trip guides in which the sedimentological aspects have been illustrated. As is generally the case, a cyclic evolution of these basins has as yet not been included in the publications, justifying thus a reinterpretation of the presented data.

In Fig. 4.1, the selected areas have been mentioned. For their interpretation, some basic publications have been consulted, as follows:

(1) Andes Fueginos – field trip guide by Olivero (1998);
(2) Patagonia and Andean foreland basins – various publications presented in congresses and symposia about South American geology;
(3) Karoo Basin in South Africa – field trip guide by Hancox et al. (2002), with annexes;
(4) Sedimentary basins of NW and Central Europe – after data presented by Ziegler (1978) and Zwart and Dornsiepen (1978);
(5) Western Ireland – field trip guides by Elliott et al. (2000) and Williams et al. (2000), respectively;
(6) Duero Basin in Spain – field trip guide by Mediavilla et al. (1998).

Besides these basic publications, additional papers about these areas have also been taken into account, when important for their interpretation. Of great help in this sense were the abstracts of the papers presented during the International Sedimentological Congresses of 1994 (Recife, Brazil), 1998 (Alicante, Spain) and 2002 (Johannesburg, South Africa), the Latin-American Sedimentological Congress of 1997 (Porlamar, Isla Margarita, Venezuela), 2000 (Mar del Plata, Argentina), and 2003 (Belém, Pará, Brazil) as well as abstracts published in the respective volumes of Annual Sedimentology Meetings, generally realized in Europe.

Cyclic Development of Sedimentary Basins
J.M. Mabesoone and V.H. Neumann, editors.

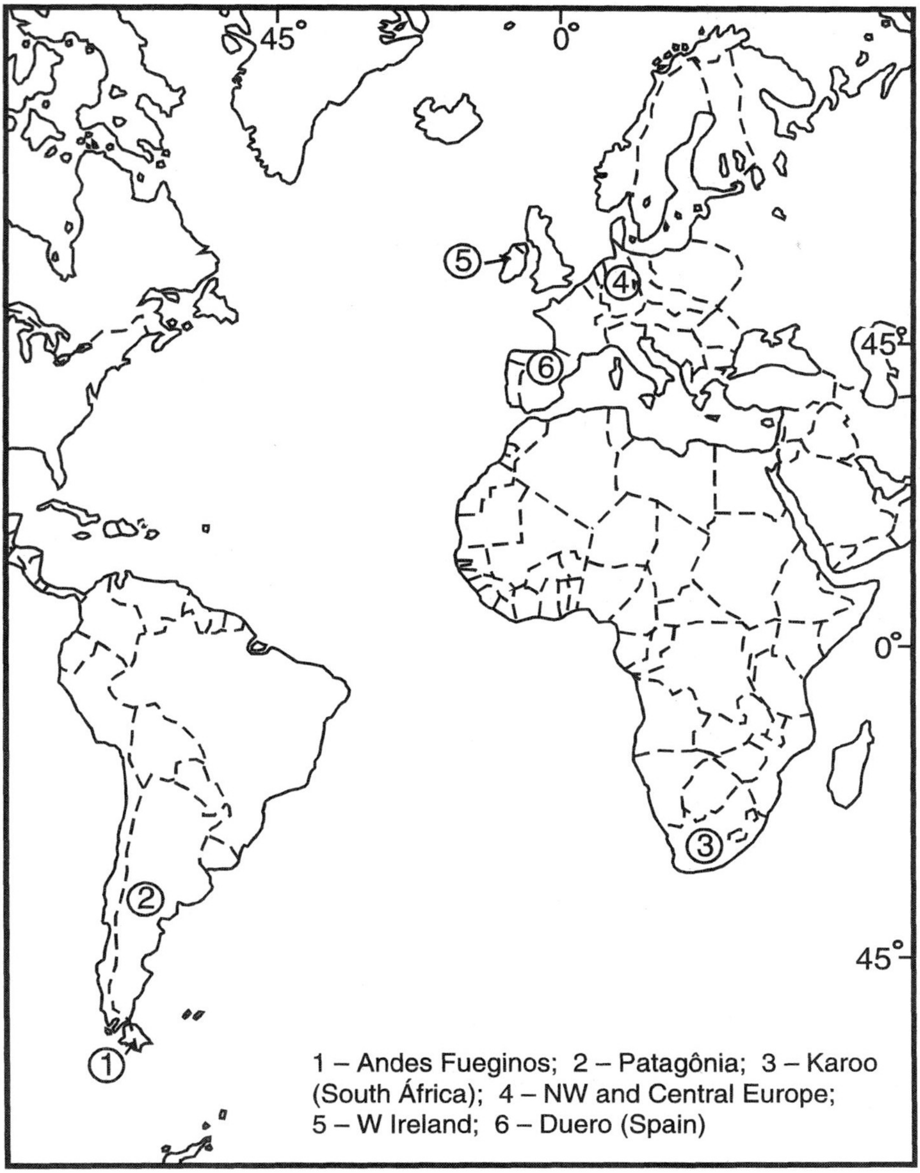

Fig. 4.1: World basins, selected areas.

4.2 ANDES FUEGINOS

4.2.1 TIERRA DEL FUEGO: MAIN GEOGRAPHY AND MORPHOSTRUCTURAL AREAS

The Andes Fueginos occupy the southernmost tip of South America on the Isla Grande de Tierra del Fuego and adjacent smaller islands. The island's territory is divided in an Argentinean and a Chilean part.

The mountain chain forms a prominent arcuate belt with a general latitudinal orientation. On the main island and its continental extension to the north, Olivero (1998) distinguishes four main morphostructural units (Fig. 4.2), from south to north:

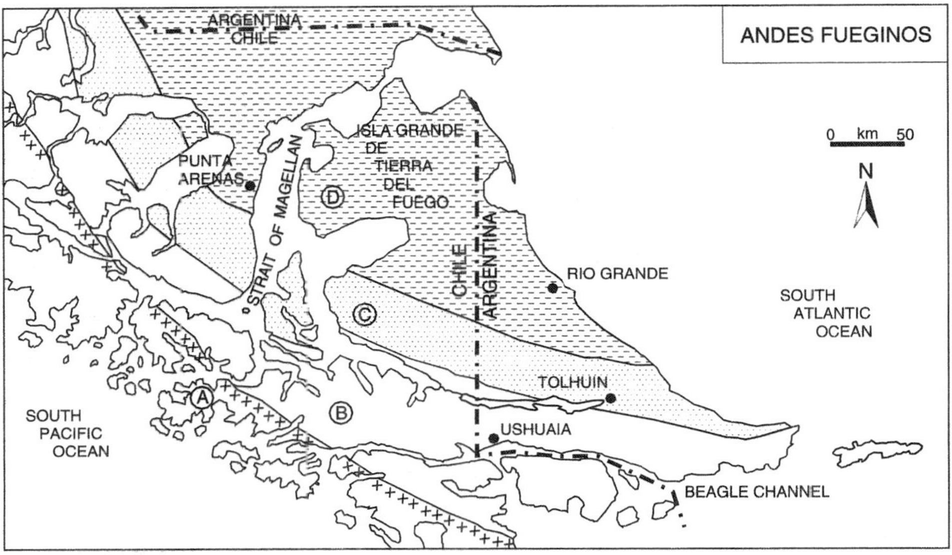

Fig. 4.2: Tierra del Fuego: geography and geological overview.

(A) Chilean archipelago, with Cretaceous-Cenozoic plutonic rocks, being the roots of a Pacific magmatic arc;

(B) Central Cordillera, north of the Beagle Channel, composed of Paleozoic-Jurassic crystalline rocks, and Late Jurassic-Early Cretaceous sedimentary and volcanic rocks of the Rocas Verdes Marginal Basin;

(C) Pre-Cordillera at the northern foothills of the mountain chain, with Late Cretaceous-Paleogene sediments, folded and thrusted in belts of the Austral and Malvinas foreland basins;

(D) Plains and low hills in the north, presenting subhorizontal Tertiary sediments covered by Quaternary glacial and fluvio-glacial deposits.

Three sedimentary basins are recognized in the area: (1) the Malvinas Basin in the NE, (2) the Austral Basin in the NW, and (3) the Rocas Verdes Marginal Basin in the S. The origin of these basins is related to Jurassic crustal stretching and rifting; however, their subsequent Cretaceous-Tertiary evolution differs markedly (Olivero, 1997).

4.2.2 STRATIGRAPHICAL SEQUENCES

4.2.2.1 Generalities

In Table 4.1 is presented a simplified and generalized stratigraphical column that also mentions the main tectonic and geological settings, and chiefly the principal Jurassic-Cenozoic events.

Table 4.1: Stratigraphical column of Andes Fueginos, and lithologic character of deposits.

Quaternary	
Glacial sequences	glacial, fluvio-glacial and littoral deposits
Paleogene	
Cabo Peña Formation and equivalent units	conglomerates, sandstones, shales and limestones, in unconformably bounded successions
La Despedida Group	
Rio Claro Formation	
Late Cretaceous	
Cerro Matrero Formation	heterolithic rhythmites (mudstone-siltstone-fine sandstones)
Policarpo Formation	mudstones, fine sandstones
Early Cretaceous	
Beauvoir Formation	black shales, marls
Yahgan Formation	volcaniclastic turbidites
Late Jurassic	
Lemaire Formation	submarine volcanic complex (tuff, breccias, conglomerates, turbidites, shales)

Olivero and Martinioni (1998) concluded that the Alpine orogenic cycle exercised a strong influence controlling the stratigraphic evolution of the region. The authors recognized seven units that reflect the main features of the evolving tectonic regimes. These units can partly be correlated with the Earth's cyclic tectonic-sedimentary episodes.

4.2.2.2 Basement

The rocks of the crystalline basement crop out along the Chilean part of the Isla Grande de Tierra del Fuego and the westernmost Argentinean side of the Beagle Channel. These rocks are chlorite-sericite and biotite-garnet schists, greenstones and amphibolites, highly deformed, and divided into two stratigraphical units separated by an important unconformity. The peak of metamorphism was reached in the Middle-Upper Cretaceous. The original basement rocks are considered to be of Late Paleozoic to Early Jurassic age and they are interpreted as an accretionary prism on the Pacific margin of Gondwana.

4.2.2.3 Rocas Verdes Marginal Basin

Lemaire Formation (Late Jurassic). Resting unconformably upon the basement sequence, the Late Jurassic Lemaire Formation is composed by a complex submarine unit of rhyolitic lavas and domes, acidic volcani-clastic breccias, tuffs, conglomerates, turbidites, shales and basaltic rocks. These Lemaire rocks are strongly deformed; their stratification has only been preserved in the coarse-grained facies whereas in the fine-grained facies a penetrative cleavage has completely obliterated the original stratification (Olivero et al., 1997). The formation was deposited in hemi-grabens and its upper contact with overlying Early Cretaceous sediments is unconformable (Biddle et al., 1986).

The rhyolitic Jurassic volcanism is interpreted as an episode of regional extension related to the initial break-up of Gondwana and the opening of the South Atlantic Ocean. In contrast, the submarine silicic-basaltic volcanism is thought to represent widespread Jurassic extension and development of a narrow deep-marine volcano-tectonic rift parallel to the Andean side of South America. A continued extension during the Late Jurassic-Early Cretaceous resulted in the opening of the Cretaceous Rocas Verdes Marginal Basin.

The Late Jurassic age for the Lemaire Formation is based on paleontological evidence collected in southern Chile. Ammonites, belemnites and bivalves point to a Kimmeridgian-Tithonian fossil association (Fuenzalida and Covacevich, 1988).

Yahgan and Beauvoir Formations (Early Cretaceous). The Yahgan lithostratigraphic unit consists of a complex association of coarse conglomerates, sandstones, sandy and silty turbidites, black tuffaceous mudstones and tuffs (Fig. 4.3). From the southern Chilean archipelago to the northern margin of the Beagle Channel, these facies types change. Part of the coarse facies bears Aptian-Albian corals in its upper section, and bivalves, and Tithonian-Neocomian ammonites and belemnites in its lower section. In the finer muddy and sandy facies, a few horizons bear a distinctive trace fossil assemblage (Olivero and Martinioni, 1996). Diverse types of basaltic rocks are intruded in this Yahgan Formation that is furthermore strongly folded, chiefly with respect to the finer facies, and shows a low-grade metamorphism. The unit is interpreted as the volcaniclastic infilling of the Rocas Verdes Marginal Basin, between the Pacific volcanic arc and the South American continent. This infilling took place in clastic wedges, with the thicker and coarser facies restricted to the S, near the volcanic arc. These coarse facies become markedly thinner towards the N, grading into fine-grained basin plain rocks and then into slope and distal platform lime-rich black mudstones and marls in the Austral Basin.

Fig. 4.3: Andes Fueginos: Beagle Channel, E Ushuaia; Early Cretaceous Yahgan Formation sandy trubidites.

The finer facies have been taken together in the Beauvoir Formation, forming a thick section of marine sedimentary rocks in central Tierra del Fuego to the north of Lago Faguano (Fig. 4.2). The rocks of this formation are very homogeneous, with the beds up to 1 m in thickness, consisting mainly of dark-coloured mudstones, lutites and tuffs. Some intervals show rhythmites of fine sandstone and mudstone. The depositional environment was marine, partly of deeper realms involving slope and basin plain deposits (Biddle et al., 1986). The age of the Beauvoir Formation is inadequately constrained due to the poorly preserved inoceramid bivalves. However, the suggested age is late Early Cretaceous, possibly Albian.

Cerro Matrero and Policarpo Formations (Late Cretaceous). The Late Cretaceous rocks in Tierra del Fuego are but poorly known. The stratigraphical column of the central part of the island consists of a thick mudstone dominated unit passing upward into a rhythmic interlayering of mudstone and fine sandstone. The lower section shows massive to poorly laminated mudstone and siltstone, analogous to the Beauvoir Formation sediments, but with a distinct younger fossil invertebrate assemblage. The uppermost section with heterolithic rhythmites is characteristic for the Cerro Matrero Formation. At other places, the unit presents silty-sandy mudstones, clayey siltstones and argillaceous, very fine sandstones. In calcareous concretions, there occur trace fossils.

Similar sequences of mudstones and fine sandstones are known from the Mitre Peninsula. There its upper part is exposed along the Atlantic coast and called the Policarpo Formation.

In central Tierra del Fuego, paleocurrent measurements point to a northward-directed sediment dispersal pattern. The Late Cretaceous rocks were deposited in a marine shelf environment below the wave base level. The fossil assemblages determined at different places, include Turonian-Coniacian, Santonian-Campanian and Maastrichtian marine species as well as some palynomorphs.

4.2.2.4 Austral Basin

Springhill Formation and other Early Cretaceous units. The Austral Basin appears on surface in the northernmost part of Isla Grande de Tierra del Fuego (Fig. 4.4), not exactly more belonging to the Andes Fueginos. The stratigraphy of the area has been presented by Cagnolatti et al. (1989). During the Early Cretaceous, when the basin suffered subsidence due to cooling, the sea invaded the area. About 700 m thick sequence of pelites and marls record the event; the deposits have been united into a number of lithostratigraphic formations. At the base, there occurs still a sequence of sandstones, siltstones and claystones of Berriasian-Valanginian age, accumulated in continental, littoral and transitional environments, and resting unconformably on top of the Late Jurassic Lemaire Formation.

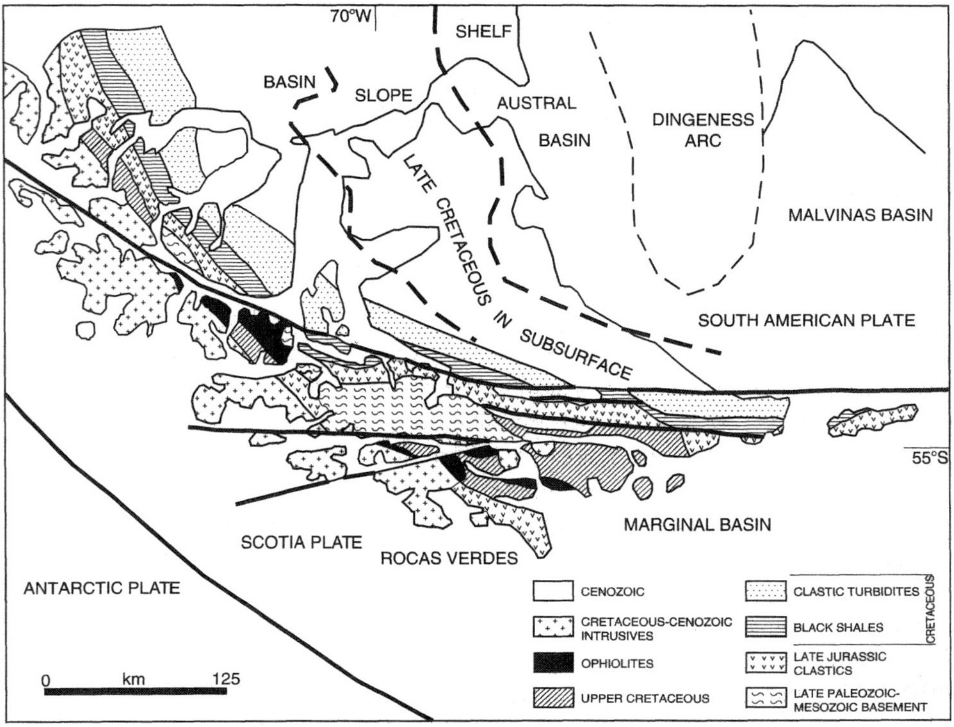

Fig. 4.4: Sedimentary basins of Tierra del Fuego.

4.2.2.5 Austral and Malvinas foreland molasse basins

During the Late Cretaceous, the compressional Patagonidic orogenic phase resulted in tectonic inversion and closure of the marginal Rocas Verdes Basin, besides metamorphism, folding and an initial uplifting of the Andes Fueginos.

Rio Claro Formation and equivalent units (latest Cretaceous-Paleocene). The Rio Claro Formation represents the first deposited molassic sediments of the foreland stage of the Andes Fueginos. At that time, the already uplifted central Andean rocks became exposed to subaerial erosion, and the lowermost clastic sequence evidences an Andean source. Typical rocks of the formation are massive sandstones with minor, alternating beds of coquinas, mudstones and conglomerates. At other places, there appear marine sandstones, coarse conglomerates, resedimented breccias, and mudstone-siltstone rhythmites. The age has been based chiefly on Late Cretaceous-Early Paleocene palynomorphs and Paleocene to Early Eocene mollusks.

The Rio Claro strata are steeply tilted and folded in rather large (over 2 km in wavelength) anticlines and synclines, striking generally in an E–W direction. The formation appears essentially in the Austral Basin, while in the Malvinas Basin the sequence is called Rio Bueno Formation (Olivero, 1997). The subsurface stratigraphy of these basins is roughly consistent with the observed surface geology (Martinioni, 1997).

La Despedida Group (Eocene). This sequence rests unconformably upon the Rio Claro Formation. The best exposures of the unit are found in the northwestern part of the island and along the Atlantic shore (Fig. 4.5).

Fig. 4.5: Tierra del Fuego, Atlantic coast, Punta Torcida: sandstones and mudstones of Paleogene La Despedida Group.

The lithostratigraphic group consists of a very thick clastic wedge, reaching a thickness up to 1600 m in the south and thinning remarkably to the north. Along the Atlantic shore, the almost complete section consists of basal mudstone-dominated marine intervals; middle glauconitic sandstone-dominated marine to estuarine intervals; and upper mudstone-sandstone marine intervals. To the south along the Atlantic coastline, there appear bryozoan-rich coquinas and limestones, of Early Eocene age. Furthermore, the La Despedida Group preserves a quite rich foraminiferal microfauna, and also calcareous plankton, diatoms and radiolarians; microfossils are also known. The unit is of a typical marine origin.

The different strata are variably tilted and folded into open or asymmetric anticlines and synclines. At the top of the unit, another angular unconformity separates it from the subhorizontal deposits of the overlying Cabo Peña-Rio Leona Formations.

Cabo Peña-Rio Leona Formations (end Eocene-Oligocene). Cropping out along the Atlantic shore and to the north of the folded Eocene deposits of the La Despedida Group, occur the subhorizontal beds of the Cabo Peña-Rio Leona Formations. The outcropping strata are chiefly friable, well-bedded mudstones with some thin sandstone intercalations, or at other places massive, poorly cemented, thick sandstone beds with alternating mudstones. The depositional environment was marine to non-marine, without more specification due to lack of detailed studies. The age is believed to be latest Eocene to Oligocene, based on scarce fossils.

With exception of a few localities where the strata are moderately dipping (less than 30°), generally found adjacent to vertical or steeply inclined faults, most of the strata are subhorizontal. In some road cuts, spectecular intruded sandstone dikes may be observed.

4.2.2.6 Geological evolution

The present-day physiographic elements of southernmost South America reflect a complex interaction of the tectonic plates of South America, Scotia and the Antarctic. Since Cretaceous, this interplay resulted in a series of different tectonic regimes, which determined the geological evolution of the Andes Fueginos. The orocline, extended eastward for about 2000 km as a piece of the partly submerged and broken Nova Scotia Ridge, is located near the diffuse triple junction of the three mentioned plates (Fig. 4.2).

The origin of the Rocas Verdes Marginal Basin, the Austral Basin and the Malvinas Basin is related to Jurassic crustal stretching within a regional extensional tectonic regime that resulted in the opening of the South Atlantic Ocean. During the rifting episode, a series of grabens originated, bounded by NW trending faults. They became filled in with silicic volcanic and volcaniclastic rocks, regionally extending into Patagonia.

The opening of the Rocas Verdes Marginal Basin took place during the Late Jurassic-Early Cretaceous continued extension. The basin is located between the South American continent and a Pacific volcanic arc, and has been filled in with thick, andesitic volcaniclastic turbidites of the Yahgan Formation, chiefly during the Early Cretaceous. The dominantly andesitic composition of these rocks points to an important volcanic activity in the Pacific arc at that time. Petrographic studies and paleocurrent measurements in the Yahgan Formation are consistent with a source for the detritus in the volcanic arc. Thickness and facies distribution suggest a clastic wedge geometry typical for a volcaniclastic apron, with thicker and coarser deposits near the arc, and thinning markedly to the north, where the formation grades laterally into the Early Cretaceous slope and basinal mudstones (Olivero and Martinioni, 1996). The Austral and Malvinas Basins passed through a subsiding sag phase, with the consequent deposition of transgressive sequences with a main sedimentary source in Patagonia.

In the Malvinas and Austral Basins, the Aptian-Maastrichtian is characterized as an interval of tectonic quiescence, with deposition of basinal, slope and platform mudstones (Biddle et al., 1986). In the Rocas Verdes Marginal Basin, the Late Cretaceous marks a compressional tectonic regime that finished with the closure of the basin and its tectonic inversion. This Andean orogenic phase is responsible for peak metamorphism, isoclinal folding and initial uplifting of the Andes Fueginos (Fig. 4.6).

There exists a clear field evidence of subaerial exposure and erosion of at least part of the present Andes Fueginos during the end of Cretaceous and beginning of Tertiary, due to this uplift. Subsurface data suggest a possible southern source

Fig. 4.6: Physiography and Main Karoo Basin.

for the Late Cretaceous detrital sediments in the Austral and Malvinas Basins. Outcrop data point to a definite source in the Andes Fueginos and a sediment dispersal towards north for the Early Paleocene (Danian). The subaerial exposure of the Andes and the accompanying propagation of thrust sheets onto the Patagonian craton started the foreland evolutionary phase of both basins.

The Paleogene foreland molasse of the Andes Fueginos consists of three main, unconformably bounded, clastic wedges represented by the Rio Claro, La Despedida, Cabo Peña and equivalent lithostratigraphic units. Exposures of these molasse sequences in three parallel belts matching the trend of the Andean axis, and the general younger direction northward, suggest a migration of the depocenters, which accompany the propagation of deformation onto the foreland.

The first clear field evidence of left lateral strike-slip faulting is found affecting Oligocene and younger rocks. Initiation of E–W strike-slip faulting seems to have a post-Late Eocene lower age limit, as is suggested by the strongly brecciated Paleocene rocks and left-lateral offsetting, in the order of 20–30 km, of Cretaceous-Eocene rocks (Olivero, 1997). Neogene drifting and the associated opening of the Drake Passage resulted in important paleogeographical changes which accompany the final separation of South America from Antarctica and the final phase of the oroclinal bending of the Andes Fueginos (Cunningham et al., 1995).

The Quaternary geology of Tierra del Fuego reveals the development of geological processes, which took place at regional and global scale. Ice ages, ash-fall deposition, lacustrine sedimentation, peat accumulation and relative sea-level variations have to be taken into consideration. Valley glaciers as well as large mountain ice sheets occur today on the Andean ranges. The Fueginian Andes were repeatedly glaciated both by mountain ice sheets and hundreds of local valley and cirque glaciers. The mountain ice sheet issued several large outlet glaciers or true ice lobes, which radiated in all directions and in many cases, following tectonically controlled depressions. Glaciers persisted after 10 Ka only as cirque glaciers and small valley glaciers in the eastern Fueginian Andes, and as relics of a mountain ice sheet in the Central Cordillera Darwin. The occurrence of ice bodies within the glaciated valleys is restricted to a minimum elevation of 700–800 m (Rabassa et al., 1998).

4.2.2.7 Conclusions

Table 4.2 presents the sequence of geological events in the Andes Fueginos, and the interpretation of the cyclic development of its basins.

The Jurassic-Paleogene sedimentary rocks, which are exposed along the eastern part of the Andes on the Isla Grande de Tierra del Fuego, reflect a complex

Table 4.2: Timing of main tectonic and geologic settings, and cyclic development of Andes Fueginos sedimentary basins (Olivera, 1998).

ages Ma				tectonic-sedimentary episode	dominant tectonic regime	structural features	main geologic settings		
0	Cenozoic	Quat.		oscillatory-emergent	strike-slip		foreland molasse drifting	Austral and Malvinas Basins	separation South America from Antarctica
	Cenozoic	Tertiary	Neogene	oscillatory-emergent	strike-slip		foreland molasse drifting	Austral and Malvinas Basins	separation South America from Antarctica
50	Cenozoic	Tertiary	Paleogene	oscillatory-emergent	compressional	clear evidence of left-lateral faulting; oligocene: cessation of thrust propagation; 2nd pulse of rapid uplifting and cooling; basement-involved thrust	foreland molasse drifting	Austral and Malvinas Basins	Rio Claro, La Despedida, Cabo Peña units and equivalents
65	Mesozoic	Cretaceous	Late	oscillatory-emergent / submergent	compressional	onset of thrust propagation onto foreland area; 1st pulse of rapid uplifting and cooling	post-rift deposits	tectonic inversion	Cerro Matrero, Policarpo units
100	Mesozoic	Cretaceous	Early	submergent	compressional / extensional	ductile deformation (isoclinal folding/and peak metamorphism	post-rift deposits	Rocas Verdes Marginal Basin	Yahgan and Beauvoir Formations
144, 150	Mesozoic	Jurassic	Late	submergent	extensional		syn-rift deposits	Rocas Verdes Marginal Basin	Lemaire Formation

basement, accretionary prism on Panthalassic margin of Gondwana.

evolution of tectonic regimes. These regimes guided the development of the sedimentary basins and also the final separation of South America and Antarctica. Three sedimentary basins developed in the area: the Malvinas Basin to the NE, the Austral Basin to the NW, and the Rocas Verdes Marginal Basin to the S. The origin of all three basins is related to Jurassic crustal stretching and rifting, but their subsequent Cretaceous-Tertiary evolution differs markedly, neatly reflecting the two different cyclic tectonic-sedimentary episodes involved: submergent during Middle Jurassic-Late Cretaceous, and oscillatory-emergent since Late Cretaceous.

The Rocas Verdes Marginal Basin evolved as a backarc basins filled in with Cretaceous volcaniclastic, deep marine rocks derived from a Pacific-facing andesitic volcanic arc. Late Cretaceous tectonic inversion and closure of the basin provoked a strong deformation and metamorphism, with an initial uplift of the Andes Fueginos and the development of an elongated foredeep flanking the northern continental front of the rising cordillera, during the submergent tectonic-sedimentary episode of that time. Within the syn-rift and post-rift deposits, a cyclic succession of small advances and retreats of the sea level is recorded.

During the following oscillatory-emergent episode from the Late Cretaceous onwards, the Austral and Malvinas foreland basins developed, with the deposition in pulses of deep to shallow marine and littoral molassic sediments, resulting in the three main unconformably bounded cyclic lithic successions. The Neogene drifting of continental blocks, associated with the opening of the Scotia Sea, brought together the northern and southern boundaries of these basins along the present Magallanes-Fagnano fault system. The Quaternary glaciation, characteristic for the episode, strongly affected the Andes Fueginos, being still present today in the region.

4.3 ANDEAN FORELAND BASINS: PATAGONIA

4.3.1 CORDILLERA DE LOS ANDES

The Andes fold belt extends over more than 10,000 km through Middle and South America. Although the cordillera has been traditionally considered as one single morphological unit, it constitutes actually various and different morphostructural elements (Fig. 4.7; Loczy and Ladeira, 1976, Chapter 28). Paleozoic, Mesozoic and Tertiary geosynclines show many interruptions and variations parallelling the Andean mobile belt. The primary orogeneses of the Andes developed since the beginning of the Paleozoic until about its end. The modern Andes started its development in the Late Mesozoic, with the first uplifts and a reactivation during the Oligocene.

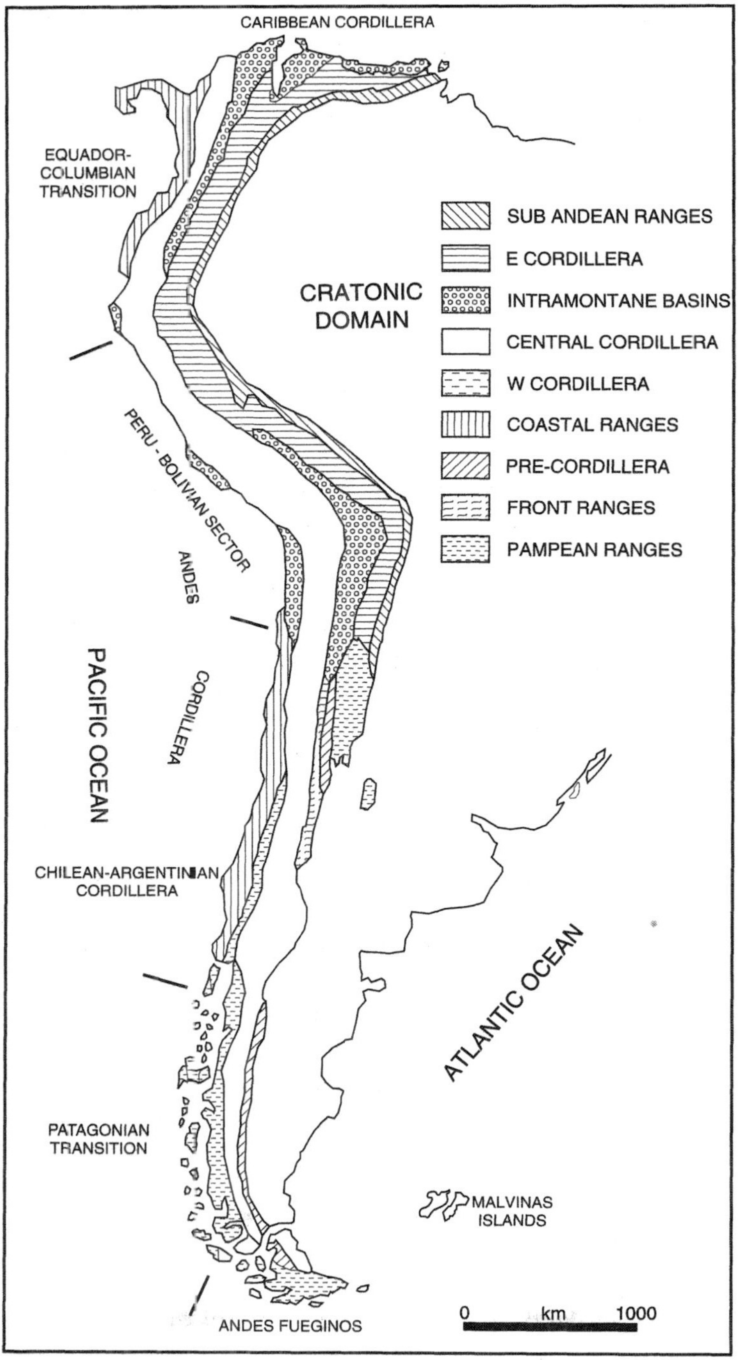

Fig. 4.7: Andes mountain chain; morpho-structural elements.

Indeed, the Cretaceous System is a key epoch to understand the tectonics of South America, due to the fact that it records the transition from an extensional to a compressional Andean-type system, similar to the present times. The Late Cretaceous drift of the continents triggered important kinematic changes. The rollover velocity of the Pacific trench switched from negative to positive, producing an important modification in the kinematics of the upper crustal deformation. The Andes were sensitive to these changes, and therefore provide a source for understanding the processes. The post-Paleozoic tectonics of South America resulted mainly from the rifting and subsequent readjustments contemporaneous to the drifting. The continent did not drift as a rigid entity, but was fragmented into mega-mosaic shield-blocks, readjusted by vertical and roto-translational movements. The derived movements would not only explain the crustal fracturing, with associated magmatism and the eastern taphrogeny, but also the counterpart events which have gone on in the western compensatory Andean Orogeny and the transcurrent faults of continental importance of South America (Rezende, 1972).

Three different stages can be recognized in the Andean orogen during the Cretaceous, which match what is also known from the Atlantic passive margin and interior basins. The first stage was characterized by a generalized extension in the Andean arc and surrounding areas. This tectonic regime controlled a widespread volcanism along and across the arc. Sedimentation showed quick variations from limited carbonate ramps to clastic and volcaniclastic deposits interfingered with volcanic episodes. The backarc basins dominated the sedimentation at this stage. Extensive carbonate and clastic platforms cover most of the pre-Andean regions. The foreland and the continental margins were also dominated by extension. A series of diachronous rift systems developed from west to east, as for instance the Salta rift basins, the Cretaceous Paraná Basin and the marginal basins of the platform.

The second stage appears to be a transitional period. It records a retreat of the Pacific sea from the orogenic areas, a continentalization of the volcanic arc, and important retroarc basins. Thermal subsidence controlled the sedimentation in these basins. Widespread clastic deposits, with volcaniclastic aprons, dominate the sedimentary facies during this stage. The foreland areas slowly changed from syn-rift facies to thermal sag facies, as for instance in the San Jorge Basin (Figari et al., 2000), although in some proximal areas normal fault reactivations have been observed. In the marginal platforms, this stage coincides with the break-up unconformity of the passive margins.

The third stage is associated with incipient shortening and the first inception of retroarc foreland basins, and progressed from south to north. This stage led to the present Andean regime, where shortening and crustal thickening became significant. The foreland areas began their tectonic inversion that continued during most of the Cenozoic.

The comparison among different segments and transects across the continent shows that the stages were diachronous, strongly controlled by crustal rheology, inherited from the pre-Cretaceous geologic history.

In the present subtitle have been considered the continental foreland basins of the Eastern Cordillera and part of the Subandean belt in southern Argentina (Fig. 4.8). The extension in Tierra del Fuego has been dealt with in detail, in another subtitle. The Patagonian part considers the cyclic development of the sedimentary basins, with the Neuquén Basin in the north and the Austral and Malvinas basins in the south, interrupted by the Pampean, Patagonian and Deseado Massifs. The northern part considers the foreland basins between the easternmost mountain chain and the Precambrain Brazilian shield and the Chaco-Paraná intracontinental syneclise.

4.3.2 MESOZOIC-CENOZOIC EVOLUTION OF SOUTHERN SOUTH AMERICA

During the Mesozoic and Cenozoic, the tectonic development of southern South America was controlled by a complicated subduction regime along its western margin and the evolving spreading centre of the Mid-Atlantic ridge along its eastern margin (Uliana and Biddle, 1988). Processes associated with activity along these two plate boundaries provided the first-order controls on subsidence and uplift in the area, and thus controlled the sites available for sediment accumulation. Both long- and short-term sea-level variations were superimposed on the tectonic framework and modified the paleogeography of the continent.

Within the cyclic tectonic-sedimentary episodes, from mid-Triassic onwards, the basin development and the sedimentary infillings have been considered.

4.3.3 PATAGONIAN FORELAND BASINS

The integrated basin analyses of this area in southern Argentina have been presented in various papers (Spalletti and Franzese, 1996; Spalletti et al., 1999a, b, 2002; Franzese and Spalletti, 2000; Spalletti, 2000). Details about different sedimentological aspects have also been mentioned, due to the intense sedimentological studies effectuated in the basins.

The Patagonian sedimentary basins occur in various geoloigcal settings. The southern part of South America represents an extensive platform area with Permo-Carboniferous and Mesozoic-Cenozoic covers upon a tectonically deformed basement of Precambrian to Early Paleozoic age. The basins found at the eastern margin of the Andes mountain chain developed on this platform

Fig. 4.8: Andes foreland basin area in Patagonia (Argentina): geography.

and basement, mainly as the consequence of the opening of the South Atlantic Ocean. The Austral and Magallanes Basins in the very south of the continent and the Neuquén Basin of south-central Argentina are polyhistory basins formed and deformed by active-margin processes (Spalletti et al., 1999a, b).

During the break-up of the Gondwana supercontinent an interaction of hot spots or mantle plumes affected the separation of continents. In the Mesozoic South America has been affected by two of these major plumes: the Karoo plume upon the SE of the region at about 183 Ma, and the Paraná-Etendeka plume at about 132 Ma. The sedimentary basins then formed resulted from all these tectonic activities, being in Patagonia Jurassic backarc basins (Neuquén and Austral Basins) and Cretaceous rift basins (at the Atlantic margin of Argentina: San Jorge and San Julian Basins; Spalletti et al., 2002).

4.3.4 EARLY PERMIAN – EARLY JURASSIC OSCILLATORY-EMERGENT EPISODE

First records of sedimentation during this episode date from the Late Triassic (Uliana and Biddle, 1988; Spalletti and Franzese, 1996). At that time, Patagonia was an almost positive land. Narrow and isolated, rapidly subsiding, continental rifts, with NNW-SSE direction, filled in with volcaniclastic sediments developed in NW Patagonia and the Deseado Massif of south-central Patagonia (Fig. 4.9). Almost everywhere these Triassic rocks rest on a sharp lithological and structural discontinuity. The abundance of volcanic and volcaniclastic rocks record significant syntectonic magmatic activity. The mid-Late Triassic successions are entirely non-marine. On top of the basal section of volcaniclastic rocks, extensive organic-rich shales accumulated in stratified rift lakes, in turn followed by a shallowing-upward package capped by sandy red beds, deposited in fluvial systems (Spalletti and Barrio, 1998). In the Magallanes-Austral and Malvinas basins, a Triassic graben fill seems also to be present (Biddle et al., 1986).

Paleogeographic reconstruction of the Late Triassic basins provides a picture that is in concert with events associated with the pre-break-up stage of the Gondwana supercontinent. Heat buildup and consequent high continental freeboard provide a good explanation for the dominance of non-marine facies shown by the southern South American sequences (Worsley et al., 1984).

4.3.5 MIDDLE JURASSIC-LATE CRETACEOUS SUBMERGENT EPISODE

Jurassic. The Jurassic period was characterized by general persistence of extensional conditions and fault-driven subsidence. All over Patagonia, the belt of

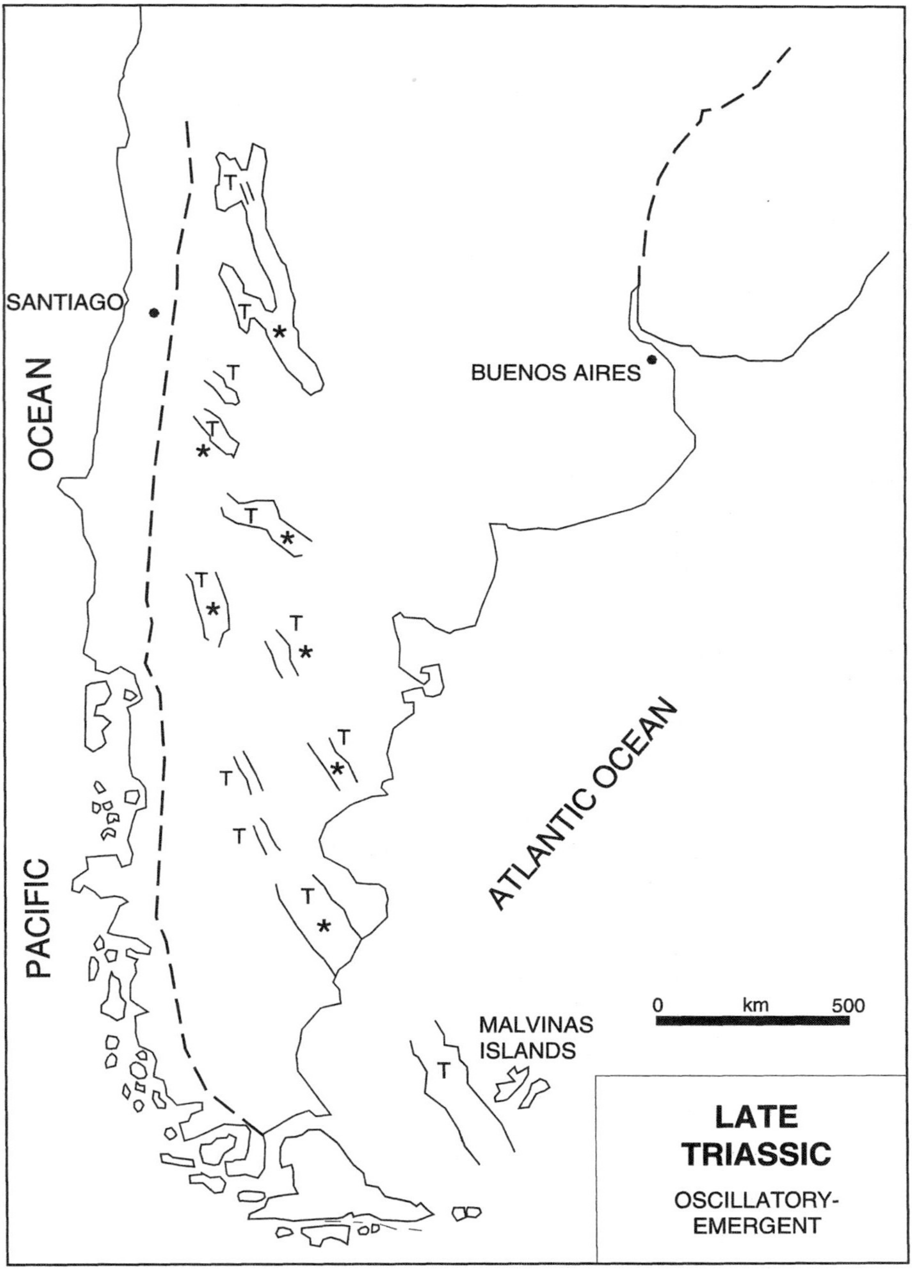

*Fig. 4.9: Patagonian Triassic basins. Symbols: T – terrigenous sediments, * – volcanics.*

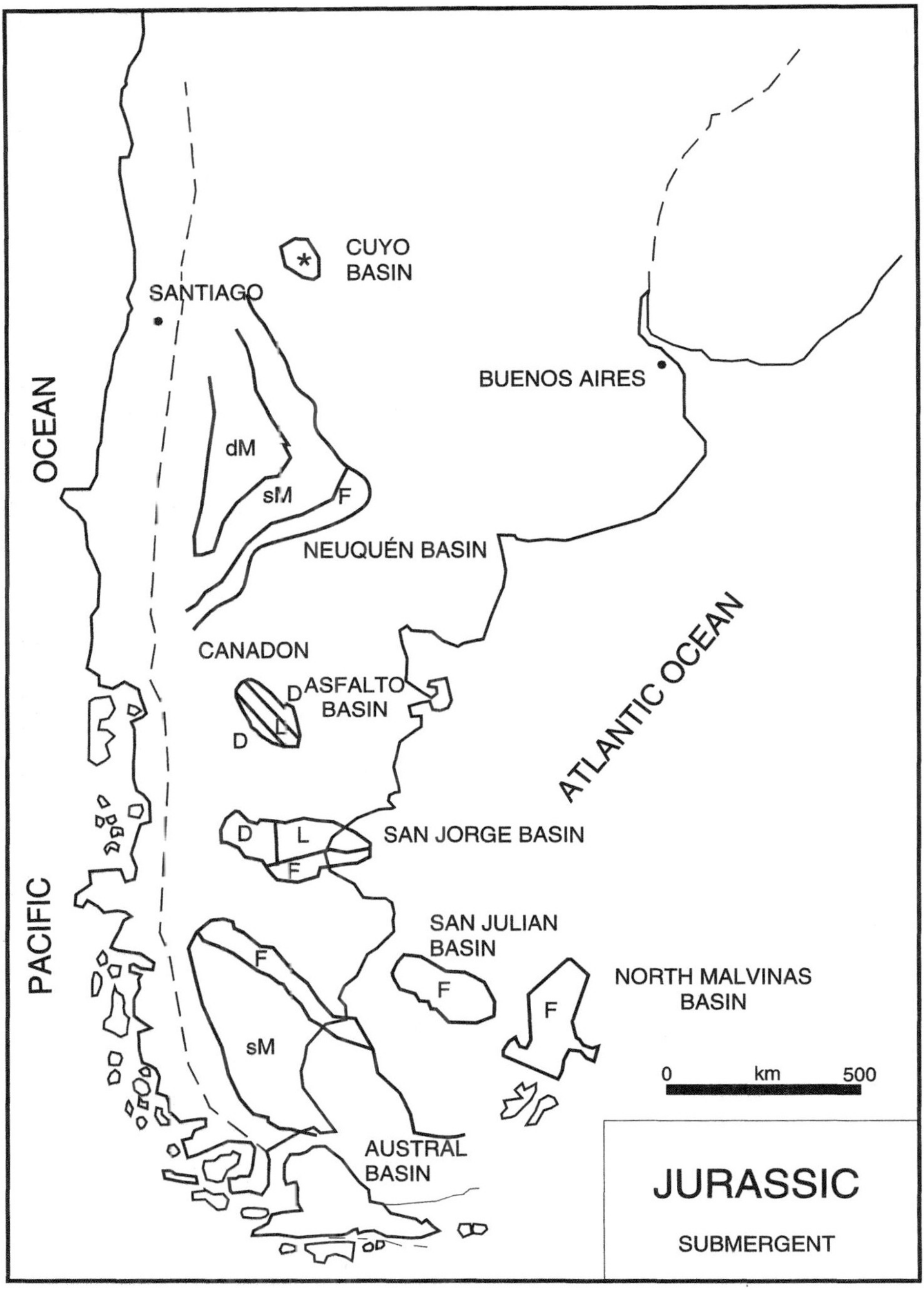

Fig. 4.10: Patagonian Jurassic basins. Symbols: C – coastal, F – fluvial, L – lacustrine, D – deltaic, sM – shallow marine, dM – deep marine.

active and fault-driven subsidence became wider by enlargement of the graben system and appearance of new fault-bounded troughs (Fig. 4.10). Due to the submergent character of the episode, a generalized subsidence of the area permitted the sea to encroach upon it.

During the Early-Middle Jurassic, the sea already advanced upon the South American plate edge and invaded the sedimentary basins from the west. In the Neuquén Basin shallow to deep marine deposits related to this transgression are recognized (Riccardi et al., 1997). Fluvial depositional systems occur at the fault-bounded basin borders, and in a small northern rift in the Austral Basin. Near the boundary between the early and middle Jurassic, a dominantly acidic volcanism covered large areas of northern Patagonia. In the Middle Jurassic the paleogeographical situation remained essentially the same. At the end of the period, most of the Patagonian region, south of the Gastre Fault System, was characterized by a bimodal volcanism. In central and southern Patagonia, more NW-SE and NNW-SSE trending grabens formed as a result of the widespread extensional tectonism. A transcurrent displacement along the Gastre Fault System controlled the depocenter of the Cañadón Asfalto Basin in north-central Patagonia, with fluvial and lacustrine facies associations (Veiga, 1998).

In the Late Jurassic, significant paleogeographical changes took place. The Andean magmatic arc reached up to 50°S and the silicic volcanism became restricted to SW Patagonia, where submarine rhyolite flows intercalate with deep-marine siliciclastics. The Rio Mayo, San Jorge and Austral basins reached their full development. The Rio Mayo Basin maintained a western connection with the Pacific Ocean, resulting in intra-arc and proximal backarc shallow marine carbonates; to the east, only periodic marine episodes are recorded, with mixed siliciclastic-carbonate sediments. The San Jorge Basin is characterized by distal lacustrine and proximal fluvial deposits along its northern and southern borders; these early rift continental deposits grade laterally into continued shallow marine sediments of the Rio Mayo Basin. Shallow marine facies in most of the Austral Basin point to the onset of widespread extension, and to the west, deep marine deposits associated to basic lavas, suggest no effective connection between this basin and the Pacific Ocean. Towards the north, the Cañadón Asfalto Basin became filled in with lacustrine and deltaic deposits prograding from its western and eastern margins. In the Neuquén Basin, there was some restriction due to the growth of the magmatic arc in the west, resulting in anoxic shales (main source rocks for the Neuquén oil fields) and marginal carbonate-siliciclastic ramp facies to the south (Spalletti et al., 1997, 1999b; Doyle et al., 1998).

Cretaceous (Berriasian-Turonian). The Cretaceous southern South American paleogeographic evolution is related with two main geotectonic features: an active margin along its western side, and the splitting of Africa and South America, and seafloor spreading in the southern South Atlantic. Permanent subduction of the Pacific plates under the South American plate favoured the development of the Andean magmatic arc. Seafloor spreading in the southern South Atlantic started at 132 $\pm$ 1 Ma. The rift subsidence in the eastern South American basins increased in the Neocomian and Barremian, and terminated (rift-sag unconformity) during the Aptian. The basin configuration remained almost the same (Fig. 4.11).

No major paleogeographic changes are recorded for the Valanginian-Hauterivian. The Andean magmatic arc covers the whole western Patagonian margin. The Deseado Massif separates the San Jorge Basin from the Austral Basin. In this latter basin, a wide marine platform is characterized by coastal and shallow marine siliciclastic sediments (Limeres et al., 2000). The platform is bounded by fluvial deposits along its NE margin. Towards the south deeper platform and slope fine-grained deposits are formed (Bertels, 1989). The Rio Mayo Basin embayment became drastically reduced; deltaic facies prevailed. To the west, periodic marine deposits confirm a connection of the embayment with the Pacific Ocean. The San Jorge Basin is reduced to a very small size; its restricted depocenter is filled with lacustrine and deltaic deposits. In the Cañadón Asfalto Basin, an enlarged continental depocenter is dominated by fluvial deposits. To the south of the Neuquén Basin, fluvial red beds are associated with lacustrine shales and evaporites; to the north, widespread mixed siliciclastic and carbonate marine ramp deposits represent the record (Matheos et al., 1994; Spalletti et al., 1997; Schwarz, 1999; Sagasti, 1998, 2000; Veiga and Spalletti, 2000). Offshore, the Colorado Basin rift formed and was filled in with fluvial and lacustrine sediments.

The Aptian was a time of transition. Continental red beds are widespread in the Neuquén and San Jorge Basins. The topographic barrier of the magmatic arc closed the Rio Mayo Basin. Towards the south, several paths through the volcanic chain connected the Austral Basin with the Pacific Ocean. A variety of sediment facies types is present in this extensive basin: in the north, from a very active source area, fluvial deposits are associated with restricted marine black shales; mudstones and shales prevail in distal platforms; in the south, widespread deep marine facies, and along the Andean magmatic arc black shales and turbidites accumulated. The San Jorge Basin shows a large depocenter that incorporated the northern Cañadón Asfalto Basin and a southern N–S oriented depression. Along the marginal areas, there occur fluvial deposits, and deltaic systems prograded from the northern and western margins into an extensive lacustrine system. With respect to the Neuquén Basin, a continental depocenter shows marginal fluvial red beds and a large continental sabkha muddy-evaporite depositional system.

The Albian panorama is characterized by a continuous, positive volcanic chain that separated Patagonia from the Pacific Ocean. Fluvial red beds are widespread in the Neuquén Basin. In the offshore Colorado Basin, the first marine

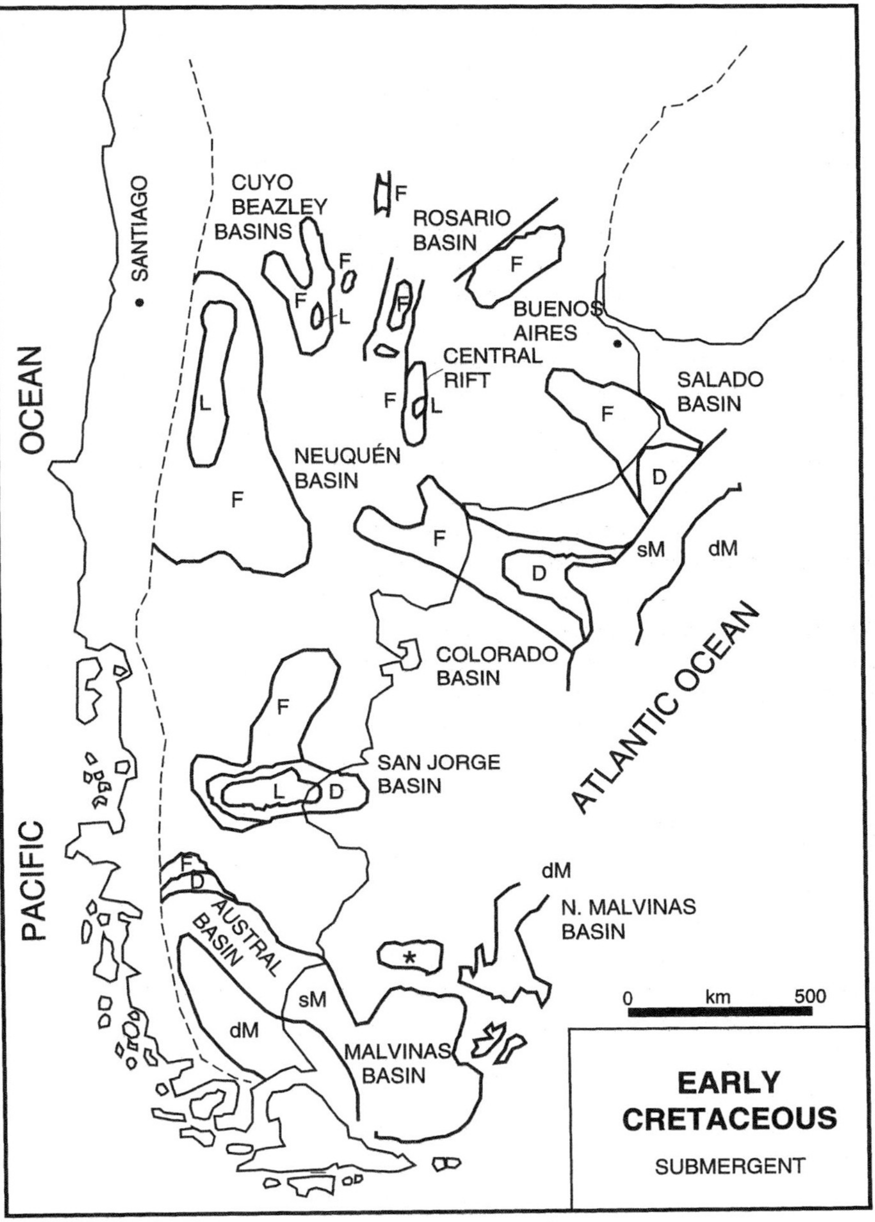

Fig. 4.11: Patagonian Cretaceous basins. Symbols: see Fig. 4.10.

deposits accumulated. In the San Jorge Basin, extended fluvial systems terminate in deltaic-lacustrine depocenters with siliciclastic sediments; in the southern domain, the fluvial system is reduced; but to the north and northeast, the systems expanded. The Austral Basin passes through an early foreland stage; the marginal basin is closed and from the northwest corner a very active deltaic system prograded towards the south where pro-deltaic and deep marine shales are associated to N–S oriented axial turbidite systems. The basin is now connected to the ocean through its southern and southeastern sides (Spalletti and Franzese, 1996; Spalletti et al., 1999a, b).

During the Cenomanian-Turonian, the Colorado rift opened again and became filled in with fluvial and subordinate lacustrine sediments. The also offshore Valdéz and Rawson rift basins developed with a continental early rift clastic infilling. These offshore basins occur on the wide continental shelf of the area. The Neuquén and San Jorge basins are integrated in a unique depocenter dominated by fluvial red beds, with subordinated shallow lacustrine and playa deposits. In the subsurface of the depression, the fluvial deposits are represented by reservoir sandstones. In the Austral Basin, the foreland stage causes a strong detrital contribution from the west and progressive migration of the depocenter towards the east. The northwest prograding zone continues with its fluvial and deltaic facies. Shallow marine sediments are represented by mudstones and fine-grained sandstones. Towards the basin centre mudstones and shales accumulated in deeper sectors of the platform and talus slope. North to south oriented axial turbidite systems remain along the western Austral Basin.

4.3.6 LATE CRETACEOUS – CENOZOIC OSCILLATORY-EMERGENT EPISODE

The most recent oscillatory-emergent episode started at about 80–90 Ma ago, still during the Late Cretaceous (Senonian). From that time onwards, the first signs of regression of the sea become evident (Vail et al., 1977; Haq et al., 1988). The episode seems to have initiated in Campanian times.

Sediment accumulation during the Senonian was characterized by a trend towards the enlargement of depositional sites and an increase in the amount of marine influence, in spite of the starting oscillatory regression. At many places, the Late Cretaceous sequences display a progressive overstepping of older terranes. At most locations earlier Mesozoic faults tend to become inactive and thickness patterns point to a regional style of subsidence, dominated by sediment loading and lithospheric cooling with a minimum of differential downwarping. More pronounced subsidence, presumably related to flexural loading of the South American crust, was only locally developed at backarc positions within the Austral and Neuquén Basins. All along the Atlantic margin the depositional framework was one of prograding wedges, facing a progressively deeper Atlantic Ocean.

Senonian. The paleogeography of this period shows a widespread oscillatory transgression embracing the Colorado Basin and the North Patagonian shallow platform. The San Jorge Basin becomes again a large and isolated continental depocenter. A general regression is recorded in the Austral Basin, caused by renewed uplift along the Andean margin and a marked NNW-SSE fluvio-deltaic progradation. The Atlantic basins suffered a late sag stage. A foreland stage started in the Neuquén Basin. Positive areas in the region were Western Patagonia with a high relief, the northeast corner of the North Patagonian Massif, the Rawson High, the Deseado Massif-Dungeness Arch and the Malvinas High (Fig. 4.12).

Details point for the Colorado and Valdéz-Rawson Basins a late sag stage with a transgression from the Atlantic Ocean, resulting in shallow marine and partially marine facies. The North Patagonian platform was also dominated by the transgression, and a marginal fluvial-lacustrine, and periodic shallow marine deposits represented by fine-grained clastics and biogenic carbonates are recorded. Towards NE, the platform was connected to the Colorado Basin. The foreland stage of the Neuquén Basin shows also some shallow marine deposits. The San Jorge Basin is characterized by extended areas of fluvial sedimentation though ephemeral lacustrine, deltaic and fluvial deposits accumulated in the central area of the depression. At the southeastern margin of the basin, a shallow trough was filled in with coarse fluvial deposits derived from the Deseado Massif. Furthermore, the Austral Basin shows a general regressive phase, with a huge embayment towards its northwest corner. Fluvial deposits and prograding deltaic-esturine clastics have been recorded. In marginal areas of the basin appear shoreface and nearshore tidal dominated sediments. Distal platform shales and mudstones are widespread in the Austral and Western Malvinas Basins, which were connected between them. The deepest sector of the Austral Basin was located at the southernmost part of continental Argentina.

An important side effect of the Late Cretaceous flooding was the decrease of the clastic influx into most basins. Shaly and marly facies are thus common, and they are locally associated with carbonates, while areas marginal to the seaways are characterized by red bed–evaporite assemblages.

Especially in the Neuquén Basin, the southern Argentinean basins are known for their dinosaur faunas (Bonaparte, 1996). In this basin, one of the biggest species was found (*Gigantosaurus carolinii*), and numerous dinosaur footprints have been preserved (Apesteguía, 2002).

Paleogene. A survey of the Early Tertiary sequences of the Andean domain shows a considerable along-strike variation in depositional regimes. A pause in

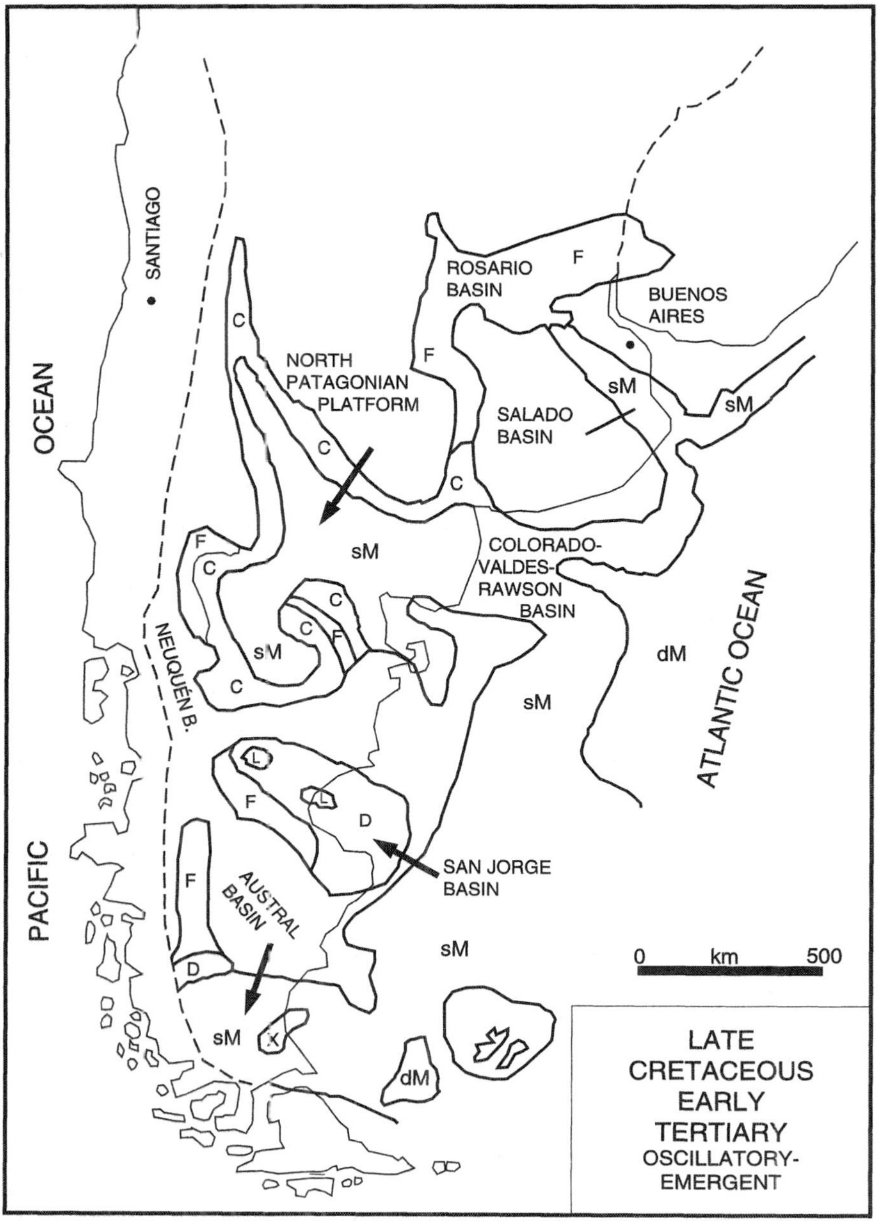

Fig. 4.12: Patagonia: basins and highs in Late Cretaceous-Tertiary. Symbols: see Fig. 4.10.

Andean magmatism and a marine retreat took place. The volcanic rocks are associated with lacustrine sediments and were deposited within a series of restricted intermontane basins.

In sharp contrast, the southernmost Andes show few indications of magmatic activity. Sedimentation in mid-plate settings continued to be mostly confined to those basins active during the Late Cretaceous. A considerable reduction in the areal extent of the depositional sites, and the diminution of the regions under marine influence, appears to be the response to base level lowering related to the Early Tertiary global eustatic sea-level fall (Haq et al., 1988). Because of the stable tectonic conditions, intraplate relief continued to be limited and dominant accumulations are fine grained. In Patagonia, many areas invaded by the Late Cretaceous epeiric flooding turned into vast loess plains made up by distal pyroclastics punctuated with numerous paleosols (Gonzalez, 1999).

Persistent drowning along the Argentine shelf continued. To a large extent, a pronounced downward shift and offlap arrangement of the Paleogene clastic wedges is attributed to reduction in shelf accommodation concomitant with global eustatic sea-level fall.

Neogene. Early Neogene paleogeographic reconstruction shows arc buildup and renewed marine flooding. In the Andean arc the most obvious change was reactivation of the main magmatic belt, after a period of relative quiescence that lasted most of the Oligocene. Over large areas, igneous activity expanded and invaded the foreland of west-central Argentina. In southern Argentina, however, the magmatism was superimposed on older arc terranes, and the increase in magmatic activity was modest. There is also no evidence of pervasive arc or backarc extensional faulting, as was the case more towards north. Most indicators point to the continuation of flexural downwarping of the Austral foredeep trough.

Over many areas of continental Argentina, the upper Paleogene and lower Neogene are missing (Biddle et al., 1986), because of a composite regional unconformity largely due to reduced accommodation-erosion beveling plus non-deposition, around the 30 Ma eustatic drop. The mid-Oligocene event of shelf exposure was followed by a change toward increased marine encroachment that lasted from the late Oligocene until the middle Miocene, parallelling the early Neogene trend of global eustatic rise. The Neogene marine inundation occurred in the same general areas as the Late Cretaceous flooding. The span of the marine beds, however, varies from place to place because of a regional onlap on an irregular surface and diachronous comeback of the shoreline as a result of variable rates of clastic influx. The large areal extent of the Miocene marine deposits reveals that mean freeboard of the plate interior was still consistently low. The dominant lithofacies types are very similar to those developed as a consequence of the Late Cretaceous drowning. Shales and fine-grained clastics with some carbonates, associated with sabkha evaporites at the fringes of the epeiric seaways covered eastern Argentina. As in the Paleogene, thin pyroclastic loess with paleosols continued to dominate the subaerial settings of Patagonia (Franchi et al., 1984).

Some 15 Ma ago, and in many areas 10 Ma or even later, the southern South American scenery began to be dominated by the processes which led to the present configuration of the Andean tectonic-magmatic belt. Among the complete sequence of partially overlapping events may be cited: regional uplift and Cordilleran morphogenesis, and east–west-directed compression and contractional deformation. The Andean uplift in itself promoted an overall increase in the magnitude of the sedimentary influx. Because of the augmented topographic relief and the late Cenozoic eustatic fall, some of those sediments were carried to the Atlantic continental margin depocenters. A sizeable proportion, however, was trapped at intra-plate locations to form the sedimentary fill of a new and diverse suite of depositional sites generated as a response to the late Neogene tectonic framework. Unlike most of the previous Mesozoic-Cenozoic interval, the sedimentary input was largely captured within basins developed under a compressional regime. East of the main Cordilleran belt, 2000–4000 m thick late Neogene clastics accumulated as asymmetric foreland wedges developed over downflexed portions of the craton in front of segments of the Andean, thin-skinned fold and thrust belt. The northern part of the Neuquén Basin, east of the Principal Cordillera and the Chilean-Argentinean Austral Basin are good examples.

With the Miocene uplift of the Andes, large areas of Patagonia were elevated above the base level and entered into a net erosional regime. Late Neogene deposits are areally restricted and consist of fluvial to eolian sandy facies in valley fills related to the fluvial systems that carried Andean detritus to the Atlantic shelf (Franchi et al., 1984). The updip edge of the late Neogene marine incursions was confined to positions close to that of the present shoreline. Farther away from the Andes, on the Malvinas Plateau, submersion continued to be dominant and sedimentation was essentially biogenic (nannoplankton and siliceous oozes), possibly reflecting vigorous upwelling associated wit the mid-Tertiary development of the Circum-Antarctic Current.

4.3.7 CONCLUSIONS

Although the time span covered by the sedimentary history of the Patagonian Andean Foreland basins, in Argentina, comprises only the three last Phanerozoic tectonic-sedimentary episodes, the cyclic development of these basins becomes clear. The two oscillatory-emergent episodes favour the development of continental interior fracture and wrench basins, as intracontinental and pericontinental rifts. The submergent episode is characterized by continental interior sag phenomena, often implanted upon the rifts of the foregoing episode. This means that the basins formed and reactivated between Triassic and Cretaceous are the same, with a rifted base and a subsided upper part. The Late Cretaceous-Cenozoic basins are commonly rifts, often independent from the earlier developed basins.

Within the tectonic and sea-level oscillations of Triassic-Early Jurassic and Late Cretaceous-Cenozoic times, marine transgressions and regressions as well as continental environment phases alternate, often as third-order non-periodic tectonic cycles, caused by changes in rate of tectonic subsidence and sediment supply (Vail et al., 1991), in this case evidenced by the formation and uplift of the Andean Cordillera. Also higher order episodic events are sometimes recorded in the sediment successions in the different basins, as presented in the numerous papers about detailed studies effectuated in the area. These latter events have also been mentioned in some sediment sequences deposited during the Middle Jurassic-Late Cretaceous submergent episode.

4.4 KAROO BASIN (SOUTH AFRICA)

4.4.1 AFRICAN SEDIMENTARY BASINS: KAROO BASIN

The Phanerozoic sedimentary basins in Africa are either internal (intracratonic) or external (pericratonic or marginal) basins (Burollet, 1982). In each group, special mechanisms resulted in a typical sedimentary behaviour, corresponding to different tectonic-sedimentary episodes of the earth's history.

The Karoo Basin is a syneclise-type intracontinental depression, subject to varying degrees of subsidence in which sedimentary material has accumulated above a cratonic platform. The basin is fully isolated in the interior of the African craton, with links to border features that are left over from the previous substratum. It is filled in with Paleozoic series, sometimes with a Mesozoic filling on top. The basin developed to the north of the Paleozoic Cape chains which underwent folding during the Devonian; it started with glacial deposits during the Late Carboniferous and sank during the Triassic and Jurassic under thick continental formations, first detritic and last volcano-detritic.

4.4.2 MAIN KAROO BASIN

4.4.2.1 Physiographical and geological aspects

Southern South Africa presents four major physiographic aspects from E to W: the High Interior Plateau, the Great Escarpment, the Great Karoo and the Cape

Coastal Mountains (Fig. 4.13). The High Interior Plateau, elevated between 600 and 1800 m above sea level, is underlain by almost flat-lying strata of the Karoo Supergroup. The Great Escarpment that forms the boundary between the Great Karoo and the High Interior Plateau, is a prominent topographic feature traceable throughout southern South Africa; it was initiated by Late Jurassic uplift and erosion, probably associated with the beginning of the Gondwana

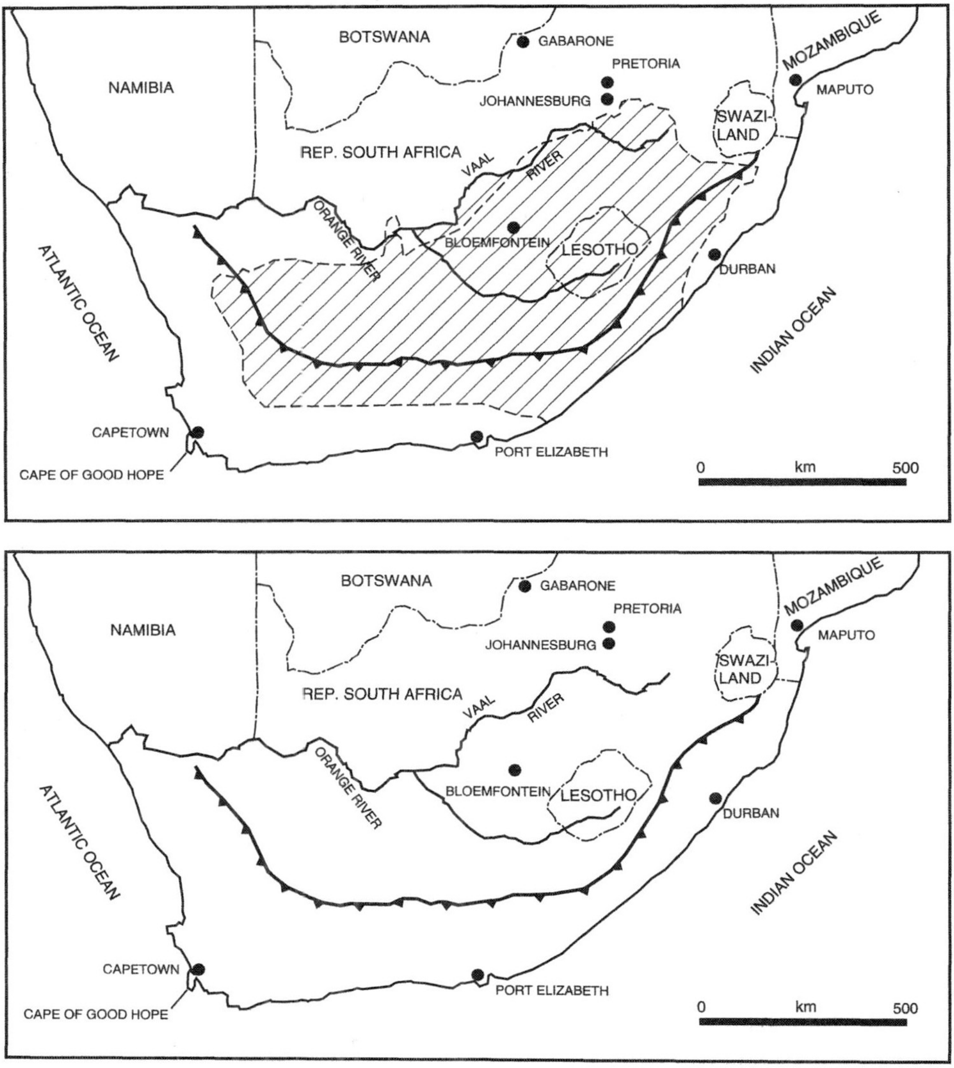

Fig. 4.13: Southern Africa: physiography.

supercontinent fragmentation. The Great Karoo is characterized by undulating ridges and plains, formed by gently folded strata. The Cape Coastal Mountains, with altitudes of over 2000 m, are dissected by steep valleys.

Geologically, the rocks which belong to the Karoo Supergroup occur in most of southern Africa, however with the best outcrops in the main Karoo Basin of South Africa (Fig. 4.14), and with equivalent units in Namibia (Stollhofen and Stanistreet, 1997) and Botswana (Modie, 2002b). The ages of the strata range from Late Carboniferous (about 300 Ma ago) to Middle Jurassic (about 183 Ma ago). The southern and southwestern margins of the basin are constituted by the Cape Fold Belt, to the east it extends beneath the Indian Ocean, and to the north and northeast the basin margins are in part structural and in part erosional. The main Karoo Basin of South Africa is a retro-arc foreland fill, developed in front of the Cape Fold Belt. The basin has a typical asymmetrical fill due to the nature of the lithospheric flexure. The sedimentary fill corresponds to the sequences which belong to the Middle Devonian-Late Carboniferous submergent, the Permian-Early Jurassic oscillatory-emergent and the Middle Jurassic-Late

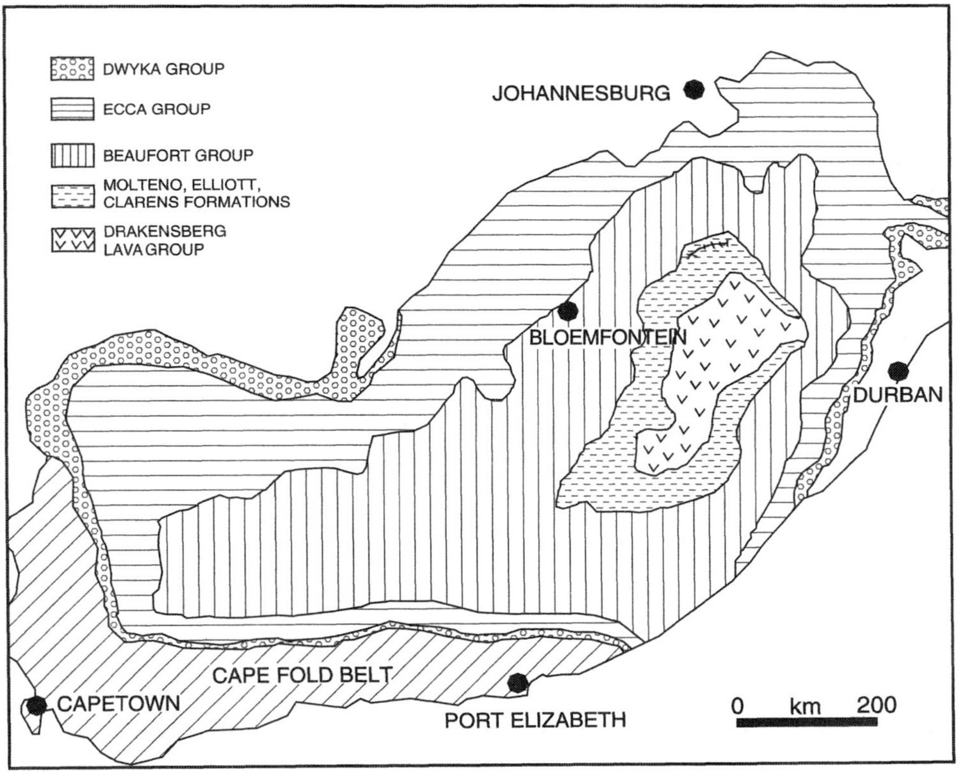

Fig. 4.14: Geology of Karoo Basin in South Africa.

Cretaceous submergent episodes, with basal and top contacts marking profound changes in the tectonic setting, from extensional to foreland, and from foreland to extensional, respectively (Catuneanu et al., 1998; Hancox et al., 2002).

4.4.2.2 Lithostratigraphy and depositional environments

In Table 4.3, the stratigraphical column of the Karoo Supergroup in the main Karoo Basin has been presented. With exception of the Drakensberg volcanics, the rest of the supergroup is composed of sedimentary rocks. A major hiatus occurs between the lower Dwyka, Ecca and Beaufort groups and the upper so-called Stormberg Group, comprising most of the Middle and part of the Late Triassic. A NE-SW section through the basin, from distal to proximal, attests a thickness increasing significantly towards the Cape Fold Belt. The proximal and distal facies are separated by a narrow transitional zone, with important lateral facies changes.

Table 4.3: Stratigraphical column of the Karoo Supergroup in the Main Karoo Basin of South africa (after Catuneanu et al., 1998).

proximal facies (S, W)		distal facies (NE)
Middle Jurassic–Late Cretaceous submergent episode:		
	Drakensberg Group (Middle Jurassic, volcanics)	
Permian–Early Jurassic oscillatory-emergent episode:		
	"Stormberg Group" (Middle Triassic-Middle Jurassic)	
	Clarens Fm.	
	Elliott Fm.	
	Molteno Fm.	
Beaufort Group (Late Permian-Middle Triassic)		
	Tarkastad Subgroup	
Burgersdorp Fm.		Driekoppen Fm.
Katberg Fm.		Verkykerskop Fm.
	Adelaide Subgroup	
Tekloof Fm. (W)	Balfour Fm. (C)	Normandien Fm. (N)
Abrahamskraal Fm. (W)	Koonap/Middleton Fms. (C)	
	Ecca Group (Permian)	
Fort Brown Fm.		Volkrust Fm.
Ripon & Collingham Fm.		Vryheid Fm.
Whitehill Fm.		Pietermaritzburg Fm.
Prince Albert Fm.		
Middle Devonian-Late Carboniferous submergent episode:		
	Dwyka Group (Late Carboniferous-Early Permian)	

N, north; NE, northeast; C, central; S, south; W, west

Dwyka Group (Late Carboniferous-Early Permian). Sedimentation in the main Karoo Basin started in the Late Carboniferous about 300 Ma ago (Moscovian/ Late Westphalian), with the deposition of the well-known glacigenic deposits of the Dwyka Group (Visser, 1987; Fig. 4.15). The sediment section consists of massive to crudely stratified diamictites, conglomerates, fluvioglacial pebbly sandstones, rhythmites and mudstones, attaining a thickness of more than 800 m. Visser (1986) distinguishes a platform facies in the south and an inlet/ valley facies in the north. The rather thick and to some extent, lithologically uniform platform facies shows chiefly layered diamictites, deposited as clast-rich tills beneath grounded ice sheets, and massive clast-poor tills, deposited from beneath floating ice sheets. The inlet/valley facies association comprises remains of supraglacial abrasion till, outwash sand and gravel deposits, and morainic and glacial lake deposits, preserved in topographic lows. Trace fossils are common in the finer-grained sediments of this latter facies association. Successive phases of ice expansion and contraction are recorded in nine, fining-upward, vertical sedimentary cycles, with thickness varying between 60 and 100 m. Each cycle displays a transition from basal terrestrial or subaqueous moraines to upper glacio-lacustrine shales (Visser, 1986). The final short-lived glacial phase covered large areas of southern Africa (Modie, 2002a), including more than half of the main Karoo Basin surface, with a marine ice sheet, prior to the

Fig. 4.15: Laingsburg (South Africa): Dwyka tillites.

amelioration of the climate, with a consequent break-up of the ice sheet and a sea-level rise of about 100–150 m. The overall pattern of southwards ice movement suggests that the surface profile during the Dwyka time was directed to the south. This explains also the diachronous age of the top of the Dwyka Group, younger in the north (Artinskian, upper Early Permian, about 262 Ma) relative to the south (Sakmarian, middle Early Permian, about 271 Ma). Thus, the continental glaciation lasted longer in the north, due to a higher altitude, than in the south where the marine environment led to a rapid melting of the floating ice. As a result of this gradual deglaciation of the continental areas during the Artinskian, the northern Dwyka succession ends with coal-bearing fluvial-deltaic sequences (Cairncross, 2002).

Ecca Group (Middle Permian). The base of the Ecca Group is diachronous, younger in the distal NE sector relative to the S and W proximal sector. The coal-bearing Dwyka sequences are overlain in places by the marine shales of the Pietermaritzburg Formation of the distal facies of the Ecca Group. These shales accumulated in a moderate to deep marine environment. Upward the realm turned gradually continental due to marine regression, and the deposits became interbedded sandstone, shale and a few coals beds accumulated in a fluvial-deltaic environment. The lithostratigraphic unit (Vryheid Formation) is correlatable with the proximal deep marine environment of the southern part of the basin. Another sea-level oscillation returned a shallow to deep-marine realm with sedimentation of dark-coloured shales with intercalated fine-grained sandstones that constitute the upper sequence of the distal Ecca Group (Volkrust Formation).

The proximal facies in the W and S of the main Karoo Basin starts with a sequence of dark greenish-grey shales with some silty layers from a deep water environment (Prince Albert Formation) that passes conformably into carbonaceous shales with chert bands and lenses, accumulated in a shallowing marine to brackish and reducing realm (Whitehill Formation). The upward following Collingham Formation represents a distal submarine fan facies of alternating siltstone and shale beds, with yellowish tuff layers. The proximal submarine turbidite facies part is composed of greywackes, siltstones and shales, arranged in Bouma sequences (Ripon Formation). The upper Fort Brown Formation of the Ecca Group accumulated in a regressive shallow marine environment with greenish-grey shales and subordinate sandstones of paraglacial deltas and braidplains.

The bounding surfaces of the proximal facies are all conformable. In the distal part, the sea-level oscillations caused unconformable bounding surfaces at the base of the Ecca Group and at the base of the uppermost Volkrust Formation due to the transgression of the Ecca Sea over the non-marine Vryheid facies. The other bounding surfaces are conformable due to the intermediate regressional oscillation phase.

Beaufort Group (Late Permian-Early Triassic). The contact between the Ecca and Beaufort Groups is characterized by a transition from subaqueous to subaerial deposition (Rubidge et al., 2000), showing the same phenomenon throughout the whole basin. However, this contact is highly diachronous. The group is the most extensively exposed stratigraphic unit in the main Karoo Basin and the most fossiliferous; it consists chiefly of mudstones and siltstones, interbedded with subordinate lenticular and tabulate sandstones, deposited in a variety of fluvial depositional systems.

(a) *Adelaide Subgroup* – By the lower Adelaide Subgroup times, non-marine conditions of sedimentation had firmly established throughout the Karoo Basin, lasting until the end of the deposition of the Karoo sedimentary sequence. The base of the subgroup is diachronous, however conformable in both the proximal and distal sections, that is a facies contact from marine to non-marine depositional systems. There exists some divergence in the opinions of Catuneanu et al. (1998) and of Hancox et al. (2002), with respect to the subdivision in formations, being different in the western, southern and northernmost outcrop areas (see Table 4.3).

In the area W of 24°E longitude, sedimentation started by a source area tectonism and the accumulation of a 2700 m thick fining-upward succession of sandstones and reddish-brown mudstones with numerous thin chert bands, and rich in reptilian faunas (Abrahamskraal and Teekloof Formations). The sequence was deposited by overbank flooding from meandering and low-sinuosity river systems, which drained an extensive alluvial plain sloping gently towards NE in the direction of the receding Ecca shoreline. The climate had turned warm and dry, from semi-arid to almost arid, with a seasonal rainfall, recorded by the presence of hematitic mudrocks, gypsum pseudomorphs (desert roses), desiccation cracks, playa lake evaporites, and pedogenic carbonate nodules. The Abrahamskraal and Teekloof Formations grade eastwards into the Koonap and Middleton Formations which are a succession of thick mudstone dominated deposits, with a thickness of 2800 m, consisting of a shallow lacustrine facies at the base, overlain by vertically stacked fining-upward sequences. These sequences comprise erosively based sandstones, siltstones and shales or mudstones, deposited in meandering river channels, flanked by crevasse splays and mud-dominated floodplains. These are overlain by a 120 m thick sandstone sequence (Balfour Formation), composed of fining-upward sandstone-dominated sequences deposited by low-sinuosity streams. In the western section, this formation seems to be absent. The northeastward occurring Normandien Formation includes interbedded sandstones and mudstones deposited by meandering rivers with channels flanked by wide semi-arid floodplains.

(b) *Tarkastad Subgroup* – A twofold subdivision of this lithostratigraphic unit has been adopted for the southern proximal and northeastern distal sections of the main Karoo Basin. The proximal facies is represented by the Katberg and Burgersdorp Formations, and the distal facies by the Verkykerskop and Driekoppen Formations, which latter however, may bear different formation names throughout the basin. Increased tectonic uplift of the Cape Fold Belt is held responsible for this new tectonic-sedimentary cycle as part of a large northward tapering clastic wedge that emanated from the south of the fold belt in response to orogenic loading.

The lower Katberg Formation of the proximal area lies unconformably on the underlying Balfour Formation. It consists chiefly of thick laterally extensive, light olive-green, coarse-grained sandstones, composed of transverse and longitudinal bar macroforms with horizontal and trough cross-bedding (Fig. 4.16). These sandstones have been deposited in a shallow braided river environment with pulsatory discharge. In abandoned channel fills and braidplains, mudstones have been preserved. The northern distal facies (Verkykerskop Formation) consists chiefly of thin, laterally extensive, medium- to fine-grained sandstones, from transverse bar macroforms with internal planar cross-stratification.

Conformably lying on top of the Katberg Formation, the proximal facies of the Burgersdorp Formation consists predominantly of thick, laterally extensive, fining-upward sandstones overlain by siltstones and mudstones. These sequences

Fig. 4.16: Wapadsberg (South Africa): Early Triassic Katberg Formation (Beaufort Group), fluvial sandstones.

are attributed to meandering river and floodplain realms. The Driekoppen Formation of the distal facies is composed of thin, fine-grained channel sandstones covered by massive to diffusely laminated siltstones and mudstones. These deposits represent suspended-load dominant meandering river sediments.

The Tarkastad Subgroup may be considered as a single fining-upward sequence. Sandstone-dominated braidplain deposits at the base, grade upwards into mudstone-dominated floodplain deposits of meandering river systems. The contact between the Tarkastad Subgroup and the overlying Molteno Formation is unconformable across the whole area of occurrence. The limit between the lower and upper formations of the subgroup is conformable with a gradual transition in both proximal and distal areas.

Late Triassic-Middle Jurassic lithostratigraphic units ("Stormberg Group"). This group includes the Molteno, Elliott and Clarens Formations. A major stratigraphic gap separates the strata from the underlying sequences. In this group, no proximal and distal facies are distinguished. The entire succession may be regarded as distal Karoo facies, not extending up to the Cape Fold Belt, but separated from it by a proximal region of syn-depositional by-pass and reworking of older Karoo sequences. The southern limit of the group's depositional area was close to the present-day preservation area.

The Molteno Formation is composed predominantly of tabular sheets of medium- to coarse-grained sandstones, internally structured by horizontally and cross-bedded macroforms. The laterally continuous sheets accumulated in braided streams on a vast braidplain. Infillings of abandoned channels and water bodies on this braidplain are represented by siltstones, mudstones and coal deposits which however, are not very abundant.

Reddish floodplain mudstones with subordinate channel and crevasse splay sandstones, representing deposition in mixed-load dominated meandering river systems, characterize the Elliott Formation. The unit is a pedogenically influenced red bed sequence. Accumulation occurred under conditions of increasing aridity, as evidenced by changing patterns of alluvial sedimentation into ephemeral and floodplain playas. Upwards eolian input of wind-blown loessic dust sediments increases, making the transition towards the upper Clarens Formation.

This Clarens Formation crops out in a restricted sub-basin around Lesotho. It is represented by pale yellow to white, well-sorted, fine-grained and carbonate-cemented, sheet-like sandstones which weather into spectecular massive cliffs (Fig. 4.17). The thickness of this sandstone varies considerably due to paleotopography and original dune heights. Four rather different facies are distinguished from the base up: (1) fine-grained wadi-playa lake facies, (2) coarse braided wadi facies, (3) mass flow facies, and (4) wind-blown dunes. The depositional realm was evidently desertic, with wind-blown dunes and shallow playa

Fig. 4.17: Clarens (South Africa): Early Jurassic Clarens Formation sandstones.

lakes. Towards the end of the depositional period, the climate became less dry, and wet desert processes of stream and sheet floods became more dominant.

Considering the stratigraphic conformity and the gradual transition between the fluvial-dominated Elliott and the eolian-dominated Clarens environments, the whole sequence may be taken together as one coarsening-upward sequence.

Drakensberg Group (Middle Jurassic). The early volcanic activity of this sequence occurred contemporaneously with the sedimentation of the Clarens Formation. At this time, the conditions of the main Karoo Basin changed from the Late Triassic transitional regime, with a tectonic inversion, into an extensional regime. A volcanic plume caused doming and resulted in the ruption of tholeiitic basalts of the Karoo Igneous Province, eventually marking the initial stages of the break-up of Gondwana. However, this volcanic event was short-lived, centred around 183 Ma (Hargreaves et al., 1997).

4.4.2.3 Importance of fossils in interpreting the development of the main Karoo Basin

The Permo-Carboniferous to Jurassic aged rocks of the main Karoo Basin are world renowned for their wealth in synapsid reptile and early dinosaur fossils

(Hancox and Rubidge, 1997). The respective assemblages allowed a detailed biostratigraphic subdivision of the Karoo Supergroup. Integrated sedimentological and paleontological studies contributed to a more precise definition of a number of problematical formational contacts within the supergroup, as well as enhancing paleo-environmental reconstructions, and basin development models. One of the most important conclusions drawn was that many contacts between the various lithostratigraphic units are diachronous trough the basin, from distal NE to proximal S and W.

4.4.2.4 Basin history

The Karoo Basin formed along the northwestern margin of the Gondwana supercontinent during the Late Carboniferous to Early Permian, due to the low angle, 25–40° dip, northward subduction of the paleo-Pacific plate beneath the Gondwana plate. The position of the plates' margin trench is uncertain, but best estimates place it some 1000–1500 km south of the main basin. During the Permian, the basin was located behind a magmatic arc and associated thrust belt (Cape Fold Belt), and can therefore be classified as a retro-arc foreland basin (Cole, 1992). The basin floor is largely composed of old stable lithosphere of the Kaapvaal Craton and younger lithosphere of the Natal-Namaqua Mobile Belt. This heterogeneity of the lithosphere had a strong influence on the nature of the flexural profile and subsequent sequence development.

Until recently the Karoo Basin was considered to be a uniform subsiding basin, created by flexural loading and dynamic subsidence. However, Catuneanu et al. (1998) showed it to be non-uniform, being composed of distinct proximal and distal sectors, which are out of phase regarding their lithic fills. In such a setting, orogenic loading provokes subsidence in the proximal sector and increase in accommodation space, and a corresponding uplift and decrease in accommodation space in the distal sector. During orogenic unloading the reverse occurs. These two tectonic settings correspond to the proximal foredeep and distal forebulge, being separated by the foreland basin hingeline.

The respective sedimentary sequences are preserved in both proximal and distal facies, which are however, contrasting and with distinct stratigraphic signatures, due to different base level changes in the two areas. The hingeline between the two facies types migrated northwards, away from the orogenic belt during the Late Carboniferous to Permian oscillatory-emergent episode, in response to thrust sheet advance. During the Triassic-Middle Jurassic submergent interval the hingeline moved south in response to the visco-elastic relaxation of the foreland lithosphere.

Tectonism in the Cape Fold Belt caused a succession of second and lower order stages of loading and unloading cycles. A first-order compressional stage defined the northward migration of the hingeline. During this stage, the Dwyka, Ecca and Lower Beaufort Groups accumulated, with a great composite thickness in the proximal area reflecting orogenic loading and foredeep subsidence. Followed a first-order orogenic quiescence stage with a southward migration of the hingeline. The resulting deposits include the Upper Beaufort and the Stormberg groups, which are more complete in the distal area and reflect orogenic unloading and forebulge subsidence.

4.4.2.5 Conclusions

The major stages in the evolution of the Karoo foreland basin system is thus: (1) Late Carboniferous-Permian – orogenic loading with the progradation of the orogenic front, leading to the progradation of the foreland system; (2) Early-Middle Triassic – orogenic loading without progradation of the orogenic front, leading to the high-rate retrogradation of the foreland system; (3) Late Triassic-Middle Jurassic – orogenic unloading with the retrogradation of the centre of weight in the orogenic belt due to the erosion of the orogenic front, leading to the low-rate retrogradation of the foreland system.

4.4.3 KAROO SEQUENCES IN NAMIBIA AND BOTSWANA

4.4.3.1 Introduction

Whereas the main Karoo Basin in South Africa records the most complete representation of the Karoo Supergroup, similar sequences are also present in the neighbouring Namibia and Botswana. In these countries, the sequences appear to be less complete. Karoo Supergroup sediments are also found in other African countries (Cairncross, 2002).

Following the Late Proterozoic Panafrican orogenic cycle, the onset of Karoo rifting during latest Carboniferous to Early Permian times, within an oscillatory-emergent tectonic-sedimentary episode, represents the first regionally tectonic event in southern Africa (Stollhofen and Stanistreet, 1997). Karoo strata preserve a record of the complex interplay between sedimentation, effusion of flood basalts and extensional tectonics that predated and accompanied the break-up of Gondwanaland. The initial plate tectonic control on Karoo basin development may be viewed in terms of the Samfrau active margin S of southern Africa that developed a major flexurally deformed foreland basin in front of the northward vergent Cape Fold Belt. The northerly trending elongate graben and half-graben structures of southern and southeastern Africa, however, define rift systems which show widespread evidence for incremental basin extension including repeated phase of syn-sedimentary faulting.

4.4.3.2 Namibia

Permo-Carboniferous Karoo deposits in southern Namibia include the glacially dominated Dwyka Group and the shelf sediments of the overlying Ecca Group, with a total thickness of about 400 m (Bangert et al., 1997). The Dwyka Group splits up into three major sequences: a first glacial phase represented by tillites and diamictites, a second interglacial phase with offshore mudstones in which tuff horizons were deposited, and a third, minor glacial phase with dropstone-bearing sections during two depositional periods, in Early and Late Permian respectively. Peat accumulation in conjunction with the non-marine, terrestrial and paralic clastic sediments, most commonly sandstones and mudstones of the Ecca Group. Rifting controlled the Dwyka glacial and Ecca coal measures sedimentation. The sediments accumulated in various tectonic-sedimentary basins, including foreland, intracratonic rift, and intercratonic grabens and half-grabens (Cairncross, 2002).

The area that was to become the Namibian continental margin developed a tectonic zonation during its long-lived history, recorded by syn-sedimentary fault systems showing varying architectural geometries and depositional style of contemporaneously deposited strata. Evidence of this syn-depositional faulting includes thickness and facies changes or variation of vertical stacking patterns across faults, well displayed within the breakaway detachment zone at various stratigraphic levels. Analysis of faults, outcrop scale fracture and facies patterns and the depositional architecture of the (1) Carboniferous-Permian, (2) Late Triassic-Eearliest Jurassic, and (3) Early-Middle Jurassic megasequences both in southern and northwestern Namibia has enabled the identification of the direction and sense of the slip along the major, contemporaneously active faults of the rift system (Stollhofen and Stanistreet, 1997).

In South Africa, the Whitehill Formation of the lower Ecca Group was defined on a lithological basis as a white-weathering, fine-grained, black carbonaceous shale unit, distinguished from adjacent units by its conspicuous white colour in outcrop. In southern Namibia, however, the black shale facies changes in outcrops into a greenish-brown, silty shale facies (Werner et al., 2002), in which at least five different facies successions can be distinguished. In this area, dark shales dominate the sequence interrupted by a silty shale band in the middle of the formation. Further to the north in Namibia, several marker horizons of various lithologies, such as dolomitic limestones, carbonate concretions and tuffs in the upper third, and a medium- to coarse-grained sandstone in the lower part, are very important tools for correlation. Especially pyroclastic fall-out ash tuffs and bentonites represent true chronostratigraphic marker horizons. The Namibian Karoo provides thus sequences allowing a precise correlation between lower Ecca basinal and marginal coal-bearing facies that is not possible in the central and northern facies transition of South Africa, due to lack of exposure or drilling.

The Omingonde Formation of north-central Namibia is a succession of mid-Late Triassic terrestrial red beds deposited in a series of NE-SW trending transfer grabens associated with the pre-break-up tectonics of the South Atlantic rift (Smith and Swart, 2002). Three facies associations are recognized in the Omingonde outcrops. They are defined by thin, distinctive lithologies, sedimentary structures, vertebrate taphonomy and paleosols, and are considered to represent a sequence of increasingly arid fluvial environments in the Omingonde Basin through time. First, homogeneous mottled dark red siltstone with interbedded conglomeratic feldspathic sandstone bodies, accumulated in alluvial fans and wet floodplains, in part within axial lakes which filled elongate depressions running parallel to the base of the fault. Second, two major amalgamated sheet sandstone bodies, up to 25 m thick, composed of numerous vertically stacked fining-upward channel fill sequences, with a conglomerate-lined basal erosion surface, deposited in braided rivers and dry floodplains. Third, laterally accreted sandstone bodies separated by pale red overbank mudrocks with well developed calcic vertisol structures, represent ephemeral meandering streams and dry floodplains with saline lakes. The rather rapid switching of fluvial style, the abundance of extrabasinal clasts and the proximity to an active fault margin suggest that tectonism controlled the fluvial style and floodplain accretion rates in the Omingonde Basin. Desiccation features and a thick bed of loessic silts point to a climatic aridity as an overriding factor in the basin development and the preservation of vertebrate remains.

4.4.3.3 Botswana

Geological and mineral exploration investigations across the geological spectrum of Botswana have revealed the occurrence of major sedimentary sequences which represent episodes of sedimentation. Accumulation and preservation of sedimentary rocks can be traced back from the Archaean through to the recent Kalahari Desert environment. Most of these sedimentary rocks have been preserved in basins of intracontinental sag and rift settings (Modie, 2002b).

The last major episode of ancient sedimentary basin development is represented by the Kalahari Karoo Basin wherein are preserved sequences of the Karoo Supergroup. This basin occupies much of central Botswana where it defines an intracratonic sag basin, although the eastern margin indicates a fault-bounded graben system (Green et al., 1980).

The Late Paleozoic glacial event has been widely reported in Botswana, where it is represented by the Dwyka Group (Smith, 1984). This glaciation resulted in the accumulation of widespread glacial deposits throughout southern Africa and led to the development of large coal reserves during Permo-Carboniferous times. In Botswana, glaciclastic deposits of the Dwyka Group occur in the centre of the Kalahari Karoo Basin (Johnson et al., 1996). Much of the record is poorly understood owing to the extensive cover by Cretaceous to recent deposits of the Kalahari Desert environment. Deposits of the Dwyka Group are accessed mainly through drilling and a few isolated exposures along the eastern fringe of the Kalahari Karoo Basin. Evidence for glaciation in Botswana consists of basal diamictites, with rare faceted and striated clasts, and siltstones with thin fining-upward lamination pulses, interpretd as tillites and varvites, respectively (Modie, 2002a). Regionally, the Late Paleozoic glaciation of southern Africa is believed to have started to form on uplifted highlands during the Cape Orogeny, with the development of a major foreland main Karoo Basin in the south and a smaller intracratonic Kalahari Karoo Basin to the north.

4.4.4 LANDSCAPE DEVELOPMENT

Geomorphologically, the multitudinal landforms of southern Africa may be classified in a rather simple system of denudational cycles that are developed uniformly throughout the greater part of the area (King, 1967, Chapter IX). The face of Africa is a polygenetic and complex feature developed over a vast interval of geologic time, with successive local phases of denudation or sedimentation that have developed in response to the attitude of the surface assumed periodically after disturbing land movements. Best preserved are remnants that have stood not upon the elevated country but in depressions where they have been buried for long intervals of time beneath covering formations that have recently been stripped to reveal the resurrected topography more or less in its pristine state. The first fossil topography that survived and is significant in the modern Karoo landscape is a Late Paleozoic denudational topography, the so-called pre-Karoo surface of Late Carboniferous time. It was a landscape of moderate relief, heavily glaciated, with striated pavements beneath the mantle of Dwyka tillites, and upwards covered by the Ecca and Beaufort Group sediments. After a mild period of epeirogenesis, an intra-Karoo landscape developed in Triassic times, with a low relief and desert features; it is covered by the sediments of the Stormberg Group. Over Jurassic Africa and following the Late Triassic sedimentation phase, smooth planation unaccompanied by significant deposition was almost ubiquitous, the Gondwana landscape. After the fragmentation of Gondwanaland, the area underwent more cyclic episodes of relief formation with tectonic interludes, until the development of the present-day topography of the area.

4.5 NORTHWEST AND CENTRAL EUROPE

4.5.1 INTRODUCTION

The tectonic structures of West and Central Europe are very complex due chiefly to the rapid succession of four important orogenic periods: Cadomian (Baikalian), Caledonian, Variscan and Alpine (Zwart and Dornsiepen, 1978). The four successively resulting mountain chains have partly occupied the same space, resulting in an intricate pattern of rocks formed or deformed during each orogeny. Due to this multiple orogenesis, the presence of older Precambrian rocks is difficult to demonstrate.

The studied here comprises most of NW Europe, including Great Britain, exception made of western Ireland (see item 5), and a part of Central Europe (Figs. 4.18 and 4.19), that had an important influence on the successively formed sedimentary basins, often grabens, present in the area.

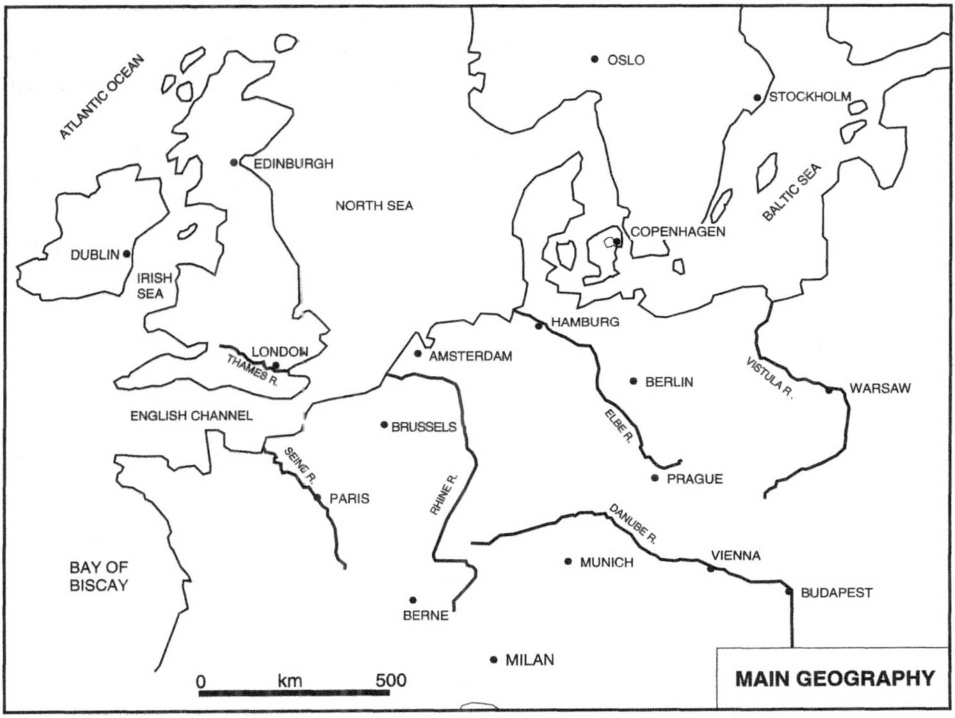

Fig. 4.18: Study area of NW and Central Europe.

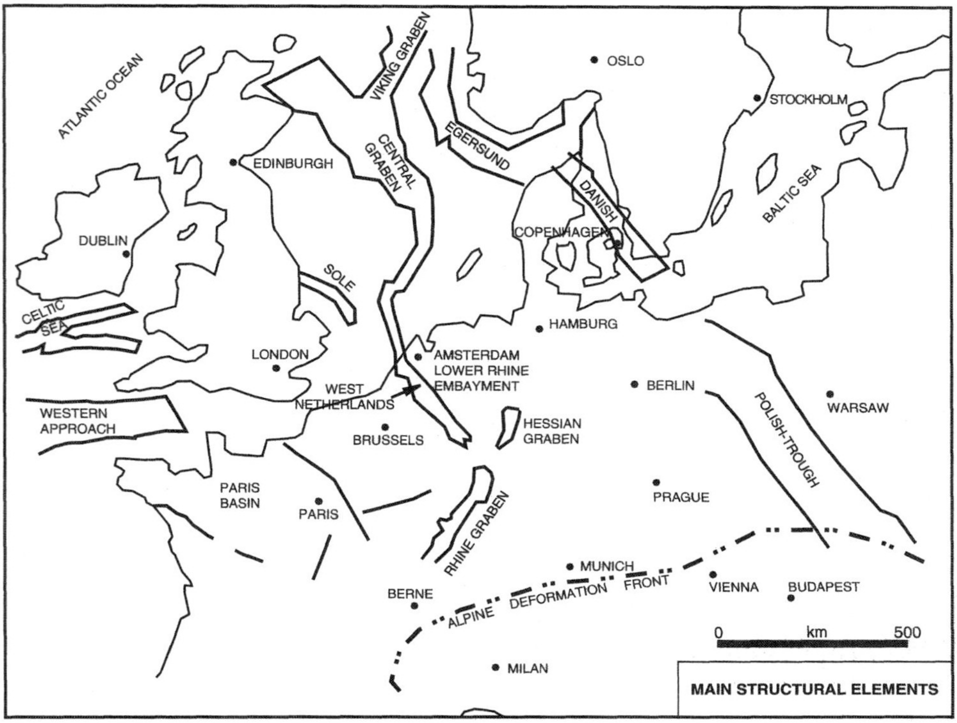

Fig. 4.19: Main structural elements of NW and Central Europe.

4.5.2 TECTONIC FRAMEWORK

4.5.2.1 Pre-Cadomian basement

Low grade Proterozoic sediments and volcanics are known from the Armorican and Bohemian Massifs, where crystalline basement rocks predating these supra-crustals have also been found. Basement rocks are also known from wells drilled in the North Sea area.

The Cadomian (Baikalian) orogeny has been demonstrated in England, the Armorican Massif and the Bohemian Massif, either by the unconformity of Cambrian strata on folded basement, or by geochronological methods. The folding is accompanied by a regional metamorphism of varying grade up to granulite facies. Post-orogenic Cadomian granitic magmatism occurs in the Armorican and Bohemian Massifs.

4.5.2.2 Cadomian orogeny

The Cadomian orogeny took place at the end of the Neoproterozoic and is contemporaneous with the Pan-African and Baikalian events. It is well dated by unconformities in the before mentioned massifs and by geochronological methods. In both cases the orogeny shows folded and metamorphosed sedimentary and volcanic rocks of Late Neoproterozoic age.

In the Armorican Massif, the rock sequence is called the Brioverian (Vendian-Late Cryogenian). Its lower section starts with coarse detrital arkoses and conglomerates, followed by volcanic rocks of spilitic and keratophyric composition. The middle and upper sections consist of graywackes, shales and sandstones with minor limestones and tillites; not very abundant are the mafic and felsic volcanics. Towards the south the Brioverian becomes thinner. At several places in Normandy and Brittany, the folded Brioverian is unconformably overlain by Cambrian rocks. At the eastern flank of the massif, a beginning rift development started the basement of the well-known Paris Basin (Pomerol, 1978).

The situation is more complex in the Bohemian Massif. The Neoproterozoic consists there of graywackes, shales and sandstones, with intercalated mafic and felsic volcanics, mainly in the lower part of the sequence. The rocks are strongly folded. Away from the unconformably overlying Paleozoic, north- and westwards, the Proterozoic grades into greenschist facies and then into intermediate pressure amphibolite facies.

In conclusion, it can be stated that the Cadomian orogeny is widespread within Central and West Europe. It involves Neoproterozoic sediments and volcanics, unconformably overlain by the Lower Paleozoic. Metamorphism grades from anchizone to granulite facies. Cadomian and post-Cadomian granites and Cambrian volcanics occur in the Armorican and Bohemian Massifs. The Cadomian rocks are almost everywhere overprinted by Variscan structures and metamorphism.

4.5.2.3 Caledonian orogeny

The Caledonian orogeny occurs besides the main belt in Scandinavia, Ireland and Wales, in the Ardennes and a zone from northern Germany towards Poland and Rumania (Fig. 4.20). South of this zone no Caledonian folding based on geological evidence, such as unconformities, can be ascertained, although numerous radiometric data on metamorphic and igneous rocks indicate a thermal event during the Caledonian orogenetic period.

The geological time span from Cambrian to Early Devonian is one of the most controversial issues in middle-European geology. Caledonia folding is known from the northern border of the Variscan belt in southern Ireland, Wales and the

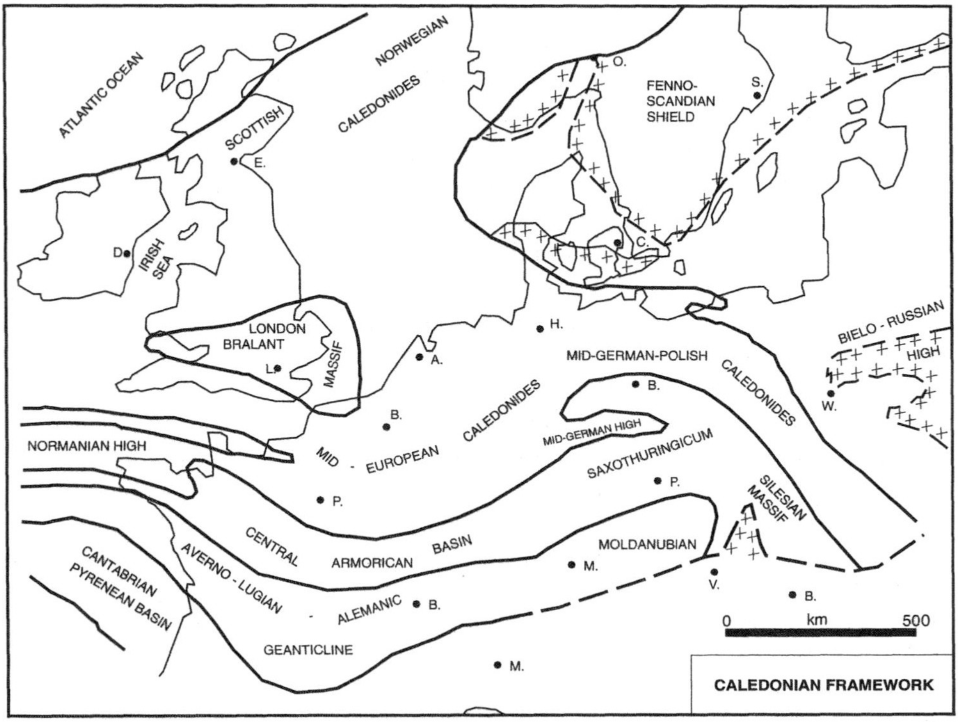

Fig. 4.20: Paleogeography of Caledonian occurrences.
(+) Precambrian basement.

Ardennes. The deformation in S Wales and southern Ireland is rather mild, mainly without metamorphism or with anchimetamorphism. In the Brabant Massif and the Ardennes, the deformation is more intense, resulting in strong folding, the development of cleavage and a greenschist-facies metamorphism. Furthermore, data from drill holes indicate the existence of a narrow Caledonian fold belt at the SW border of the east European platform running from the North Sea over Denmark and northeastern Germany to Poland. There may also exist a connection between the Caledonides of the Brabant Massif and the North Sea-Poland belt, hidden by younger sediments in northern Germany, or obscured by Variscan structures.

4.5.2.4 Variscan orogeny

The Variscan orogeny is much better known although many problems still remain unsolved. Folding, metamorphism, and granitic activity started in the Devonian, but the most widespread and intense folding, low-grade metamorphism, and granitic intrusion are of Late Carboniferous age. The orogeny was active in almost all Central and Western Europe. It affects the Cadomian and a small part of the Caledonian basement, and a more or less complete Paleozoic sequence from Cambrian to Devonian and Carboniferous. The reworking of the Caledonian basement at the northern border of the Variscan belt is weak because the intensity of the Variscan deformation decreases towards the north. The reworking of the Cadomian basement in the internal zones of the Variscan belt is very intense. The strong deformation, together with a high-grade metamorphism, has obliterated many traces of the pre-Variscan structures and metamorphism.

At many places in the Variscan belt the folding is autochthonous as in most of the Rhenohercynian and Saxothuringian zones and does not involve nappe structures. In general, the folding in unmetamorphosed and low-grade regions is quite simple. Nappe structures in low-grade rocks occur in the Harz Mountains, the southern part of the Rheinische Schiefergebirge and the Montagne Noire. In high-grade regions, as in the Moldanubian zone, folding is more complex with three or more fold generations.

There are four folding phases during the Variscan orogeny. The first two are dated between 380 and 360 Ma, that is Middle Devonian (Acadian). The last two phases took place around 320–300 Ma during the Carboniferous (Sudetic). In conclusion, it can be stated that the Acadian-Bretonic folding and metamorphism occurred in the central part of the Variscan chain. The later Carboniferous events have folded and metamorphosed the outer parts of the Variscan belt as well as the earlier formed more internal zones.

In NW Europe the Variscan, orogeny came to a close in the late Westphalian. During the Permian post-orogenic uplift and partial collapse of the Variscan fold belt went parallel with the subsidence of two new intracratonic basins. Rifting in the Arctic-North Atlantic led to the opening of a first seaway between the Arctic and NW Europe during Late Permian. The Triassic development of NW Europe, Arctic-North Atlantic and Tethys was dominated by regional crustal extension at the onset of the split-up of the Pangean continent. The Triassic is characterized by an orogenic volcanism.

4.5.2.5 Alpine orogeny

The Alpine orogenic period started with the Jurassic split-up of Pangea, leading to the opening of the Tethys and a progressive rifting in the Arctic-North Atlantic; this was accompanied by the development of major rift systems in NW Europe. Of these, only the North Sea graben experienced a short-lived volcanic phase. During the Cretaceous rifting, transform faulting and progressive subsidence of the Tethys basin continued, with some seafloor spreading in the Neocomian, and minor folding phases during the Aptian-Albian and latest Cretaceous (Senonian).

With the Early Tertiary onset of seafloor spreading in the northern North Atlantic and in the Norwegian-Greenland Sea, the Mesozoic rifts of NW Europe became inactive and started to subside regionally. The main Alpine orogenic phases took place during the Paleogene and early Neogene. Alpine suturing of Eurasia and Africa at that time, was accompanied by compression and inversion of the NW European Mesozoic troughs, located at distances of up to 800 km to the north of the Alpine front. However, the Atlantic seaboard as well as the North Sea Basin continued to subside during the Tertiary. The late-orogenic minor folding phases of the Alpine orogeny occurred in the Late Neogene and Quaternary. Concomittant with these last phases, a new set of rifts developed in the Alpine domain and its immediate foreland. Their subsidence is correlative with the collapse of the Mediterranean basins.

4.5.3 SEDIMENTARY BASIN DEVELOPMENT

4.5.3.1 Cadomian (Late Cryogenian-Vendian) oscillatory-emergent episode

From the sedimentological point of view, detailed studies of the Late Neoproterozoic sedimentary rocks have not yet been made. Metamorphosed sediments, such as metagreywacks, slates, shales and sandstones are known from the Armorican and Bohemian Massifs, respectively. Research about tectonics and metamorphism suggests these sediments to be of Late Cryogenian-Vendian age (Zwart and Dornsiepen, 1978), and thus may be attributed to the Late Neoproterozoic oscillatory-emergent episode.

4.5.3.2 Caledonian (Cambrian-Early Devonian) submergent and oscillatory-emergent episodes

Cambrian-Early Ordovician. During the Cambrian-Early Ordovician submergent episode the accumulated deposits are essentially of marine origin. These deposits occur chiefly in E Ireland, Wales, S Scotland, the Armorican Massif, and some parts of the London-Brabant Massif and in Central Germany. Two groups of marine sediments may be distinguished: (1) conglomerates and sandstones of littoral origin, showing many erosional hiatuses, with the material derived from the Precambrian shield areas; (2) fine-grained clastic sediments, of shallow to rather deep marine realms. The thickness of up to 3000 m of these latter deposits suggests an overall deposition in subsiding troughs. The climate seems to have become temperate, after the Late Neoproterozoic glacial events, due to drift of the area towards lower latitudes. Many of the uppermost sections of the Cambrian-Early Ordovician sediment sequences have later been eroded due to the Caledonian orogenetic events. A more detailed sedimentological study of a Cambrian succession has been made in the Middle Cambrian Bray Group of SE Ireland (Strogen and O'Byrne, 2000).

At the southern flank of the London-Brabant Massif, in the French-Belgian border area, there appear some outcrops of Cambrian deposits consisting of an alternation of slates and pyrite-containing sandstones with a scarce fossil content, folded during the Caledonian orogeny.

Late Ordovician-Early Devonian. Between Late Ordovician and Early Devonian, with special emphasis on the Silurian, the accumulated sedimentary sequences show many erosional interruptions and hiatuses, due to the character of the episode, being oscillatory-emergent. During the Late Ordovician and Early Silurian most of NW Europe was still covered by the sea. The occurrences in Wales are composed of dominantly littoral to shallow marine sandstones, with a considerable thickness of a few thousands of metres, suggesting a continued subsidence of the area. Towards SE the sediments become finer, composed chiefly of slates and shales, with abundant graptolites.

Upwards in the stratigraphic section it appears that the subsidence decreased in intensity as to become almost equal to the sedimentation rate. The sea turned less deep and regressed, with a sediment accumulation becoming coarser towards sandstones. This type of deposition continued into the Early Devonian as proved by the conformable position of the Late Silurian to Early Devonian strata appearing in Normandy. Finally at the end of the episode, most of NW Europe had turned continental, starting the epoch of the so-called Old Red Sandstone. Especially towards the SE of the Old Red continent, marine influences continued to exist. For instance after Wehmann et al. (2000), an alternation of small transgression and regression phases enabled the accumulation of quartzitic sandstones and sandy shales in the studied area of the southern Rhenish Massif in the Mosel region of Germany. The depositional environments have been interpreted as marine to brackish littoral to shallow marine, with frequent emergence.

4.5.3.3 Early Variscan (Middle Devonian-Late Carboniferous) submergent episode

From the Middle Devonian onwards, the sedimentary sequences in NW Europe become more frequent. They have been deposited in a number of basins from SE

Ireland to the Polish trough (Ziegler, 1978; Pozaryski and Brochwicz-Lewinski, 1978; Figs. 4.21–4.23).

Northern British Isles. In the domain of the Arctic-North Atlantic Caledonides strong post-orogenic uplift coupled with transcurrent faulting led during the Devonian to the rapid subsidence of the intramontane Old Red "Molasse" basins of Ireland, Scotland and Western Norway. In the North Sea area, subidence in the Midland Valley Graben and the Orcadian and Northumberland basins started in the Early Devonian. This was accompanied by extrusion and intrusions of alkaline volcanics and granites. Continental and in part lacustrine series, with dominantly sandstones and shales reaching up to thickness of 6000–8000 m, were deposited in the fault-bounded Old Red basins of the British Isles and western Norway. During the Carboniferous, the Midland Valley, the

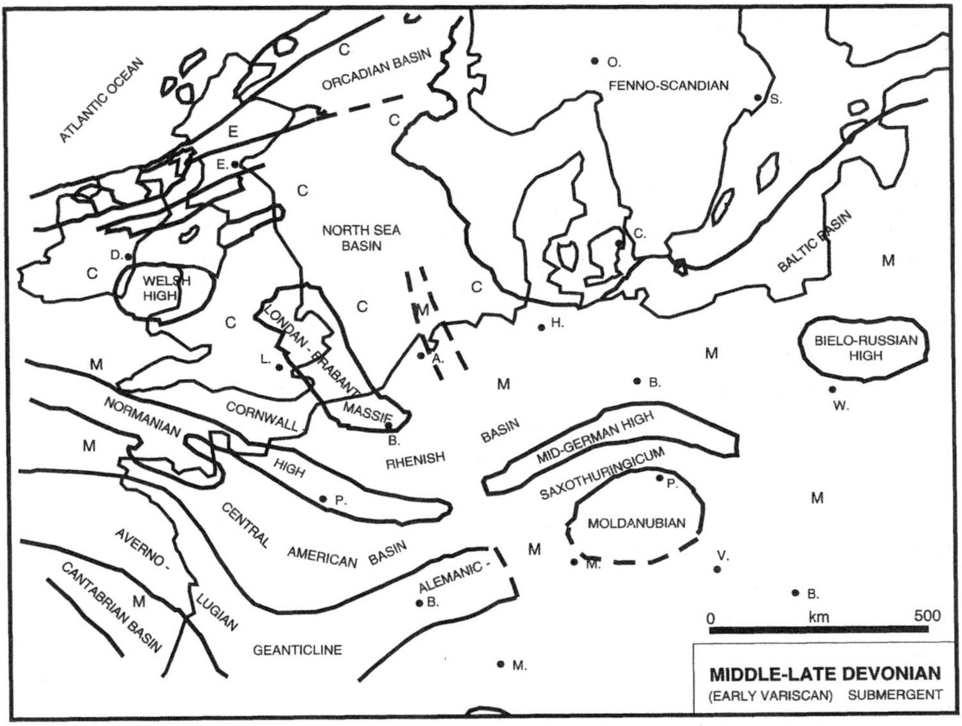

Fig. 4.21: Paleogeography of Variscan occurrences, Devonian. Symbols: M – marine (limestones), M(c) – marine clastics, C – continental sandstones, P – continental sandy shales, i – inversion areas, v – Lower Tertiary volcanics, × – Upper Tertiary volcanics.

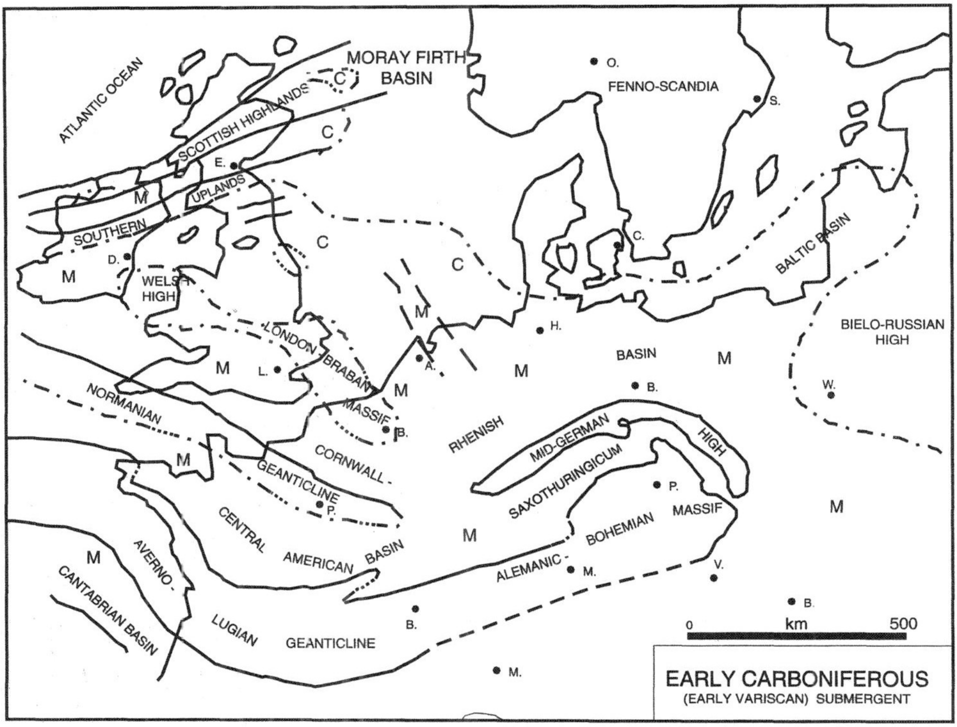

Fig. 4.22: Paleogeography of Variscan occurrences, Early Carboniferous. Symbols: see Fig. 4.21.

Northumberland-Solway Basin and the Dublin Trough continued to subside rapidly, with accumulation of chiefly limestones in the Early Carboniferous and more sandy-shaly deposits in the Late Carboniferous. All these sequences have also been found in the central parts of the North Sea.

Cornwall-Rhenish Basin. During Devonian times a large sedimentary basin developed also, under a tensional setting of the rapidly degraded Mid-European and North German-Polish Caledonides: the Cornwall-Rhenish Basin with fine clastic sediments and limestones. A marine transgression originating presumably from such areas of continuous Silurian-Devonian marine deposition in the east as the Harz and southern Poland, and possibly similar areas to the west, invaded this basin during the Gedinnian and Siegenian. During this time, the proper basin continued its subsidence, consequence of the submergent tectonic-sedimentary episode.

In the Early and Middle Devonian, clastics were shed into this basin from the Old Red continent to the north. Shallow-water sands accumulated in shelf areas

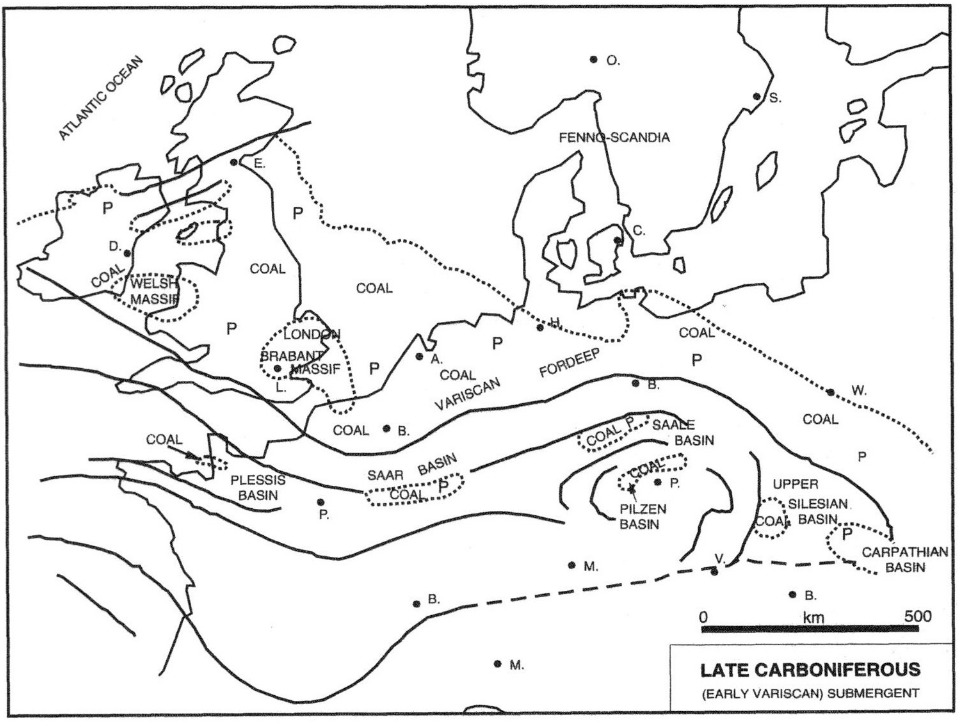

Fig. 4.23: Paleogeography of Variscan occurrences, Late Carboniferous. Symbols: see Fig. 4.21.

where they were in part trapped in local sub-basins. In the southern deeper parts of the basin, Early and Middle Devonian series are represented by thick black shale sequences. Increased tectonic activity during the Middle Devonian resulted in block faulting of the basin floor and the accentuation of a number of fault-bounded sub-basins and intervening swells. This was accompanied in Southern England and in parts of the Rheinisches Schiefergebirge by a strong, partly submarine, bimodal volcanism that emphasizes the intracratonic tensional nature of the Cornwall-Rhenish Basin which may be regarded as a complex rift belonging to the Devonian North Atlantic fault system. A possible extension of the rift system into the North-German and Polish lowlands is masked by a thick Late Paleozoic and Mesozoic cover.

The Middle Devonian tensional phase went parallel with a regional marine transgression that reached far to the north, shutting off much of the clastic supply to the deeper part of the Cornwall-Rhenish Basin. During the Givetian, large reef-fringed carbonate platforms established in shelf areas and on local heighs, extending into the central North Sea area as a rather narrow, fault-controlled sea arm.

Extension tectonics persisted throughout Late Devonian into Early Carboniferous. These tectonic disturbances triggered density currents, slumping and even olistostromes on the flanks of the Mid-German High. In the axial part of the Cornwall-Rhenish Basin, carbonate deposition continued, with only minor amounts of clastics. Calcareous turbidites were derived from shelf areas and local highs. Volcanic activity decreased considerably at that time.

During the Early Carboniferous, the northern shelf of the Cornwall-Rhenish Basin was occupied by the Waulsortian reef and "Kohlenkalk" platform that extended from Ireland to Poland. In the southern parts of the basin, carbonate deposition on local highs was gradually drowned out by continental subsidence. In the same time a peak of basalt volcanic activity took place, interpreted as another major rifting phase in the southern part of the basin. This volcanism correlated with the onset of a similar phase in England and Ireland.

The onset of the Variscan orogeny is suggested by the end Devonian-beginning Carboniferous accentuation of the Mid-German High, as illustrated by an increasing influx of clastic material. Culm graywackes started appearing, first in Moravia, later in the other part of the basin. An underthrusting along the margins of the Normanian and Mid-German highs turned the Cornwall-Rhenish Basin strongly polarized to the south, thus assuming the geometry of a classical foredeep. Only in the early Namurian the Culm flysch reached its peak. The Kohlenkalk shelves that occupied the northern parts of the basin were drowned by the Namurian transgressive marine shales which reach thickness of over 2000 m in the northern Netherlands and Germany. In the Central North Sea Basin, these shales grade northwards into red beds.

By Late Namurian times, sedimentation exceeded subsidence in the southern, proximal parts of the Variscan Foredeep, as is illustrated by the basin-wide change of the depositional regime from a deeper water flysch type to a shallow-water to continental molasse type, with sandstones and shales.

During the Westphalian continental paralic conditions prevailed throughout the basin; these, however, were periodically interrupted by short-lived marine incursions at the onset of a new oscillatory-emergent tectonic-sedimentary episode. The Westphalian coal measures reach maximum thickness of 3500 m in the North German lowlands.

The Carboniferous geological history of the NE German foreland basin has been studied by McCann (1998). This basin is situated between the Precambrian Baltic Sea/Scandinavian shield in the north, and the tectonically influenced areas during the Cadomian, Caledonian and Variscan orogenies in the south. The Lower Carboniferous sediment sequences are chiefly carbonate rich and of marine origin. In the Namurian, continental realms developed at the northern basin margin, with turbidite sedimentation in the south. By Upper Carboniferous times, the sequence had become wholly continental and clastic.

The sedimentary evolution of the basin is explained by the structure and evolution of the Variscan thrust front in the south and local tectonics towards the north.

The Gondwana glaciation of the southern hemisphere had also wide ranging effects in low equatorial latitudes (Lemon and Tucker, 2002). The mid-Carboniferous strata in Europe are cyclic, with a range of lithologies and thicknesses in the sedimentary rhythms through the succession and across the region. The respective cycles are composed of mixed clastic-carbonate to wholly carbonate sections. Within the succession, local major sharp-based channel sandstones occur with fluvial-dominated, fining-upward into estuarine facies. These are interpreted as a consequence of short-lived glacial advances and associated rapid sea-level falls, as incised-valley fills.

The evolutionary tendency in all these basins appears to fit rather well into the general idea of the mid-Paleozoic submergent tectonic-sedimentary episode.

4.5.3.4 Late Variscan (Permian-Early Jurassic) oscillatory-emergent episode

In Europe, the Variscan orogeny came to an end with the Late Carboniferous compressional phase. In the Early Permian the development of the Variscides and their northern foreland was characterized by the emplacement of a complex fault system and an intense post-orogenic volcanism. In NW Europe the mapping of the fault pattern is hampered by a thick Mesozoic and Tertiary sedimentary cover and has to be made by geophysical information and well data. A number of rather small-sized rift basins developed in the region in the beginning of the episode. Areas with maxima of volcanic thicknesses may be associated with pull-apart structures at the termination of subsidiary wrench faults. A modification of plate movements during the last suturing phase of the Laurasia supercontinent apparently caused intense fracturing of the European Variscides and their foreland. This had wide implications for the subsequent evolution of NW Europe, with a repeated reactivation of the faults.

Permian basins. During Middle and Upper Permian, subsidence generated two large basins in the Variscan foreland (Ziegler, 1978; Figs. 4.24 and 4.25). One of these is the so-called Southern Permian Basin that extends from Poland to England over a distance of about 1500 km. It overlays the Variscan foredeep basin, encroaching in the east upon the Variscan fold belt, where the Polish Trough started its full development (Pozaryski and Brochwicz-Lewinski, 1978). During the Early Permian (Rotliegendes), clastics derived from the rapidly denudated and eroded Variscan Mountains were deposited in this basin under

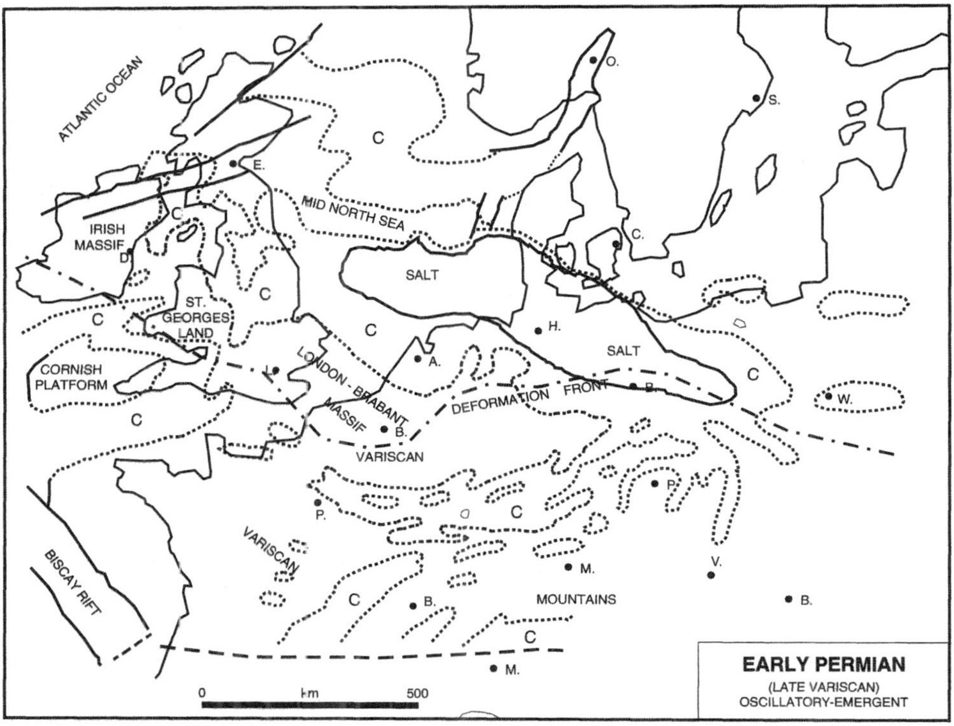

Fig. 4.24: Paleogeography of Early Permian basins. Symbols: see Fig. 4.21.

eolian and sabkha conditions as the classical sequence of Rotliegendes red beds (Glennie, 1972). The respective clastic sequences are composed of conglomerates, sandstones and shales, deposited in desertic alluvial fans and proximal to distal ephemeral streams, with a widespread eolian influence. In the central part of the basin, the red beds are interbedded with massive halite sequences.

The other large basin is called Northern Permian Basin (see Fig. 4.25) that is separated from the Southern Permian Basin by a Mid-North Sea-Denmark trend of highs. This northern basin extends from northern Scotland (Moray Firth) to the Oslo Graben. It overlays the central North Sea Devonian Old Red Basin in the west and transgressed over Lower Paleozoic sediments and metamorphics in the east. The basin showed only a moderate subsidence tendency during the Early Permian. The accumulated deposits consist of red beds ranging from conglomerates to shales, reaching a maximum thickness of some 600 m.

In both Permian basins, subsidence exceeded the sedimentation rate during the Early Permian. This led to the development of large topographic depressions which were located well below the sea level of the Permian oceans. Continued rifting and an eustatic sea-level rise led to the onset of a seaway extending from

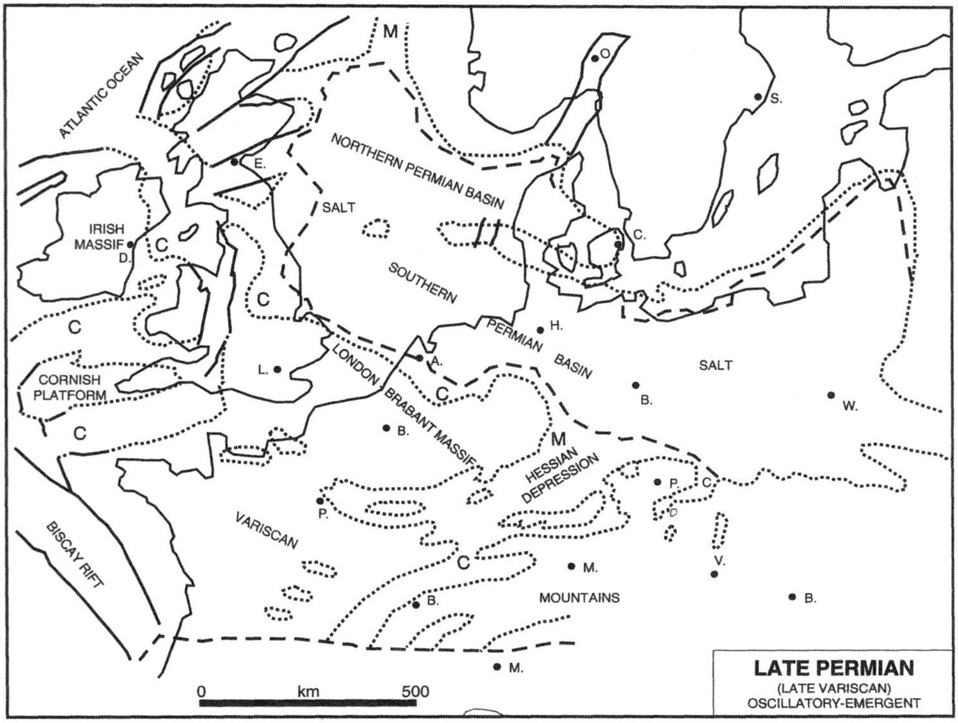

Fig. 4.25: Paleogeography of Late Permian basins. Symbols: see Fig. 4.21.

the Arctic southwards. From this seaway the Late Permian (Zechstein) sea transgressed into the two NW European Permian basins under possibly catastrophic conditions.

In the very southwest of the region, the Paris Basin also started its development (Pomerol, 1978). This basin is situated to the south of the London-Brabant High and the Rhenish Massif which form the southern limit of the Permian basins (see Fig. 4.19). After the erosion of the Hercynian massifs and the infilling of the formed depressions with continental detrital sediments, the Paris Basin began its individualization. In these rather small depressions an about 900 m thick succession of conglomeratic to medium-grained sandstones accumulated.

The Late Permian (Zechstein) seas entered the area through the northern North Sea. The rapid transgression was followed by the deposition of thin deep-water limestones and laminated anhydrite in basinal centres, and thick shallow-water carbonate and sulphate banks on the basin margins and on local highs. An arid climate, a restricted influx of sea water and oscillatory eustatic sea-level changes led to the rapid infilling of the two basins with the cyclical Zechstein evaporite series, with thickness of some 1000 m in the Northern and some 1500 m in the Southern Permian Basin. The excellent correlation of these depositional cycles in the two basins point to a free communication between them. The Zechstein sea also transgressed southward through the incipient Hessian depression deep into the Variscan fold belt (Fig. 4.21).

Triassic-Early Jurassic basins. At the transition from the Permian to the Triassic the marine connection between the Arctic and the NW European Permian basins was interrupted. The proper basins continued to subside; however, their structural framework was gradually modified by the emplacement of a new set of grabens and flexure-bound troughs that transected in part the Permian tectonic elements, connecting also the Paris Basin to the central basinal areas (Fig. 4.26).

During the Triassic, regional tensional stresses caused the rapid subsidence of the fault-bounded Polish Trough and the North Danish Basin in which Triassic sequences reach maximum thickness of 3000 and 6000 m, respectively. In the northern and central North Sea, over 1000 km long Viking-Central Graben

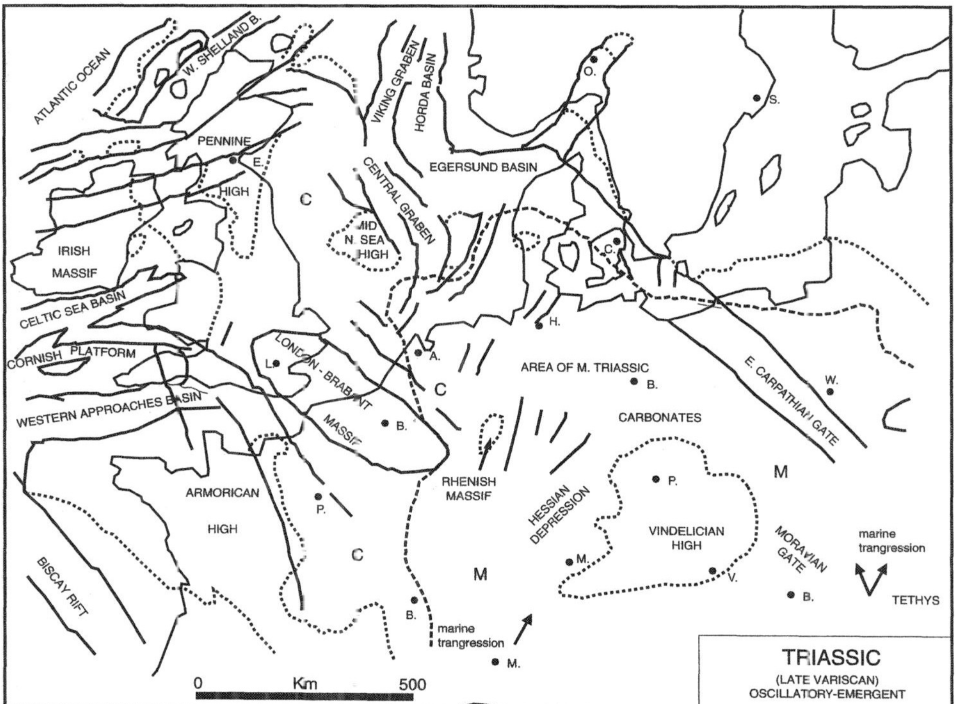

Fig. 4.26: Paleogeography of Triassic basins. Symbols: see Fig. 4.21.

system came into evidence. This rift that transects the Northern Permian Basin, cuts across the separating ridge, and reaches the Southern Permian Basin, presumably developed from the north to the south by fracture propagation. Furthermore to the west, various rift graben basins developed over Britain, containing appreciable thickness of Triassic sediments. Along the southern flanks of the Southern Permian Basin a number of essentially NNE-SSW striking troughs came into evidence during the Triassic, being from east to west: the Thuringian-Westbrandenburg, the Hessian and the Emsland depressions. In NW Europe, an overall absence of intense Triassic rift volcanism is very striking, suggesting that the emplacement of the European Triassic rifts is caused by regional crustal extension only.

During the same period a gradual sea-level rise was accompanied by the progressive overstepping of the Permian basin margins. Continued subsidence provoked a link-up between the NW European basins, making also a connection with the Tethyan area. A renewed regional regression marked the onset of the Late Triassic Keuper. In the northern North Sea the entire Triassic is represented by continental sandstone red beds, with a thickness of up to 3000 m. In the southern area the Triassic occurs in its classical tripartite Germanic facies, composed of Early Triassic (Scythian) red sandstones, Middle Triassic (Muschelkalk) marine limestones, and Late Triassic (Keuper) sandy claystones with anhydrite and gypsum lenses; all form chiefly continental and transitional, more rarely marine environments.

Triassic sedimentation rates generally remained in balance with subsidence rates. By Late Triassic times, the Variscan fold belt was degraded to the point that only disjointed low relief highs emerged from the immense, monotonous tidal flats to floodplain realms that covered much of NW Europe.

During the Jurassic, the complex Triassic graben system of NW Europe underwent a polarization to a few major rifts, most of which remained active till the Late Cretaceous. Thus, a number of Triassic grabens became inactive. Main elements are the North Sea Rift (Viking Graben and Central Graben), the Polish-Danish troughs, the Western Approaches Basin, connected with the Paris Basin, and the graben-shaped Celtic Sea-Bristol Channel Basin (Fig. 4.27). The stratigraphic record of these basins reflects a number of more or less synchronous tectonic events that are correlative with major rifting phases in the North Atlantic. In this context, the grabens of NW Europe can be considered as an integral part of the Arctic-North Atlantic rift system that gradually opened during the Mesozoic.

At the transition to Early Jurassic (Liassic) times, an extensional tectonic pulse affected the entire rift system. This pulse preceded the Early Liassic transgression that inundated much of NW Europe. A new seaway connecting the NW European basins with the Arctic basins opened through the Norwegian-Greenland sea area. The earliest Alpine tectonic phase caused a mild accentuation of the major positive elements from which clastic material was shed into the down-warped intervening lows. In the northern North Sea axial parts of the Viking Graben, massive fluvial sands were deposited. Moreover, in the Danish Trough sands are well developed. In the Polish Trough, brackish-water clastics transgress unconformably over Middle and Upper Triassic sequences. In northern Germany and in the southern North Sea, marine shales and minor sands overlay conformably the Keuper series. During the Sinemurian, fully marine conditions were established in much of the whole area with transgressions coming from the Tethys as well as from the Arctic-North Atlantic seaway. In the North Sea area the Liassic sequences are largely developed in a cyclic open-marine shaly facies. The Polish Liassic series of fresh- to brackish-water clastics were transported westwards into northern Germany where they interfinger with marine shales. In the Paris Basin and the connected Aquitaine Basin, medium-grained clastics are essentially lacking with shales and limestones being the dominant lithology. In the western England grabens marine shales with minor carbonates prevail.

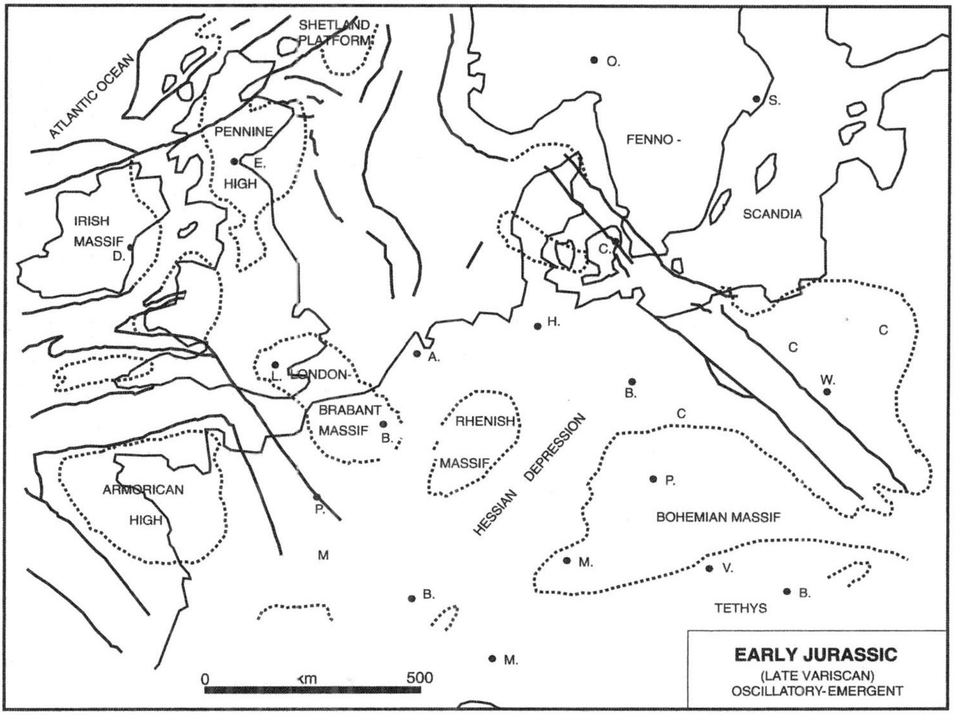

Fig. 4.27: Paleogeography of Early Jurassic basins. Symbols: see Fig. 4.21.

The more recent studies made in the areas with Permian-Early Jurassic sediments deal more with certain sedimentological details than with integrated basin studies.

4.5.3.5 Early Alpine (Middle Jurassic-Late Cretaceous) submergent episode

At the onset of the following submergent episode in the Middle to Late Jurassic, a number of discrete rifting phases affected the entire metastable platform of NW Europe.

In the Central North Sea, a large high encompassing the Mid North Sea and the southern Danish highs was uplifted during the beginning of the Middle Jurassic. This uplift separated again the northern North Sea Basin from the southern one, and can be considered as a rift dome. Its crestal parts were transected by grabens. At the triple junction of these grabens, a major alkaline volcanic centre was emplaced. The grabens at the NE side of the high and the troughs at the SE flank became inactive. The Liassic seaway that extended through the Irish Sea was interrupted by regional uplift. On the other hand, the Celtic Sea-Bristol Channel and the Western Approaches basins continued to subside differentially. This was accompanied by a minor volcanic activity as illustrated by the Fuller's Earth of southern England. The Polish Trough also continued its subsidence whereby a seaway between the eastern part of the NW European Basin and the Tethys was reopened.

Middle-Late Jurassic basins. In the area of the central North Sea rift dome, Early Jurassic and older sediments were subjected to profound truncation. Erosion products were shed northwards into the continuously subsiding Viking Graben, Danish Embayment and Horda-Egersund basins (Fig. 4.28) where they were deposited under paralic and deltaic conditions, as sandstones and shales, also with some coal formation. Clastics shed southwards from this dome accumulated in the incipient Sole Pit, West Netherlands, Lower Saxony and Altmark-Brandenburg basins in which marine conditions prevailed. These basins are generally referred to as the "Marginal Troughs". The sediments deposited herein are sandstones near the border of the dome and silt and claystones more distally. Right-lateral wrench movements between the Danish-North German block to the north and the Variscan massifs to the south were caused by crustal distension in the North Sea rift and in the Polish Trough. In this trough, sandstones, claystones and some limestones accumulated.

The Sole Pit Basin connected towards the south through the English Midlands towards the Paris Basin surrounding the London-Brabant Massif. The Liassic is still characterized by a variety of rocks comprising sandstones, shales and

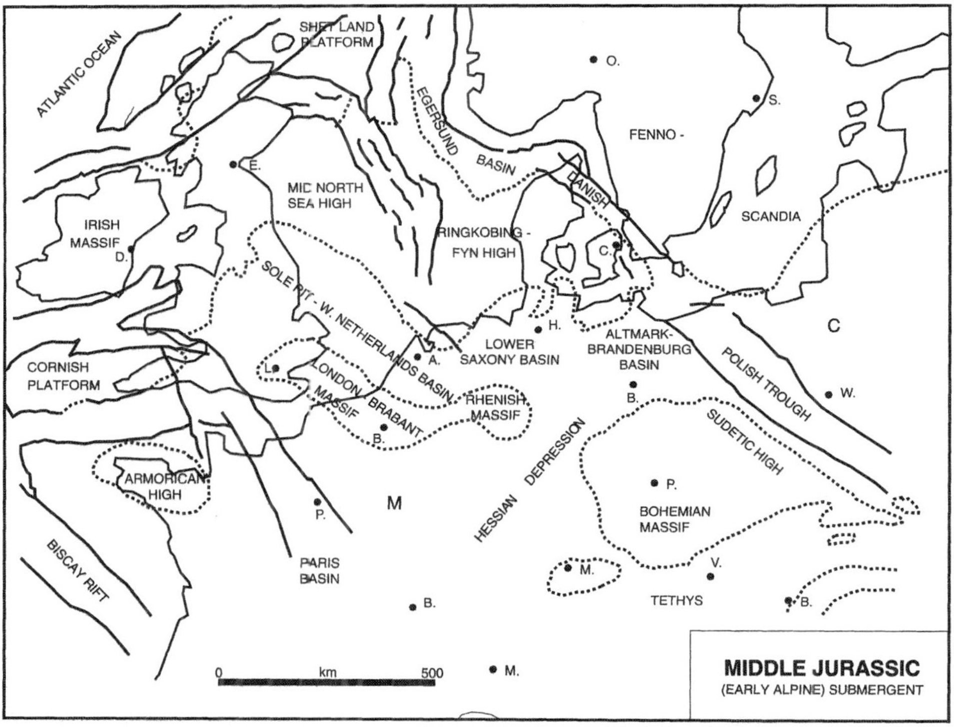

Fig. 4.28: Paleogeography of Middle Jurassic basins. Symbols: see Fig. 4.21.

limestones, while the Middle Jurassic sequences are dominantly marls and limestones with a minor sandstone component. The rocks seem to have deposited in a number of local sub-basins developed within the area. In the Paris Basin, limestones dominate in shallow-sea to platform facies.

Continued differential subsidence of the Viking Graben and regional subsidence of the central North Sea rift dome were accompanied by the progressive transgression of the Jurassic seas which reached their maximum extent in the Late Jurassic (Fig. 4.29). By this time, deep-water conditions were established in the Viking and Central grabens. The Late Jurassic sequences consist in these areas of highly organic shales locally interbedded with sandy turbidites. During the Kimmeridgian, a renewed marine connection established between the northern and southern North Sea basins. In the latter, the Late Jurassic sequences display a wide facies variety ranging between carbonates and evaporites in the Altmark-Brandenburg and the Lower Saxony basins towards the Polish Trough. Paralic series were deposited in the West Netherlands Basin and shales with minor carbonates and clastics in the Sole Pit Basin. The differential subsidence of the North Sea rift and the Polish Trough went parallel with an accentuation of

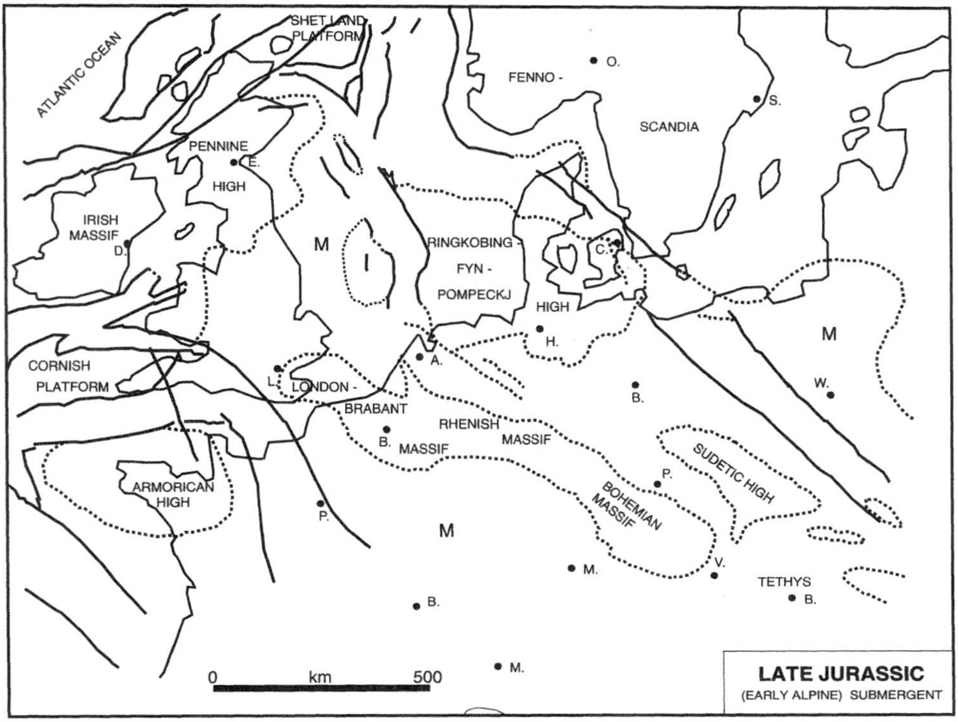

Fig. 4.29: Paleogeography of Late Jurassic basins. Symbols: see Fig. 4.21.

the Marginal Troughs and the uplifting of the London-Brabent, the Rhenish and the Bohemian massifs, with a contemporaneous closure of the Hessian Strait. Furthermore, a Sudetic Strait opened separating the Sundetic High from the Bohemian Massif.

The Late Jurassic Sea continued its extension through the English Midlands, connecting with the western Celtic Sea-Bristal Channel and Western Approaches basins, and towards SE with the Paris Basin. The respective basins were filled in with a variety of sediments, generally sandstones around the tectonic highs, and becoming finer to shales, marls, and limestones towards the basin centres. In some of these basins, anhydrites have also been found.

Cretaceous (Berriasian-Turonian) basins. At the beginning of the Cretaceous a major rifting phase affected the entire Arctic-North Atlantic rift system, coinciding with a pronounced eustatic sea-level drop. This rifting caused seafloor spreading in different parts of the North Atlantic that correlates with phases of relative tectonic quiescence and gradual rise of the sea level.

The early Alpine tectonic phase affected the entire North Sea area. In the Viking Graben, the rifting impulse resulted in the accentuation of the existing seafloor topography, causing a strongly block-faulted submarine relief of up to 1000–2000 m. In the Central Graben rift tectonics is less obvious due to the interference of an intense Zechstein salt diapirism.

The downfaulting was accompanied by a temporary uplifting and emergence of the rift flanks. The Marginal Troughs became more accentuated, parallel with the uplifting and emergence of the Pompeckj Platform and the London-Brabant-Bohemian Massifs (Fig. 4.30) from which clastics were shed into the adjacent basins. A break in sedimentation was followed in the emergent areas by a rapid transgression with a consequent decrease in the clastic influx into the Viking and Central grabens. During the Early Cretaceous the topography of this rift system became infilled by up to 1200 m thick, rather deep-water shales. Along the Fennoscandian Borderzone, chiefly sands were deposited in the Danish-Polish Trough.

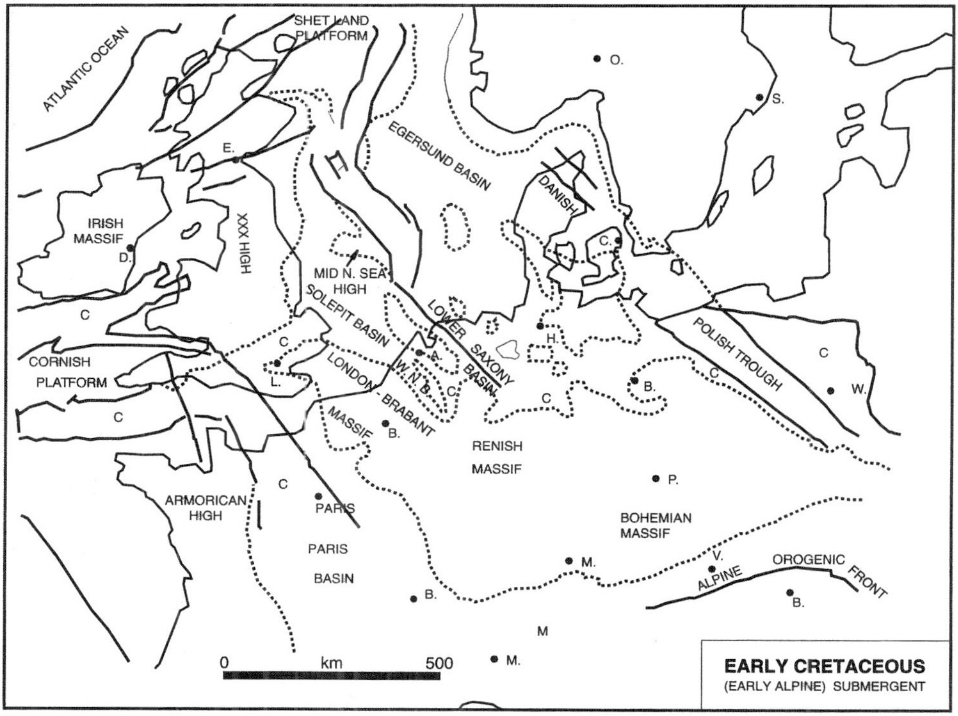

Fig. 4.30: Paleogeography of Early Cretaceous basins. Symbols: see Fig. 4.21. WNB – West Netherlands Basin.

The early Aptian rifting phase was again followed by a short-lived emergence of the flanks of the North Sea rift, and accompanied by a renewed sharp accentuation of the Marginal Troughs. In the Lower Saxony Basin the tectonism was accompanied by a basic volcanism. In the Albian the basinal marine sediments are represented by the marls from the deep part of the basin. Towards the end of the transgressive sedimentation, the deposits become finer grained with no increase of the clay fraction (Fenner et al., 1998). Throughout the North Sea area the early Aptian tectonic phase was followed by a regional transgression that culminated in the Senonian, already during the onset of the next oscillatory-emergent tectonic-sedimentary episode.

Due to the opening of the Bay of Biscay, a major rifting and transform-faulting event caused the uplifting of its margins. Towards E clastic material was shed into the Celtic Sea Trough and the Hampshire and Paris basins. The Celtic Sea and the Western Approaches troughs subsided differentially; in these basins the Early Cretaceous clastic sequences reach thickness of over 1000 m. A regional unconformity corresponding to the early Aptian phase marks the base of the so-called Greensand Series in Britain. By this time the Celtic Sea-Bristol Channel and the Western Approaches troughs had almost become inactive. Similarly to the North Sea area, carbonate deposition set in on the Celtic Sea shelf with the onset of the Late Cretaceous. Subsidence was relatively uniform in both troughs where maximum chalk thickness is of the order of 600–800 m near the shelf edge. Sequence stratigraphical criteria can be applied to the Aptian-Albian sediments of SE England (Eyers, 1994). The classical interpretation of one fully marine deposition, with the commencing of nearshore sands, developing in offshore sandwaves with a subsequent transition into deeper marine clays, appears to be more complex. The studied sequence revealed estuarine, shoreline, tidal flat, shoal and sandwave sand facies within the lower section, through transitional silty beds into deep marine clays with numerous phosphatized hiatus surfaces in the uppermost section. The system tracts appear to be complete and consist of six sequences for the whole section.

4.5.3.6 Late Alpine (Senonian-Holocene) oscillatory-emergent episode

Late Cretaceous and Early Tertiary tectonics in the central and northern North Sea was purely extensional. This contrasts with the Marginal Troughs, the Danish and Polish troughs and the southern part of the North Sea rift in which the Late Alpine tectonism was of a tangential compressive nature. These basins became mildly inverted during the Senonian, with a major inversion during the Early Tertiary. In this process of inversion, the previously extensional basins were deformed by wrench movements whereby their basin fill was folded and uplifted above the erosional base level. At the same time, the Rhenish-Bohemian Massif was dissected and in part uplifted along a number of wrench and steep reverse faults. These inversion movements coincide with major orogenic events in the Alpine domain, due to the continent-to-continent collision between the Eurasian and African plates.

The regional importance of the Late Cretaceous to Early Tertiary inversion tectonics in NW Europe is best illustrated by comparison of the paleogeographic maps of both periods (Figs. 4.31 and 4.32). However, it should be borne in mind that in areas surrounding the Rhine Graben the present-day distribution of Mesozoic series has been strongly influenced by Eocene and younger erosion across the Rhine Graben rift dome. Therefore, it seems likely that Senonian series originally extended unbroken from the Paris Basin to eastern Bavaria, upper Austria and into the Alpine domains. Strong inversion in the Polish Trough resulted in the upwarping of the Polish Anticlinorium (Pozaryski and Brochwicz-Lewinski, 1978). This major inversion structure extends southwards

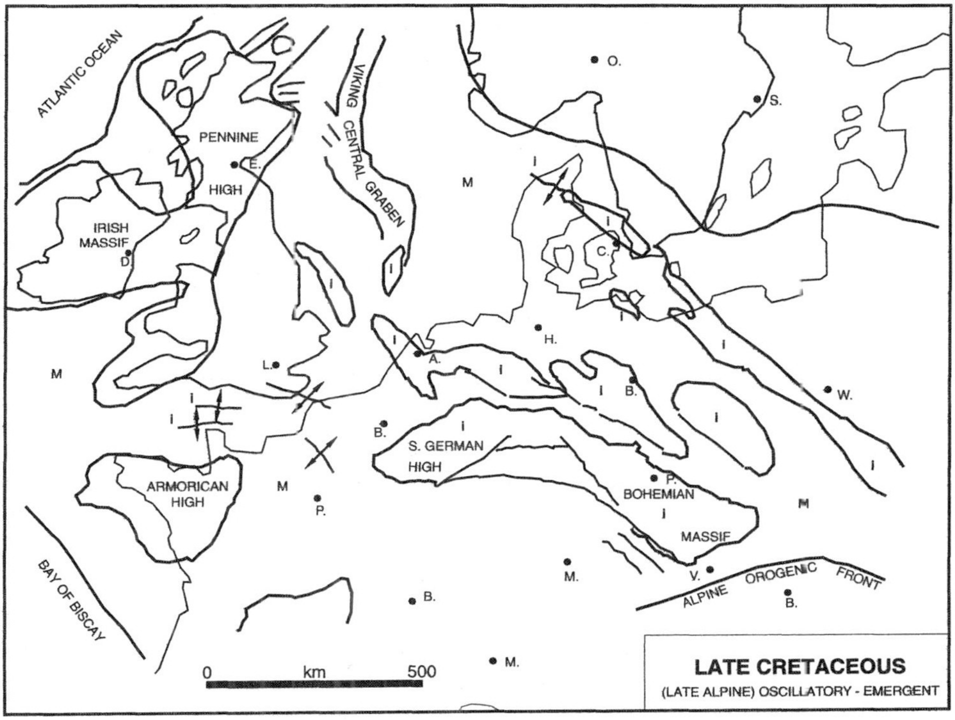

Fig. 4.31: Paleogeography of Late Cretaceous (Senonian) basins. Symbols: see Fig. 4.21.

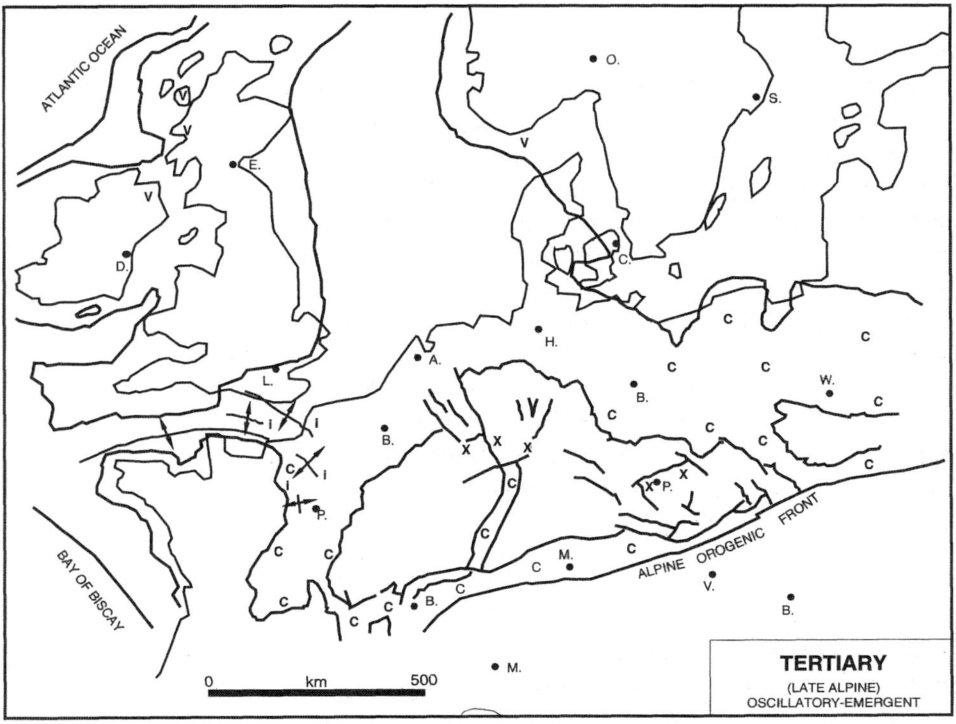

Fig. 4.32: Paleogeography of Tertiary basins. Symbols: see Fig. 4.21.

under the Carpathians and northwards into the Rönne Graben located to the east of Bornholm that itself was uplifted at that time.

Inversion in the Channel area and the Western Approaches Basin was mild during the Early Tertiary. Main inversions took place there during the Miocene at a time, when the inverted troughs of the NW European Basin were already stabilized. This discrepancy in timing of the main inversion movements between the two areas is probably related to a modification of plate movements during the Late Alpine orogenic phases. Updoming of the Pays de Bray Anticline in the Paris Basin is also thought to have taken place during the Miocene.

During the Senonian, at the onset of the Late Alpine oscillatory-emergent tectonic-sedimentary episode, an extensive sea covered most of NW Europe, in which the individual sedimentary basins became mere local depressions, with exception of the Viking Central Graben that deserves some distinction (Fig. 4.31); even the existing intrabasinal highs became inundated. In almost the whole area limestone and marl deposition prevailed. Only at the northern and southern flanks of the still continental South German High and the Bohemian Massif, the sediments have a more sandy character. In the Viking Central Graben, the deep-sea deposits have a rather clayey character.

From the beginning of the Tertiary onwards the sea started to regress from most of the present-day continental NW Europe due to the uprising of the Alps. The largest Cenozoic basin is the NW European Basin that extends over a distance of more than 2000 km from the northern North Sea to Poland. Subsidence of this megabasin was all but uniform. In comparison, the dimensions of the Tertiary Alpine and Carpathian foreland basins, the Rhine and Rhône grabens, the Paris Basin and the Atlantic shelf basins of France, Ireland and Scotland are much smaller (Fig. 4.32). The tectonic origin of the various Cenozoic basins of NW Europe differs considerably. The development of the North Sea Basin and the Atlantic shelf basins was strongly influenced by the onset of seafloor spreading in the Norwegian-Greenland Sea during the Eocene. The foredeep basins of the Alps and the Carpathians strongly subsided during the Tertiary due to the uprising of the Alpine mountain chains. The Rhine and Rhône grabens and the Vienna Basin form part of the Late Tertiary and Quaternary collapse system that affected the entire Alpine fold belt and its forelands.

The gradual, oscillating sea-level drop caused profound paleogeographic changes in the North Sea area. Apart from the renewed differential subsidence of the North Sea rift, the Shetland platform was uplifted and tilted eastwards resulting in an eastward-directed drainage system. As a result, an outbuilding clastic foreset wedge became deposited during the Paleocene and Eocene. Slope failure of this foreset sequence triggered density currents which supplied sands to the downfaulted rift valleys of the Viking Central Graben in which water depths were in the order of several hundreds of metres. During the Eocene, the supply of clastics from the Scottish Highlands gradually diminished. In the central and southern North Sea, the Paleocene and Eocene sequences are represented mainly by silts and clays. For instance, the Belgian sector of the southern bight of the North Sea is an example of a shallow marine siliciclastic intracratonic sedimentary basin, with a ramp-type margin setting characterized by low subsidence rates, relative tectonic quiescence and a low gradient relief (Jacobs, 1994; Jacobs et al., 1998). Here two major transgressive–regressive cycles can be discerned during the Eocene. The Lower and Middle Eocene basal sediments are clayey, displaying a fining upward trend, deposited on a mud shelf. Upward the deposits become more sandy, contain calcarenite horizons, coarsen upward and are chiefly of tidal and lagoonal origin. The upper cycle consists of an alternation of sandy and clayey deposits of a distal origin, gradually fining upward, and passing during the uppermost Eocene in tidal flat sands.

A regional hiatus marks both the base and the top of the Oligocene. These breaks in sedimentation were probably caused by the eustatic sea-level changes of the oscillatory-emergent tectonic-sedimentary episode. The Oligocene series consists predominantly of clays and silts attaining thickness of up to 1000 m in the central North Sea. In the Lower Rhine Embayment of western Germany marine shelf deposits in the northwest of the basin were interfingering with more continentally influenced sediments in the south (Nickel and Schäfer, 1997). The sedimentation during the Rupelian (Early Oligocene) was not tectonically disturbed and therefore related to the pre-Oligocene relief, with sandy and clayey deposits accumulated in a shallow marine to a more coastal environment. This pattern changed in the Chattian (Late Oligocene) when syn-sedimentary NW-SE striking faults became activated. In the centre of the subsiding embayment, divided into several tectonic blocks with different stratigraphic successions and sedimentary thicknesses, more than 1000 m of clastic and lignitic sediments were deposited in coastal and near-coastal continental realms. The lignite deposits accumulated mainly during the Miocene. The conversion of the Lower Rhine Embayment from an original peat accumulation area into a landscape dominated by clastic sand and clay sedimentation has been interpreted as the consequence of the origin of a primitive Rhine River that is the fluvial system cutting through the Rhenish Slate Mountains (Rheinisches Schiefergebirge) from the Upper Rhine graben to the Lower Rhine Embayment (Gliese and Hager, 1978). Alkaline volcanism associated to this Rhine rifting persisted till subrecent times.

In contrast to the North Sea rift system the inverted Mesozoic troughs and grabens and their surrounding areas remained stable throughout the Tertiary. From this it must be concluded that during their inversion these basins reached a state of thermal and isostatic equilibrium that persisted throughout the Cenozoic. This is particularly evident in the case of the Polish Trough, but also applies to the Marginal Troughs and to the post-Miocene history of the Western Approaches and Celtic Sea basins.

The Paris Basin suffered a new transgression of the sea after a long emersion period at the end of the Cretaceous (Pomerol, 1978). This invasion of the sea came from the Atlantic Ocean in the west as well as from the North Sea in the north, with a predominance of this latter seaway. The accumulated sediments are chiefly carbonates between Paleocene and Middle Eocene (Lutetian), when an anticline uplifted in the north reconstituting the ancient London-Brabant Massif. The Paris Basin became thus isolated from the north and turned a mere sea arm of the western English Channel. The deposits from the Upper Lutetian onwards are from west to east marine and lagoonal limestones, lacustrine limestones and calcareous to quartz sands, respectively. The long distance from the open sea explains a certain closure culminating in the precipitation of gypsum. In the Oligocene, an ultimate transgression connected the Paris Basin with the west as well as with the south, turning Normandy and Brittany into islands. From this time onwards the accumulated sediments are chiefly sands and clays, ranging in the beginning from marine to later fluvial-deltaic realms. Finally, the basin became uplifted in the Pliocene and Quaternary turning an area of denudation and erosion.

The southern North Sea Basin that comprises the present mainland of the Netherlands and the adjacent part of the southern North Sea shows a pattern of basin development in the Quaternary distinctly different from that in Neogene times. Considerably greater amounts of sediment were accumulated per time unit during the Quaternary as compared to the Neogene (Zagwijn and Doppert, 1978). At the transition from the Miocene to the Pliocene, the sediments deposited by the Rhine River changed from coarse-grained to predominantly fine-grained. A thick series of clays with interbedded lignites and some beds of sand and gravel were laid down in Early Pliocene times. These fluviatile deposits interfinger with a series of marine sediments at the margin of the North Sea. A widespread hiatus at or close to the Plio-Pleistocene transition is noticeable in marine as well fluvial sedimentary sequences. This hiatus probably resulted from erosion subsequent to tectonic uplift in the basin, possibly in addition to glacio-eustatic sea-level changes related to the first glacial episode of the Quaternary. In the beginning of the Pleistocene marine beds accumulated in the same depositional areas as those in Neogene times. A swift regression took place at the onset of the Günz (Nebraskan) glaciation, at 1.8 Ma ago. In the later part of the Middle Pleistocene and during the Late Pleistocene, tongues of marine deposits related to interglacial marine transgressions were formed. However, fluvial sedimentation became predominant, with a supply by the rivers coming from the south. Furthermore, two other regional unconformities were formed in the Quaternary. The first occurred at the transition form the Early to Middle Pleistocene (Günz/Nebraskan) glaciation and the second in a later phase of the Middle Pleistocene (Mindel/Kansan) glaciation. The beds above the latter unconformity contain elements indicating the presence of inland ice from Scandinavia close to or partly covering the southern North Sea Basin. The particular character of this basin in the Quaternary as compared to that in the Neogene has been evidently the result of epeirogenetic uplift of the hinterland together with increased subsidence in the Netherlands and the adjacent offshore area.

The Rhenish Massif is located at the northwestern margin of the Variscan fold belt of central Europe. From Tertiary times onwards it has been subjected to uplift movements while the adjacent Lower Rhine Embayment was subsiding into the updoming Rhenish Massif. It appears that this configuration developed by crustal collapse due to extensional movements (Wuestefeld et al., 1997).

4.5.4 NW AND CENTRAL EUROPE: COMPLEX INTERPLAY OF OROGENIC PHASES AND RIFT FORMATION

The structural and stratigraphic complexity of NW and Central Europe is the result of a long geological evolution during which orogenic events leading to plate suturing alternated with periods of rifting causing fragmentation and disintegration of the newly formed plate assemblages, this all in a cyclic sequence of events. From Paleozoic to Cenozoic times three such orogenic and rifting cycles have been recognized. These cycles correspond neatly to the following pairs of submergent and oscillatory-emergent tectonic-sedimentary episodes as follows:

Cenozoic-Late Cretaceous (Senonian)	oscillatory-emergent	Alpine orogeny
Late Cretaceous (Turonian)-Middle Jurassic	submergent	
Early Jurassic-Early Permian	oscillatory-emergent	Variscan orogeny
Middle Devonian-Late Carboniferous	submergent	
Middle Ordovician-Early Devonian	oscillatory-emergent	Caledonian orogeny
Cambrian-Early Ordovician	submergent	

During the Caledonian diastrophism the Laurentian-Greenland shield was sutured with Fennoscandia, forming the Laurasian continent. Devonian-Early Carboniferous rifting along the Caledonian suture zone indicates the instability of this plate assembly. However, its break-up was prevented by the Variscan orogeny, during which Laurasia was welded together with Gondwana to form the Pangean Megacontinent. The instability of such super-plate caused its disintegration during the Mesozoic to Tertiary rifting phases. The following Alpine orogeny resulted in a renewed suturing of Eurasia and Africa. However, the onset of a new rifting stage can be witnessed by a new rift emplacement during the collapse of the Mediterranean basins. Each of these megatectonic cycles left their marks on the geology of NW and Central Europe.

During the development of the sedimentary basins of the area, many different geotectonic processes were active. In many areas basins of various geotectonic origins stacked on top of one another, resulting in very great sedimentary thicknesses. The accumulated sediment types reflect, besides climatic influences, the periodicity in such sedimentation phases, as presented in Chapter 3 (see Mabesoone, 2003a).

4.6 IRELAND

4.6.1 IRELAND'S STRUCTURE AND SCENERY

The surface features of Ireland are closely connected in their underlying structure and geological origin (Fig. 4.33).

A vast stretch of Carboniferous limestones covers the greater part of the heart of the island, reaching the sea in a number of places between masses of varying shape and size, of older rocks which are arranged in a discontinuous ring. The central stretch of Carboniferous limestones is also interrupted in places by outcrops of older rocks which are clearly inliers, and in places by outcrops of younger rocks forming outliers. The Quaternary of Ireland shows boulder clays occupying at least half of the country as well as very large stretches of peat. Over the heart of the country, this cover is so complete that the underlying Carboniferous limestone is rarely exposed at the surface. Even the upland masses of the rim are literally plastered with drift. It becomes clear, therefore, that the whole of Ireland was covered by ice at one time or another during the Quaternary.

The centre of the island is a plain, or rather a low plateau, with a surface at only about 50 m above sea level, so that there is but little natural fall. The underlying rock of most of this plain is limestone, and owing to solution of the rock and the development of underground watercourses, the surface drainage is naturally irregular or ill defined. This has been made far worse by glacial interference and the irregular deposition of the boulder clay masses, sands and gravels. Furthermore, the present-day climate of Ireland is one of almost constant moisture that hinders the drying of the surface and promotes the growth of mosses and the development of bogs. These bogs which in part represent the later phase of glacial lakes, act as a sponge, holding up moisture and rendering still worse the already poor drainage.

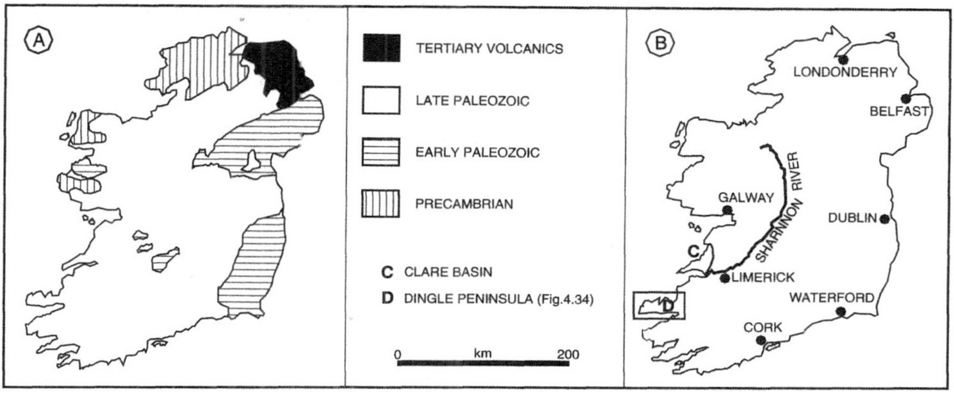

Fig. 4.33: Main geological structures of Ireland, Quaternary omitted.

4.6.2 MAIN GEOLOGY

Apart from the large sheets of Tertiary basaltic lavas in Antrim, the thin representatives of chalk preserved under them, and the large Tertiary granite mass of the Mourne Mountains, practically the whole of Ireland is built up of Paleozoic rocks, except for the Pleistocene drift deposits. These Paleozoic deposits are included in the Cambrian-Early Ordovician submergent, Late Ordovician-Early Devonian oscillatory-emergent, and Middle Devonian-Late Carboniferous submergent tectonic-sedimentary episodes.

Broadly speaking, though various rock groups and structural units may be traced from Britain into Ireland, their distinctive features are softened by the Irish climate. In the north, the three major divisions of Scotland – the Highlands, the Central Lowland, and the Southern Uplands – are continued across the narrow Irish Channel. The Higlands appear as the mountains and hills of County Kerry, of Donegal, where conditions most resemble those of the Scottish Highlands, and of County Mayo, of Connemara. The Southern Uplands of Scotland reappear as the uplands of County Down and County Armagh. The Irish continuation of the Central Lowlands is obscured, because they merge westwards into the central Irish plain, the boundary faults die out and the rift structure becomes less obvious. Besides, but even more important, the huge sheets of Miocene lava stretch right across the area of Antrim.

That part of Ireland which is now Northern Ireland has thus a varied and interesting scenic and geological heritage. Eire or the Republic of Ireland has likewise great contrast between the central plain and the surrounding uplands. The central plain not only boasts the great bogs, but many of its glacial hollows form lakes, which mostly drain into the Shannon River. In outline, these lakes are extraordinarily tortuous where their waters have invaded the boulder clay hollows between drumlins, and thus are studded with innumerable islands and islets. A greater contrast is offered where in the west, the Carboniferous limestone emerges at the surface from under the mantle of boulder clay. In County Clare, there appear the limestone pavements, associated with sheep pastures.

On both sides of the central Lough Derg lake rise anteclinal masses of Devonian Old Red Sandstone and Silurian rocks. The so-called Golden Valley is a Carboniferous well-drained limestone plain.

Southeastern Ireland, roughly the counties of Wexford and Wicklow, consists largely of Cambrian and Ordovician rocks, with a structural resemblance to central Wales. The Wicklow Mountains form a large granite mass, showing the typical rounded forms associated with the granite, with deep secluded valleys on the east.

The southwest of Ireland is essentially Armorican and consists of long parallel ridges of Old Red Sandstone separating drift-filled valleys excavated in Carboniferous limestone. The ridges and valleys trend towards WSW, and where they run into the Atlantic Ocean, they provide the classic example of a coastline with drowned valleys or rias. The Mountains of Kerry, developed on Old Red Sandstone, include the magnificent group of Macgillycuddy's Reeks with Carantonhill reaching 1041 m, the highest point of Ireland, overlooking the lakes of Killarney on the Carboniferous limestone. Immediately to the NE there occur Westphalian age rocks, however without seams or streaks of coal.

Western Ireland together with the greater part of Scotland and with Scandinavia, belongs to the Caledonian orogenic belt, and has not been considered in the foregoing item about NW Europe.

4.6.3 EXAMPLES OF WESTERN IRISH SEDIMENTARY SEQUENCES

4.6.3.1 Dingle Basin

Generalities. The Dingle Peninsula is the northernmost of three peninsulas which make up the County Kerry in the SW corner of Ireland (Fig. 4.34). The peninsula is a mountainous area, about 40 km long and 24 km wide.

The Dingle Basin occupies most of the Dingle peninsula. The rocks show an Ordovician to Carboniferous succession. The oldest part of the Siluro-Devonian

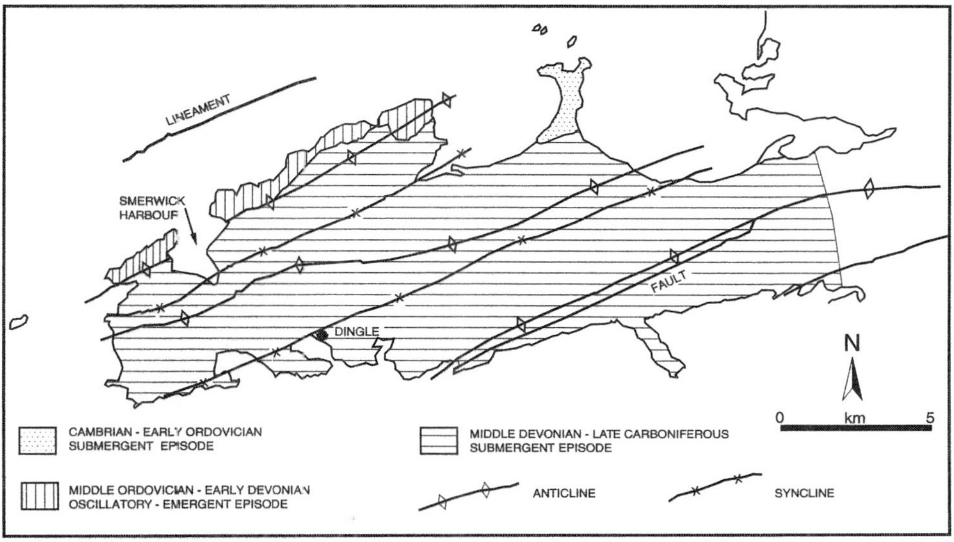

Fig. 4.34: Geology of Dingle Peninsula (W Ireland).

sequence was deposited in the Dingle Basin that formed in a transgressive tectonic regime on convergence between Avalonia and Laurentia. The younger sequences were deposited in basins on the northern margin of the Middle-Late Devonian Munster Basin of South Kerry and Cork (Williams et al., 2000).

Stratigraphy and Lithologic Sequences. The stratigraphical column presented in Table 4.4 recognizes the distinguished units in the Dingle Basin and their respective sedimentary sequences. Structurally the basin is constrained between two fundamental, ENE-trending lineaments to the south of the putative trace of the Iapetus Suture Zone. The northern limit of the peninsula is formed by the North Kerry Lineament and the southern one by the Dingle Bay Lineament that

Table 4.4: Cronostratigraphy of Dingle Basin.

Middle Devonian-Late Carboniferous submergent episode:
Slieve Mish Group (Famennian)
post-rift extensional half-graben
Ballyroe Group (Frasnian-Famennian)
syn-rift extensional half-graben
Carrigduff Group (Frasnian)
syn-rift extensional half-graben
hiatus (Givetian)
Pointagare Group (Eifelian)
post-Caledonian, pre-rift extensional graben
Caherbla Group (Eifelian)
post-Caledonian, pre-rift extensional graben
hiatus (Late Emsian)
Middle Ordovician-Early Devonian oscillatory-emergent episode:
Smerwick Group (Siegenian)
Late Caledonian, pre-rift, strike-slip pull-apart/extensional half-graben
Dingle Group (Pridolian-Gedinnian)
Late Caledonian, pre-rift, strike-slip load-induced foreland basin
Duquin Group (Wenlockian-Ludlovian)
intra-arc basin
hiatus (Llanvirnian-Llandoverian)
Cambrian-Early Ordovician submergent episode:
Annascaul Formation (Tremadocian-Arenigian)
Early Caledonian, strike-slip basin

continues into the country as the South Ireland Lineament. These inherited Caledonian lineaments formed long-lived, basin-bounding faults, and had an important influence on the structural and sedimentary evolution of the Dingle Basin. The area was subjected to compressive and transgressive tectonic regimes during Early and Late Caledonian deformation, and Late Carboniferous Variscan deformation. These two orogenic periods were separated by a period of Middle to Late Devonian extension.

The Cambrian-Early Ordovician submergent tectonic-sedimentary episode shows still a remainder represented by the strike-slip basin Annascaul Formation of the Annascaul inlier. The unit is composed of deep-water clastic and volcanic rocks.

The Middle Ordovician-Early Devonian oscillatory-emergent episode is the most represented in the basin, comprising the Late Caledonian orogenic phase. Due to the tectonic oscillations, a number of lithostratigraphic groups can be distinguished, with marine and continental sediments.

The Silurian Dunquin Group (Fig. 4.35) records an intra-arc basin phase with shallow marine, shelf, littoral, and ephemeral-fluvial siliciclastic depositional systems as well as volcanic rocks (acid pyroclastics and andesitic lavas), deposited around volcanic islands. Three phases of basin evolution are involved: (1) active subduction, (2) major volcanic episode triggered by stress

Fig. 4.35: Dingle Basin, Ballyferreter (Co. Kerry): Silurian Dunquin Group.

relaxation associated with the ending of subduction, (3) post-subduction thermal subsidence.

Follows a phase of strike-slip fault-controlled subsidence at the end of the Silurian, represented by non-marine sandstones, claystones and minor conglomerates of the lower Dingle Group, accumulated in lacustrine, lake margin and succeeding fluvial systems. Upward the middle-upper Dingle Group sedimentation culminated in the deposition of laterally extensive sandy braided axial river systems which continued into the lowermost Devonian. Alluvial fan conglomerates occur also in this sequence, in a load-induced foreland basin.

At the end of the episode, in the NW Dingle domain, the Early Devonian Smerwick Group documents sandy and gravelly ephemeral-fluvial and erg-margin processes on an ancient terminal fan, under Late Caledonian pre-rift, strike-slip, pull-apart to extensional half-graben basin conditions.

The following Middle Devonian-Late Carboniferous submergent episode comprises a post-Caledonian, pre-rift, extensional graben passing upwards into syn-rift extensional half-grabens, each phase recorded by one or two lithostratigraphic groups. After inversion of the Dingle and Smerwick sequences under a sinistral transgressive regime, the region became increasingly influenced by N–S extension. The Dingle Bay Lineament developed as a topographical high of exhumed basement, already stripped of cover sequences in Late Silurian-Early Devonian times. In the late Emsian-Eifelian, Pointagare and Caherbla Group sedimentations started, and to the S-SE the Munster Basin iniciated. The North Kerry Lineament and the Dingle Bay Lineament were reactivated as normal faults with southward and northward displacements, respectively, demarcating the basin margin to a large graben.

The Pointagare Group is found in the NW Dingle Domain, outcropping in high sea cliffs along the coast. It rests with an angular unconformity upon the Smerwick Group. The lower section is represented by a greyish conglomerate sequence as a fining-upward succession of sandy conglomerate, pebbly sandstone and sandstone sheets deposited in a low-sinuosity to braided river system and in alluvial fans. Upwards there appear eolian cross-stratified sandstones which form the base of the upper section. From base to top occurs a fluvial-dominated sequence to switches into an eolian-dominated sequence. The respective depositional environments are perennial and ephemeral braided rivers and alluvial fans, passing into an erg to erg-marginal realm with a fully developed dune field.

In the SE of the Dingle Domain, the Caherbla Group is found. Its lower section is again coarse-clastic, with a thickly bedded, massive breccio-conglomerate, with clasts up to boulder size, within a sandy matrix. Interfingering with and overlying the conglomerate section appears a fluvial-eolian sandstone with large-scale cross bedding. The respective depositional environments were the same as conclude for the Pointagare Group.

Both groups illustrate the inherent tectonic control on fluvial drainage patterns, the location of erg accumulation, fluvial-eolian facies distribution and ultimate sequence preservation.

The Late Devonian is represented by three lithostratigraphic groups, with a maximum thickness of 1550 m. In central and southern Dingle Peninsula, the period is only represented by the post-rift Slieve Mish Group, with sandstones deposited in perennial low-sinuosity and braided river systems and alluvial fans. The syn-rift Carrigduff and Ballyroe Groups occur in the NW Dingle Domain. The sequences are also chiefly sandstones deposited in isolated syn-rift half-grabens. The accumulation took place in perennial braided rivers and alluvial fans in the lower Carrigduff section, and in perennial and ephemeral braided rivers, alluvial fans and erg margins, with tidal incursions in the upper Ballyroe section. Both groups record two oscillation periods of subsidence separated by uplift and tilting, indicative for a punctuated, active tectonism on the northern rift margin.

In the Early Carboniferous, a deposition still took place in the eastern part of the Dingle Basin, represented by the Tralee Group, at the end of the Old Red Sandstone phase.

4.6.3.2 Munster Basin

The Munster Basin is located to the SE of the Dingle Basin. It started its development in about the beginning of the Middle Devonian, with rifting. Sandstones of the Dingle Basin Carrigduff Group are also apparent in parts of the Munster Basin fill.

Quin (2000) presented a detailed study about a shallow marine succession in the South Munster Basin. The author distinguishes a transgressive sequence in rocks of Late Devonian-Early Carboniferous age. From the onset of marine conditions during the Late Devonian, a shallow marine sequence developed across the basin, in the east sand dominated, in the west mud dominated. At the Devonian-Carboniferous boundary, a transgressive mudstone unit covered the underlying section. Following the transgressive phase, a sand-dominated sheltered shelf facies accumulated in the west, and a muddy, open-shelf sequence in the east. The sinks for sand and mud deposition had thus reversed compared to the Late Devonian sequence. Oscillations in relative sea level and basin architecture are held responsible for this environmental change.

4.6.3.3 Clare Basin

The Atlantic coast of County Clare provides spectecular exposures of turbidite, slope and deltaic depositional systems which accumulated in a Namurian basin (Elliott et al., 2000; Fig. 4.33), during the Middle Devonian-Late Carboniferous submergent tectonic-sedimentary episode. The Clare Basin was an intracratonic basin formed on continental crust that underwent active extension during Late Devonian and Early Carboniferous. The basin was deformed during the Late Carboniferous-Early Permian by Variscan deformation. In the south of Clare, around the Shannon estuary, the strata are folded into rather tight folds. To the north, the intensity of folding decreases and broad, open folds prevail.

Climatically, northern Europe was equatorial during the Carboniferous, and the Clare Basin experienced a humid tropical, non-seasonal climate. The Early Carboniferous marine transgression into the basin is thus chiefly recorded by limestones which occur over great extensions throughout Ireland. Towards the mid-Carboniferous, clastic material supply increased, and mixed carbonate-clastic high-frequency sequences are common, and studied in detail in northern England (Tucker, 1998). Visean carbonate banks, the dominant mud-mound type at that time, are characterized by multi-component, highly structured carbonate muds. The overlying limestones suggest a significant northward shallowing in the area (Devuyst and Lees, 2000). Sea-level fluctuations in the Late Carboniferous were dominated by a glacial eustatic component related to the Gondwana ice sheet (Lemon and Tucker, 2002). One of the peaks of this glaciation was in the Namurian. The marine bands were produced during periods of glacio-eustatic transgression. Widespread sequence boundaries formed during periods of falling sea level. However, the Clare Basin was located distant from coeval oceans, but remained a rather deep depression of sufficient size to generate deep-water waves.

Sediment supply to the basin via fluvial systems was substantial and comprised a mixed load of clay, silt and sand, with the latter no coarser than medium-sized sand. This sediment was derived from the weathering and erosion of older sedimentary rocks. It is likely that the hinterland supplying the sediment, was tectonically quiescent during the Namurian. Sedimentation rates were generally high, except during periods of rising sea level. The prevalence of these high sedimentation rates combined with the fine-grained nature of the sediment load probably accounts to a large degree for the abundance of syn-sedimentary deformational features in the turbidite, slope and deltaic systems.

The lithostratigraphy of the Namurian section in the Clare Basin recognizes two groups of strata: (1) Shannon Group, with Clare Shale, Ross Sandstone and Gull Island formations; (2) Central Clare Group, comprising five major cyclothems. Correlation and reconstruction of the sedimentary history of the basin fill is aided greatly by the highly refined biostratigraphic subdivision of the Namurian, based on the occurrence of discrete fossiliferous horizons, the marine bands. The Clare Shale Formation comprises condensed black shales containing some fossiliferous marine bands, accumulated in a deep-water basin under reducing circumstances (Collinson et al., 1991). The Ross Sandstone Formation records a sandstone-dominated, axial deep basin turbidite system which had its maximum development in the southern axial zone of the basin, and is absent to the north. The dominantly fine- to medium-grained classical turbidites have been deposited in a regional northeasterly flow direction, varying appreciably, parallel to the inferred trend of the basin axis (Lien and Walker, 2000). The Gull Island Formation is interpreted as a fine-grained, unstable slope system that prograded broadly eastwards into the basin. In the north, away from the basin axis, the formation is represented by predominantly undisturbed, stable siltstones, and occasional thinner turbidite packages. The Central Clare Group cyclothems represent fine-dominated, unstable river delta systems, overlying transitionally the basin slope Gull Island deposits. At the base, there occur rhythmically banded siltstones which provide a marked facies in contrast with the siltstones of the underlying slope. The group is composed of five cyclothems defined by widespread faunal concentrate marine bands. Each cyclothem broadly coarsens upwards and has been interpreted as the product of a prograding delta system. The cyclothems are sequences bound by erosional unconformities.

4.6.4 CONCLUSION

Also in the case of western Ireland, the sedimentary basin development points to a cyclicity in accordance with the cyclic tectonic-sedimentary episodes involved.

4.7 DUERO BASIN (SPAIN)

4.7.1 CHARACTERISTICS OF THE BASIN

The Duero Basin in Spain occupies the major part of the NW Iberian Peninsula, with a surface of about 50,000 km^2 (Fig. 4.36; Santisteban et al., 1996), being the largest Cenozoic basin in Spain. It is surrounded by high-relief mountains: to the north (Fig. 4.37), of Paleozoic and Mesozoic sedimentary rocks; to the south and west, by Paleozoic igneous and metamorphic rocks; and mainly to the east, by Mesozoic siliciclastic and carbonate rocks. These borders formed during the Alpine orogeny. The Late Hercynian structure of NW Spain influenced considerably the structural and geodynamic evolution of the basin. The main structural lineaments of the basement reacted under the new tectonic conditions of the Alpine orogeny, but also new fault lines appeared. The sedimentary record of the Duero Basin shows that the borders of the basin developed independently (Fig. 4.38).

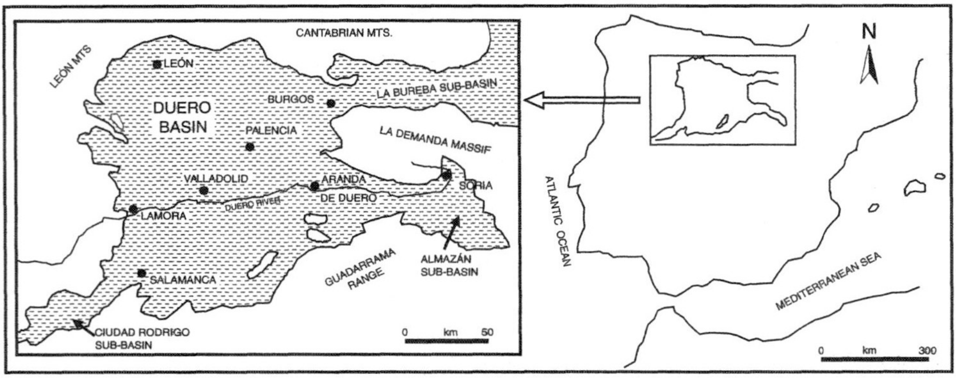

Fig. 4.36: Geography of Duero Basin (Spain).

Fig. 4.37: Duero Basin, Cantabrian Mountains in background.

The basin has a roughly quadrangular shape, with three rather narrow sub-basins protruding near the corners: the Ciudad Rodrigo Basin in the SW corner, the Almazán Basin extending between the Iberian Range and the Central Mountain system of Spain (e.g. the Guadarrama Range), and the La Bureba Basin Corridor in the NE corner, linking the Duero and Ebro basins (see Fig. 4.36).

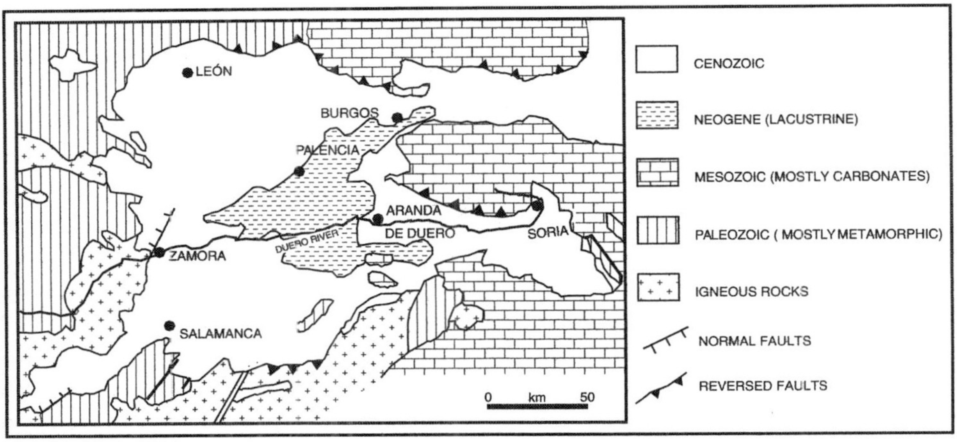

Fig. 4.38: Main geology and tectonics of Duero Basin (Spain), after various authors.

4.7.2 SEDIMENTARY RECORD

The stratigraphic framework of the Duero Basin has been studied and described since the nineteenth century, by various authors. The ideas of one of them (Hernández-Pacheco, 1915, 1930) still remain the basis of the present-day stratigraphical research. This author studied the stratigraphy, "sedimentology" (of a rudimentary sort, but including the first description and interpretation of fossil point bar deposits in the world), paleontology and tectonics of a large part of the basin. Mabesoone (1959) made a detailed sedimentological study of a part of the basin, distinguishing five facies types, however without entering in major stratigraphical aspects.

Most of the sedimentary fill of the Duero Basin accumulated in terrestrial sedimentary environments, with the outcropping of a great variety of lithofacies. Siliciclastic sediments are widespread, ranging from gravels to clays, with a well-marked dependence on their source area. The sediments derived from the north and east are lithic, those derived from the south and southwest are arkosic to lithic. Carbonates are also abundant, with their largest volumes near the basin centre. Evaporites occur restricted to the eastern half of the basin, towards its centre and northeast.

Due to the numerous previous works, an abundant stratigraphical nomenclature was generated, rich in local names, with a general overview by Portero et al. (1982). Consequently, attempts to establish a valid stratigraphical framework for the whole basin failed, also because it is difficult to trace particular stratigraphical units across the basin. Because the proposed units were based on lithostratigraphy, almost without any tectonic analysis, no general basin model could be established. A general facies distribution shows fans along the borders and lacustrine realms in the centre.

Santisteban et al. (1996) have divided the sedimentary record of the Duero Basin in three tectonic-sedimentary complexes (TSC), composed of several tectonic-sedimentary units (TSU), bounded by unconformities of tectonic or climatic origin (Table 4.5). Each one of these complexes refers to a stage of tectonic evolution of the basin, developed under particular climatic conditions which determined their overall mineralogical composition. These stages reflect oscillations during the Late Cretaceous-Cenozoic tectonic-sedimentary episode, the only one recorded in the Duero Basin of Spain.

Tectonic-sedimentary complex A (TSC-A) is of Late Cretaceous to Paleocene age. The unit consists of siliciclastic, carbonate and evaporite deposits arranged in a fining-upward sequence. The TSC-A deposits occur in stratigraphical continuity of the Upper Cretaceous in the N, E, and SE borders. In the W and SW borders, they rest unconformably upon a thick lateritic weathering profile on the Paleozoic basement rocks. The complex was deposited in environments ranging from terrestrial towards W, to marine towards E.

The following TSC-B is Eocene to Oligocene in age. It is composed of essentially siliciclastic sediments and a few carbonates, with exception of the Almazán sub-basin where the carbonates attain a considerable thickness. The whole sequence has a general coarsening-upward character. The TSC-B deposits form a fringe near the basin borders, resting there unconformably upon rocks of the underlying complex and the pre-Tertiary basement. The complex splits into three units bounded by unconformities. Most of the deposits accumulated in terrestrial alluvial fan and river environments; marine deposits occur only in the northeastern corner.

The tectonic-sedimentary complex C (TSC-C) is of Miocene to Holocene age. It consists of siliciclastic, carbonate (Fig. 4.39) and evaporite deposits in a fining-upward sequence. It is best represented in the central and NW parts of the basin where it covers the sediments of the previous complexes. The Neogene to Quaternary complex has been divided into five units which progressively onlap previous units and the borders. These rocks formed in alluvial fan, fluvial and lacustrine realms. Tectonic stability favoured the development of weathering profiles, both in the basin margins and the borders.

The well-known gravel sheets, called *rañas*, cover large parts of the basin borders, extending over large areas towards the basin centre. They seem to have formed at a number of different times since the Oligocene-Miocene, following episodes of fluvial incision (Martín Serrano, 1991).

Near the northern basin border the three tectonic-sedimentary complexes are disposed in relation to a highly active mountain front. Tectonics in the fast-

Table 4.5: Tectonosedimentary complexes Duero Basin

Ma	ages			tectonic complexes	tectonic units	sediments	climate	tectonic trends	
0–2	quaternary								
2	Neogene	Plioc.	Late	TSC-C		siliciclastic, carbonate, evaporite deposits, fining upward sequence	mediterranean humid		overall +
	Neogene	Plioc.	Early	TSC-C					
	Neogene	Miocene	Late	TSC-C	N5			+	
10	Neogene	Miocene	Late	TSC-C	N4			–	
	Neogene	Miocene	Middle	TSC-C	N3			+	
20	Neogene	Miocene	Early	TSC-C	N2			–	
	Neogene	Miocene	Early	TSC-C	N1		arid	+	
24,6	Paleogene	Oligocene	Late	TSC-B	P3	siliciclastic sediments scarce carbonates, coarsening upward sequence	arid – subtropical	–	overall ↑
30	Paleogene	Oligocene	Early	TSC-B	P3			+	
40	Paleogene	Eocene	Late	TSC-B	P2			–	
50	Paleogene	Eocene	Middle	TSC-B	P2			+	
	Paleogene	Eocene	Early	TSC-B	P1			–	
60	Paleogene	Paleocene	Late	TSC-A	MC	siliciclastic, carbonate, evaporite deposits, fining upward sequence	seasonal tropical	+	overall +
	Paleogene	Paleocene	Early	TSC-A	MC				
70–90	Cretaceous	Late	Senonian	TSC-A	MC				

Fig. 4.39: Duero Basin, E Torremormojón (prov. Palencia): Miocene fluvial clastic sediments (base) and lacustrine limestones (hills).

moving borders is a major control that masks the climate imprint on sediments by means of constant rejuvenation of the source area. The complexes are characterized by a homogeneous lithological composition. They are bounded by tectonic unconformities and their trends and geometrical disposition record different stages in the border evolution. The Complex A deposits are mainly polymictic conglomerates of limestone and quartzite clasts, cemented by carbonate. They arrange in tabular to lens-shaped beds of metric thickness and alluvial origin. The deposits rest, with a low-dip tectonic unconformity, upon Maastrichtian clays and dolomites. The Complex B sediments are also polymictic conglomerates of the same type as those of Complex A, however separated from these by an angular unconformity. These conglomerates alternate with red clay-sandy beds. The continuity in sediment characteristics is interpreted as a result of the homogeneous nature of the source area in the Cantabrian Mountains. Complex B deposits show syn-sedimentary folds. The sediments of the Complex C overlay unconformably those of Complex B. They are also polymictic conglomerates in tabular and channelized bodies, and red clays and occasionally sandstones. The continuous similarity in composition between the deposits of these complexes is due to the proximity of the source area and the high erosion rates in this area.

In the basin centre prevail Complex C deposits, split up into several units bounded by sedimentary breaks related to oscillating tectonics, that is mostly recorded by patterns of migration of environments, forming sequences. Lateral relations of these sequences reveal depocenter shifts related to subsidence pulses. Units N3, N4, and N5 characterize the basin centre. N3 and N4 sediments are very similar: (1) sands and clays accumulated in proximal and distal river floodplains; (2) green and black clays deposited in marshes; (3) nodulized and brechified marls and limestones of marginal lacustrine to palustrine origin, laterally related to (4) alternances of laminated limestones and marls deposited in lake centres, far away from alluvial inputs. Siliciclastic inputs reached the lakes without marsh fringes; there they have been reworked and mixed with lacustrine fauna, and the sediments became alternances of sands and clays. The deposits of both units appear arranged in sequences of several orders (third to fifth). At other places, a sequential arrangement of N4 unit sediments occurs in a section composed of gypsum and carbonates (gypsarenites and dolomites), with to the top, stromatolitic limestones, dolomitic marls and clays. A tectonic pulse took place after the N4 unit sedimentation, generating symmetrical and asymmetrical folds. A planation surface cuts these tectonic structures coeval to N5 unit sedimentation. These are the first incised sediments in the basin centre, and composed of conglomerates and sandstones of Late Miocene age, according to mammal fossils (Santisteban et al., 1997).

The southwest border is one of low uplift rates during the entire Cenozoic, so all tectonic-sedimentary complexes onlap it. However, in Complex B times a higher uplift rate and shortening, conditioned sedimentation in this area, resulting in a coarsening-upward trend for this complex. The Complex A sequence in this area begins with a lateritic soil profile developed on Paleozoic metasediments. The deposits resting upon this profile share their mineralogical composition with it; they are kaolin-rich, quartz-arenites interpreted as braided river deposits. Upward the sandstones turn finer grained, from a progressively increasing width of the braided river valleys and a floodplain development. The Complex B sediments in this area are arkoses to litharenites of fluvial origin, cemented by dolomite. Finally, the Complex C is characterized by the scarcity of deposits which mainly belong to the N1 unit. The sediments are chiefly red clays with some gravel levels interpreted as alluvial fan and river terrace deposits.

4.7.3 BASIN HISTORY

During Mesozoic times, the area occupied by the present-day Duero Basin, was a marine and terrestrial area, open to the north and east, under an extensional regime. To the west and south, the neighbouring emergent Hesperic Massif supplied sediment to this basin. The emergent massif underwent intense weathering under a tropical climate that generated lateritic soil profiles, tens of metres deep.

At the end of the Paleocene, the compressional phase of the Alpine orogeny started to cause uplift of the basin borders, and consequently a retreat of the marine environments towards E and NE. This phase induced progradation of alluvial river systems towards the basin centre. Progressive uplifting along the margins of the basin provoked deformation of the alluvial sediments. Major changes in paleogeography took place, changing from a rather flat landscape, with small tectonic-induced highs, to a well-differentiated sedimentary basin. Tectonic movements resulted in the opening of the Ciudad Rodrigo sub-basin, as well as several small basins in other areas. All these basins share a similar record of sedimentation and weathering.

The Early Neogene landscape of the basin was about similar to the present-day one. However, a long-lasting modification started when the Atlantic fluvial network captured some of the endorheic fluvial systems in the SE corner of the basin. The resulting, new exorheic rivers initiated the process of draining the basin towards W, and transporting enormous volumes of sediment to the Atlantic Ocean, with a contemporaneous enlargement of the area connected to the exorheic drainage.

Meanwhile, fault-related subsidence favoured the still continuing lake deposition in the remaining endorheic central and northeastern basin realms. In these areas, marginal alluvial fans fed fluvial systems flowing into the central lakes. Tectonic stability in the basin allowed the development and preservation of thick Mediterranean weathering profiles. Furthermore, the basin fill onlapped the eroded borders of the basin.

The coexistence of the continuously expanding fluvial network in the SW areas, and the rather restricted lacustrine realms on the opposite site, continued until the drainage of the whole basin was captured, probably only near the end of the Tertiary. At that point, the whole basin was connected to the Atlantic Ocean base level through the ancestral Duero River, and the last lacustrine environments had disappeared from the basin (Mediavilla et al., 1996).

The sedimentary history of the Duero Basin occupies only one episode of the cyclic development of sedimentary basins, that is the Late Cretaceous-Cenozoic oscillatory-emergent tectonic-sedimentary episode. However, within this period the oscillating tectonic pulses of the Alpine orogeny, are neatly recorded in the sedimentary basin infilling. These pulses constitute lower order cycles in many of the sediment sequences, and are also reflected in the migration of the depositional environments.

It may thus be concluded that even in the case of short-term tectonic-sedimentary events, a cyclic development of sedimentary basins becomes evident.

5. THE SEDIMENTARY RECORD OF FORELAND-BASIN, TECTOPHASE CYCLES: EXAMPLES FROM THE APPALACHIAN BASIN, USA

F.R. ETTENSOHN

5.1 INTRODUCTION

On the regional scale of North American foreland basins and their surrounding cratonic environs, second- and third-order, unconformity-bound, sedimentary sequences are generally widespread, and some coincide with Sloss sequences. Moreover, because the sequences recur in the same order several times at more or less irregular intervals, they are clearly cyclic at timescales generally greater than 10^7 years. The association between these cyclic sequences and lithospheric flexure, however, was not made until 1985 in two papers on the Catskill Delta complex (Ettensohn, 1985a, b). In these papers, Ettensohn made the connection between repeated episodes of subsidence and related black-shale deposition and phases of major structural deformation at various locations along preserved parts of the Acadian orogen as noted by Boucot et al. (1964) and Johnson (1971). Four unconformity-bound, foreland-basin, sedimentary sequences of similar composition were shown to correspond temporally and spatially to major phases of the Acadian orogeny, each of which was sequentially localized at promontories, structurally controlled prominences left on a continental margin after rifting (e.g. Thomas, 1977), along the former eastern margin of Laurentia. Moreover, Ettensohn (1985a) used Johnson's (1971) term *tectophase* to reflect each phase of the orogeny and showed that each tectophase produced a third-order, unconformity-bound sequence of sediments, the mapped distribution of which, both parallel and perpendicular to foreland-basin trend, can be used to track the progression of an orogeny. Similar patterns subsequently emerged from the foreland-basin deposits of the Taconian, Salinic, and Ouachita orogenic events (Ettensohn, 1991, 1993, 1994; Ettensohn and Pashin, 1993; Ettensohn and Brett, 2002). Despite different ages and tectonic circumstances of the Taconian, Salinic, and Ouachita sequences contains a similarity ordered succession of lithologies, which suggests that each lithologic type in the sequence broadly reflects a specific flexural responds to a particular phase of deformational loading or lithospheric relaxation during the tectophase. Moreover, the facts that these sequences occur

Cyclic Development of Sedimentary Basins
J.M. Mabesoone and V.H. Neumann, editors.

in different orogenies in widely different orogenic belts, recur during any one orogeny, and concur in time and space with otherwise determined deformational phases, suggest that orogenies do indeed occur in distinct tectophases and that these phases are responsible for the cyclic sequences that characterize most foreland basins. The rationale behind the flexural control of these sequences and their typical stratigraphic manifestation is discussed below.

5.2 FLEXURAL RATIONALE AND STRATIGRAPHIC MANIFESTATIONS

The association of ancient clastic wedges with former orogenies and their deformation-related orogenic highlands has been long understood. Less well understood, however, is the fact that the same orogenic highlands that produce the clastic sediments also contribute to foreland-basin subsidence via deformational loading (e.g. Walcoot, 1970; Price, 1973; Beaumont, 1981), thereby generating much of the accommodation space necessary for sediment accumulation. Modeling has shown that the deformed orogenic highlands, or fold-thrust belts, produces surface and subsurface loading that depresses or flexes the lithosphere cratonward of the orogen to form migrating foreland basin (Fig. 5.1A; Price, 1973). The resulting flexural models (e.g. Beaumont, 1981; Jordan, 1981; Quinlan and Beaumont, 1984) suggest that lithospheric loading produced by convergence-related crustal shortening generates or relaxes stresses that cause adjacent parts of the craton to rise or subside in long-wavelength deformation, which can influence relative sea-level fluctuations and cratonic sedimentary sequences across broad areas. The nature of predicted lithospheric responses in these models necessarily depends on the rheology assumed for the lithosphere (Quinlan and Beaumont, 1984). However, comparison of model predictions with sedimentary sequences in the Appalachian Basin and elsewhere seems to best support the loading and unloading of a temperature-dependent, viscoelastic lithosphere as modeled by Quinlan and Beaumont (1984), Beaumont et al. (1987, 1988), and Jamieson and Beaumont (1988). Moreover, while all of the above deals with supracrustal loading, more recent evidence suggests that subcrustal loading, related to mantle dynamics near subduction zones, may also generate cratonic subsidence (Mitrovica et al., 1989; Gurnis, 1990, 1991, 1992; Kominz and Bond, 1991; Coakley and Gurnis, 1995; Moresi and Gurnis, 1996) and that its effects may be similar to and concurrent with those from supracrustal loading (Gurnis, 1992). Nonetheless, comparison of model predictions with actual, foreland-basin sedimentary sequences seems to validate the usefulness of the models of Beaumont et al. (1987, 1988) and Jamieson and Beaumont (1988) in explaining the origin of the foreland-basin sequence cycles. The application of these models has been described in some detail by

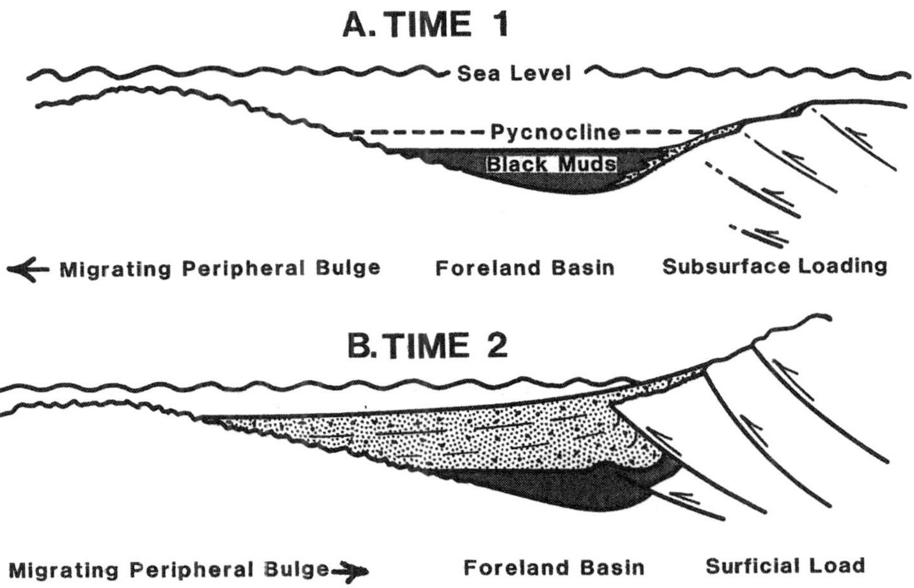

Fig. 5.1: Schematic diagrams showing possible relationships between foreland-basin generation and infill and deformational loading. (A) Basin-bulge formation and migration with subsurface loading and little clastic influx during the first three parts of a typical cycle. (B) Part four of a typical cycle showing "loading-type" relaxation resulting from a static, surficial load that supplies coarser clastic sediments to a subsiding basin (see Fig. 5.2). The pycnocline is a zone of thermohaline density gradation in deeper basins with decreasing O_2 content. Dark stipple: dark, organic-rich muds; large stipple: coarse clastic sediments; wavy lines: unconformities (adapted from Ettensohn and Brett, 2002).

Ettensohn (1991, 1994, 2004); therefore, only a summary, mostly extracted from these papers, will be described below.

5.1.1 CYCLE ORIGINS

The typical foreland-basin sedimentary cycle consists of seven parts produced as surface (nappes, thrusts, and folds) and subsurface (buried, obducted blocks, and flakes) deformational loads accumulate on the cratonic margin, severely loading the adjacent lithosphere. To isostatically compensate for the load, the adjacent lithosphere deforms into a downwarped flexural or retroarc foreland basin just cratonward of the deforming orogen and an uplifted peripheral bulge on the cratonward margin of the basin (Fig. 5.1A). The stacking of thrusts and folds supplies most of the load, but a subordinate component is also produced by sediment loading (Beaumont, 1981; Tankard, 1986). As orogeny proceeds and thrust loads shift cratonward, the foreland basin and peripheral bulge also shift away from the advancing load (Fig. 5.1A). Most of the loading and accompanying basin-and-bulge migration will advance cratonward in a direction nearly perpendicular to the strike of the orogen, but if orogeny is diachronous along its length, basin-and-bulge migration will shift parallel to the strike of the orogen (Ettensohn, 1985a, 1987). Once active loading ceases, the lithosphere responds by relaxing stress through a series of stages that ultimately give rise to "post-orogenic" clastic wedges (Ettensohn, 2004). Inasmuch as each orogeny typically progresses in a series of pulses or tectophases (Jamieson and Beaumont, 1988), the complete loading-relaxation cycle of each tectophase generates an idealized cycle of lithofacies (Fig. 5.2) over the course of several millions to tens of millions (10^6–10^7) of years, based on the chronology of Gradstein and Ogg (2004). The major parts of each cycle and their likely origin (Fig. 5.2) are briefly discussed below.

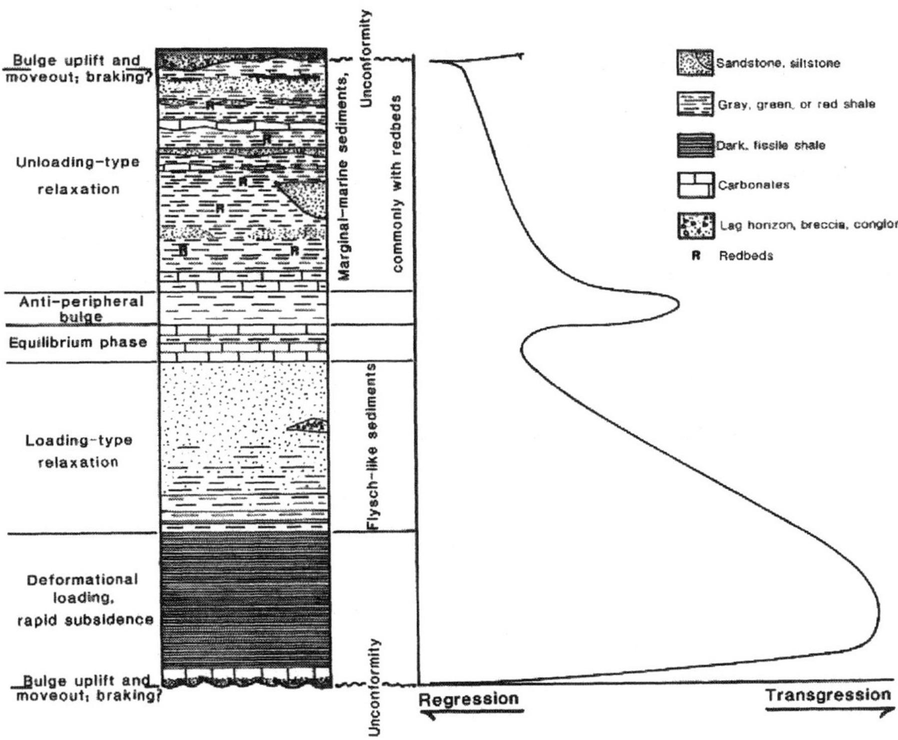

Fig. 5.2: Generalized tectophase cycle, showing sequence of flexural events, typical sequence of lithologies, and transgressive–regressive curve for early subduction-

5.1.2 BASAL UNCONFORMITY

The initial result of loading is bulge moveout and uplift, which results in diachronous uplift of the foreland (Quinlan and Beaumont, 1984). The resulting bulge will generally be one to two times the width of the adjacent foreland basin, and unconformity formation proceeds from the foreland basin cratonward due to subaerial exposure or elevation into wave base as the bulge migrates. Because unconformity formation migrates cratonward toward areas of decreasing subsidence and deposition on or near intracratonic highs, these unconformities typically "open up" cratonward, such that the space-time value of the lacuna on the unconformity increases in that direction. The distribution of the resulting unconformity is generally localized to parts of the foreland basin and adjacent craton next to the locus of tectonism and will be approximately parallel to the strike of the associated orogen (Ettensohn, 1993, 1994). In the larger Appalachian orogen, where most tectonism was localized at continental promontories that were subject to greater shortening and resulting deformation (Dewey and Burke, 1974; Dewey and Kidd, 1974; Ettensohn, 1994), the distribution of unconformities is generally asymmetric toward the involved promontories, even within the foreland basin. However, in parts of the foreland basin most proximal to those promontories, unconformities may not develop. In these areas, deformational loading is so persistent and intense at the adjacent promontory that resulting subsidence readily offsets the effects of bulge uplifts or braking. Thus, unconformity formation in time and space can be an important indicator of tectonic influence (Ettensohn, 1991, 1993, 1994; Ettensohn and Pashin, 1997).

The unconformities that define the initiation of tectophases are typically regional in nature (Ettensohn, 1993, 1994), but some of the same unconformities developed into the much more extensive, continent-wide, or possibly global, surfaces that Sloss (1963) used to define his sequences. Such unconformities generally seem to represent the initiation of significant new episodes of tectonism that mark brief periods of major plate reorganization. At these times, the response of the continent to new subduction or collision seems to have been one of impedance, resulting in the broad, large-scale, upward lithospheric deflections of large parts of the continent during few-million-year episodes, which Sloss and Speed (1974) called "emergent". Dickinson (1974, p. 22) called such a response "braking" inasmuch as subduction is initially retarded or braked by the immobile continent being subducted. However, if the continents at any one time were largely surrounded by interconnected subduction zones as already suggested, plate reorganization on one margin may have enabled nearly synchronous tectonism and related flexure on other margins as well, generating interregional, and perhaps global, unconformities in the process.

5.1.3 SHALLOW-WATER TRANSGRESSIVE DEPOSITS

After bulge moveout, rapid subsidence of the foreland basin begins, and shallow transgressive seas initially move across the unconformity surface. Subsidence is typically so rapid that the resulting shallow-water deposits are very thin or condensed. Nonetheless, in subtropical areas with little previous clastic influx, transgressive carbonates typically develop. In areas with immediately preceding tectonism, residual clastic debris on the surface, sometimes exposed long enough to have been reworked by eolian processes (Grabau, 1932, 1940; Cecil et al., 1991), will generate very mature, thin, shoreface sand deposits in this position. Whether clastic or carbonate, however, rapid subsidence and the consequential deepening ensure that any such deposits are short lived.

5.1.4 DARK-MUD SEDIMENTATION

Subsidence in the foreland basin is largely an isostatic response to deformational loading caused by crustal shortening and load transfer along a steep basement ramp. Early in the development of an orogen, it is likely that such ramps may have been able to accommodate up to 20 km of vertical deformation without creating major subaerial topography or source areas (Jamieson and Beaumont, 1988) so that initially much of the deformation must have occurred in the subsurface or in subaqueous environments, generating little or no subaerial relief (Karner and Watts, 1983; Fig. 5.1A). As a result, no major source of externally derived sediment is available to begin basin infilling. Hence, after the shallow-water deposits are drowned, the foreland basin experiences sediment starvation, and in the absence of major clastic influx, organic matter from the water column and suspended silts are the predominant sediment input into the early basin. At the same time, however, the basin continues to undergo rapid subsidence with which the scant sedimentation cannot keep pace. Consequently, the deepening water column becomes stratified, and the organic matter is buried and preserved as dark to black muds in resulting oxygen-deprived (dysoxic or anoxic) environments. Although dark muds typically characterize the rapidly subsiding proximal and central parts of a foreland basin, at the same time in more distal parts of the basin and on adjacent parts of the craton, reduced subsidence will typically produce coeval, transgressive deposits of carbonates or light-coloured shale overlying the basal unconformity. Nonethless, the black or dark shales are the most distinctive part of each sedimentary cycle, and mapping their successive distribution in time and space is probably the best way of tracking the cyclic sedimentary response of foreland basins to tectonic cues (Ettensohn, 1994, 1998).

The intense deformational loading and concomitant foreland-basin subsidence at this time clearly reflect ongoing plate convergence and subduction. However, yet another manifestation of subduction at this time should be arc volcanism, and although volcanics are very rare in foreland basins, the presence of a volcanic arc may be reflected in the foreland basin through the presence of wind-blown ash, preserved as bentonites. Consequently, although bentonites may occur in any part of the foreland-basin sequence, they are typically more common in the shallow-water, transgressive, and early dark-mud, depositional phases of a foreland-basin cycle, because this is when subduction is most active.

5.1.5 FLYSCH-LIKE SEDIMENTATION

Dark-mud deposition will occur throughout central parts of the foreland basin as long as active orogeny and deformational loading continue, but once active thrust movement declines and tectonic quiescence ensues, the deformational load becomes static. The lithosphere responds to the now static load by relaxing stress, as a result of which the foreland basin subsides and the peripheral bulge is uplifted and shifts toward the subsiding load (Fig. 5.3A). By this time, substantial subaerial relief has been generated by emplacement of a surficial load (fold-thrust belt), and surface drainage nets have had time to develop. As a consequence, coarser grained clastic debris is eroded and transported into the foreland basin in the form of deeper water deltaic deposits, turbidites, contourites, and debris flows, while nearshore clastic sediments commonly become redistributed as tempestites by storms (Fig. 5.1B). While the subsiding foreland basin is filled with deeper water, flysch-like sediments, the adjacent bulge is uplifted and migrates basinward, generating a regional unconformity that may truncate previously deposited flysch-like sediments and/or a regressive carbonate sequence atop the bulge depending upon the relative disposition of sea level. Ettensohn (1994) has called the flexural process at this stage "loading-type relaxation", and it results in a regressive, foreland-basin infill (Fig. 5.2). Moreover, as the bulge migrates back toward the load and sediments overflow from the filled basin, adjacent cratonic sequences will also appear to be regressive in nature.

In subduction-type orogenies, the foreland or retroarc basin on stationary lithosphere is commonly "balanced" as sediment by a peripheral, forearc, or pro-foreland basin on converging lithosphere (Dickinson, 1974; Willett et al., 1993; Johnson and Beaumont, 1995). Such "double-sided" or "symmetrical" orogens theoretically assure the presence of two "sinks" for the dispersal of debris eroded from the subaerial load. Where one of these sinks is absent, as for example in the loss of basins on converging crust in some collisional orogens, or where climatic factors favouring weathering and erosion are focused more on one side of an orogen than on the other, the other sediment sink or basin may fill more rapidly and aggrades upward into marginal-marine or terrestrial settings that would not normally be expected during this stage.

type orogenies. Sequences are unconformity-bound, but may be incomplete or eroded from the top.

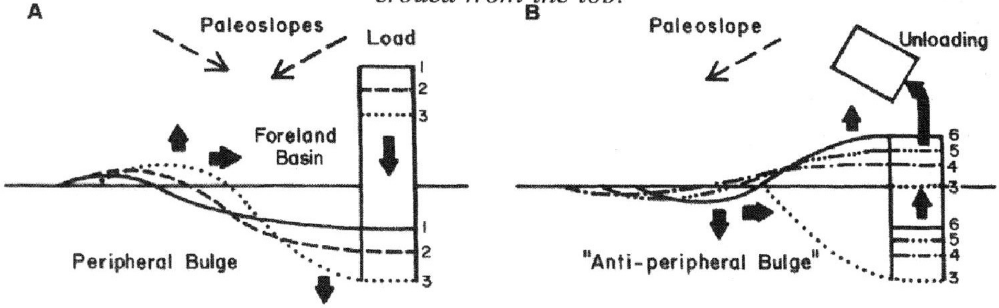

Fig. 5.3: Schematic diagram showing two types of flexural response to lithospheric stress relaxation (redrawn from Beaumont et al., 1988). (A) "Loading-type" relaxation – thrust migration ceases and the resulting static load causes lithospheric relaxation with major basin subsidence in part 4 of a typical cycle.

5.1.6 THIN, REGRESSIVE, SHALLOW-WATER, CARBONATE OR SHALE BLANKET

As the adjacent surface load is eroded, the rate of siliciclastic influx eventually exceeds basin subsidence rates, and the foreland fills or overflows with siliciclastic sediment. This infilling of the foreland basin, combined with greatly lowered source areas and a waning supply of clastic sediment, may set the stage for deposition of an extensive, but thin, blanket of shallow-water carbonates or mixed carbonates and shales. If the basin sits in arid, subtropical latitudes, carbonates typically predominate, however, if the basin occurs at rainy, equatorial, or temperate latitude, shales or mixed carbonates and shales are more likely. These carbonates or shales mark the culmination of shallowing and regression that began with the cessation of loading and the beginning of loading-type relaxation (Fig. 5.2). The carbonate or shale blanket may become very widespread at this time because the filled basin and lowered source areas briefly approach the same elevation, enabling shallow seas to spread widely.

5.1.7 THIN, TRANSGRESSIVE, MARINE, PRECURSOR SEQUENCE

The above phase of "elevation equilibrium" is short-lived, because the area of the former orogen and foreland basin begins to rebound upward in isostatic response to the lost load (Fig. 5.3B). During this "unloading-type relaxation", previously eroded upland areas and adjacent parts of the foreland basin rebound and a compensating "anti-peripheral bulge", called a peripheral sag by Ettensohn (1994) forms and moves toward the rebounding area (Beaumont et al., 1988; Fig. 5.3B). A short-lived, transgressive sequence of open-marine shales or shales and carbonates is deposited in the peripheral sag (Fig. 5.2), but the sequence is really only a thin, marine precursor to the thick wedge of prograding siliciclastic sediments that will follow.

5.1.8 MARGINAL-MARINE AND TERRESTRIAL CLASTIC WEDGE

The final part of the cycle is characterized by a cratonward-prograding wedge of predominantly marginal-marine and terrestrial siliciclastic sediments, which have at time been described as "post-orogenic", "molasse-like", or "deltaic" sediments (Fig. 5.2). Because the rebounding load consists of formerly beveled highlands and previously deposited foreland-basin sediments, sediments eroded from the rebounded area will be primarily fine-grained, composed mainly of siltstone, silty shale, shale, mudstone, or shaly carbonate (Ettensohn, 1994, 2004). Because the flexural process begins from a state of approximate elevational equilibrium at or near sea level, a single cratonward-dipping paleoslope becomes established (Fig. 5.3B). In this setting, red beds, paleosols, coals, and even thin carbonates, may be common. Red beds are especially common, and because of their association with terrestrial parts of deltas, many Appalachian Basin red bed units in this part of the cycle, like the Queenston, Bloomsburg, Catskill, Bedford-Berea, and Pennington-Mauch Chunk, have been called "deltas". In reality, they are delta complexes or tectonic delta complexes, which may reflect several different marginal-marine environments (Friedman and Johnson, 1966; Ettensohn, 2004) that prograded far beyond the foreland basin, giving the appearance that the foreland basin has "overflowed" onto the craton. In the process, proximal parts of the foreland basin may be cannibalized and generate an unconformity that "opens up" toward the tectonic highlands (Goodman and Brett, 1994). Moreover, any erosion that accompanies the bulge-moveout phase of the next tectophase or orogeny will readily subsume such an unconformity and parts of the underlying succession. Hence, an incomplete cycle of lithologies may reflect erosion during the succeeding tectophase or the advent of a new tectophase before the sedimentary expression of the previous one is complete (Ettensohn, 1994, 2004).

5.1.9 MODEL SUMMARY AND IMPLICATIONS

As Fig. 5.2 indicates, the typical, tectonic, foreland-basin cycle consists of seven parts in ascending order: (1) basal unconformity; (2) shallow transgressive carbonates or shoreface sands; (3) dark shales; (4) flysch-like clastic sequence; (5) thin blanket of shallow-water carbonate or carbonate and shale; (6) thin, open-marine shale sequence; (7) a thick, marginal-marine, clastic sequence with red beds. Because orogenies occur in pulses or tectophases (Jamieson and Beaumont, 1988), one or more of these cycles will constitute the sedimentary record of an orogeny in its foreland basin. As an example, the cyclic sedimentary record of two or three tectophases from the Ordovician-Silurian Taconian orogeny in the Appalachian Basin (Fig. 5.4) is shown in a section parallel to basin strike (Fig. 5.5) and in two sections perpendicular to strike (Figs. 5.6 and 5.7). Individual cycles, however, may not exhibit a complete sequence of lithologies because of erosion at the base of the succeeding cycle (Fig. 5.8), because the succeeding tectophase began before the sedimentary expression of the previous was complete (Fig. 5.6, tectophase 3), or because the area of the section during deposition was too distant from the major locus of tectonism and resulting subsidence (Fig. 5.7, Blountian tectophase). It is also possible for subcycles of black shale and flysch-like clastics to repeat several times in a single tectophase (Fig. 5.8), each subcycle perhaps reflecting movement of major thrust systems during the tectophase. Based on current observations, the simplest noted cycles consist of an unconformity, dark shales and a flysch-like sequence (Fig. 5.7, tectophase 3). However, one consistent feature of any foreland-basin cycle or subcycle is that the sedimentary record of each successive tectophase migrates farther cratonward (Figs. 5.6–5.9), or farther along strike during oblique convergence or in transgressive orogenies (Figs. 5.5 and 5.9) than did the previous one (Ettensohn, 1985a, 1991, 1994, 1998). Moreover, the best way to observe this progressive perpendicular-to-strike or parallel-to-strike movement of the cycles is to map the distribution of the dark-shale part of each cycle (Ettensohn, 1985a, 1994, 1998; Fig. 5.9). The dark shales represent not only the time of maximum subsidence and transgression in each cycle, and hence its greatest extent, but they are also probably the most distinctive and easily recognized of the cycle lithologies.

The development of such cyclic sequences in foreland basins seems to be most characteristic of subduction-type orogenies during early parts of the convergence history of an adjacent continental margin. During early parts of convergence, mass transfer from one plate to another and intraplate shortening mainly occur along a steep basement ramp that defines the rifted continental margin. Based on modeling in the Appalachians, it is likely such a ramp may have accommodated up to 20 km of vertical deformation without generating a major subaerial load (Jamieson and Beaumont, 1988). Much of this load, moreover, would have been concentrated narrowly on or near the basement ramp, and the isostatically compensating foreland basin would have been correspondingly narrow and deep. However, in successive tectophases, it is likely that the deformational

(B) "Unloading-type" relaxation – erosional unloading results in rebound and erosion near unload areas in parts 5–7 of a typical cycle.

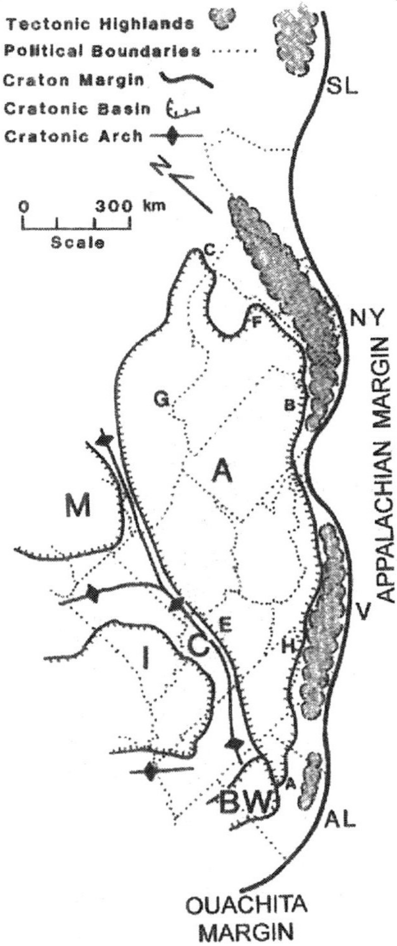

Fig. 5.4: Schematic representation of the southeastern margin of Laurentia during the Taconian orogeny, showing major structural features. The promontories were apparently mediated successive tectophases during each orogeny; promontories: SL, St. Lawrence; NY, New York; V, Virginia; AL, Alabama. ABFC is the line of section in Fig. 5.5, EH is the line of section in Fig. 5.6, and GF is the line of section

front would have surmounted the continental margin and developed sufficient topography to advance substantially onto the craton as a surficial load (Jamieson and Beaumont, 1988). Because the deformational load is now more expansive and spreading itself across a greater area, the resulting lithospheric flexure

in Fig. 5.7. *A*, Appalachian Basin; *M*, Michigan Basin; *I*, Illinois Basin; *BW*, Black Warrior Basin; *C*, Cincinnati Arch.*

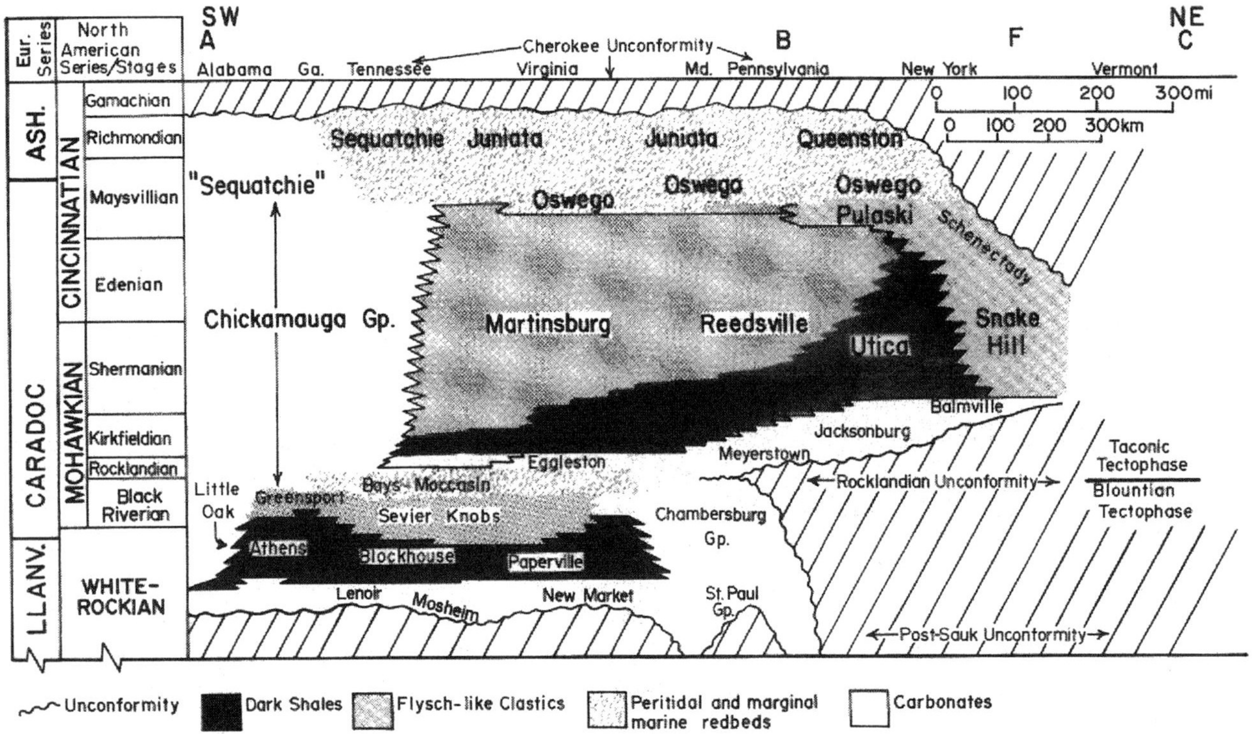

Fig. 5.5: Schematic Middle-Upper Ordovician section (line ABFC in Fig. 5.4), parallelling the strike of the Appalachian Basin and showing the repetition of foreland-basin tectophase cycles for the Blountian and Taconic tectophases. Note the northeastward migration of tectophase sequences in time, reflecting the diachronous northeastward migration of Taconian tectonism on the southeastern margin of Laurentia. No vertical scale intended (adapted from Ettensohn, 1994).

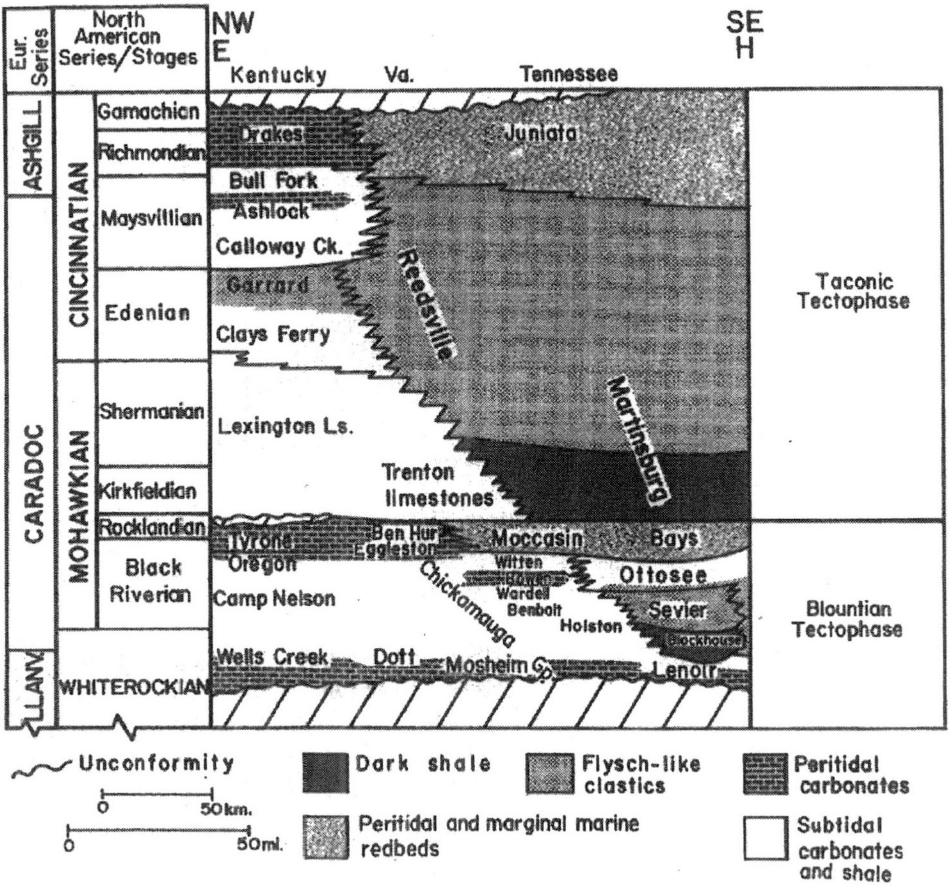

Fig. 5.6: Schematic southeast-northwest section (line EH in Fig. 5.4), nearly perpendicular to the strike of the foreland basin, showing the cratonward migration of foreland-basin tectophase cycles in time. No vertical scale intended (adapted from Ettensohn, 1991).

generates a broader, shallower foreland basin (Karner and Watts, 1983), and the successive broadening of foreland basins during any one orogeny is a common pattern. During the Taconian orogeny, for example, not only do successive foreland basins move cratonward, reflecting the advancing load, but also they become greater in area (Fig. 5.9) and the contained cyclic sequences, especially dark-shale and flysch-like parts of the cycles, overall become shallower with time. In fact, except for shale colour, in the later-formed foreland basins of an orogeny, foreland-basin cycles may become diffuse and difficult to distinguish from normal cratonic sedimentation, pointing out the fact that the same flexural

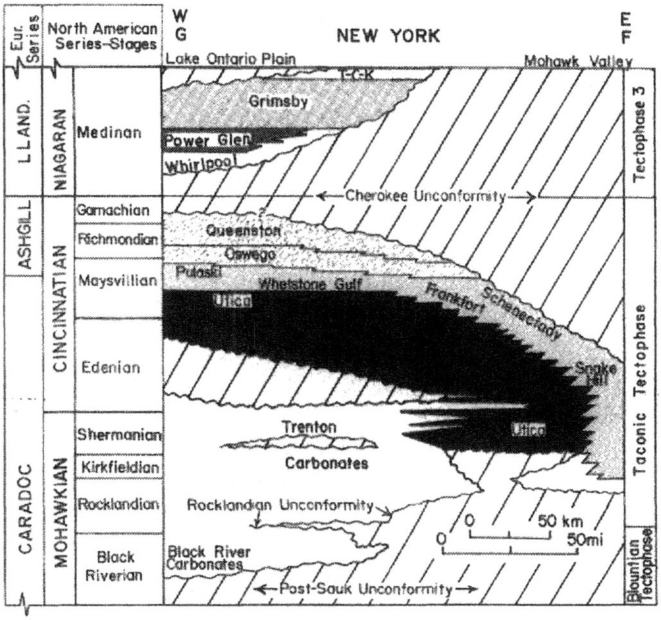

Fig 5.7: Schematic east–west section, nearly perpendicular to the strike of the foreland basin (line GF in Fig. 5.4), showing the nature and disposition of the three Taconian tectophases. The Black River carbonates, though unconformity-bound, reflect an atypical and very poorly developed tectophase sequence, probably because of the great distance from the Virginia locus of the Blountian tectophase.

stresses controlling foreland-basin sedimentation may also influence patterns of cratonic sedimentation some distance from the foreland basin. In addition, if a continental margin like the Appalachian margin experiences multiple orogenies, patterns of foreland-basin shallowing and broadening not only continue, but may be amplified by the increasing flexural rigidity of the lithosphere with time (Watts et al., 1982). By the latest of such orogenies when the deformational front has advanced far onto the craton, the resulting foreland basin may be so shallow and distant from the sea that typical cycles will not develop, marginal-marine and terrestrial clastic sediments predominate, and these sediments overflow the foreland basin to form a siliciclastic blanket that can spread hundreds of kilometres across large parts of the adjacent craton (Ettensohn, 1994, 2004).

Note the cratonward migration of the final two tectophases. Symbols as in Fig. 5.6; no vertical scale intended (adapted from Ettensohn, 1991, 1994).

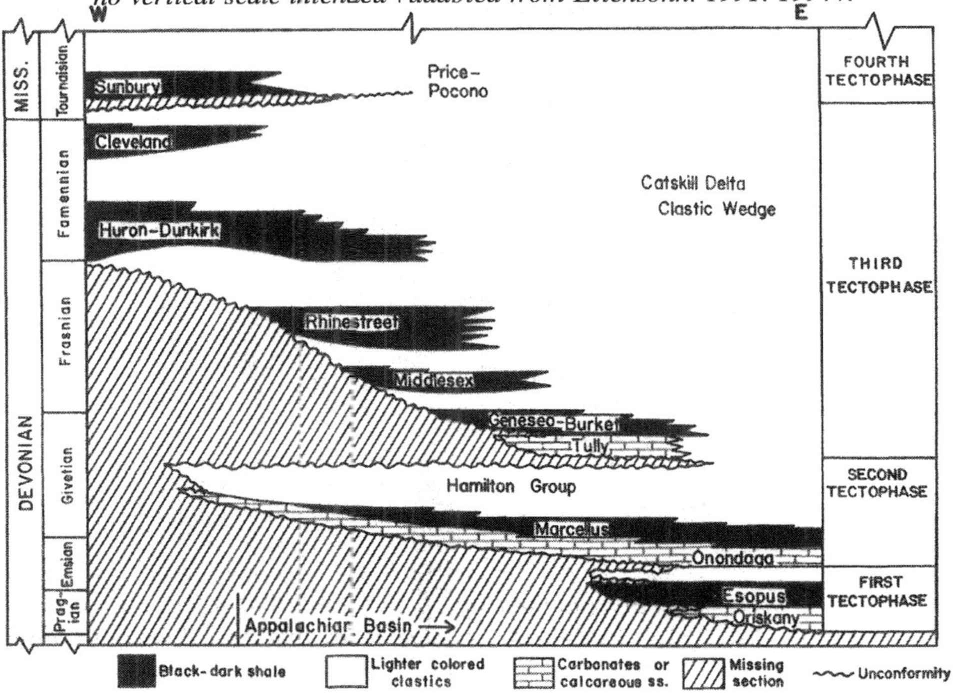

Fig. 5.8: Highly schematic, composite, east–west section from east-central New York to north-central Ohio in the northern Appalachian Basin. Parts of four tectophases are shown, but only the final Mississippian sequence went to completion, but is not completely shown here. Each of the early three tectophases apparently did not go to completion before the next one started, and erosion from the succeeding tectophase partially destroyed the sedimentary record of the previous one. Note the cratonward migration of succeeding tectophase sequences and the fact that the third

5.3 APPALACHIAN BASIN FORELAND-BASIN CYCLES

5.3.1 TACONIAN OROGENY

The first North American, Paleozoic, foreland-basin cycles began about 461 Ma ago with the onset of the Taconian orogeny, which marks initial closure of the Iapetus Ocean and the first major sedimentary and tectonic differentiation of Laurentia since its Late Proterozoic formation; the Taconian orogeny also

tectophase is composed of subcycles that may reflect the movement of individual thrust complexes. No scales intended (from Ettensohn, 1987, 1994).

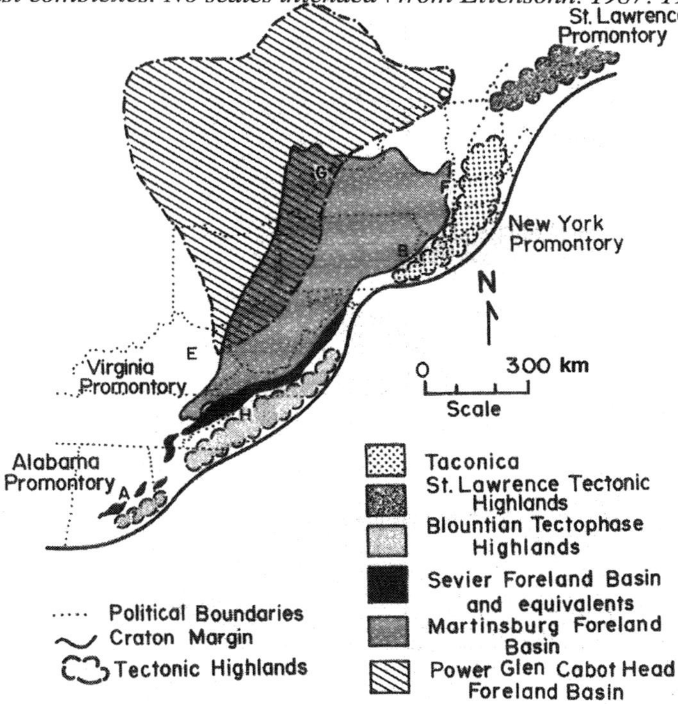

Fig. 5.9: Schematic map of the Appalachian Basin area showing the relative positions of Taconian tectonic highlands, dark-shale foreland basins, and continental promontories during Late Ordovician to Early Silurian time. Note the northwestward migration of dark-shale foreland basins in time and the asymmetry of basins toward respective promontories. Each of these basins more or less represents the maximum aerial distribution of the respective tectophase sequence. The Sevier Basin was the product of the Blountian tectophase, the Martinsburg Basin was the product of the Taconic tectophase, and the Power Glen-Cabot Head was the

marks the beginning of the Caledonian orogenetic cycle on the Appalachian margin (Fig. 5.10). The earliest cycle is defined at its base by the post-Sauk unconformity of Sloss (1963) of late Middle Ordovician age (late Llanvirn; Whiterockian), which defines the boundary between his Sauk and Tippecanoe sequences (Fig. 5.10). The lacuna on the unconformity may extend upward in places beyond the Middle-Late Ordovician (Llanvirn-Caradoc; Whiterockian-Mohawkian) transition into early Mohawkian (early Caradoc) time, and the unconformity is so widespread across North America that is has several different

product of Early Silurian tectophase 3. The lettered lines indicate locations of the sections in Figs. 5.5–5.7 (adapted from Ettensohn, 1991, 1994).

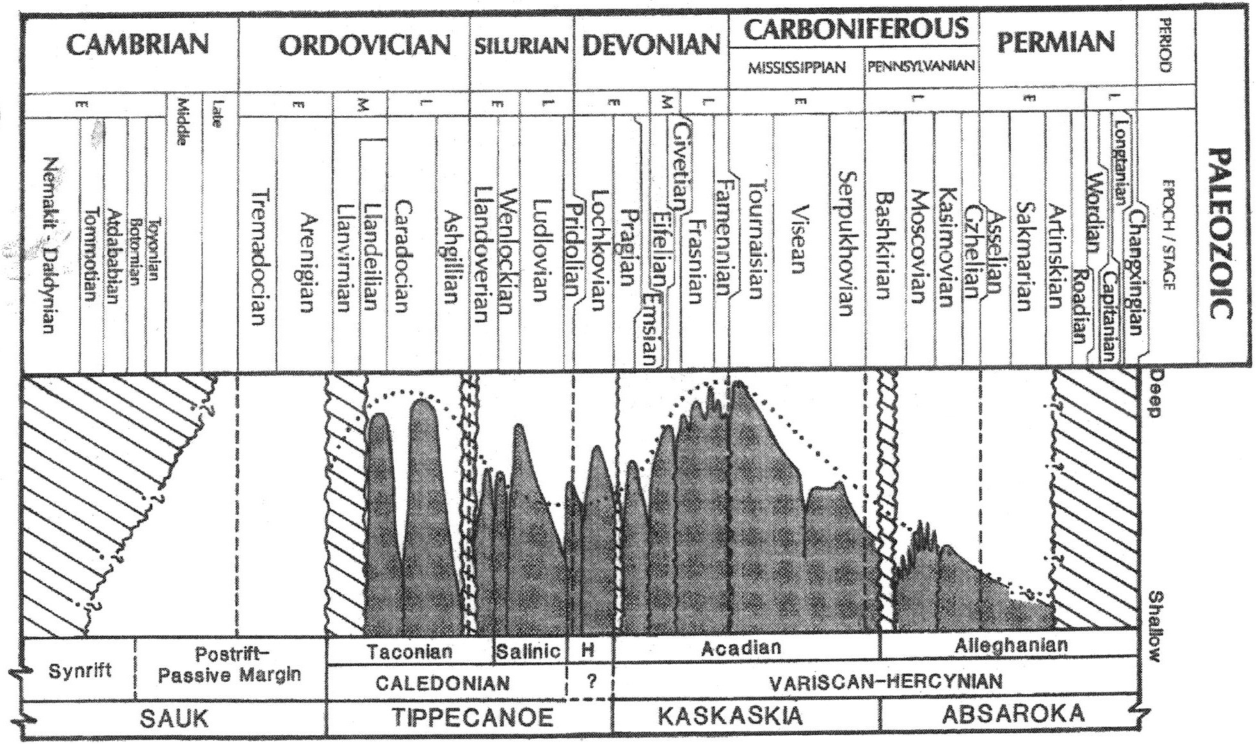

Fig. 5.10: Paleozoic geologic timescale, showing the occurrence and relative duration of 13 third-order, tectophase cycles in the Appalachian Basin as a relative sea-level curve. In the columns on the right side of the diagram, from left to right, the respective orogenies, orogenic cycles, and Sloss sequences related to each cycles are shown. The Helderberg interval (H) on the column could represent a separate, but minor, event involving the Carolina terrane, or could

names (e.g. St. George unconformity, Owl Creek Discontinuity, unconformity C, sub-Tippecanoe unconformity, Knox unconformity).

In fact, the duration and extent of the unconformity probably preclude a wholly bulge-related origin and argue for a larger cause like major continental breaking with the inception of Taconian convergence. Although the causes of Taconian convergence were probably heterogeneous in kinematic style at different places and times, it is clear from deformation and from the development and migration history of the cycles (Figs. 5.5 and 5.9) that there was a northeastward progression of convergence along the eastern Laurentian margin in space and time and that the successive tectophases were localized at continental promontories (Dewey and Burke, 1974; Dewey and Kidd, 1974; Ettensohn, 1991). Alhtough it is now realized that Taconian deformation continued northward during Early Silurian (early Llandovery; Alexandrian or Medinan) time (e.g. Van Staal, 1994; Van Staal and de Roo, 1995; Ettensohn and Brett, 2002), the very widespread Cherokee unconformity (Dennison and Head, 1975) separates the last two cycles (Fig. 5.7). Although this unconformity clearly has tectonic components near the Appalachian Basin, its duration and widespread nature beyond the Appalachian Basin (e.g. McKerrow, 1979) area are almost certainly related to eustatic drawdown accompanying the peak of Ordovician-Silurian glaciation (Dennison, 1976; Hambrey, 1985; Caputo and Crowell, 1985; Bjorlykke, 1985). The last Taconian cycle is wholly Silurian in age and reflects a third tectophase focused on the St. Lawrence Promontory, although there is also evidence of continental accretion at the Virginia promontory (Hibbard, 2000; Hibbard et al., 2002). Although the extent of the dark-shale phase of the cycle has been reconstructed (Fig. 5.9), only distalmost parts of the cycle (Fig. 9.7) are preserved in the northern Appalachian Basin. Hence, as a result of the Taconian orogeny, three foreland-basin, tectonic cycles with duration of approximately 7, 10, and 6 Ma respectively, corresponding to the three tectophases, are preserved in the Appalachian Basin (Fig. 5.10); only the first two of these cycles apparently went to completion (Figs. 5.5–5.7).

5.3.2 SALINIC DISTURBANCE

The final, Early Silurian, Taconian cycle is also immediately followed by an apparently unrelated Early Silurian (latest Llandovery-early Wenlock; Clinton) cycle in northern parts of the Appalachian Basin. The cycle reflects the first part of a two-cycle, orogenic event, called the Salinic Disturbance (Boucot, 1962; Fig. 5.10), evidence of which is best preserved in the northern Appalachian Basin. Unlike Taconian tectophases, the cycles are more carbonate-rich and migrate southward in time and space (Ettensohn and Brett, 1998, 2002; Figs. 5.11 and 5.12). The Salinic Disturbance apparently reflects the southward

be associated with either of the orogenies that bound it. The superimposed dotted curve reflects the two second-order orogenetic cycles (Timescale from Prospectors & Development Assoc. of Canada, B. Grant compiler, 1999).

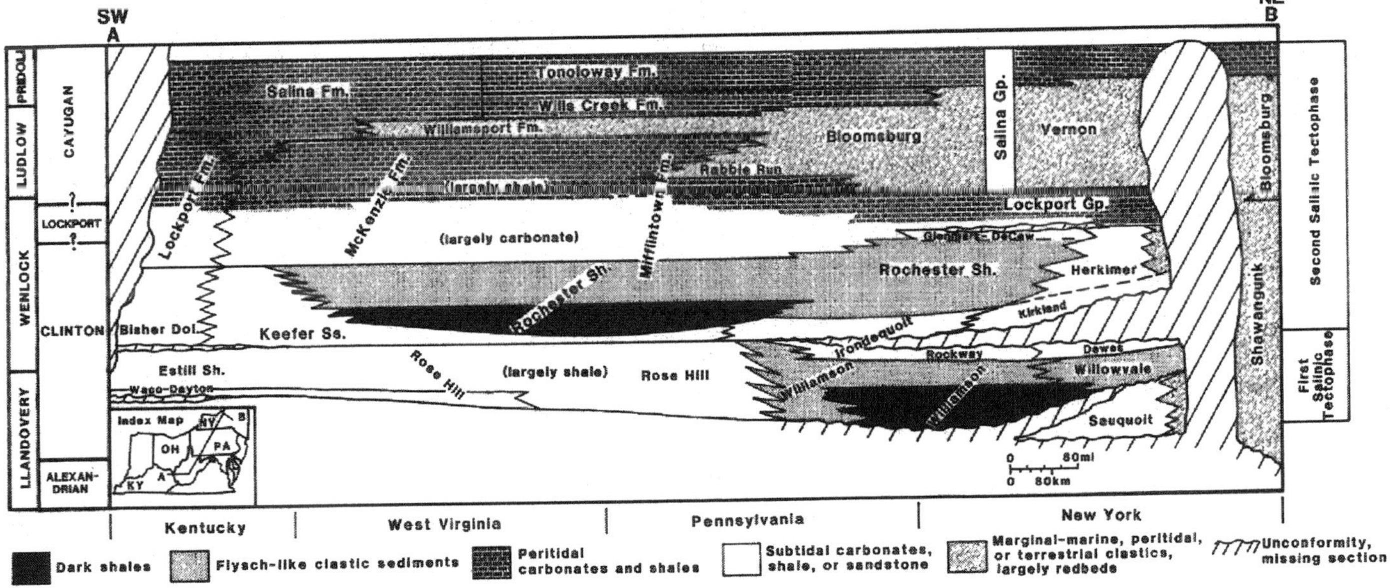

Fig. 5.11: Schematic Silurian section partially parallel to basin strike showing two tectophase sequences related to the Salinic Disturbance. Note the southwestward migration in time that reflects a similar trend in Salinic convergence (Fig. 5.12). The first sequence is not complete and contains no subaerial components, but the second sequence went to completion and ends with Bloomsburg-Vernon red beds and evaporites. No vertical scale intended (after Ettensohn, 1994).

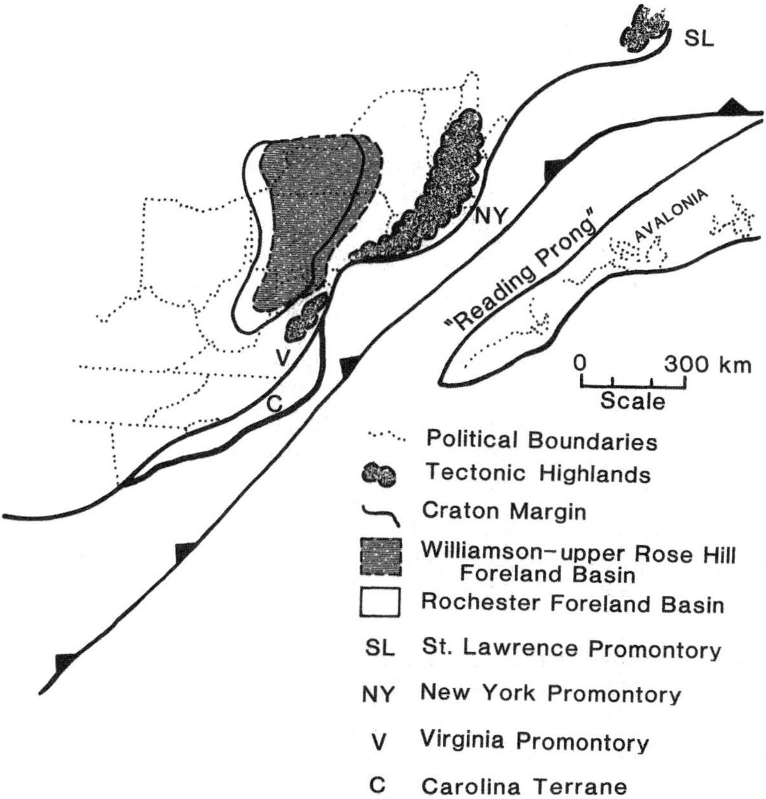

Fig. 5.12: Silurian tectonic setting showing convergence of the "Reading Prong" with the northern Appalachian margin to generate the Salinic disturbance. Note the southwestward migration of Salinic foreland basins based on the distribution of dark-shale basins (see Fig. 5.11) that tracked the progress of Salinic convergence (adapted from Ettensohn and Brett, 1998).

extension of Caledonian tectonism during the docking of Avalonian terranes in the "Reading Prong" of Baltica with the northern Appalachian margin of Laurentia (Ettensohn, 1994; Ettensohn and Brett, 1998; Fig. 5.12). Although Rodgers (1987) concluded that the disturbance was concentrated near the St. Lawrence Promontory, tectophase cycles from the Appalachian Basin indicate that the effects of convergence were prominent as far south as the Virginia Promontory. The disturbance, with two tectophases of approximately 2 and 4 Ma duration respectively, persisted from latest Llandovery to late Pridoli (mid-Clinton to late Cayugan) time (Figs. 5.10 and 5.11). The earlier tectophase cycle is incomplete, but the second tectophase cycle went to completion and is well known for its

Bloomsburg red beds and related Salina evaporites that formed during unloading-type relaxation (Ettensohn, 2004; Fig. 5.11).

5.3.3 HELDERBERG INTERVAL

Conformably to unconformably overlying the Bloomsburg-Salina sequence in the central and northern Appalachian Basin is a mixed-carbonate-clastic sequence of latest Silurian-Early Devonian age (Pridoli-Lochkovian-early Pragian; latest Cayugan-early Ulsterian), called the Helderberg Group. The predominance of carbonates in the group has led most workers to conclude that deposition occurred during a time of tectonic quiescence between the Taconian (or Salinic) and Acadian orogenies (e.g. Dorobek and Read, 1986; Smosna, 1988). However, evidence for structural reactivation in the area (Goodmann, 1988; Linn et al., 1990) and for synchronous Silurian-Devonian magmatism in the area (Horton et al., 1995; Samson and Secor, 2000; Hibbard et al., 2002) suggests coeval tectonism that generated upland sources for small clastic wedges in the sequence. Helderberg facies are very complex and have been further complicated by superimposed, small-scale eustatic cycles of likely glacial origin (Saltzman, 2002), but two, third-order, transgressive–regressive cycles of about 2 and 4 Ma respectively, centred on small, local, dark-shale basins in east-central parts of the larger Appalachian Basin, have been identified (Smosna, 1988; Fig. 5.10). Although the cycles may be tectophase, foreland-basin cycles, the tectonic origin of the loading is uncertain. The loading, however, may reflect the continued southward extension of Salinic tectonism or some interaction between accreting Avalonian terranes and the recently accreted Caroline Terrane (Fig. 5.12).

5.3.4 ACADIAN OROGENY

While the Salinic Disturbance apparently reflects the actual docking of Avalonian terranes with the southeastern margin of Laurentia and possible development of a subduction zone (Fig. 5.12), the Acadian orogeny seems to represent a diachronous, transgressional convergence between these terranes and the continent margin in a north-to-south direction (Ettensohn, 1987). Four, foreland-basin cycles, some with pronounced subcycles (Figs. 5.8 and 5.10), developed during periods of approximately 12, 9, 28, and 42 Ma respectively, and migrated southward in time and space (Ettensohn, 1985a, 1994, 1998; Fig. 5.8). Initiation of the orogeny is marked in the Appalachian foreland basin by the pre-Oriskany or Wallbridge Discontinuity, which also defines the base of Sloss's (1963) Kaskaskia sequence and the beginning of the Variscan-Hercynian orogenetic cycle (Fig. 5.10). The first tectophase was focused at the St. Lawrence Promontory in northern New England and in the Canadian Maritime Provinces in late Early Devonian time (Pragian-Emsian; mid-Ulsterian). The second tectophase reflects Middle Devonian (Eifelian-early Givetian; Erian) collision with the New York Promontory, whereas the third tectophase represents Middle-Late Devonian (late Givetian-Famennian; late Erian-Chatauquan), southward migration of oblique convergence between the New York and Virginia promontories; it is characterized by at least five pronounced subcycles. This progression of tectophases is clearly recorded in the southward migration of foreland-basin tectophase cycles (Ettensohn, 1985a, 1994, 1998), a scenario that parallels Rodger's (1987) early observations of a northeastwardly to southwestwardly shifting orogeny. Only the fourth tectophase went to completion, and it effectively took the entirety of the Mississippian Period and probably extended into the earliest Pennsylvanian time, persisting for about 42 Ma; it was the longest of the Appalachian tectophase cycles (Ettensohn, 1994, 2004; Fig. 5.10).

5.3.5 ALLEGHANIAN OROGENY

Throughout most of the Appalachian Basin, Mississippian and Pennsylvanian sections are separated by the so-called "Mississippian-Pennsylvanian", or sub-Absaroka, unconformity of Sloss (1963). Despite its Mississippian-Pennsylvanian appelation, it is actually Early Pennsylvanian in age (Englund et al., 1979; Chesnut, 1989, 1992; Pashin et al., 1991) and in large part probably reflects bulge uplift and migration accompanying inception of the Alleghanian orogeny (Quinlan and Beaumont, 1984; Beaumont et al., 1987, 1988; Ettensohn and Chesnut, 1989), but a major component of continental braking related to collision of Gondwana with southeastern margin of Laurussia may also be involved. The initial timing of the collision is poorly constrained, but if the age of arc development (Sinha and Zietz, 1982; Dallmeyer, 1986) and the age of unconformity development (Englund et al., 1979; Chesnut, 1989, 1992; Pashin et al., 1991) are interpreted to approximate the inception of orogeny, then the Alleghanian orogeny began in latest Early Pennsylvanian (late Namurian or Bashkirian; Morrowan) time.

The clockwise convergence of Gondwana toward Laurussia (e.g. Ziegler, 1989; Scotese, 1998), the age and distribution of clastic wedges in the foreland basin (Chesnut, 1989; Patchen et al., 1985a, b), and flexural modeling (Beaumont et al., 1987, 1988), all suggest that the orogeny proceeded from south to north, but interpretation of the sedimentary record is still uncertain about the manifestation of tectophase cycles. Based on structural criteria, Geiser and Engelder (1983) suggested that there were two Alleghanian tectophases, the earliest of which they called the Lackawana tectophase. In contrast, Donaldson and Eble (1991) suggested the presence of three tectophases, but their earliest tectophase include lowest parts of the Lower Pennsylvanian section that are gradational with underlying Mississippian rocks, and hence, according to Ettensohn and Chesnut (1989) and Ettensohn (1994, 2004), were parts of the final Acadian tectophase (Fig. 5.10); their final two tectophases, however, are divided at nearly the same point suggested later in this chapter.

The Lackawana tectophase may reflect early development of a west-directed subduction zone below the newly accreted southeastern margin of Laurussia and development of a magmatic arc forward of the Alabama and Virginia promontories during Early and Middle Pennsylvanian time (Bashkirian-Moscovian, Morrowan-Desmoinesian; Fig. 5.13). Concomitant loading developed a foreland basin throughout southern and central part of the larger Appalachian Basin, but the extent of individual units and the very shallow-marine to terrestrial nature of the included sediments suggest that the basin was broad and

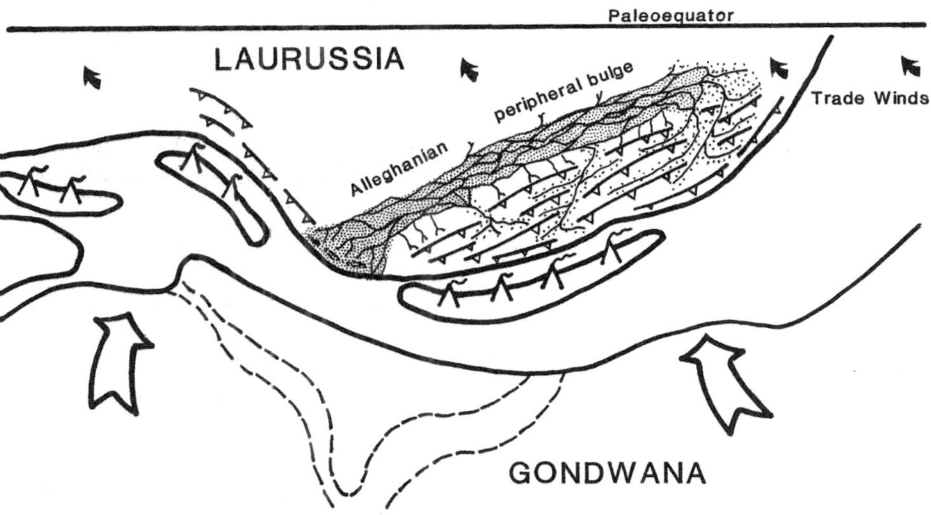

Fig. 5.13: Schematic paleogeographic diagram of the collision zone between Gondwana and the southeastern margin of Laurussia during Early Pennsylvanian time. The compact stipple represents one of the braidplain sandbelts running longitudinally along the northwest margin of the foreland basin; this sandbelt has truncated some of the earlier, transverse drainage between it and the thrust belt to the southeast. As suggested in the figure, most of the major drainage (loose stipple) from the thrust belt probably entered the braidplain from the north and northeast. During early parts of each fourth-order cycle (see Fig. 5.14), the foreland basin (braidplain and transverse drainage) would have been flooded with marine waters from the south forming an elongate estuary/tidal complex.

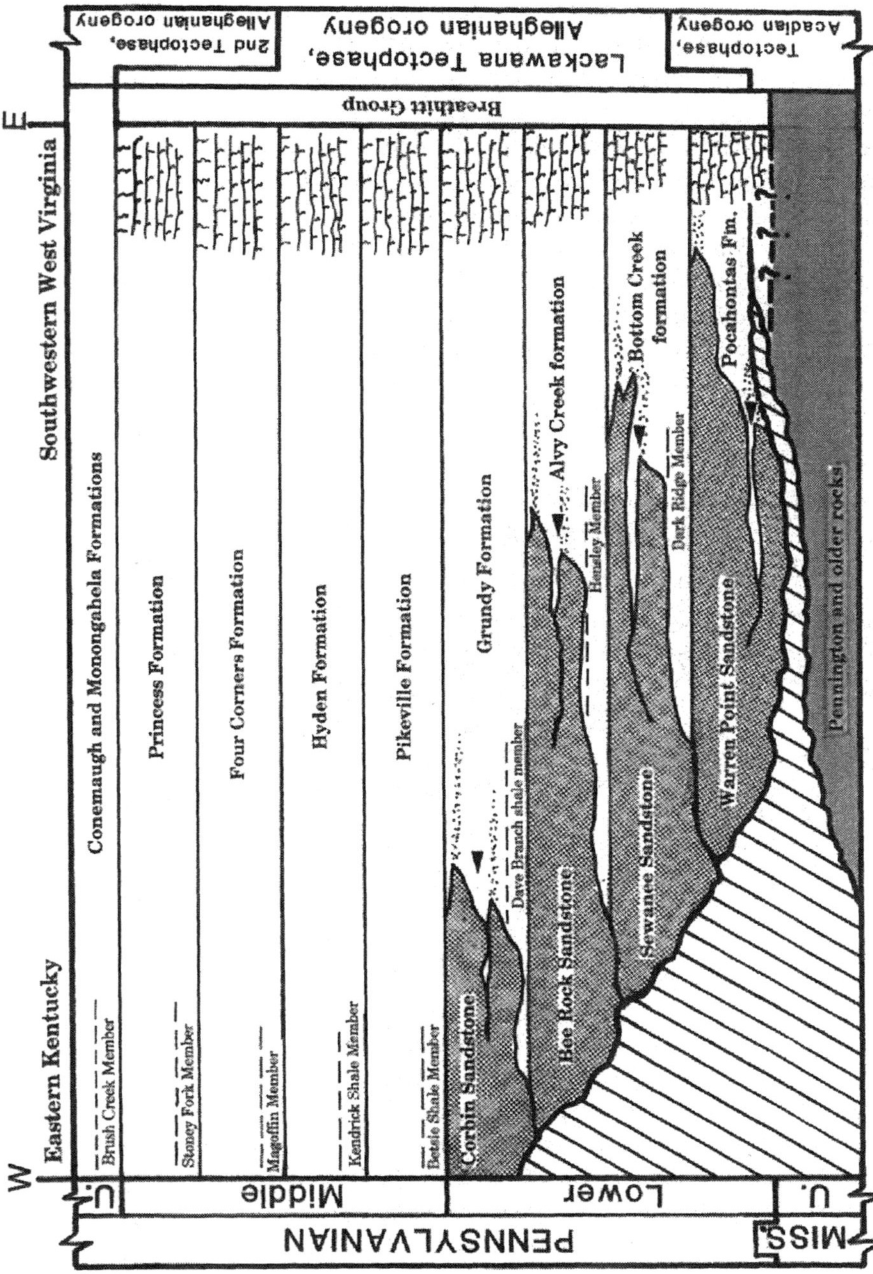
W
E
Eastern Kentucky
Southwestern West Virginia
Brush Creek Member
Conemaugh and Monongahela Formations
Stoney Fork Member
Princess Formation
Four Corners Formation
Magoffin Member
Hyden Formation
Kendrick Shale Member
Pikeville Formation
Betsie Shale Member
Grundy Formation
Corbin Sandstone
Dave Branch shale member
Bee Rock Sandstone
Alvy Creek formation
Hensley Member
Sewanee Sandstone
Bottom Creek formation
Dark Ridge Member
Warren Point Sandstone
Pocahontas Fm.
Pennington and older rocks
Breathitt Group
2nd Tectophase, Alleghanian orogeny
Lackawana Tectophase, Alleghanian orogeny
Tectophase, Acadian orogeny
U.
Middle
Lower
U.
PENNSYLVANIAN
MISS.

Fig. 5.14: Schematic diagram showing the eight, fourth-order, Lower Pennsylvanian cycles of the Breathitt Group, which largely comprises the Lackawana tectophase. Each cycle begins with a named marine zone that represents marine flooding of the foreland basin. The basin then infilled with five to seven, fifth-order, coal-clastic cycles of likely glacio-eustatic origin represented on the right by successive coal beds (undulating lines with rooting). In the lower four cycles, the coal-clastic cycles were truncated by quartzarenite, sandbelt complexes representing braidplains that migrated westward in time with each new cycle. The arrowheads in the four lower cycles point to mid-formational, marginal-marine shales that may reflect further subdivision of the lower four cycles. The trailing, loose stipple at the eastern termination of each sandstone body represents the possibility of some eastern, transverse streams feeding the braidplain channels. While the lower four cycles probably represent "loading-type" relaxation, the upper four cycles, lacking major quartzarenite channel sands, apparently prograded beyond the limits of the former foreland basin and probably represent "unloading-type" relaxation. The overlying, Upper Pennsylvanian Conemaugh, and Monongahela formations reflect a new tectophase and may be separated from the Princess Formation by a subtle unconformity. The diagonally lined area at the base of the diagram is the Early Pennsylvanian, sub-Absaroka or "Mississippian-Pennsylvanian" unconformity (adapted from Chesnut, 1992).

shallow, perhaps reflecting the fact that deformation had advanced more expansively cratonward than during previous orogenies. Sediments deposited during the tectophase are included in the Breathitt Group (Chesnut, 1992, 1994, 1996; Fig. 5.14) and exhibit a pronounced, fourth-order, transgressive–regressive cyclicity of thin marine horizons with overlying coal-bearing sequences of largely estuarine, tidal or alluvial-plain origin, each of which contain groupings of five to seven, fifth-order, coal-clastic cycles (Greb et al., in press; Fig. 5.14). The fourth-order cycles are defined by marine flooding surfaces at their bases and reflect ~2.5 Ma of time; they were probably tectonic in origin and include the fifth-order cycles with ~0.4 Ma periodicities of likely glacio-eustatic origin (e.g. Chesnut, 1992, 1994, 1996; Fig. 5.14). In the four earliest of the fourth-order cycles, the cratonward margin of coal-bearing sequence is truncated by a channel complex of quartz-pebble-bearing quartzarenites (Warren Point, Sewanee, Bee Rock, and Corbin sandstones; Fig. 5.14), interpreted to represent braided-stream sandbelts that ran longitudinally along the distal margin of the foreland basin on a south- to southwest-dipping paleoslope (Chesnut, 1994; Greb et al., in press). Each of the sandstones, moreover, appears to be further divided by a mid-formational shale, which also seems to reflect a marine flooding surface that probably divides each of the cycles containing distal quartzarenites into two fourth-order cycles (Greb et al., in press; Fig. 5.14, arrowheads). The thicker, more basinward, coal-bearing parts of each fourth-order cycle are characterized by litharenites and sublitharenites in channels that trend westward or northwestward, transverse to the foreland basin and the longitudinal trends of the quartzarenite sandbelts. The quartzarenite sandbelts appear to truncate most of the transverse channels and adjacent parts of the coal-clastic cycles containing them. The exception probably occurred at the top of each cycle below the next marine flooding event, where the transverse channels must have been gradational into the longitudinal sandbelt and provided at least some of the quartz sands and pebbles for the longitudinal belts; this implies that there must have been different sources for transverse streams in the earlier coal-bearing parts of the cycle than for the distal, longitudinal streams and their feeder streams in later parts of each cycle (Greb et al., in press). The western or cratonward boundary of each longitudinal sandbelt was apparently the peripheral bulge (Chesnut, 1994), and the fact that the bulge, sandbelts, and related coal measures shifted progressively westward in time and space (Fig. 5.14) suggests a tectonic origin in response to a shifting deformational load (Greb et al., in press). As has been suggested for similarly shifting Devonian-Mississippian black-shale subcycles (Fig. 5.8), each of these fourth-order cycles may reflect the cratonward movement of a major thrust system (Tankard, 1986; Greb et al., in press). Overall, the lower four, Lower Pennsylvanian (Bashkirian; Morrowan) cycles are unique in the presence of shifting quartzarenite belts and in the less marine, more estuarine nature of their truncated, basinward, coal-bearing parts (Fig. 5.14).

The overlying, Middle Pennsylvanian (Moscovian; Atokan-Desmoinesian), forth-order cycles are similarly bound by regionally extensive, marine zones, but they lack the distal, truncating, quartzarenitic, channel sands, suggesting the absence of a major, quartz-sand, sediment source. In addition, the marine zones are much more extensive, and the intervening coal-clastic cycles are more tidal in origin up to the level of the peak, marine-flooding event (Magoffin Member), after which the cycles become more fluvially dominated, each of these cycles, moreover, apparently prograded beyond the limits of the foreland basin (Fig. 5.14), indicating a filled to overflowing foreland basin (Chesnut, 1994; Greb et al., in press). The overlying Upper Pennsylvanian (Kasimovian; Missourian) units of the Conemaugh Group may unconformably overlie the Middle Pennsylvanian units just discussed and apparently represent a different depositional framework with different sources in a different basin, and hence, probably represent a new tectophase.

Clearly, the sedimentary sequence and lithologies present in the Lower-Middle Pennsylvanian Breathitt Group appear to be vastly different than those that characterize earlier tectophases. These differences no doubt reflect the collisional nature of the Alleghanian orogeny and the resulting broad, shallow foreland basin, cyclic glacio-eustatic influences, and presence in the humid equatorial belt (e.g. Cecil and DuLong, 2003; Cecil et al., 2003, 2004). Nevertheless, elements of the Breathitt sequence still parallel typical tectophase development, and suggest that the entire sequence does represent a distinct tectophase. It begins with the major sub-Absaroka unconformity (Figs. 5.10 and 5.14), reflecting at least in part major bulge uplift and moveout, and if each of the four (or eight) Early Pennsylvanian cycles with distal quartzarenites is interpreted to represent cratonward movement of a major thrust system, then each of the cratonward-shifting cycle sequences (Fig. 5.14) is similar to Devonian-Mississippian black-shale subcycles (Fig. 5.8). The marine flooding event at the base of each cycle, which is represented by dark shales and/or dark limestones, apparently reflects basin subsidence accompanying thrust moveout and loading. The overlying series of coal-clastic cycles probably represents loading-type relaxation with the westerly or southwesterly directed marginal-marine and terrestrial clastic sediments taking the place of flysch-like clastic sediments in a shallow foreland basin. The fifth-order, coal-clastic cycles that nearly fill each basin at this stage are merely responses to climatic and eustatic controls caused by alternating glacial (coal-rich) and interglacial (clastic-rich) episodes (Cecil and DuLong, 2003; Cecil et al., 2003, 2004). The muds, litharenites and sublitharenites that predominate in this phase of each cycle largely represent the sedimentary unroofing of each thrust complex and probably filled, and even overfilled, the shallow foreland basin, especially toward its eastern margin, which is proximal to source areas; and indeed, all cycles do thicken in this direction (Greb et al., in press). This pattern of filling or overfilling toward eastern source areas may have also been reinforced by climatic asymmetry, inasmuch as the Pennsylvanian paleoequator and west-directed trade winds that encroached upon it from the southeast and transported weather systems toward the basin, were aligned on the foreland-basin side of the Alleghanian highlands (e.g. Scotese, 1998; Cecil et al., 2003; Fig. 5.13). With low westward gradients into the nearly filled foreland basin, the thrust complexes, now unroofed to the level of Early Paleozoic sandstones or basement igneous and metamorphic rocks, would have experienced higher gradients to the east, and north and generated high-gradient streams carrying quartz-rich sediments in those directions (Fig. 5.13). Some of the transverse, westward drainage may have also transported quartz-rich sediments westward, and there is evidence for this in later parts of each cycle, but most paleocurrent evidence from the quartzarenite sandbelt complexes support transport from the north and northeast (Greb et al., in press). Because continental collision effectively prevented the symmetrical development of major sedimentary sinks to the east and southeast, much sediment generated on the eastern flanks of the mountains must have emerged from the mountains toward lower areas in the north that had not yet experienced Alleghanian convergence. These streams emerged from the mountains in the north and northeast with their quartz-rich bedloads to form the large, low-gradient, braided, trunk streams that then flowed southward along the lowest, westward margin of the foreland basin toward the deepest, most rapidly subsiding parts of the foreland basin in what is now Alabama (Fig. 5.13). The peripheral bulge no doubt acted as a westward limit for the braided stream complexes (Chesnut, 1994), but as successive thrust-sheet loads migrated westward, so did the bulge and successive braided-stream belts (Fig. 5.14). Cratonward migration of the braided-stream sandbelts occurred at least four, and possibly eight, times before the thrust complexes ceased moving and were weathered and eroded to near elevational equilibirum with the filled foreland basin, as indicated by the absence of major quartzarenite units in subsequent parts of the Breathitt Group. These segregated quartzarenite units would seem to be unusual manifestations of loading-type relaxation, even under a collisional regime, but in the Appalachian Basin they probably represent the unique coincidence of climatic and tectonic asymmetry (e.g. Johnson and Beaumont, 1995), and their absence in overlying units probably signals the beginning of unloading-type relaxation near the end of the tectophase.

By Middle Pennsylvanian time, distal, sandbelt development had ceased, the foreland basin had filled, and four, successive, fourth-order cycles defined by major marine zones (Pikeville, Hyden, Four Corners, and Princess formations; Fig. 5.14) apparently overflowed cratonward beyond the basin (Chesnut, 1994; Greb et al., in press). The lower three cycles are defined by marine zones (Betsie, Kendrick, and Magoffin members; Fig. 5.14) that probably represent the most widespread, Pennsylvanian, marine flooding events in the Appalachian Basin (Greb et al., 2002). In the two, intervening, coal-clastic sequences, coals are widespread, and the immature clastic sediment therein are tidally dominated and largely reflect a west-dipping paleoslope (Chesnut, 1994; Greb et al., in press). The two overlying cycles, defined by the marine, Magoffin and Stoney Fork members at their bases, are more sandy and fluvial-dominated, and in the final cycle, the direction of thickening begins to change from the southeast into the Central Appalachian Basin to the east and north into the Northern Appalachian or Dunkard Basin (Fig. 5.15).

Although continued presence of fourth-order cycles in the Middle Pennsylvanian section seems to reflect some continuation of cyclic loading, the absence of major quartzarenite-sandbelt sedimentation probably reflects the lowering of major source areas and the establishment some degree of elevational equilibrium between these areas and the largely filled foreland basin, which marks the onset of unloading-type relaxation. Transition to this stage of loading is also supported by the predominance of a west-dipping paleoslope and apparent overflow of the filled foreland basin. The widespread nature of the three earliest Middle Pennsylvanian marine zones, moreover, may indicate influence by the anti-peripheral bulge or peripheral sag described earlier, during which thin marine horizons become very widespread (Ettensohn, 1994).

Hence, Lower and Middle Pennsylvanian units of the southern and central Appalachian Basin seem to be best described in terms of a single, foreland-basin, tectophase cycle. The general pattern of the sequence cycle still appears to be present, but it is overprinted by the effects of a shallow foreland basin, which insure the predominance of marginal-marine to terrestrial sedimentation in every part of the cycle; by the effects of tropical climatic asymmetry, which force the predominance of clastic or organic sediments in every part of the cycle; and by the effects of glacio-eustatic and tectonic cycles, which impart other levels of cyclicity across all parts of the larger tectophase cycle.

By Late Pennsylvanian, and possibly Early Permian (Kasimovian-Artinskian; Missourian-Leonardian), time, foreland-basin sedimentation shifted northward into the northern Appalachian or Dunkard Basin (Fig. 5.15), apparently tracking the similar progression of Alleghanian orogeny. The sequence of resulting sediments, comprising Conemaugh, Monongabela, and Dunkard groups, may unconformably overlie the Middle Pennsylvanian Breathitt Group and its equivalents, and probably represents the second and final Alleghanian tectophase. In contrast to deposition during the Early and Middle Pennsylvanian

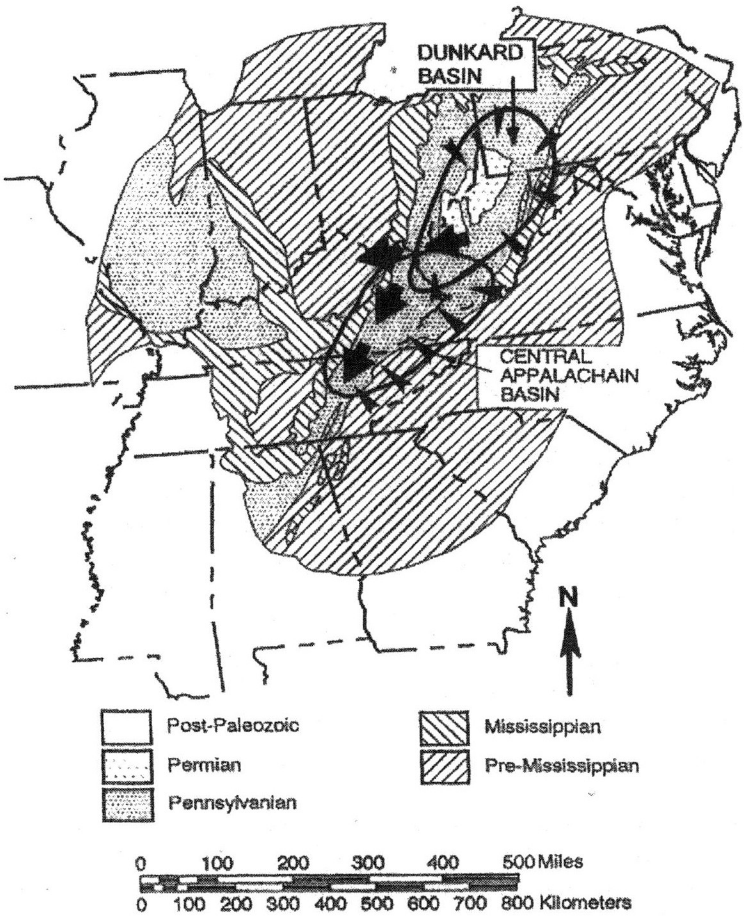

Fig. 5.15: Map of east-central United States showing general locations of the Central Appalachian and Dunkard (Northern Appalachian) basins, which in large part represent foreland basins for the Lackawana and second Alleghanian tectophases, respectively. Large dark arrows represent major paleoslopes in each basin, whereas the smaller arrowheads represent secondary paleoslopes. The northward

Lackawana tectophase, in which marine flooding entered the southern and central Appalachian foreland basin from the south, during early parts of the final tectophase. Subsidence in the north opened the Dunkard foreland basin to brief episodes of marine flooding from the Illinois Basin, perhaps through structurally related sags in the Cincinnati Arch like the Sebree Trough and

Cumberland Saddle (see Chesnut, 1994; Ettensohn et al., 2002). By this time, Ouachita and early Alleghanian orogenies had closed seaways to the south, and deformational loading to the north during the new tectophase apparently lowered the Cincinnati Arch and focused foreland-basin subsidence in the area of the Dunkard Basin (Fig. 5.15). The tectophase began with exposure along a possible unconformity and subsequent marine inundation, represented by the Brush Creek Limestone (Fig. 5.14). The Brush Creek, together with five, overlying, cyclic, marine horizons, may represent an early period of overall transgression from the west in response to cyclic thrust loading in the orogen. By late Conemaugh time, deltaic and alluvial-plain sediments, prograding from sources around the basin margin during loading-type relaxation (Fig. 5.15) effectively closed the basin to further marine inundation and generated a period of starved-basin infilling by chemical (lacustrine limestone) or organic (coal) deposits in large fresh-water lakes and swamps, represented by the Monongahela Group (Kovach, 1979; Donaldson and Shumaker, 1981; Donaldson and Eble, 1991). The presence of distinct basin geometry with multiple paleoslopes suggest continuation of loading-type relaxation. By latest Pennsylvanian and possibly Early Permian time, however, the basin had apparently filled with alluvial-plain sediments of the Dunkard Group, prograding into the basin from the south and east (Kovach, 1979; Donaldson and Shumaker, 1981; Milici and de Witt, 1988). This progradation probably marks the inception of unloading-type relaxation and is the final phase of Appalachian sedimentation. The foreland-basin tectophase cycle manifest by this tectophase is even less distinct than the previous one, but still shows early transgression followed by two different episodes of relaxation-related progradation. Although cyclicity is present, the larger fourth-order cycles appear to be absent or indistinct, perhaps reflecting the fact that the final Alleghanian tectophase was probably the product of transgressional or strike-slip movement due to escape tectonism following the convergence and rotation of Gondwana against the southern margin of Laurussia (e.g. Gates et al., 1986).

5.4 DISCUSSION AND CONCLUSIONS

The Appalachian Basin is a composite foreland basin formed during the ocean-basin-closure phase of a Wilson cycle (Wilson, 1966; Dewey and Burke, 1974). The closure occurred in two, phased orogenetic cycles that appear to have had worldwide, relative, sea-level manifestations. In the Appalachian Basin, these cycles are apparent as generalized, second-order, sea-level cycles (Fig. 5.10). The earlier, Ordovician-Silurian, Caledonian orogenetic cycle on the Appalachian margin of Laurentia represents closure of the Iapetus Ocean through subduction involving island-arc complexes and Avalonian terranes during formation of the new continent, Laurussia; it was accomplished during two orogenies – the earlier Taconian, progressing from south to north, and the later Salinic, progressing from north to south. The following, Variscan-Hercynian orogenetic cyle, in contrast, reflects closure of the Rheic Ocean during collision with Gondwana and the related transgressional adjustment of Avalonian terranes to form Pangea; it occurred on the Appalachian margin of Laurussia during two orogenies – the earlier Acadian, progressing from north to south, and the later Alleghanian, progressing from south to north.

Each of the orogenies, moreover, occurred in a series of two to four shorter tectophases, which on the Appalachian margin of Laurentia/Laurussia were apparently mediated by the occurrence of continental promontories, which focused convergence and at the same time restricted its along-strike continuation. Appalachian tectophases range in duration from about 2 to 42 Ma with an average of 12 Ma, based on the chronology of Gradstein and Ogg (2004), and each reflects an episode of deformational loading followed by crustal relaxation. Flexural modeling suggests that each complete tectophase can be described in terms of a seven-part model, each part of which leaves a distinct record in the foreland basin. The complete or incomplete sedimentary record of each tectophase is bound by unconformities and is manifest in the foreland basin as a generalized, third-order, transgressive–regressive cycle of lithologies, called herein a *foreland-basin tectophase cycle* or *sequence*. Inasmuch as the Appalachian margin of Laurentia/Laurussia was subject to nearly continuous convergence from Middle Ordovician to Permian (?) time during closure of the Iapetus and Rheic oceans, the Appalachian foreland basin during this time developed 13, stacked, third-order, tectophase cycles from four orogenies, arranged in two second-order, orogenetic cycles (Fig. 5.10). Eleven cycles from the first three orogenies all exhibit at least lower parts of the typical, sedimentary manifestation that includes three major lithologies in ascending succession: a basal dark marine shale, a flysch-like clastic sequence, and a molasse-like sequence of marginal-marine to terrestrial, clastic sediments with red beds. The dark shales form the most distinctive unit in each cycle and can be mapped in space and time (Fig. 5.9) to show the along-strike progression of tectophases during orogeny. During the ocean-closure phase of a Wilson cycle, this type of tectophase cycle is typical of early, subduction-type orogenies, in which deformational loads must initially mount a continental-margin ramp, and as a result, end up accumulating across short distances at the margin, especially near promontories. The concentrated load results in deeper isostatically compensating foreland basins in which the sediment-starved conditions that favour dark shales can easily develop. The early orogenies are also typically "symmetrical" orogenies with sediment sinks in the subduction zone on one side and in the foreland basin on the other, and, in the Appalachian Basin, the early orogenic belts were more or less perpendicular to climatic zones and largely developed in arid, subtropical conditions (Scotese, 1998). All of these situations militated against maximum clastic influx to the foreland basin, resulting in a basin that was rarely overfilled, in periodically common sediment-starved conditions, and in nearly continual marine inundation. Hence, in the 11 early cycles, dark marine shales are common and marine sediments prevail.

The last two Alleghanian cycles, however, are vastly different in that molasse-like, marginal-marine to terrestrial, clastic sediments predominate, marine sediments are uncommon, and a prominent fourth- to fifth-order cyclicity overprints each tectonic cycle. Although the same general sequence of flexural events seems to have generated the last two cycles, the differences largely reflect the late stage, collisional nature of the orogeny and the facts that the orogeny was tectonically and climatically "asymmetrical". Inasmuch the Alleghanian orogeny reflects final closure of intervening seas through collision of continents, the deformational load had already surmounted the marginal ramp and advanced far enough onto the craton that the more dispersed deformational load could only generate broad, shallow foreland basins incapable of supporting deeper water, marine regimes. Moreover, collisional highlands to the south and east prevented the symmetrical distribution of clastic debris to the ocean side of the orogen, and most of this sediment must have eventually found its way via northern routes and transverse draining into the foreland basin. In addition, due to continental rotation during collision, the foreland basin and source areas more or less paralleled the equator in the humid, tropical zone. As a result, clastic influx into the Alleghanian foreland basin was probably far greater than that during earlier orogenies, and the already shallow basin was usually close to overfilling, or overfilled, with clastic debris throughout much of its history. Hence, cycle character and composition were largely controlled by orogeny type and timing in the larger ocean-closure cycles and by disposition relative to climatic belts. Lower level, superimposed cyclicity, on the other hand, was probably controlled by tectonism at the fourth order and by glacial eustasy at the fifth order.

The tectophase cycles and their bounding unconformities are best considered to be regional phenomena, although some of the unconformities have interregional, and perhaps, global extent. Three of the unconformities coincide with bounding unconformities of Sloss's (1963) sequences. Two of these unconformities exhibit major lacunas in the Appalachian Basin, but all three represent major tectonic turning points. Both the post-Sauk and sub-Absaroka unconformities reflect major lacunae, the former representing the initiation of convergence with the newly formed Laurentian landmass and its resulting resistance or braking and widespread emergence in response, and the latter reflecting collision with Gondwana, probably with similar braking and emergence. Between these two is the Wallbridge discontinuity, which defines the base of the Kaskaskia sequence, but lacks the great extent and missing time interval of the other two; it records the beginning of the Variscan-Hercynian orogenetic cycle and the beginning of the Acadian orogeny. There is growing evidence that these major sequence-bounding unconformities are interregional, and possibly global (Ronov et al., 1969; Sloss, 1972, 1976; Soares et al., 1974), and similar evidence has been accumulating for some of the Silurian and Devonian tectophase cycles (Johnson, 1971; Ettensohn and Brett, 1998). If, in fact, orogenic, epeirogenic, and eustatic processes are closely related (e.g. Gurnis, 1992) and if tectonic reorganization on one margin may condition activity on connected subduction zones elsewhere, then there is a very real possibility that some third-order tectophase cycles may also be interregional or global. Clearly, regional attributes like the presence and distribution of continental-margin promontories, the angle of convergence, and paleoclimate have exercised important control on the thinning, duration, and nature of tectophase cycles. However, in the absence of such controls, the possibility of similar orogenetic and tectophase cycles in widely separated foreland basins may be a powerful tool in determining ultimate causes and in developing interregional correlation.

6. GEOLOGICAL HISTORY OF THE GREATER AMAZONAS BASIN IN BRAZIL

J.M. MABESOONE AND V.H. NEUMANN

6.1 INTRODUCTION

The E–W trending Greater Amazonas Basin in Brazil is an extensive low-lying depression of about 125,000 km^2 (Fig. 6.1), becoming its relief higher than 200 m only in the extreme west, near the frontier with Peru. It is drained by the Amazon River, running in its axis. The numerous affluents from the south as well as from the north turn the Amazon into a perennial stream with a high-volume runoff. Most of the area is covered by a dense equatorial forest that hinders the appearance of fresh rock outcrops.

The proper basin has been implanted upon the Amazon Craton, dividing this Precambrian area into a Guyana shield in the north and a Central Brazilian or Guaporé shield in the south (Fig. 6.2). After Loczy and Ladeira (1976, p. 447) the basin formed as a rift furrow (in the sense of Bucher, 1933), maybe already in Archean times, separating both shields. However, in spite of the rift's E–W trend, the main traits of the former contiguous crustal provinces of the Amazon Craton remained evident in the subdivision of the sedimentary basin into four sub-basins, each with its proper identity (see Fig. 6.2, and Cordani and Brito Neves, 1982), separated by transverse ridges. In this way, the westernmost Acre Basin, currently part of an Andean foreland basin, is separated from the Solimões or Upper Amazonas Basin by the Iquitos arch. The Purus arch divides the Solimões Basin and the Middle-Lower Amazonas Basin, actually called simply Amazonas Basin (Eiras et al., 1994). The Gurupá arch divides this Amazonas Basin from the Amazon Mouth Basin that exists at the Atlantic coast. Therefore, the four distinguished sub-basins are considered separately.

6.2 ACRE BASIN

6.2.1 GENERALITIES

The Acre Basin is formed to the west of the Iquitos arch and extends into Peru and Equador (Fig. 6.3) as an Andean foreland basin. In Brazilian territory, it

Cyclic Development of Sedimentary Basins
J.M. Mabesoone and V.H. Neumann, editors.

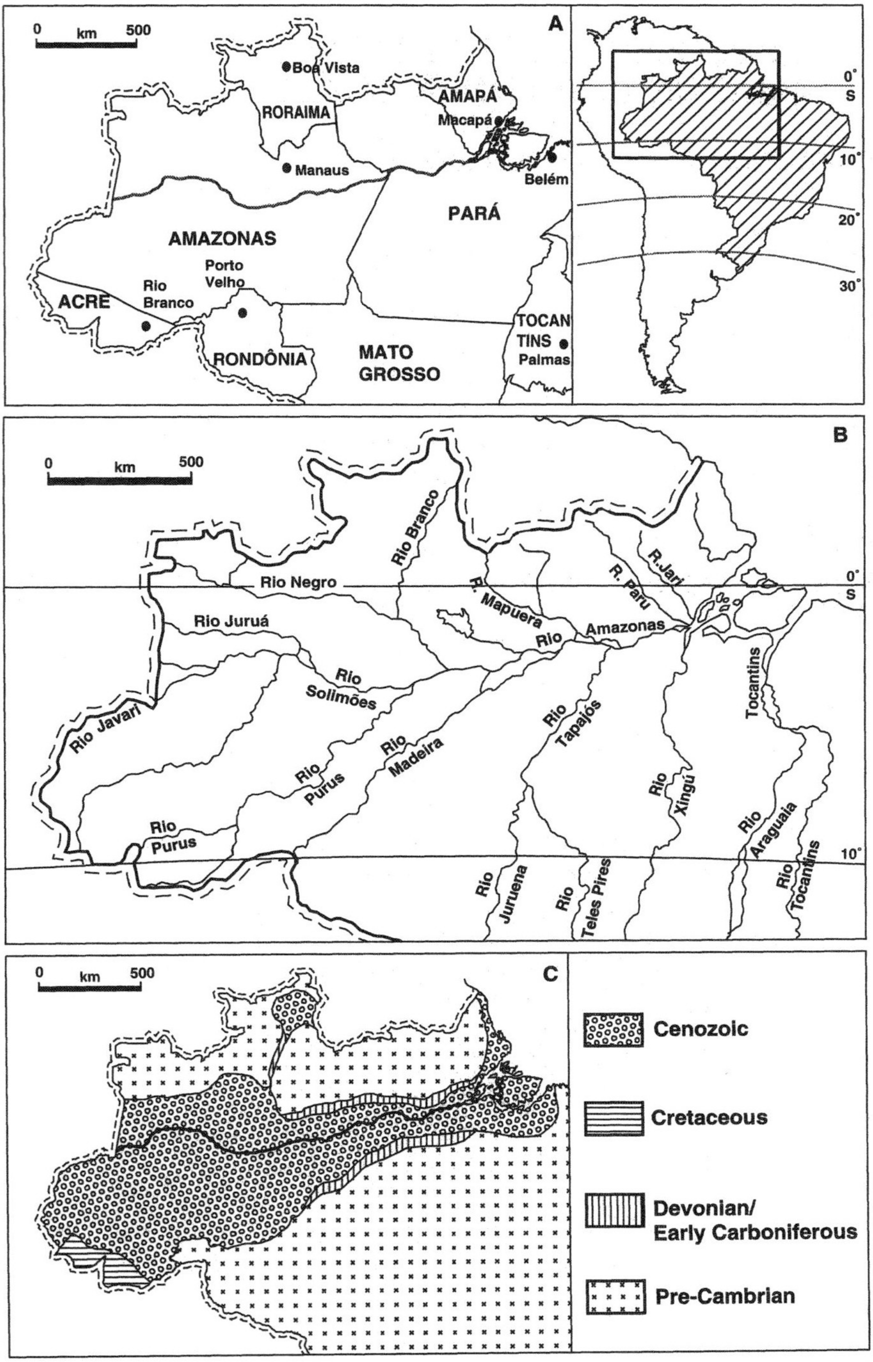

Fig. 6.1: Greater Amazonas Basin (Brazil): (A) geography of Amazon area; (B) hydrography of present-day Amazon River and principal affluents; (C) main geology.

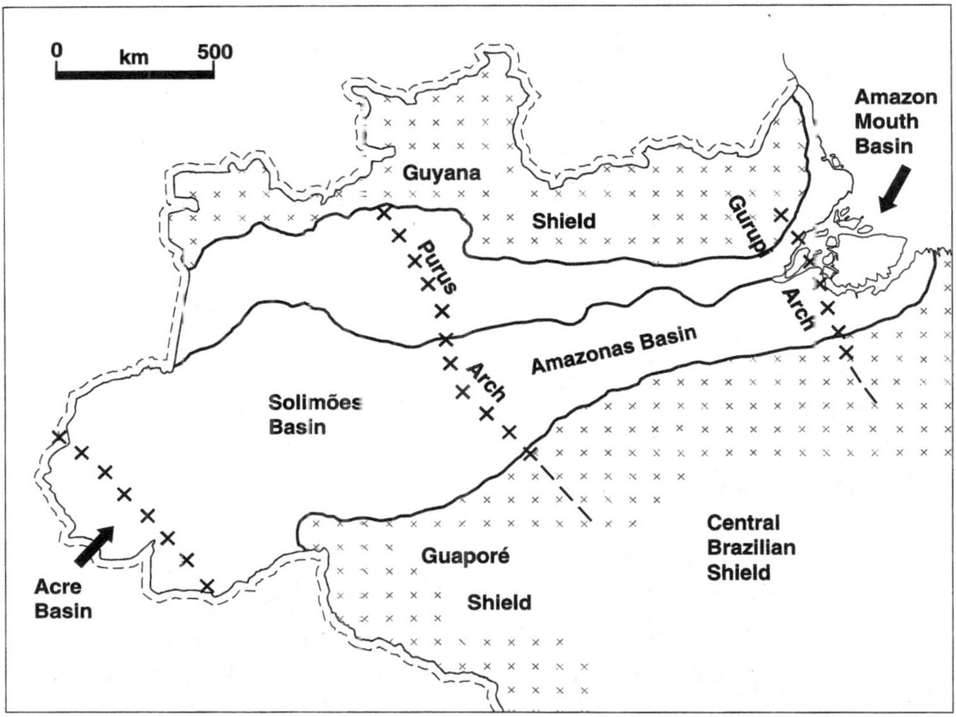

Fig. 6.2: Sub-basins of Greater Amazonas Basin.

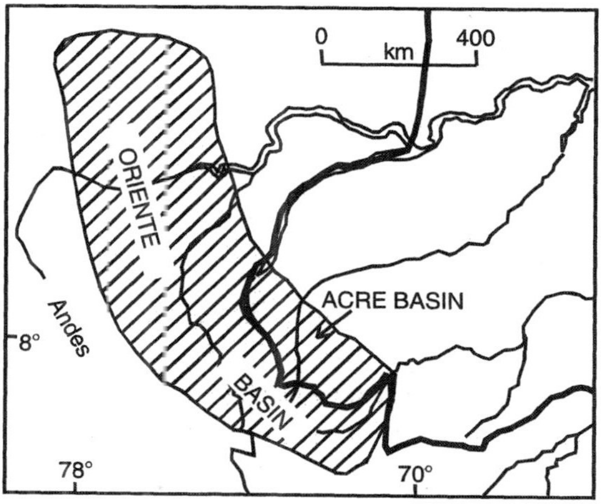

Fig. 6.3: Andean foreland basin in Brazil (Acre Basin) and Peru and Ecuador (Oriente Basin).

occupies a surface of about 230,000 km^2 in the westernmost parts of Acre and Amazonas States, and is drained by the Javari and Juruá rivers with their affluents flowing into the Solimões River, affluent of the Amazon.

6.2.2 STRATIGRAPHY AND SEDIMENT SEQUENCES

The Acre Basin suffered the influence of the Andes uprising and shows therefore from the Middle Mesozoic onwards a sequence different from that of the other Amazonian sub-basins. The stratigraphical column is presented in Table 6.1 (Miura, 1972; Feijó and Souza, 1994), together with the sediment types of which it is composed. The sequence is incomplete due to the Andean tectonics that caused the erosion of the greater part of the Paleozoic units. The known sequence is chiefly known based on well sections.

Table 6.1: Lithostratigraphic sequences in the Acre Basin.

Cenozoic oscillatory-emergent episode, since Cenomanian	
Solimões Formation	cs
Jaquerana Group	
Ramon Formation	ca, sh
Divisor Formation	ss
Rio Azul Formation	cs, sh
Moá Formation	ss
Jurassic-Cretaceous submergent episode, until Cenomanian	
Juruá-Mirim Formation	ss, st, ev, bas
Late Carboniferous-Triassic oscillatory-emergent episode	
Rio de Moura Formation	ss, st, cl
Cruzeiro do Sul Formation	st, cs, ca, ev
Apui Formation	cgl
Older tectonic-sedimentary episodes	
no record	

Abbreviations of lithotope symbols: cgl, conglomerate; sh, shale; ev, evaporites; diam, diamictite; ls, limestone; si, silex; ss, sandstone, cr, calcirudite; bas, basalt; st, siltstone; ca, calcarenite; cs, claystone; cl, calcilutite.

The tectonic-sedimentary episodes from Cambrian to Early Carboniferous left no record in the basin. The Late Carboniferous-Triassic oscillatory-emergent episode shows a sequence of conglomerates at the base, and limestones, dolomites, some evaporites, shales and sandstones at the top, united into three formations correlative with corresponding Solimões Basin units, and with ages confirmed by Daemon and Contreras (1971).

From the Jurassic-Cretaceous submergent episode onwards the Acre Basin assumes its proper character when the Andean orogenic phases became evident and stronger. The Jurassic and Early Cretaceous are recorded by the Juruá-Mirim Formation, with some sandstones, chiefly siltstones and evaporites in the Jurassic; a diabase intrusion has also been determined. The age has been confirmed by fossils (Uesugui et al., 1994). Upward the Late Cretaceous-Cenozoic oscillatory-emergent episode got most of the Andean influence, presenting a thick section of over 3000 m of a variety of sandy and microclastic deposits. The Late Cretaceous (Cenomanian) to Paleocene section is composed of four formations, distinguished after their lithic character, and united in the Jaquirana Group. The basal Moa Formation is essentially a fine-medium grained sandstone, the Rio Azul Formation is characterized by claystones and shales, the Divisor Formation is again sandstone, and the upper Ramon Formation is composed of shales and calcarenites. These units lie conformably one upon the other. Separated by an unconformity follows the Eocene-Pliocene Solimões claystone formation, as was defined for the Solimões Basin.

Correlate deposits of most of these sequences are also found in the Pastaza-Oriente Basin of Peru and the Oriente Basin of Equador (see Fig. 6.3).

6.2.3 DEPOSITIONAL ENVIRONMENTS AND BASIN HISTORY

The first record that appears in the Acre Basin belongs to the Late Carboniferous-Triassic oscillatory-emergent episode. The basal Apuí Formation conglomerates have been accumulated in alluvial fans. The Cruzeiro do Sul Formation fine clastic deposits, calcarenites and some evaporites, and the Rio do Moura Formation sandstones, siltstones and calcilutites represent a shallow marine to neritic, sometimes partly restricted environment under possibly transgressive conditions. The regressive section is absent, probably due to later erosion.

From the Jurassic-Cretaceous submergent episode only a Jurassic section is present, with chiefly clastic sediments deposited in fluvial and lacustrine realms, with a few evaporites suggesting an inland to coastal sabkha environment that may point to a certain marine influence.

During the Late Cretaceous-Cenozoic oscillatory-emergent episode, the Acre Basin received the detritus of the rising Andes chain, although in a more distal environment. The clastic sequences of the Jaquirana Group have been deposited in alluvial fans, rivers and deltas, with some marine contribution. The fossil assemblages determinative of age, consist chiefly of palynomorphs and very few small mollusks (Uesugui et al., 1994). In this way, the basin became a real continental foreland basin that continued its development during the Cenozoic. The Tertiary Solimões Formation clayey sediments have been apparently deposited in meandering rivers, marshes and shallow lakes.

6.3 SOLIMÕES BASIN

6.3.1 GENERALITIES

The Solimões, formerly called Upper Amazonas Basin extends for about 40,000 km^2 in Amazonas State and is limited in the north by the Guyana shield and in the south by the Central Brazilian or Guaporé shield. The western and eastern limits are geologically defined by the Iquitos and Purus arches, respectively (Fig. 6.2). Within the basin, two sub-basins may still be recognized, that of Jandiatuba and that of Juruá, separated by the Caruari arch which exercised a strong control on the chiefly pre-Carboniferous sedimentation. The Juruá sub-basin is well known because of intense research for oil. The Jandiatuba sub-basin is poorly known, partly due to restrictions on prospect in native American-Indian areas.

6.3.2 STRATIGRAPHY AND SEDIMENT SEQUENCES

The stratigraphical framework of the Solimões Basin, presented in Table 6.2 with its composing sediment types, has been based chiefly on the papers of Caputo et al. (1972), Silva (1988) and Eiras et al. (1994).

The first record comes from the Cambrian-Early Ordovician submergent episode, with the sandstones and shales of the Benjamin Constant Formation, of Middle Ordovician age as suggested by acritarchs and chitinozoans.

The Late Ordovician-Silurian oscillatory-emergent episode is represented by shales and sandstones taken together into the Jutaí Formation, of Late Silurian to earliest Devonian age.

The submergent episode of Devonian-Early Carboniferous age is also presented by a thin section. The Marimari Group is composed of two lithological formations. The basal Ueré Formation is characterized by chiefly siliciclastic sediments with silex and silicious fossils. The unit is laterally interdigitated with the Jandiatuba Formation with shales and siltstone and sandstone intercalations. A small occurrence of calcirudites has also been determined. The age of the formations ranges between Middle Devonian (Eifelian) and Early Carboniferous (Tournaisian).

During the oscillatory-emergent period that lasted from Late Carboniferous to Permian in the Solimões Basin, the representative Tefé Group shows a variety of sediment types. The sequence starts with the chiefly sandstone section of the Juruá Formation, passing conformably into the carbonate–evaporite section of the Carauari Formation that also passes conformably into the siliciclastic sequence of siltstones and shales with a few sandstone intercalations of the Fonte Boa Formation. Within the whole section basalt intrusions are present. The group is dated as Late Carboniferous to earliest Permian.

Table 6.2: Lithostratigraphic sequences in the Solimões Basin.

Cenozoic oscillatory-emergent episode, since Cenomanian	
Javari Group	
Solimões Formation	cs, ss
Alter do Chão Formation	ss
Jurassic-Cretaceous submergent episode, until Cenomanian	
no record	
Late Carboniferous-Triassic oscillatory-emergent episode	
Tefé Group	
Fonte Boa Formation	ss, st, bas
Caruari Formation	sh, cl ev, bas
Juruá Formation	ss, ev
Devonian-Early Carboniferous submergent episode	
Marimari Group	
Jandiatuba Formation	sh, ss, diam
Uerê Formation	sh, si
Late Ordovician-Silurian oscillatory-emergent episode	
Jutaí Formation	sh, ss
Cambrian-Early Ordovician submergent episode	
Benjamin Constant Formation	ss, sh

Lithotope symbols: see Table 6.1.

The next record is that of the Late Cretaceous-Cenozoic oscillatory-emergent episode, being the intermediate submergent episode sediments absent. The sequence is represented by two formations, Alter do Chão and Solimões, of the Javari Group. The lower sandstone unit has been dated as Late Cretaceous. The upper claystone section, present in the Acre Basin, is separated from the underlying sandstones by an erosional unconformity; the unit has a Miocene-Pliocene age.

The sediment sequences of the various episodes are separated between them by erosional unconformities, often followed by long-lasting hiatuses in sediment deposition.

6.3.3 DEPOSITIONAL ENVIRONMENTS AND BASIN HISTORY

The history of the Solimões Basin started in the Early Paleozoic in part as a platform cover of the Guaporé shield in the south, upon which the sea encroached during submergent episodes and high sea-level oscillations during the other episodes.

The Ordovician section belongs still to the first Phanerozoic submergent episode and its sediments became accumulated in shallow marine and wave-dominated littoral environments passing upward gradually into a tidal realm.

The Late Silurian oscillatory-emergent episode sediments show a transgression of the sea over a narrow wave-dominated platform. The sea proceeded from the west towards the east.

The transgression–regression phase of the submergent Devonian-Early Carboniferous episode lasted from Eifelian until Tournaisian. The sediment sequences of the Marimari Group have an evidently neritic to shallow marine character, passing into transitional and littoral realms. The fauna explosion of sponges, found as spicules, has been caused by upwelling of cooled ocean water against the Carauari arch (Silva, 1988).

The oscillatory-emergent episode of Late Carboniferous and Early Permian record an initial fluvial-deltaic environment with eolian reworking. Passing upward the marine influence becomes ever more evident with various oscillations, being sometimes rather restricted as proved by the evaporites. The upper formation has a typically regressive character passing from shallow marine to littoral sabkha and continental desertic environments.

The last sedimentation period also under oscillatory-emergent circumstances during Late Cretaceous and Tertiary is marked by continental realms, in the beginning of alluvial fans and low-sinuosity river plains, and later of meandering rivers, marshes and lakes developed in abandoned river channels.

6.4 AMAZONAS BASIN

6.4.1 GENERALITIES

The Amazonas Basin, formerly known as the Middle and Lower Amazonas sub-basins, is a rather narrow trench that occupies a surface of approximately 500,000 km^2, covering parts of Amazonas and Pará States. It is limited in the north by the Guyana shield and in the south by the Central Brazilian shield. In the west the basin is geologically limited by the Purus arch, separating it from the Solimões Basin, and in the east by the Gurupá arch, separating it from then Amazon Mouth Basin (Fig. 6.2). The sedimentary deposits crop chiefly out alongside the geomorphologically well marked northern and southern basin margins.

6.4.2 STRATIGRAPHY AND SEDIMENT SEQUENCES

The stratigraphical column of the Amazonas Basin and the various sediment types present in it, are mentioned in Table 6.3, based on the publications of

Table 6.3: Lithostratigraphic sequences in the Amazonas Basin.

Cenozoic oscillatory-emergent episode, since Cenomanian	
Javari Group	
Solimões Formation	cs
Alter do Chão Formation	ss
Jurassic-Cretaceous submergent episode, until Cenomanian	
no record	
Late Carboniferous-Triassic oscillatory-emergent episode	
Tapajós Group	
Andirá Formation	sh, st, ss, bas
Nova Olinda Formation	ev, bas
Itaituba Formation	ls, bas
Monte Alegre Formation	ss, st, sh
Devonian-Early Carboniferous submergent episode	
Curuá Group	
Faro Formation	ss, bas
Oriximina Formation	ss, st, bas
Curiri Formation	sh, st, diam, bas
Barreirinha Formation	sh, bas
Urupadí Group	
Ererê Formation	ss, st, sh
Maecuru Formation	ss, st
Late Ordovician-Silurian oscillatory-emergent episode	
Trombetas Group	
Manacapuru Formation	ss, st, bas
Pitinga Formation	sh, diam, bas
Nhamundá Formation	ss
Autas-Mirim Formation	ss, sh
Cambrian-Early Ordovician submergent episode	
no record	
Late Cryogenian-Vendian oscillatory-emergent episode	
Purus Group	
Acari Formation	ls, sh
Prosperança Formation	ss, cgl

Lithotope symbols: see Table 6.1.

Caputo et al. (1972) and Cunha et al. (1994). Biozones and ages have been determined by Beurlen (1970), Daemon and Contreras (1971), and Dino and Uesugui (1994).

The Proterozoic basement composed of granitic and metamorphic rocks has been covered, probably during the Late Cryogenian-Vendian oscillatory-emergent episode, by various types of clastic sediments and some limestones of the Purus Group, with two formations (Prosperança and Acari), determined in wells. No record is present of the Cambrian-Early Ordovician submergent episode.

The following Late Ordovician-Silurian period is recorded by the clastic lithostratigraphic Trombetas Group, subdivided into four formations, all composed of sandstones, siltstones, and shales. Within the Llandoverian-Ludlovian Pitinga Formation there occur some diamictites. One shale section contains Silurian graptolites and mollusks (Beurlen, 1970). Ages have been determined chiefly by chitinozoans (Grahn and Paris, 1992).

The submergent episode of Devonian-Early Carboniferous age presents a rather great number of formations established after their different lithotypes. These formations have been taken together into two groups (Urupadi and Curuá). All sequences are composed of clastic deposits of various grain sizes. A few basalt intrusions occur too. The fossil content is rather rich and varied, with mollusks, echinoderms, and foraminifers.

During the oscillatory-emergent period that lasted from Late Carboniferous to Triassic, deposition of sandstones, microclastics, evaporites, and a few limestones occurred until the Late Permian, again with a subdivision into four formations, taken together into the Tapajós Group. There does not exist any record of Triassic sediments.

From the Jurassic-Cretaceous submergent episode also no record exists in the Amazonas Basin. Only from the Late Cenomanian onwards, two formation, the lower one with chiefly sandstones and a few intercalations of coarser and finer clastic deposits, and the upper one with essentially fossiliferous clays, represent the last part of the Cretaceous submergent episode passing conformably into the most recent Cenozoic oscillatory-emergent period. The uppermost unit shows only scarce outcrops.

The formation of each episode are separated between them by erosional unconformities and hiatuses in the sedimentation; within the proper episodes the lithostratigraphic units lie conformably one upon the other, distinction made either through changes in lithology, or by different ages. Most age determinations have been made through palynological zonation (Daemon and Contreras, 1971).

6.4.3 DEPOSITIONAL ENVIRONMENTS AND BASIN HISTORY

The middle, rather narrow segment of the present-day Amazon River drainage basin that is occupied by the Amazonas Basin shows its first sediment accumulation record still in the Precambrian, during the Late Cryogenian-Vendian oscillatory episode. Upon the basement rocks, the Purus Group sediments have been deposited in alluvial fan and river environments, passing upward into a tidal realm. Because the sequence is incomplete, no more details could be obtained.

Anyhow, the basin records chiefly the Paleozoic sedimentation periods. The first signs of Paleozoic deposits are found at the westernmost flank of the basin where some outliers of the Silurian marine shales, belonging to the Trombetas Group, occur. Glacial and marine sediments, with the transgression proceeding from the east, attained in onlap the Purus arch. The various clastic sediments have been deposited in neritic and glacial-marine environments.

The sediments of the Devonian-Early Carboniferous submergent episode have been divided into two lithostratigraphic groups, of which the lower one shows in the beginning fluvial and deltaic deposits covered by neritic siltstones. After a short regression, the sea returned to invade the area causing the deposition of the upper lithostratigraphic group. Its base is recorded by a dark-coloured shale unit representing an anoxic marine environment, passing upward into a more open marine realm and finishing with a rather thick regressional section of lagoonal, deltaic and alluvial plain fluvial provenance. The marine deposits are rich in mollusks, chiefly found in the siltstones, and in the regressive sections there occur rests of plants and spores (Daemon and Contreras, 1971). The marine invasion lasted from Eifelian until Tournaisian, and the whole geological situation suggests that the Amazonas Basin Devonian is nothing more than an outlier of the important transgression more evident towards the east as, for instance, in the Northeast Brazilian Parnaíba Basin (see Chapter 7).

From the Late Carboniferous-Triassic oscillatory-emergent episode, a sedimentary record is only present until the Late Permian. The basin seems to have remained a topographically low-lying depression, with fluvial and lacustrine environments (Beurlen, 1970). In this depression, the sea could invade in short-lived transgressions in Late Carboniferous and Early Permian times. Due to these oscillations some parts of the depression turned to be closed small basins for a while, permitting the precipitation of evaporites during the then reigning arid climate. The changing salinity rates are confirmed by the fossil fauna of mollusk species resistant to such changes. The whole section is represented by sandstones, shales, limestones, and evaporites, chiefly anhydrite. It has been subdivided into four lithologically different units of which the uppermost Andirá Formation is characterized by fluvial sandstones with which the regression became completed. Then, during the Triassic the area became one of non-deposition and erosion. The link with the eastward occurring Parnaíba Basin seems to have continued as suggested by the similarity of their respective correspondent sediment sequences.

There seems to exist no record of the Jurassic-Cretaceous submergent episode in the Amazonas Basin. Only from Cenomanian onwards clastic sediments have been determined, apparently representing already the beginning of the Cenozoic oscillatory-emergent period. The Alter do Chão Formation covers extensive areas at the surface with fluvial clays, siltstones, and sometimes, silicified sandstones. The lower Cretaceous section has been dated by palynomorphs and the upper Tertiary section yielded teeth and bone fragments of vertebrates. The age of this formation reaches up to the Pliocene. At the westernmost flank near the Purus arch there occurs still a small representation of the clayey Solimões Formation, of Miocene-Pliocene age, as a distal record of the Andean foreland detritus. The Quaternary is represented by fluvial infillings of the present-day Amazon River that turned its drainage towards the Atlantic Ocean due to the Andes uprising in the west.

6.5 AMAZON MOUTH BASIN

6.5.1 GENERALITIES

The easternmost part of the Greater Amazonas Basin is represented by the Amazon Mouth Basin that extends onto the continental shelf and includes the big well-known Amazon River cone, with a surface of about 280,000 km^2. It developed since the latest Cretaceous when the Equatorial Atlantic Ocean started its opening (Szatmari et al., 1987).

The basin is separated from the eastward occurring Amazonas Basin by the Gurupá arch and in the southeast from the Parnaíba Basin by the Tocantins arch. Due to the rifting processes which determined the opening of the ocean, the basin is characterized by a central graben, called Marajó basin, with a SW-NE branch, the Mexiana sub-basin and a NW-SE branch, the Limoeiro and Cametá sub-basins (Fig. 6.4) in which most of the sediments accumulated.

6.5.2 STRATIGRAPHY AND SEDIMENT SEQUENCES

In Table 6.4, the stratigraphical column and the sediment types of the different formations have been presented, after data from Schaller et al. (1971), Rezende and Ferradaes (1971) and Brandão and Feijó (1994). The respective ages, as far as they are known, have been suggested by Dino (1994). Due to the important lithological differences three areas have been distinguished: coast, central part – shelf, and deep-sea. The sediments have been deposited unconformably either upon the Precambrian basement, or upon sedimentary outliers of the Amazonas Basin Paleozoic sequences.

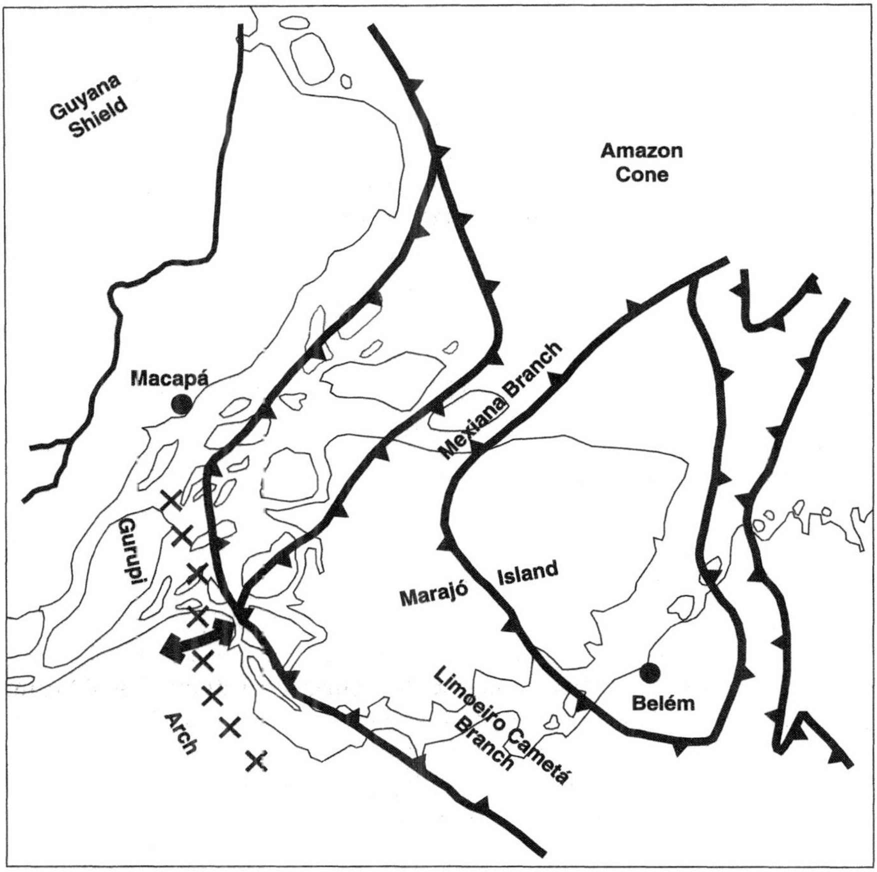

Fig. 6.4: Main structural framework of Amazon Mouth Basin.

The oldest found section in the Amazon Mouth Basin is the Calçoene Formation, composed of tholeiitic basalts, intercalated with fine to medium-grained sandstones. The sediments are non-fossiliferous, but the basalts show ages between 186 and 222 Ma, that is end Triassic to beginning Jurassic. Thus, the sequence belongs still to the Late Carboniferous-Triassic oscillatory-emergent episode.

From the Jurassic-Cretaceous submergent episode derive the shales and sandstones of the Caciporé Formation of Aptian-Albian age. The unit occurs only in the grabens of the shelf and deep-sea area, deposited during the taphrogenical active period thus attaining a considerable thickness of up to 7000 m.

The Late Cretaceous sequence that already seems to belong to the Cenozoic oscillatory-emergent episode, is composed of one unit that occurs only on the shelf and in the deep-sea area. The section presents a basal clayey complex

Table 6.4: Lithostratigraphic sequence of Amazon Mouth Basin.

Cenozoic oscillatory-emergent episode, since Cenomanian		
	Pará Group	
coast	Tucunaré Formation	ss
shelf	Pirarucu Formation	st, ss, cs
deep-sea	Orange Formation	st, cs, sh
coast	Marajó Formation	ss
shelf	Amapá Formation	cr, ca, cl
deep-sea	Travosas Formation	sh
shelf, deep-sea	Limoeiro Formation	cs, st, ss
Jurassic-Cretaceous submergent episode, until Cenomanian		
shelf, deep-sea	Caciporé Formation	sh, ss
Late Carboniferous-Triassic oscillatory-emergent episode		
shelf	Calçoene Formation	ss, bas
older tectonic-sedimentary episodes		
	no record	

Lithotope symbols: see Table 6.1.

passing upward in the shelf area into sandstones and siltstones. Biostratigraphical data suggest an age between Cenomanian and Paleocene (Regali et al., 1974). Upward in conformable succession follows an intermediate sequence, Paleocene to Miocene in age, with the common trend of coastal sandstones, stratified detrital limestones, massive biohermal limestones, marls, and calcirudites (talus detritus) on the shelf, and deep-sea clays to shales. This section shows a different formation name for every lithotype (see Table 6.4). The uppermost Late Miocene-Holocene sequence that rests conformably on top of the foregoing section, shows a trend different from that which occurs commonly on the Atlantic coast. From the continent seaward, one finds coarse sands in the coastal area, fine sands, silts, and clays in the middle shelf area, and chiefly clays in the deep-sea on the slope of the Amazon cone. This trend has also been subdivided into three formations after the trend's lithology.

6.5.3 DEPOSITIONAL ENVIRONMENTS AND BASIN HISTORY

The sediments of the lowermost Triassic section determined in the basin, representing an oscillatory-emergent period, join fluvial, lacustrine and eolian facies types which reflect the arid to desert climate that characterizes the Brazilian sedimentary basins of that time. The unit seems to be an extension of the geocratic continental realm of the Parnaíba Basin to the east, and thanks its preservation to subsidence in a fault depression.

The proper Amazon Mouth Basin developed since the Late Cretaceous in connection with the already open North Atlantic Ocean and the then opening Equatorial Atlantic Ocean. Gravity faulting was the principal cause of the subsidence, which created the structural framework of the basin. This framework comprises a chain of major grabens among which the Limoeiro-Cametá and Mexiana grabens are the most prominent (Fig. 6.4), with marginal step faults suggesting a rifting process that started at the end or after the last basic magmatic episode in the area (Rezende and Ferradaes, 1971).

The taphrogenic processes of the ocean opening started in the Aptian-Albian (Szatmari et al., 1987), still within the Jurassic-Cretaceous submergent episode. That is why the correspondent Caciporé Formation is represented only by clastic sediments deposited in fluvial, lacustrine and deltaic systems, with the contribution of turbidites in the subsiding rifts. The about 125 Ma old basalts (Thomaz Filho et al., 1974) constitute the base of the sediment section.

After a short erosional period sedimentation started again in the Late Campanian during the regression at the end of the Jurassic-Cretaceous submergent episode and its transition into the Cenozoic oscillatory-emergent episode (Aguiar et al., 1966). The correspondent Limoeiro Formation is composed of some conglomerates and conglomeratic sandstones deposited in alluvial and fault-scarp fans, occurring near the great fault lines. The more distal part is represented by various-sized sandstones with some claystone intercalations representing typical fluvial accumulations in braided and low-sinuosity streams, passing oceanward into neritic siltstones and deep-sea claystones.

Follows conformably on top of the foregoing section the Marajó, Amapá, and Travosas formations, these distinguished on the base of lithological differences. The basic components of the Marajó Formation are sandstones, accumulated in low-lying river plain, littoral, and shallow marine realms as is confirmed by the fossil content. The influence of continental supply is rather important due to the drainage of the present Amazon River that started its eastward course at that time. The clastics pass laterally towards the continental shelf into the detrital carbonate sequence of the Amapá Formation. The offshore Travosas Formation shales accumulated on the continental slope and in the deep-sea of the prograding Amazon cone.

Conformably on top of this sequence there occur still three lithostratigraphic units, the Tucunaré Formation with a sandy sequence, the Pirarucu Formation composed of argillaceous sediments, and the Orange Formation with silts and clays. Many of these sediments have a carbonate component. From the coast oceanward the environments grade one into the other in the following trend: fluvial – paralic-clastic shelf – fine clastic slope and deep-sea.

The carbonate deposition finished definitively at the end of the Miocene. From this time onwards clastic sections became deposited, chiefly due to the great influx of clastic material from the Amazon River. This sedimentation was the cause of an active cutting and infilling of many submarine canyons and channels. The sediment transported along these channels makes up the bulk of the Amazon cone that seems to be composed of numerous coalescent fans.

6.6 INTEGRATION OF DATA AND CONCLUSIONS

From the foregoing it may be concluded that the present-day Amazon drainage basin has very little to do with the geology of the Greater Amazonas Basin that appears to be neither a continuous rift valley nor a syneclise. Although the first rifting processes which finally opened the basin, could have occurred already in Archean times (Loczy and Ladeira, 1976), the more intense rifting started only in the beginning of the Phanerozoic, when in the Cambrian the Late Proterozoic supercontinent broke up, and North America drifted away from the remaining Gondwana supercontinent in the south (Valentine and Moores, 1974). Herewith the southern Guaporé or Central Brazilian shield drifted more towards west than the northern Guyana shield by differential processes with transcurrent faulting. Then the rift became gradually deepened and wider, transforming it into an important sedimentary basin.

As Beurlen (1970) observed already, the Greater Amazonas Basin presents geotectonically three completely different segments.

The westernmost, very large segment comprises in Brazil the Acre and Solimões basins, limited in the E by the Purus arch, and extending westward into the present Andean foreland basins of Peru and Equador (Fig. 6.3). This area formed in the Paleozoic and Early Mesozoic the northwestern part of the Guaporé shield, not really belonging to the Paleozoic sedimentation area of the Amazonas Basin, exception made during the Late Carboniferous, an oscillatory-emergent period.

The rather narrow E–W trending furrow that is represented by the Amazonas Basin *s.s.*, between the Purus and Gurupá arches is characterized by mainly Paleozoic deposits.

The easternmost segment is the Amazon Mouth Basin that started its development when the Equatorial and South Atlantic Oceans opened since the Late Mesozoic.

In fact, the Paleozoic sedimentary depression of northern Brazil shows no relation whatever with the present-day Amazon basin, such that, as sometimes is ascertained, a Paleozoic Amazonas syneclise did never exist. The Amazonian Paleozoic is nothing more than a preserved part of a bigger area belonging to the extensive epicontinental transgression and the wide-spanning epeirogenetic subsidence which resulting deposits have been preserved in the low-lying Amazon depression. During the Mesozoic, this depression did not suffer a subsidence worth mentioning and almost no sedimentation occurred. The Early Mesozoic is completely absent, whether due to non-deposition, or to later erosion, being uncertain. At the time of the Jurassic-Early Cretaceous basalt outpours (Thomaz Filho et al., 1974; Almeida, 1986), the depression suffered most probably some doming. The actual Amazon depression is because of that, a young structure showing a young geological development as is recorded by the Cretaceous lithostratigraphic units, with a deposition from the Early Cretaceous onwards in the Acre Basin and from the Campanian onwards in the other parts of the depression.

The Acre and Solimões basins represent the platform covers of the western Guaporé shield and got most of their sediments derived from the relief of that time. Only during the Cambrian-Early Ordovician submergent episode, some marine deposition took place at the end of the period. The oscillatory-emergent Late Ordovician-Silurian episode is recorded by a sequence of clastic continental sediments, such as is common for this type of period.

The Devonian-Early Carboniferous submergent episode is present in great extensions in the Solimões and Amazonas basins, everywhere in clastic facies types and an absence of carbonate sequences. The thalassocratic phase lasted from Emsian to Frasnian. Towards the end of the Devonian and during the Early Carboniferous an uplift of the Guaporé arch, between the Amazonas and Amazon Mouth basins, changed the environment into a regressional continental area, with interior lake basins. Due to the geographical position of South America in that time, cold climates prevailed as is suggested by lithotypes and fossil assemblages.

The Hercynian orogenetic period has been felt in two phases during the Late Carboniferous-Triassic oscillatory-emergent episode, in the beginning and at the end. In between a sediment sequence accumulated composed of clastic deposits, some limestones, and important anhydrite and halite deposits. The episode being a geocratic phase shows continental realms of rivers and lakes, having occurred also periodic ingressions. The marine basins became restricted and closed, with strong evaporation in closed basins and sabkhas. The climates were dry when the South American continent drifted northward, passing through the arid climate belt of middle latitudes. The transgressions appear to have proceeded from the west, while the eastern part of the Amazonas Basin remained a topographically rather high area. Between the Early Permian and Late Triassic, a first volcanic cycle with basalt extrusions affected the whole region and has been associated to the opening of the North Atlantic Ocean and the separation of North America from Gondwana (Thomaz Filho et al., 1974).

Due to the emergent character of the episode, many older sequences seem to have been eroded, while the area turned rather flat and low-lying.

A second volcanic cycle with basalt intrusions and outpours took place during the Jurassic and Early Cretaceous, already within the Jurassic-Cretaceous submergent episode. This cycle has been associated with the opening of the Equatorial and South Atlantic Oceans and the separation of South America from Africa. In the Acre Basin in the west the first signs of the Andean uprising became felt in about the mid-Cretaceous. In the Solimões and Amazonas basins, no record of this episode has been registered, meaning that probably the area had no connection with the sea, being separated from the Atlantic by the Gurupá arch and the doming of this region caused by the basalt intrusions. Only in the Amazon Mouth Basin area, the beginning of taphrogenesis in Aptian-Albian times resulted in a clastic sediment fill.

Very important for the sedimentation in the Amazon furrow is the still lasting Late Cretaceous-Cenozoic oscillatory-emergent episode. The present-day Amazon River and its affluents began to turn their drainage courses towards E, due to the uprising of the Andean mountain chain. The Acre Basin became a distal foreland basin reached by the medium to fine-grained clastic detritus coming from the west. Towards E, the Solimões and Amazonas basins received also part of this Andes detritus, supplied by the Amazon and its affluents, still as sandy deposits in the Tertiary and as clayey sediments from the Pliocene onwards. But it is the Amazon Mouth Basin that got the greater part of this detritus that bypassed the upper and middle Amazon River course, as well as the detritus supplied by the affluents from the crystalline shield areas in the north and in the south which formed the extensive Amazon cone. Beginning already in the Cenomanian, increasing during the Early and Middle Tertiary, most of the sediment accumulated from the Late Miocene onwards. These passive margin sequences show the common system tracts from the continent oceanward as follows: river and fan-delta clastics – shelf carbonates – slope and deep-sea clays and silts. Some sandy turbidites are also present in this last realm. Rossetti et al. (2004) considered the Plio-Pleistocene history of the basin as a hypothesis for explaining its actual biodiversity.

Integration of data show for the Greater Amazonas Basin and its individual sub-basins the perfect cyclic sequence of tectonic-sedimentary episodes as presented in Chapter 3. The individual depositional sequences are always separated by erosional unconformities. The present-day Amazon River basin is a young phenomenon established during the uprising of the Andes mountain chain, forcing the drainage towards E and producing a lot of detritus which greater part constructed the well-known Amazon cone in the Atlantic Ocean.

7. SEDIMENTARY BASINS OF BRAZILIAN BORBOREMA AND SÃO FRANCISCO TECTONIC PROVINCES SINCE MESOPROTEROZOIC

J.M. MABESOONE AND V.H. NEUMANN

7.1 INTRODUCTION

The Borborema and São Francisco tectonic provinces are found in the northeastern and eastern regions of Brazil and form part of the South American craton that occupies more than half of the continent (Figs. 7.1 and 7.2), including the whole of Brazil. The Borborema Province is the westernmost section of a Pan-African mobile belt that extends from the Amazon craton in the west up to the Arabian-Nubian shield in the east. As mobile belt, it is very extensive and composed of various terranes amalgamated since the Paleoproterozoic. The São Francisco Province constitutes the western part, in Brazilian territory, of the big São Francisco-Congo/Kasai-Angola shield. In both geotectonic units three main types of sedimentary basins are distinguished: large-size syneclises and small-size centroclines (after the definition of Bates and Jackson, 1987) as well as rifts which developed in cyclically repeated periods of the earth's geological history, probably since the beginning of the Proterozoic.

7.2 TECTONIC HISTORY

For a better understanding of the geological history of the sedimentary basins within and at the margins of the here studied areas, the successive mergers and disruptions of supercontinents through time, in which the South American cratonic areas were involved, need some further consideration. As the cyclic development of earth's geological phenomena suggests, also the formation of supercontinents is cyclic, where individual continents amalgamate preferentially during oscillatory-emergent tectonic-sedimentary episodes, and these thus formed supercontinents disrupt during subsequent submergent episodes. However, it appears that not every merger produces big supercontinents, at least not until the Neoproterozoic (Rogers, 1996). In the same way, dismemberment of

Cyclic Development of Sedimentary Basins
J.M. Mabesoone and V.H. Neumann, editors.

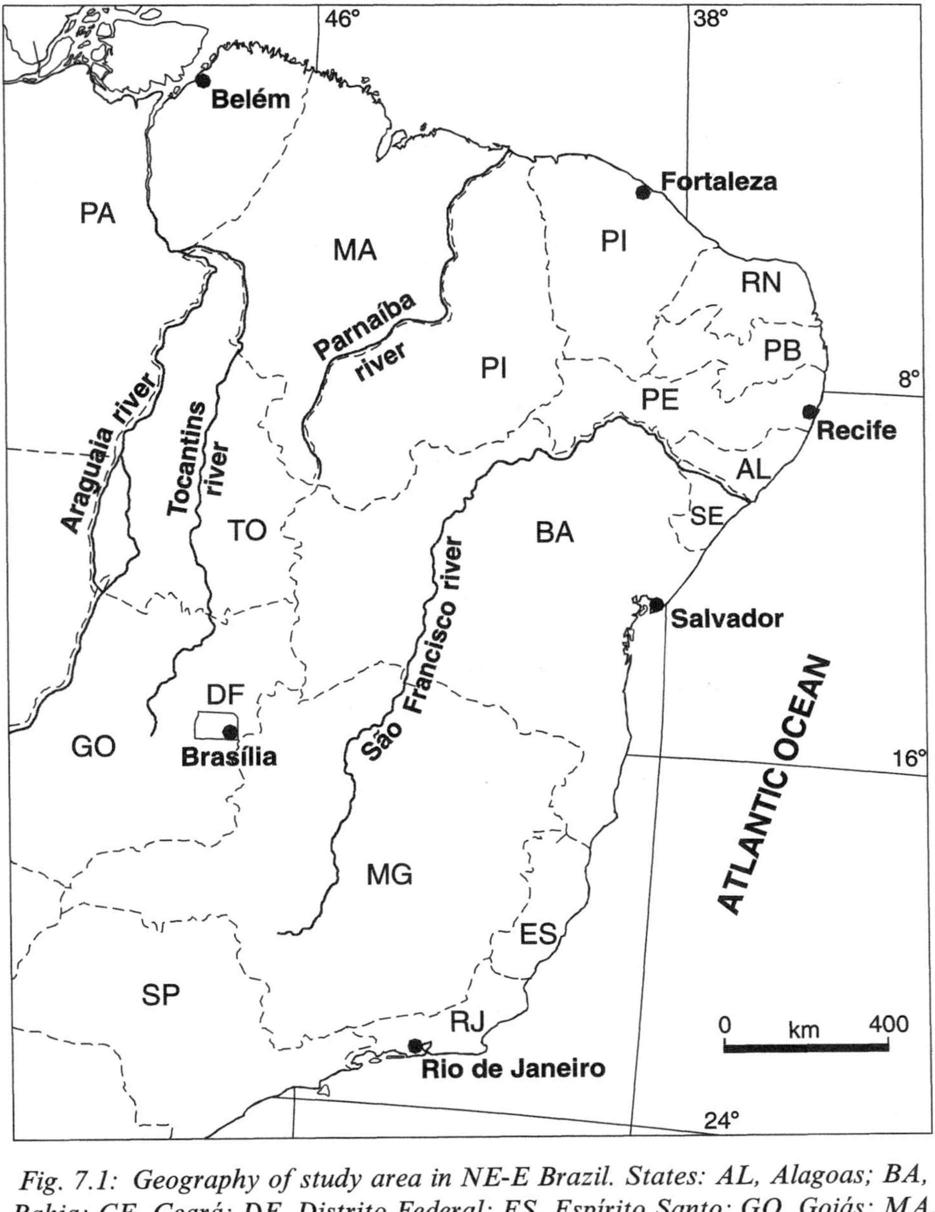

Fig. 7.1: Geography of study area in NE-E Brazil. States: AL, Alagoas; BA, Bahia; CE, Ceará; DF, Distrito Federal; ES, Espírito Santo; GO, Goiás; MA, Maranhão; MG, Minas Gerais; PA, Pará; PB, Paraíba; PE, Pernambuco; PI, Piauí; RJ, Rio de Janeiro; RN, Rio Grande do Norte; SE, Sergipe; SP, São Paulo; TO, Tocantins.

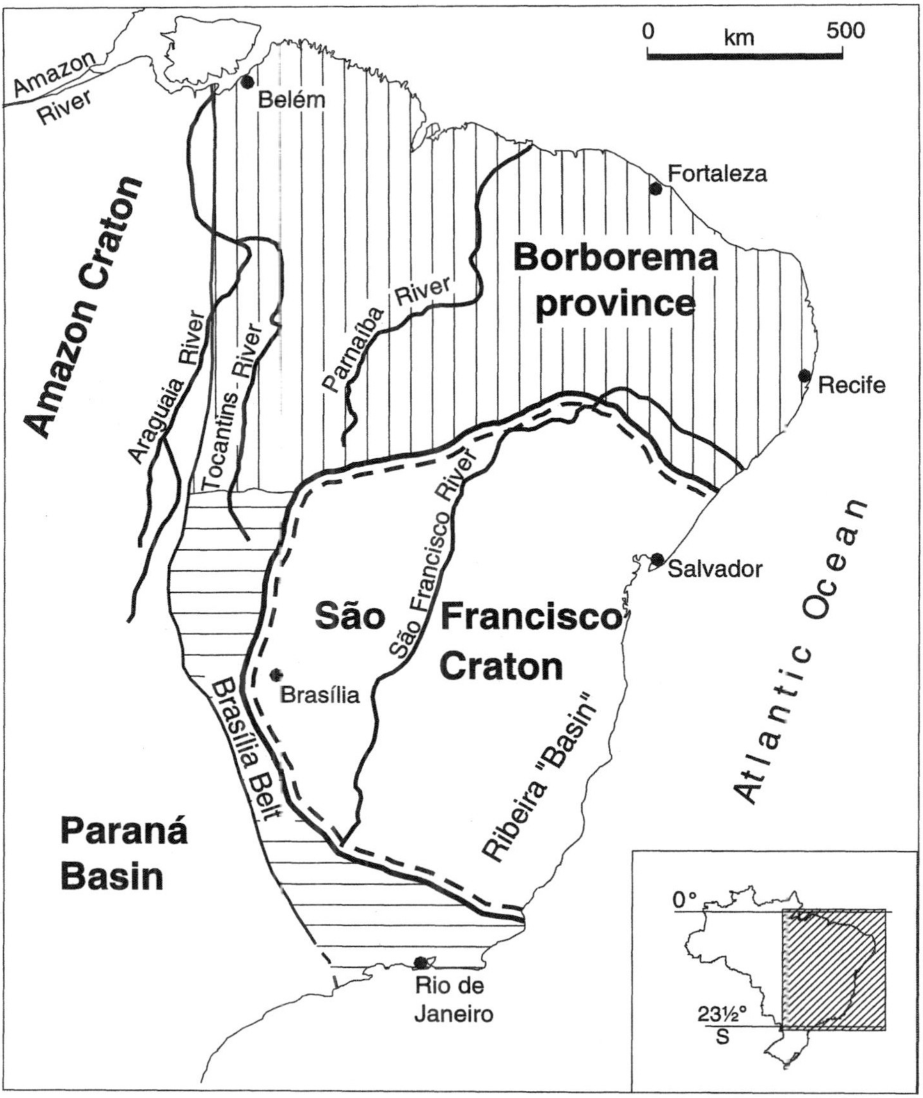

Fig. 7.2: Geographical position of Borborema and São Francisco tectonic provinces within the South American platform of Brazil.

greater units does not occur at every submergent episode. Some of this history may be concluded from the mobile belts between the shields. Their sedimentary infillings and their respective ages could suggest at what time the amalgamation of individual continents or shields, and dismemberments of the bigger units took place.

With respect to the names of the supercontinents, their assembly and their disintegration, supposed to have occurred during the Proterozoic, there exist still many uncertainties, conjectures, and inconsistencies. Rogers (1996), and Rogers and Santosh (2004) dealt with the problem in some extension, but seemingly did not arrive at a coherent conclusion. Therefore, here will be treated only the packing and rifting of the Gondwana and Pangea supercontinents and their effects on the sedimentary basin formation. The South American continent became a separate and independent unit only in the Late Mesozoic, when the Gondwana part of Pangea split up into individual fragments. At the eastern border of South America, the Atlantic marginal belt formed at that time.

7.3 GEOLOGY OF SÃO FRANCISCO AND BORBOREMA PROVINCES

The São Francisco Province is a cratonic area consolidated during the Paleoproterozoic. Its basement does not crop out at many places, but where it does, there occur granitic-gneissic complexes, strongly migmatisized. This basement has been dissected by numerous intrusions. Supracrustal structures of limited extent, affected by the rheomorphic remobilization of the basement, locally cover the gneissic-migmatitic complexes. These structures are composed of various types of metasediment, metavolcanics, and metamorphosed mafic and ultramafic intrusives in schist to low-grade amphibolite facies.

The Paleoproterozoic Transamazonian tectonic cycle (2.3–1.8 Ga; see Table 7.1) acted upon restricted parts of the province, developing fold belts which remainders have been recognized at various places. A sequence of 4000 m of sediment accumulated upon the basement, later folded, tectonized and metamorphosed. The upper stage of the Precambrian cover on the craton is represented by the deposits accumulated during the Neoproterozoic Caririano and chiefly Brasiliano tectonic cycles. The Phanerozoic cover has been preserved in Mesozoic coastal rift valleys, related to the opening of the Atlantic Ocean. In the interior part of the province there subsist remainders of an extensive but thin continental sediment cover of Late Cretaceous to Tertiary age, deposited due to relief formation, as well as Quaternary sediments of river floodplains.

The uncommon configuration of the São Francisco craton in its relation to the Congo craton, and the so-called Ribeira (Cordani et al., 1984) and West Congo belts between them, suggests that this belt could be nothing more than a Mesoproterozoic sedimentary basin implanted upon the entire craton. This supposition is supported by the Espinhaço rift that continues without interruption from the craton into the belt.

Table 7.1: Ages, orogenetic phases, supercontinent formation and glaciation epochs, since Paleoproterozoic.

Ga ±10 Ma	age: era	age: period	episode	orogenetic phase	stage	dismemberment merger	other phenomena
0	Phanerozoic				"modern"		glaciation
		Quaternary-Tertiary	O	alpine			
0.08							
		Cretaceous-Jurassic	S			dismemberment	
0.18							
		Triassic-Late Carboniferous	O	variscan (hercynian)		formation Pangea	strong glaciation
0.28							
		Early Carboniferous-Devonian	S				
0.38							
		Silurian-Late Ordovician	O	caledonian			glaciation
0.48							
		Early Ordovician-Cambrian	S				
0.58	Neoproterozoic						
		Vendian	O	brasiliano baikalian		formation Gondwana	strong glaciation
0.68							
		Late Cryogenian	S				
0.78							
		Early Cryogenian	O	caririano			glaciation
0.88							
		Tonian	S				
0.98	Mesoproterozoic				transition		
		Late Stenian	O	sunsas			glaciation
1.08							
		Early Stenian	S	grenvillian*			
1.18							
		Late Ectasian	O	espinhaço-uruaçuano			
1.28							
		Early Ectasian	S	elzevitian*			
1.38							
		Late Calymmian	O	rondoniano			
1.48							
		Early Calymmian	S	kilarnean*			
1.58							
		Late Statherian	O	rio negro			
1.68							
		Early Statherian	S	hudsonian*			
1.78	Paleo-proterozoic				stabilization		
1.88				moranian*			
1.98		Orosirian		trans-amazonica			
2.08							

Episodes: O – oscillatory-emergent
S – submergent

* after Canadian shield

The Borborema Province coincides with the so-called Nordeste fold belt, chiefly reactivated during the Brasiliano tectonic cycle. This region presents a complex arrangement, in mosaic, of different linear fold systems, mutually separated by elevated parts of the basement, related or not with faults; the province continues well into Africa, for instance in the Nigerian belts. The province became amalgamated during the Paleoproterozoic Transamazonian

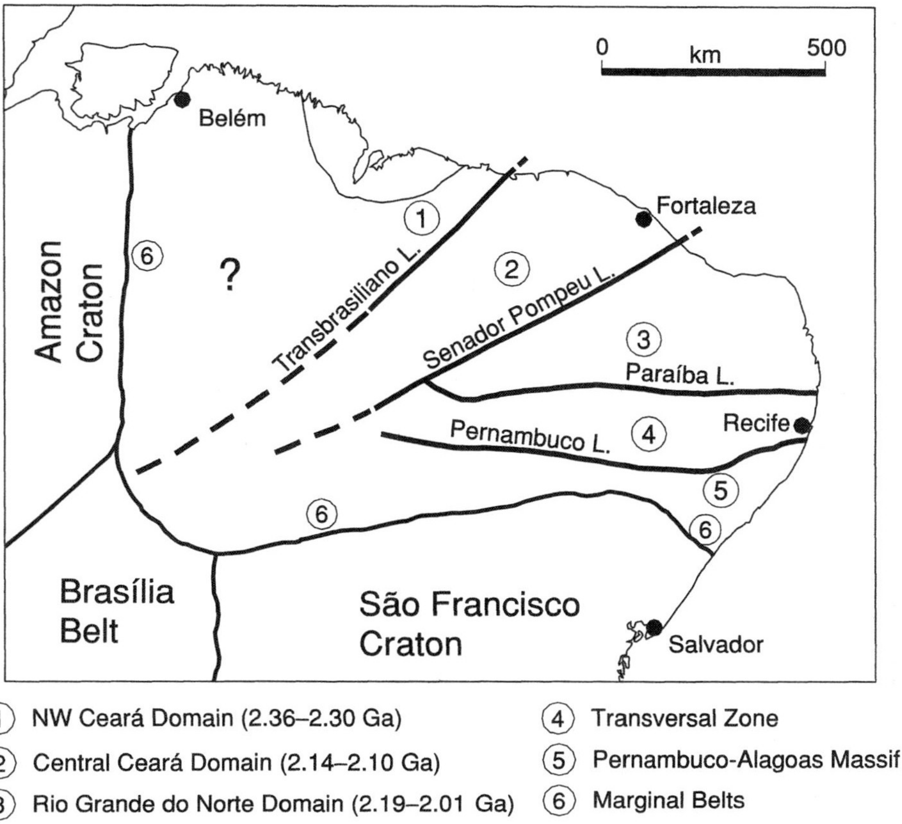

Fig. 7.3: Growth of Borborema Province during Paleoproterozoic: (1) NW Ceará domain (at 2.36–2.30 Ga), (2) Central Ceará domain (at 2.14–2.10 Ga), (3) Rio Grande do Norte domain (at 2.19–2.01 Ga), (4) Transversal zone, (5) Pernambuco-Alagoas massif, (6) marginal belts (after data from Fetter et al., 2000, and Silva Filho, 1995).

orogeny, with the merger of crustal blocks of probable Archaean age (Fig. 7.3). The intraplate events of the Mesoproterozoic maintained the thus formed geotectonic unit as a whole. Even during the more intense orogenic phases of the Neoproterozoic, of which the Brasiliano orogeny has been extremely strong, the Borborema area did not break up into smaller individual terranes (Fig. 7.4; Neves, 2001). However, the proper constitution of the province as a patchwork of terranes with different lithologies separated by important faults and lineaments, triggered the formation of a great number of sedimentary basins within the area, being those formed during the Mesoproterozoic of smaller size than those developed from the Neoproterozoic onwards.

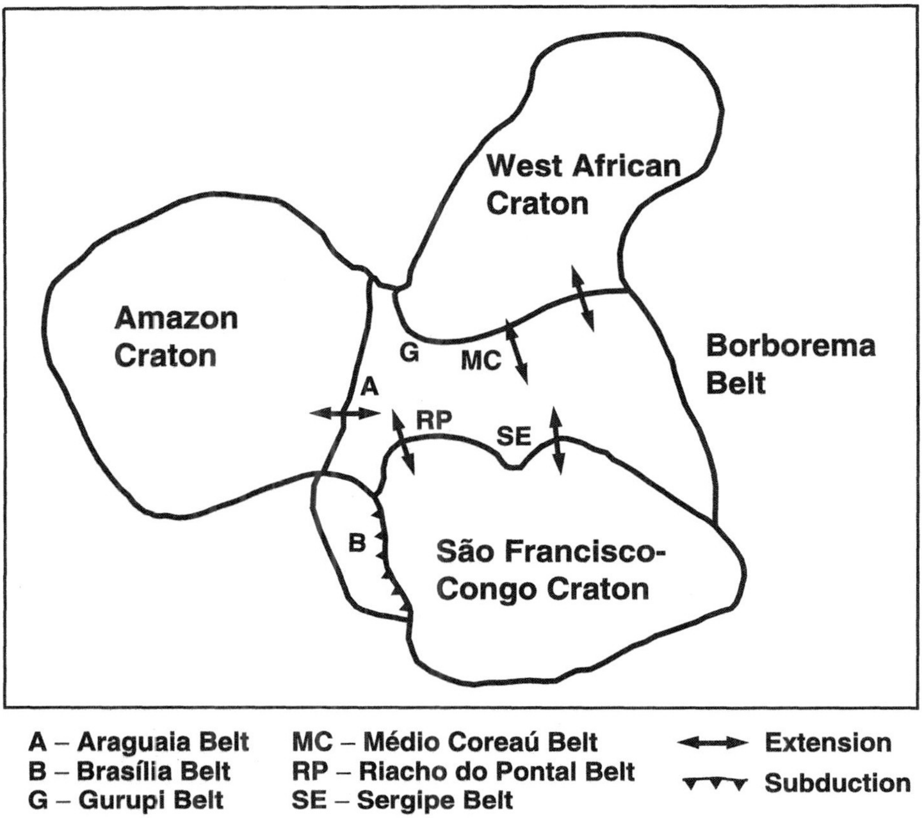

Fig. 7.4: Schematic presentation of supposed "Atlantica" supercontinent and marginal basins of Borborema and São Francisco provinces (inspired after Neves, 2001).

Only after the separation of Brazil and Africa in Cretaceous times, a new series of sedimentary basins developed along the Atlantic coast. Sedimentation during the Cenozoic took still place in reactivated older basins, as covers on plateaus, in river valleys and in coastal plains.

In the northern part of the Borborema Province there occurs still a fragment of the West African craton, called in Brazil the São Luís craton, however without any sedimentary record in its interior. Almeida and Hasui (1984) published a general overview of these areas.

Limiting both provinces with adjacent, chiefly cratonic areas, there occur marginal belts which, after their initial formation, became various times reactivated during later orogenic periods. These marginal areas have been presented in Fig. 7.5.

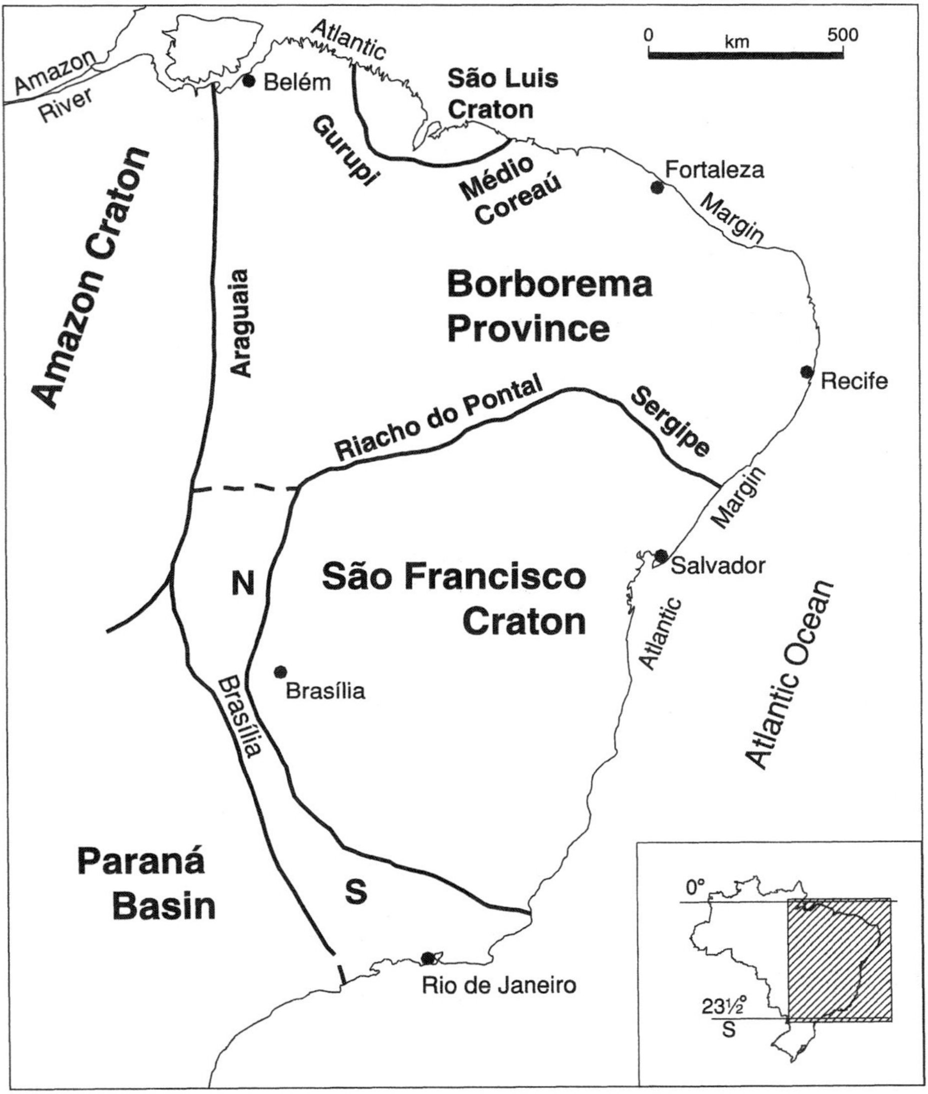

Fig. 7.5: Marginal belt basins of Borborema and São Francisco provinces.

7.4 SEDIMENTARY BASINS

7.4.1 BASIN CLASSIFICATION

In the Borborema and São Francisco tectonic provinces, two main classes of sedimentary basins may thus be distinguished: marginal and intracratonic/intra-continental basins.

With respect to the different basin types, the existing classifications are too much detailed for the purpose of this presentation, which emphasizes more the cyclic development than the proper basin characteristics. Therefore, the simple, although not perfect, basin classification of Selley (1988) is used here combined with that adapted from Kingston et al. (1983; Table 7.2). This latter classification relies very much on depositional sequences, which are basically depth-dependent across a variety of tectonic domains, and in which a single depositional cycle represents a completed single basinal class. Because most basins show repeated cycles of depositional sequences during a long-term subsidence event, the somewhat modified codification of Kingston et al. (1983) is used here to present the basin history of the study area since Mesoproterozoic. The basins found in the Brazilian Borborema and São Francisco provinces and their marginal belts have all been implanted upon continental crust, as their sedimentary sequences show, so that the basins which develop on oceanic crust, are not taken into account here.

7.4.2 BASIN TYPES

The basins formed in the Borborema and São Francisco provinces and their marginal belts developed either as rifts or as crustal sag basins. Surprisingly, it

Table 7.2: Adopted basin classification and preservation potential, after various authors.

modified after Selley (1988)				Ingersoll and Busby (1995)	adapted from Kingston et al. (1983)
generating process	basin type	plate tectonic setting	crustal type*	preservation potential**	Codification
crustal sag	intracratonic basin	intraplate collapse	C	high	IS - interior sag
	epicratonic basin		C	high	MS - marginal sag
tension	epicratonic downwarp	passive plate margin	C,T	high	ES - epicratonic sag
	intracratonic rift	sea floor spreading	C,T	high	IR - intracratonic rift
	intracontinental rift		C	high	IF - interior fracture
	post-orogenic intracratonic rift			medium	IFLL - interior fracture/wrench
compression	trench	subduction	O	low	TA - associated rift
	fore-arc	(active plate margin)	O	medium	OS(FA) - ocean sag / fore-arc
	back-arc		O	low	OS(BA -) ocean sag / back-arc
	remnant ocean basin		O	low	T - oceanic trench
wrenching	strike-slip	lateral plate movement	C	high	LL - continental wrench

*C = continental, T = transitional, O = oceanic

** high = 50–1000 Ma, medium = 5–300 Ma, low = 3–200 Ma (approximate)

appears that there exist certain periods when the basins developed chiefly as rifts and other periods when they formed by intraplate subsidence, and that these periods also occurred alternately in a cyclic succession. Besides, due to the existence of a supercontinent that included both provinces since the Mesoproterozoic, only a limited number of basin types is present, remaining almost entirely restricted to intracratonic and intracontinental types, rarely evolving into epicratonic basins at an oceanic margin.

The marginal belt basins (see Fig. 7.5) are those of (1) the Médio Coreau belt between the Borborema Province and the São Luís Craton (Santos and Brito Neves, 1984); (2) the Gurupi belt between the São Luís and Amazon cratons (Hasui et al., 1984a), respectively; the Araguaia belt between the Amazon Craton and the Borborema Province (Hasui et al., 1984b); the Sergipe and Riacho do Pontal belts between the Borborema Province and the São Francisco Craton (Santos and Brito Neves, 1984); the extensive Brasília belt at the western flank of the São Francisco Craton in the north, and with the possibly existing Paraná cratonic nucleus in its central and southern sections (Marini et al., 1984; Hasui and Oliveira, 1984; Paciullo et al., 1998). These marginal basins amalgamated the Borborema, São Francisco, and Amazon geotectonic units at the end of the Paleoproterozoic during the Transamazonian tectonic cycle, when a supercontinent (Atlantica?) formed. Many of these basins have been reactivated during orogenetic events following their initial formation. In the Late Mesozoic, the most recent marginal belt developed at the time of the opening of the Atlantic Ocean during the dismemberment of the Gondwana supercontinent. The basins formed in this area show only the initial development of a marginal basin sequence.

The intracontinental basins are those formed within the Borborema mobile area (Santos and Brito Neves, 1984), chiefly throughout the tectonically active submergent episodes (Fig. 7.6). Due to the trend of the structural units of the province, the basins show commonly an elongated shape in a SW-NE direction. The Mesoproterozoic basins appear to be of smaller size than those formed during the Neoproterozoic and later. Basin formation continued also during the Phanerozoic, especially due to the tectonic forces triggered by the opening of the South Atlantic rift.

The intracratonic basins formed within the São Francisco Craton (Mascarenhas et al., 1984; Schobbenhaus, 1996; Martins-Neto et al., 2001), in Brazilian territory (Fig. 7.7). Due to the cratonic basement structures, these basins are chiefly rifts and some sag basins upon a rifted basement. This is also the case with what is termed the Ribeira Basin (Almeida and Litwinski, 1984). Anyhow, this basin started its formation as an epicratonic basin during the Transamazonian orogenic period. Finally in the Late Mesozoic, caused by the forces triggered when the Atlantic rift opened, a rift arm developed with an S–N trend at

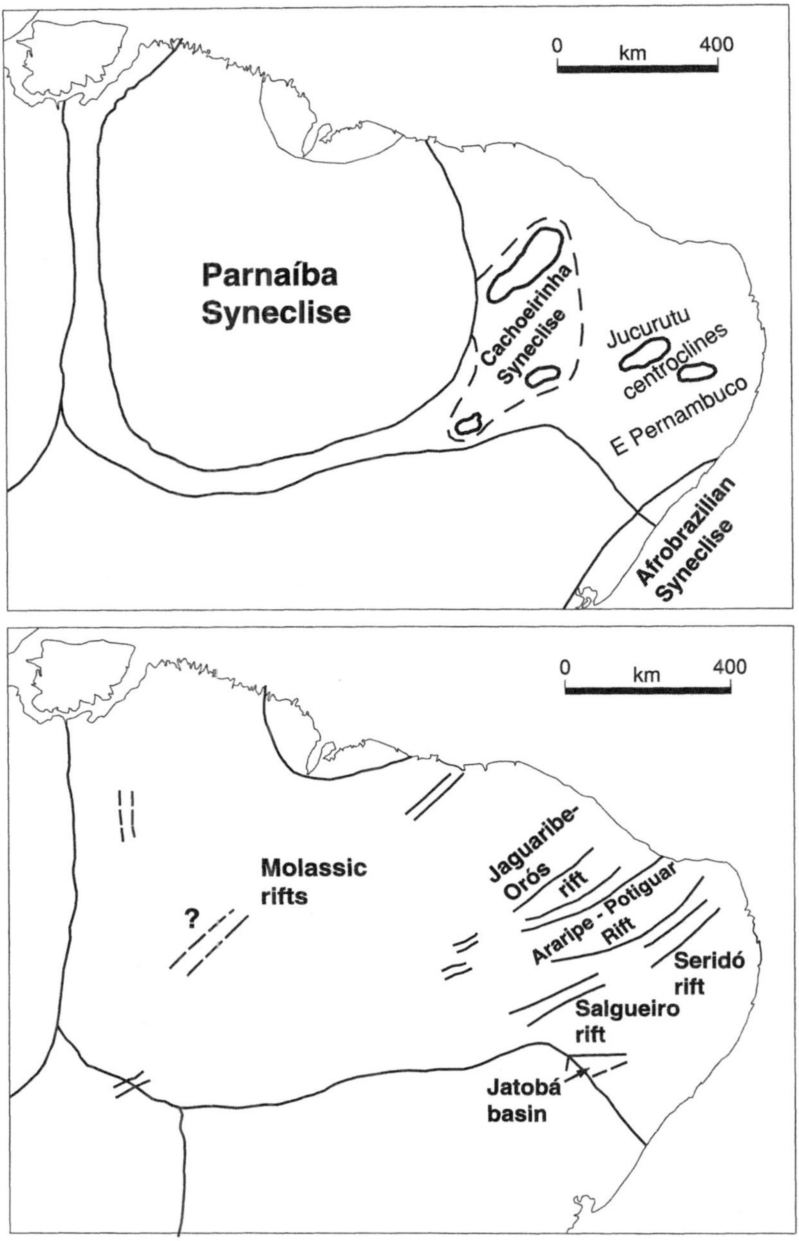

Fig. 7.6: Intracontinental basins of Borborema Province: (a) syneclises and centroclines, (b) rifts.

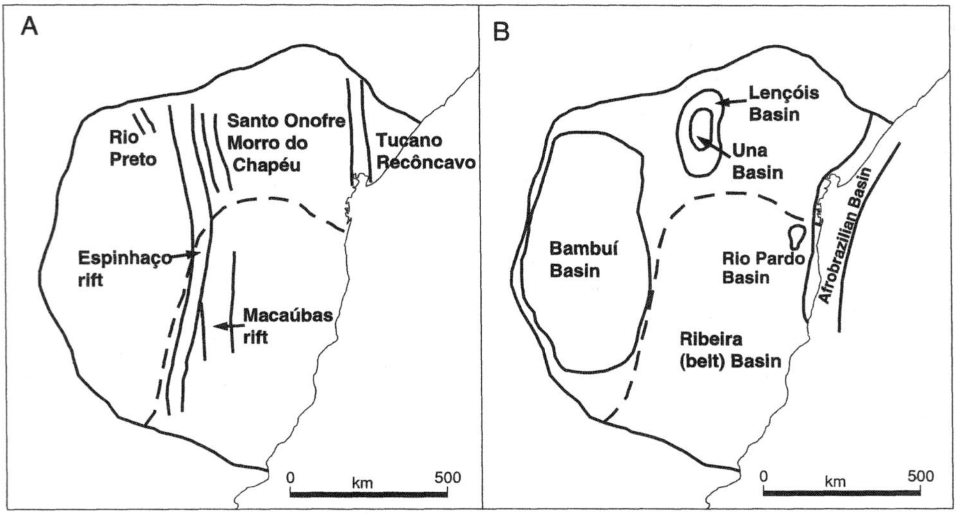

Fig. 7.7: Intracratonic basins of São Francisco Province: (a) rifts, (b) syneclises, centroclines, and Ribeira epicontinental basin.

the eastern craton border. This arm, however, became aborted in the Borborema Province where it reached the Pernambuco lineament, becoming thus an aulacogen. This rift is known as the Recôncavo-Tucano-Jatobá Basin, with its latter part in the Borborema area. Special mention has to be made to the so-called Afrobrazilian Depression (Ponte et al., 1972; Dias, 1991) that formed at what is the present northeastern margin of the Ribeira Basin, but was during its development in Middle Paleozoic times, an intracratonic basin of the then existing supercontinent. It extends northward into the Borborema Province.

In Tables 7.3 and 7.4, the basin types found in the Borborema and São Francisco provinces are mentioned together with the stratigraphy of their corresponding lithic infillings.

7.5 BASIN EVOLUTION AND SEDIMENT FILL

7.5.1 GENERALITIES

Taking into account the limited number of basin types present in the study area (see Tables 7.3 and 7.4), also their infillings show rather limited sequence successions.

The interior fracture basins (*IF*) pass in the beginning through a so-called pre-rift phase which refers to the sedimentary sequence deposited during the initial stretching period that precedes immediately the syn-rift period. Its sediments are those eroded from the existing continental surface and are generally fine clastics derived from soils. The following syn-rift phase gets its sedimentary infilling only after a sufficient relief had been formed along the subsiding trough (Magnavita, 1996). The sequence begins with a basal conglomerate deposited in fault-scarp fans, passing upward into fluvial sandstones. With the decreasing rate of rifting movement near the end of the phase, the sediments blanket them and the rift turns a gently sloping elongate or subcircular basin. The environment becomes lacustrine under humid climatic circumstances, with fine-grained clastic and carbonate deposits. In arid climates, the centre of the rift may contain ephemeral lakes and inland sabkhas, with evaporites, generally gypsum or anhydrite. In case the basin floor descends to sea level, coastal sabkhas with evaporitic facies and associated limestone reefs may develop under arid climatic conditions. When the climates are humid, however, coastal marshes and swamps may develop in which organic clays and peat become deposited, evolving later often into coal. When wrenching accompanies the interior fracture basins (*IFLL*) as in the case of post-orogenic intramontane rifts, the sediments remain continental clastics, grading from base to top, between conglomerates, sandstones, and clays.

At passive margins the intracratonic fracture basins evolve generally through intercratonic rift into epicratonic downwarp sag basins (*IF/IR/ES*), as is, for instance, the case with the development of the Atlantic margin basins. Following the pre-rift and syn-rift phases, the rift becomes intercratonic or intercontinental. With continued subsidence, the sea will invade the trough although still with restricted access and circulation; this favours the development of thermally or salinity induced density layering in the sea water. The subsequent sediment deposition is that of limestones and, under dry climates, evaporites, besides the preservation of organic matter in the organic-rich muds, which favours subsequent petroleum generation. As subsidence continues and the continents drift away from each other, the basin becomes filled in with open marine sediments. It turns in this way an epicratonic downwarp, with the following depositional systems tract, from the continent oceanward; (1) fan-delta complexes, with their proximal, medium, distal and pro-fan facies, and with clastic material derived from the continent; (2) carbonate shelf complex with two facies: bedded limestones and dolomites of the proper shelf, and biolithites of reef-like mounds found at the edge; (3) fine-clastic slope system, with a thick shale sequence and some intercalations of turbidites at the base of the slope.

Where the intracontinental fracture zone enters an epicratonic basin at the continental margin (*IF/ES*), the intermediate intercontinental rift basin will not exist. The other sediment sequences remain essentially the same.

Table 7.3: Basin types and lithic infillings of marginal basins of Borborema

Ga ±10 Ma	age: era	age: period	episode	orogenic phase	Médio Coreaú		Gurupi		Araguaia		Sergipe
0	Phanerozoic	Quaternary-Tertiary	O	alpine							
0.08		Cretaceous-Jurassic	S								
0.18		Triassic-Late Carboniferous	O	variscan							
0.28		Early Carboniferous-Devonian	S	(Hercynian)							
0.38		Silurian-Late Ordovician	O	caledonian							
0.48 0.58		Early Ordovician-Cambrian	S		Jaibaras Gr.	IFLL	Piriá Fm. Igarapé de Areia Fm.	IFLL	Rio das Barreiras Fm. Monte do Carmo Fm.	IFLL	Juá Fm. Palmares Fm.
0.68	Neoproterozoic	Vendian	O	brasiliano							
0.78		Late Cryogenian	S		Ubajara Gr.	IS					Vaza-Barris Gr. (d) Estância Gr. (m)
0.88		Early Cryogenian	O	cariríano							
0.98		Tonian	S		Martinópole Gr.	IF	Gurupi Fm.	IF	Tocantins Gr.	IF	Miaba Gr.
1.08	Mesoproterozoic	Late Stenian	O	sunsas							
1.18		Early Stenian	S						Estrondo Gr.	IS	Macururé Gr.
1.28		Late Ectasian	O	espinhaço							
1.38		Early Ectasian	S	uruaçuano							
1.48		Late Calymmian	O	rondoniano							
1.58		Early Calymmian	S						Garotiré Fm. Tucuruí Fm.	IS	
1.68		Late Statherian	O	rio negro							
1.78		Early Statherian	S		São Joaquim Gr.	IF			Natividade Gr.	IF	Jirau Gr.
1.88 1.98 2.08	Paleo-proterozoic	Orosirian		transamazônico							

Episodes: O – oscillatory-emergent (d) distal
S – submergent (m) marginal

References: Hasui et al., 1984a; Santos and Brito Neves, 1984; Hasui et al., 1984b; Marini et al., 1984; Hasui and Oliveira, 1984; Parmello et al., 1998; Almeida and Hasui, 1984 (book)

During certain phases of basin formation only intracratonic or intracontinental sag basins (*IS*) will develop as syneclises or centroclines. In these basins, sedimentation spans a range of environments including fluvial and marine sands, reefal carbonates, evaporites, and sub-wave pelagic muds. The sedimentary facies in these basins, though diverse in lithology and environment, are seldom indicative of deep water or abrupt subsidence of the basin floor. Deposition takes place close to sea level. Subsidence is thus gradual, sometimes cyclically erratic, with sedimentation being sufficiently rapid to keep the basin nearly filled at any point of time. The deposited sequences correspond almost entirely to the so-called geosynclinal sedimentary sequences of Beloussov (1962) and the depositional cycles and stages of Kingston et al. (1983), presented in Table 7.5.

It will be seen that the marginal, intracontinental and intracratonic basins of the Borborema and São Francisco provinces in Brazil, record the above-sketched types of sedimentary infillings. During the Mesoproterozoic, Neoproterozoic, and Phanerozoic eras, the respective basins show these types cyclically *and são Francisco tectonic provinces.*

	Marginal Basins										
		Riacho do Pontal		N. Brasília		C. Brasília		S. Brasília C. Mantiqueira		Atlantic Margin	
										Various groups	IF/IR/ES
	IFLL			Três Marias Fm.	IFLL	Três Marias Fm.	IFLL	Pouso Alegre Fm.	IFLL		
	IS	Vargem Grande Gr.	IS	Bambuí Gr.	IS	Bambuí Gr.	IS	São Roque Gr.	IS		
	IF	Monte Orebe Fm.	IF	Vazante Fm.	IF?	Vazante Fm.	IF?	Andrelândia Gr.	IF		
	IS	Casa Nova Gr.	IS								
				Paranoá Gr.	IF			Carandaí Gr.	IF/ES		
	IF	Jirau Gr.	IF	Araí Gr. (M) Serra da Mesa Gr. (D)	ES	Ibiá Fm., Canastra Fm. (M) Araxá Gr. (D)	ES	São João del Rei Gr.	ES		

during their history and evolution. This fact has already been suggested by Ghignone (1972) for the Late Neoproterozoic and Phanerozoic, recognizing five so-called "sequences" between Vendian and Middle Cretaceous.

7.5.2 MARGINAL BASINS

Two different areas of marginal basins have been recognized since the beginning of the Mesoproterozoic, considering the then existing supercontinent: the intracontinental marginal belts and the belts at the southwestern margin of the continent. The basins of the latter belt (North, Central and South Brasília) show an epicratonic character (*ES*). In all other belts the basins are formed by interior fracture processes, have the character of rift basins (*IF*) and their corresponding sedimentary infillings. In Table 7.3 have been mentioned the names of the various units as they are known in Brazilian geological history.

In the Early Statherian basins within this supercontinent, the identified sediment sequences, all strongly metamorphosed, are not always complete in all

Table 7.4: Basin types and lithic infillings of intracratonic and intracontinental basins of São Francisco craton and Borborema belt.

Ga ±10 Ma	age: era	age: period	episode	orogenetic phase	São Francisco Craton, intracratonic basins: craton interior		São Francisco Craton, intracratonic basins: Ribeira Basin		Borborema belt, intracontinental basins: W		Borborema belt, intracontinental basins: E	
0	Phanerozoic	Quaternary-Tertiary	O	alpine								
0.08		Cretaceous-Jurassic	S		Recôncavo-Tucano rift	IF			Cretaceous formations		Araripe-Potiguar rift	IF
0.18		Triassic-Late Carboniferous	O	variscan (hercynian)			Afrobrazilian Syneclise		Balsas Gr.			
0.28		Early Carboniferous-Devonian	S					IS	Canindé Gr.	IS		
0.38		Silurian-Late Ordovician	O	caledonian					Serra Grande Gr.			
0.48		Early Ordovician-Cambrian	S						molassic basins	IFLL	molassic basins	IFLL
0.58	Neoproterozoic	Vendian	O	brasiliano					·			
0.68		Late Cryogenian	S		Bambuí Gr.; Una Gr.	IS	Rio Pardo Gr.	IS			Cachoeirinha Gr.	IS
0.78		Early Cryogenian	O	cariríano	Santo Onofre Gr.; Rio Preto Gr; Morro do Chapéu Fm.							
0.88		Tonian	S			IF	Macaúbas Gr.	IF			Seridó Gr.	IF
0.98	Mesoproterozoic	Late Stenian	O	sunsas	Tombador Fm; Caboclo Fm.							
1.08		Early Stenian	S			IS?					Jucurutú Gr.	IS?
1.18		Late Ectasian	O	espinhaço								
1.28		Early Ectasian	S	uruaçuano							Salgueiro Gr.	IF
1.38		Late Calymmian	O	rondoniano								
1.48		Early Calymmian	S								Leste Pernambuco Gr.	IS
1.58		Late Statherian	O	rio negro	Espinhaço Gr.		Espinhaço Gr.					
1.68		Early Statherian	S			IF		IF			Jaguaribe-Orós Gr.	IF
1.78 1.88 1.98 2.08	Paleo-proterozoic	Orosirian		transamazônico								

Episodes: O – oscillatory-emergent
S – submergent

References: Mascarenhas et al., 1984
Santos and Brito Neves, 1984
Mabesoone, 2001
Almeida and Litwinski, 1984
Schobbenhaus, 1984
Almeida and Hasui, 1984 (book)

Table 7.5: Sedimentary sequences in intracratonic and intracontinental sag basins [adapted from Beloussov (1962) and Kingston et al. (1963)].

From top to base	
Molassic sequence	chiefly terrigenous sediments, ranging between conglomerates and clays from base to top; in the beginning may be marine, finishing as continental
	– regressive sequence
Lagoonal sequence	only under favourable climates: evaporites, including dolomites
Upper terrigenous sequence	terrigenous flysch: cycles of sands and clays, with occasional limestones and marls; greywackes
	subsequences with organic matter-rich sandstones and shales
Carbonate sequence	organic limestones, calcilutites, reefal limestones, varying often from one basin to another
	– marine to lacustrine sequence
Lower terrigenous sequence	uniform thick deposits of clays and sands, without fossils, sometimes cyclic stratification; at base coarse continental deposits of conglomerates and sandstones
	– transgressive sequence

marginal basins, however, they are surprisingly uniform. In a general way, the sandstone units dominate as quartzites, whereas the basal conglomerates are often absent, maybe due to the petrographic character of the source rocks. Only in the Sergipe belt the rift seems to have become almost completely filled in, because only there metamorphic limestones have been found. In the Brasília belt, a few ophiolites in orogens suggest the presence of distal oceanic crust at the belt's margins.

Upwards the interior sag (*IS*) phase of the Early Calymmian has only been identified in the Araguaia belt basin, with different-type and different-sized sandstone units, all metamorphosed. In the Early Ectasian, no sedimentary basins have been reactivated, possibly due to the rather strong Espinhaço-Uruaçuano orogenic cycle, during which older sedimentary sequences became strongly metamorphosed.

During the next Early Stenian basin formation phase, a reactivation took place in the marginal areas between the Borborema Province and the São Francisco and Amazon cratons, respectively, due to local extensional conditions (Neves,

2001, 2003), but only resulting in intracontinental subsidence (*IS*), with a correspondent sedimentary infilling pattern of sandstones, shales, and upward some limestones, again metamorphosed.

With the beginning of the Neoproterozoic, more modern evolutionary stage of the earth, the Tonian-Early Cryogenian Caririano orogenic cycle, that was essentially of taphrogenic character, with interior rift (*IF*) formation, all marginal basins within the then existing supercontinent became reactivated. The sedimentary infilling reflects neatly this type of tectonism, recorded by sandstone sections at the base, grading upwards into siltstones and shales, and limestones, some of these with a marine character indicating a transgression of the sea into these areas. Again, most but not all sedimentary sequences have been metamorphosed. The Late Cryogenian submergent episode shows one more reactivation of some of the marginal belt basins, with only intracontinental subsidence (*IS*) and the deposition of thick sedimentary sequences of lower terrigenous sandstones, followed by microclastic deposits, and often marine limestones, the well-known Bambuí limestones. Due to the very strong Brasiliano orogeny and the incorporation of the then existing supercontinent as West-Gondwana into the extensive Gondwana supercontinent, the sediments were again metamorphosed but at a rather low grade. The formation of this supercontinent with a mountainous relief triggered the development of glaciations, as is recorded by their corresponding, chiefly tillite diamictites.

During the following Early Paleozoic transitional stage towards the stabilization of the South American Platform, post-orogenic intramontane rifts developed by interior fracture and wrenching (*IFLL*) in which the so-called molassic terrigenous clastic sediment sequences were deposited, also in most of the marginal belt basins, now definitively turned intracontinental, finishing their independent behaviour.

A somewhat different history shows the Brasília belt, which, during most of the Mesoproterozoic constituted the southwestern margin of the then existing continent. In the Early Statherian, these continental margins became epicontinental downwarps (*ES*), bordering an oceanic area as is suggested by the sedimentary sequences deposited alongside these borders. Two facies types can be distinguished: marginal and distal, which can be easily mapped (Marini et al., 1984). The marginal facies in the North and Central Brasília belts are composed of terrigenous clastics, sandstones, siltstones, and shales, all metamorphosed; even a few limestones have been found in the northern area. The distal facies show strongly metamorphosed microclastic deposits, often with a carbonate component. The sequence found in the South Brasília belt is comparable with the other ones, only that different facies types have not been distinguished (Paciullo et al., 1998). Upward in the stratigraphical column, the Brasília Basin where it borders the Amazon craton, records the same basin infillings as the above mentioned other marginal basins. And, with respect to the Central Brasília Basin, the following sedimentary record dates from the sequences deposited during the Neoproterozoic assemblage of the Gondwana continent and its corresponding orogenic phases. The South Brasília Basin shows during the other periods of the Mesoproterozoic only some record of Early Ectasian age, with interior rift (*IF*) terrigenous sandy sediments, passing oceanward into the fine clastic, marly and calcareous deposits with some turbidites, of the epicratonic marginal downwarp (*ES*). During the Neoproterozoic the corresponding sediments as found in the rest of the Brasília belt, are also recorded here. Furthermore in the whole belt, there occur the so-called coarse terrigenous molassic sediments in the few basins there present (*IFLL*).

A special case present the Phanerozoic Atlantic margin basins, which formed during the dismemberment of the Gondwana supercontinent since Late Jurassic. This means that the sedimentary sequences deposited in this marginal belt, represent only one cycle: the most recent. Four phases of development are distinguished (Fig. 7.8; Mabesoone, 1994) and are generally accepted: pre-rift, syn-rift, proto-oceanic gulf, and open-ocean drift. Furthermore, there exists a subdivision into three segments: Southern Atlantic, Pernambuco-Paraíba-Rio Grande do Norte area, Equatorial Atlantic. In the South Atlantic segment the whole sequence is present, in the Equatorial Atlantic segment the sediment sequence starts with the syn-rift phase. In the area in between of eastern Northeast Brazil, which became separated from Africa only in the Late Cretaceous (Mabesoone, 1994), the basin evolution begins with a flexural phase, passing directly into the open-ocean drift phase. The deposited sediment sequences are rather uniform and as follows:

pre-rift: fluvial and lacustrine sandstones and microclastics;

syn-rift: fault-scarp conglomerates, fluvial, lacustrine, deltaic continental clastics;

proto-ocanic gulf: South Atlantic only, with limestones and evaporites of a restricted marine environment;

flexure: NE Brazil only, with chiefly continental clastic deposits;

open-ocean drift: development from narrow-ocean shallow marine limestones, into wider-open ocean with a depositional system tract, from the continent oceanward – fan-delta sandstones, carbonate shelf, fine-medium clastic slope, and some turbidites (Fig. 7.8).

This last phase continues until into the present Holocene, with increasingly sandy components, derived from the relief development of the area.

Finally, the lithic infilling of the end Jurassic-Cretaceous Recôncavo and Tucano aulacogen basins is that typical for such rift, with medium- to fine-grained terrigenous clastic sequences and fault-border conglomerates, deposited

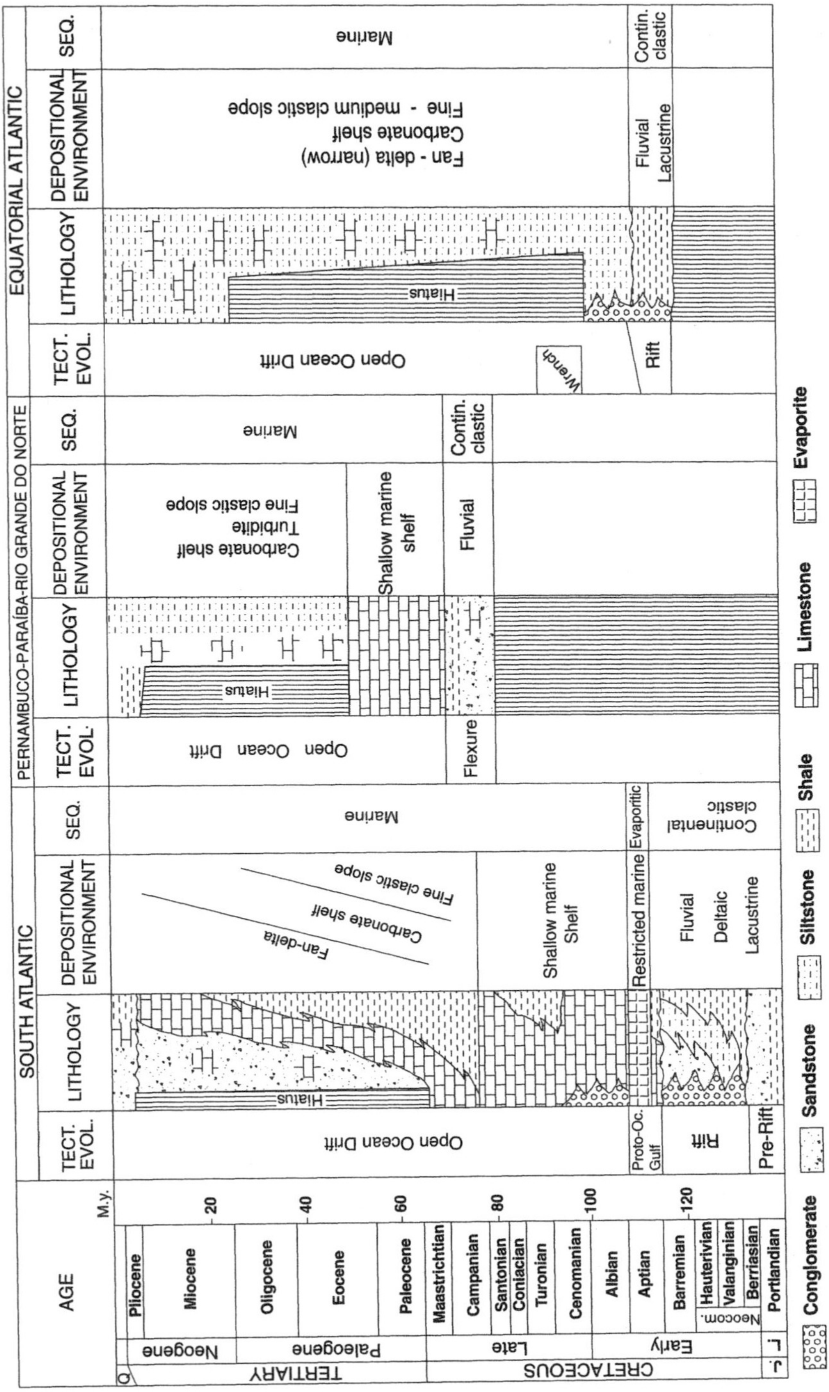

Fig. 7.8: Schematic stratigraphical succession of Atlantic margin sedimentary basins (after Mabesoone, 1994).

in continental environments. The sequence is well known, because in the southern part petroleum has been found and is being exploited.

7.5.3 INTRACRATONIC AND INTRACONTINENTAL BASINS

7.5.3.1 São Francisco Province

Although the São Francisco cratonic area became consolidated during the Paleoproterozoic, its inner strength was not enough to remain unaffected by later orogenic events. In the basement, the structural trend is orientated about N–S and along this direction later taphrogenesis subsidence formed still some intracratonic basins. The sedimentary infilling of these basins is similar to that of the marginal belt basins. The Statherian taphrogenic phase formed a rather narrow extensive N–S trough through almost the whole province (*IF*), with a thick, chiefly arenitic and microclastic infilling that became tectonized and metamorphosed in greenschist facies, essentially during the Espinhaço-Uruaçuano orogenic cycle (Table 7.4). At the end of the Mesoproterozoic and during the Neoproterozoic, the craton suffered again from tectonic activity and accompanying basin formation during three phases (*IS-IF-IS*). The lower interior sag phase resulted in rather fine sandy and microclastic sequences with a few limestones, later also metamorphosed. The Tonian interior fracture sequence reflects rift basin fills with medium- and fine-grained clastic sediments, also metamorphosed during the strong Brasiliano tectonic event. During this latter epoch, only basins of intracratonic subsidence formed in which have been deposited the lower terrigenous clastic sequence and the middle limestone sequence with the Bambuí marine limestones in platform facies (Mabesoone, 2002). Martins-Neto et al. (2001) distinguish more or less the same sequences, calling them the Espinhaço, Macaubas, and Bambuí megasequences, respectively.

The Ribeira (belt) basin is here considered as an epicratonic basin (*ES*). This seems to be confirmed by the extension of the Statherian taphrogenic N–S belt of the São Francisco craton into this basin. Only in the Neoproterozoic, other secondary small-sized sub-basins formed within this basin, in rift and maybe centrocline facies. The sedimentary infillings in the rift are chiefly psammitic; the basin must not have been very deep because on top there occur stromatolitic limestones (Macaubas Basin). The centrocline sag basin fill of Late Neoproterozoic is recorded by basal conglomerates, passing upward into sandstones and shales, with intercalated dolomites, all metamorphosed (Rio Pardo Basin).

During the Middle Paleozoic, submergent episode evolved the so-called Afrobrazilian Depression as an interior sag syneclise (*IS*), preceding the opening of the South Atlantic rift (see Fig. 7.7). Only part of the lower sedimentary sequences, chiefly medium- to fine-grained terrigenous clastics, with a Devonian marine invasion, is present in it.

7.5.4 BORBOREMA PROVINCE

The Borborema Province, not being a cratonic but an ancient mobile belt amalgamated during the Transamazonian orogenic cycle in the Paleoproterozoic, and incorporated into the then existing supercontinent, has been subject to tectonic activity during all orogenic events of the Mesoproterozoic and Neoproterozoic and the lower Phanerozoic, with the consequent formation of many intracontinental basins. In the western part of the province, only the record of the molassic basins (*IFLL*) of Early Paleozoic times and the deposits of the Parnaíba Syneclise (*IS*) from Late Paleozoic onwards have been identified. Because the basement of the syneclise is almost unknown, no older sediment sequences have been found until now. The molassic basins as well as the Parnaíba Basin show the sedimentary sequences expected for the respective basin types, more preserved and better known (Góes and Feijó, 1994). Due to the geographical position of the area in the Paleozoic, in cold climate zones (Mabesoone, 1975), almost no limestone sequences have been deposited.

In the eastern part of the province the Phanerozoic sedimentary sequences are much scarcer. The sedimentary basins formed chiefly during all orogenic phases of the Mesoproterozoic and Neoproterozoic in a cyclic succession of interior fracture or rift (*IF*) and interior sag of centrocline and syneclise (*IS*) basins. The rifts are filled in with essentially terrigenous clastic deposits, while in the sag basins the terrigenous clastics are generally finer, and also limestones are present (Santos and Brito Neves, 1984; Sá et al., 1997). The few small-sized Paleozoic molassic basins (*IFLL*) have been filled in with rather coarse terrigenous clastics, sometimes metamorphosed through contact metamorphism. The Phanerozoic interior sag phase is almost absent in the eastern part of the province, and has been treated within the Afrobrazilian Depression in the Ribeira belt basin. During the Mesozoic rift phase again an interior fracture basin (*IF*) formed, extending into Africa, and filled in subsequently with terrigenous clastic deposits in alluvial fan and fluvial facies, finer clastic sediments in rift lake facies, evaporites (anhydrite and gypsum), and a few marine limestones during the mid-Cretaceous invasion of the sea (Lima Filho et al., 1999; Mabesoone et al., 1999). Still a small fragment of the before mentioned Recôncavo-Tucano-Jatobá aulacogen, that is the Jatobá Basin, implanted upon an arm of the Afrobrazilian Depression, is present in the province. The basin is filled in with the lowermost sequence of rift basin infillings, chiefly fine sandstones and shales deposited in continental realms, accumulated on top of clastic sediments which represent a short-lived marine invasion during the Devonian high sea level stand. Younger sedimentary deposits of terrigenous fluvial origin refer to the Cenozoic relief correlated sequences.

7.5.5 EXAMPLES OF BASIN DEVELOPMENT

7.5.5.1 Introduction

Three sedimentary basins have been selected as examples of their cyclic development. (1) The Sergipe Basin, a marginal basin between the Borborema Province and the São Francisco Craton, is rather well known, although most of its lithic infilling has been metamorphosed; this basin started its formation in the beginning of the Mesoproterozoic (see Table 7.3) and finished its history as such in the beginning of the Phanerozoic. (2) The Parnaíba intracontinental basin, within the Borborema Province, during the Phanerozoic (see Table 7.4), has been studied in more detail by Petrobrás in search for oil (however, without success), and is therefore well known with respect to its sedimentary infilling. (3) The Afrobrazilian Depression, a former intracontinental basin between the South American and African continents within the São Francisco craton (see Table 7.3), probably developed since the end of the Neoproterozoic, and in which the South Atlantic rift opened. An incomplete sedimentary record of its infilling can be traced at both sides of the ocean.

7.5.5.2 Sergipe Basin

Generalities. The Sergipe belt basin, a marginal basin between the Borborema Province and the São Francisco Craton, started its development history in the beginning of the Mesoproterozoic when the so-called Atlantica continent amalgamated. During later tectonically active orogenic periods, chiefly in the Neoproterozoic, the basin became cyclically reactivated. With exception of the last-deposited sequence, the other sedimentary fill successions have all been metamorphosed, and the oldest ones stronger than the youngest. The different sediment sequences, however, remain recognizable concerning their depositional environments, which have been essentially continental with minor marine occurrences.

Because the NE Brazilian Borborema Province is one of the best studied and better known of the country's tectonic provinces, the Sergipe belt has also been extensively treated although not often with respect to the character of its sediments. Among the most complete general considerations may be mentioned the publications of Silva Filho et al., (1978a, b) and Santos and Brito Neves (1984). Later papers have not changed anymore the stratigraphical subdivision presented by the cited authors.

Stratigraphical sequences. In Table 7.6, have been presented the stratigraphical sequences determined in the basin as they are generally accepted. They show the respective tectonic-sedimentary successions deposited during the submergent episodes since the origin of the basin development.

The belt system exhibits a marked geological difference, divided into two zones of distinct folding. The Sergipe zone in the south represents the younger sedimentary sequences (Neoproterozoic and Phanerozoic) and the Alaogas zone in the north the oldest sequences (Mesoproterozoic), separated by an intermediate strip (Fig. 7.9). The belt basin is cut through by the Cretaceous Tucano Basin, part of the Recôncavo-Tucano-Jatobá aulacogen.

Table 7.6: Lithostratigraphic sequences in Sergipe belt basin.

Cambrian-Early Ordovician	molasse deposits	Palmares and Juá Formations
Late Cryogenian		
	marginal – Estância Group	Lagarto Formation
		Acauã Formation
		Juetê Formation
	distal – Vaza Barris Group	Frei Paulo/Ribeirópolis Formation
		Olhos d'Água Formation
		Capitão/Palestina Formation
Tonian		
	Miaba Group	Jacoca Formation
		Jacarecica Formation
		Itabaiana Formation
Early Statherian		
	Macururé Group	Traipu-Jaramataia Formation
		Santa Cruz Formation
Early Stenian		
	Jirau Group	

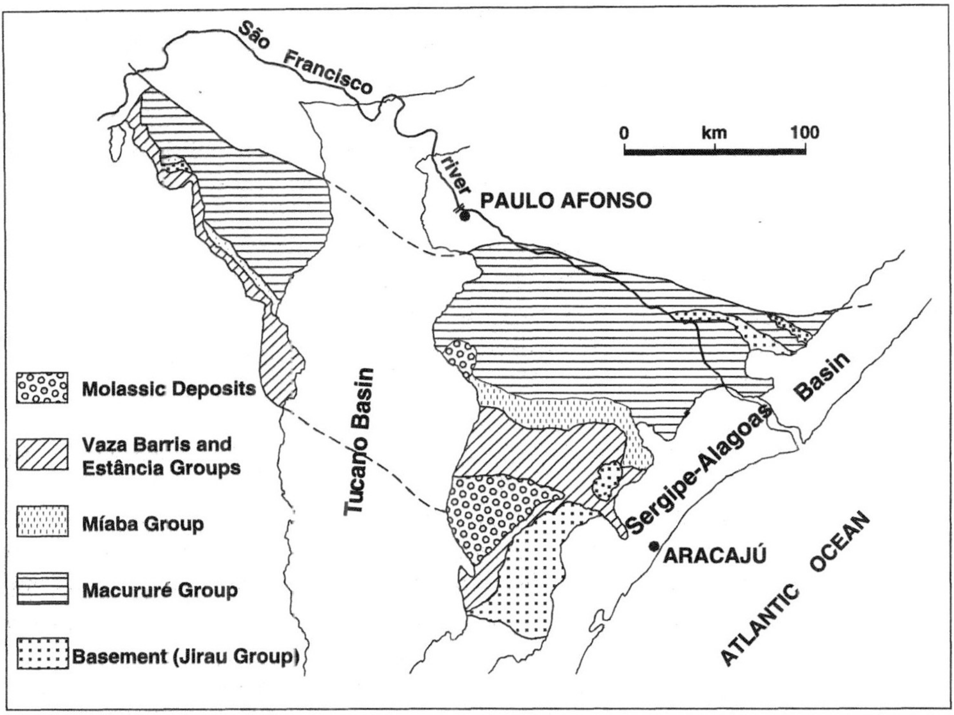

Fig. 7.9: Lithostratigraphic sequences of Sergipe belt basin.

The lower sequence represented by the Jirau Group is considered as being the basement of the basin. It developed in the earliest Mesoproterozoic episode of the Early Statherian, and is an interior fracture basin with some epicratonic sag character. It has not been subdivided into lithostratigraphic formations.

During the next epochs of the Mesoproterozoic (Calymmian and Ectasian), no other sequences seem to have been deposited, possibly due to the effects of the strong Espinhaço-Uruaçuano orogeny.

The next sedimentary sequence has been accumulated during the Late Mesoproterozoic Early Stenian epoch; represented by the Macururé Group subdivided into two formations, as an interior sag type basin.

Under the Caririano orogenetic episode of the Early Neoproterozoic, basin formation conditions prevailed as interior fracture types. Its record has been presented as the Miaba Group of Tonian-Early Cryogenian age, with three different lithostratigraphic units.

The Sergipe belt basin became strongly affected by the Brasiliano orogeny of Late Neoproterozoic times (Late Cryogenian-Vendian). Two types of sedimentary sequences became deposited in it: the Vaza-Barris Group with three formations in distal environment, and the Estância Group also with three formations in a marginal environment, both under interior sag conditions.

The last, not very important sequences developed in the Sergipe belt as an independent marginal basin, are represented by two molassic formations deposited in interior fracture to wrench basin types. This occurred already during the first Phanerozoic submergent episode in Cambrian-Early Ordovician times.

The various sediment successions are separated by basin-wide erosional and often tectonically influenced unconformities. The ages attributed to these sequences are partly confirmed by absolute age determinations. In Fig. 7.9, the approximate occurrences of the respective lithostratigraphic groups in the basin have been presented.

Sediments and their depositional environments. The basement is represented by the Jirau Group with the occurrence of sediments metamorphosed in high grade. Gneisses and migmatites dominate, but also metamorphosed limestones and quartzites have been found. Detailed studies about these sediments have not yet been made, but they must reflect the depositional conditions during the amalgamation of a supercontinent in the Early Statherian, probably as a closing ocean basin.

After a fairly long interval during which the strong Espinhaço-Uruaçuano tectonic cycle affected much of the existing lithic groups, a basin reactivation took place at the end of the Mesoproterozoic during the Early Stenian. The epoch is recorded by the Macururé Group. Its lower Santa Cruz Formation is characterized by quartzites of various grain sizes, with the intercalation of a few schist lenses. Upward there appears the Traipu-Jaramataia Formation, composed of a great variety of metasedimentary rocks, chiefly clastic, subordinate carbonate, as well as a sequence of metavolcanics. Generally, the grade of metamorphism decreases from north towards south. The depositional environments were evidently continental, but marine invasions have also taken place. However, no indication of oceanic crust in the sedimentary sequences has been detected. Both units are present in the northern zone of the belt.

During the Neoproterozoic "modern" period, the Sergipe belt basin suffered various times from extensional processes with basin reactivation, which permitted the deposition of many sedimentary successions. The Tonian-Early Cryogenian sequence, accumulated under the influence of the Cariríano orogeny, is called Miaba Group and has been subdivided into three formations. The lower Itabaiana Formation shows a basal conglomerate upon which feldspathic quartzites with large cross bedding are found. The next Jacarecica Formation presents rather restricted occurrences and is represented by conglomeratic greywackes and polymictic conglomerates, with rare intercalations of red claystones and siltstones. Both formations have neatly been deposited in continental environments, chiefly through erosion of a surrounding area with high relief. The upper Jacoca Formation, composed of limestones, dolomites and dark-coloured claystones, was deposited in a marine environment. All sediments have suffered a medium-grade metamorphism.

The Late Neoproterozoic was a tectonically very active period due to the strong Brasiliano orogeny that affected the greater part of the Brazilian Precambrian. The Sergipe belt suffered from rather strong extensional processes (Neves, 2003), which resulted in two groups of sedimentary sequences, one marginal, the other more distal. However, no record of an eventual presence of oceanic crust has been detected. The distal realm is represented by the Vaza-Barris Group, with its sediments low-grade metamorphosed, and the marginal realm by the Estância Group which sediments underwent only anchimetamorphic alterations (Mello, 1977). The Vaza-Barris Group has been subdivided into three lithostratigraphic units which appear to present the lower terrigenous or transgressive sequence, the middle carbonate or marine sequence, and the upper terrigenous sequence, as is presented in Table 7.5. The Capitão/Palestina Formation shows metamorphosed greywackes, quartzose sandstones, siltstones, claystones, and shales deposited in continental environments. The Olhos d'Água Formation indicates the marine facies of the sequence, recorded by metamorphosed dark-coloured limestones, and some siltstones and claystones. Moreover, in the upper terrigenous sequence of the Frei Paulo/Ribeirópolis Formation, metamorphosed fine sandstones, greywackes, siltstones, claystones and shales accumulated again in fluvial and lacustrine realms. The marginal Estância Group shows the same type of successions, also in three formations. The Juetê Formation is represented by arkosic and quartzose sandstones, with intercalations of shales, some cherts and matrix-supported conglomerates, deposited in alluvial fans which pass through braided and meandering river systems into alluvial plains and more rarely into lakes and swamps. The middle Acauã Formation is composed of oolitic and pisolitic limestones, pure to carbonaceous limestones and dolomites, few shales, with at the top often pelites, and with the presence of algal stromatolites; this sequence has been deposited in a rather shallow marine environment with restricted circulation and oscillating sea level. The sequence finishes with the Lagarto Formation with fine arkosic and micaceous sandstones and greywackes, accumulated in deltaic realms passing into meandering fluvial systems, under regressive circumstances. After deposition, deformational forces affected the sequences in various degrees (Mabesoone, 1994).

The molassic sequences, deposited in fault depressions and in the foreland of a more vigorous relief formed during the post-Brasiliano epeirogenetic uplift, accumulated in continental alluvial fan and chiefly braided river systems. The Juá Formation has been deposited in an interior fracture to wrench zone, and shows rather coarse clastic sediments among which polymictic conglomerates and conglomeratic greywackes and arkoses stand out. The Palmares Formation has more the character of a foreland deposit, with finer greywackes and quartzose sandstones, with rare intercalations of breccias and conglomerates; these sediments were evidently deposited in alluvial fans and braided to low-sinuosity rivers.

Herewith the sedimentation in the proper Sergipe belt finishes. The Mesozoic Tucano and Sergipe-Alagoas basins belong to the Atlantic marginal basin history.

Basin history. The Sergipe belt basin history is an example of the cyclic development of a Precambrian basin in a marginal setting. Its development started in the Early Mesoproterozoic and finished in the beginning of the Paleozoic, although not in all tectonic-sedimentary episodes deposition took place. Besides, none of the accumulated sequences shows any record of oceanic crust. The basin reactivation occurred chiefly at the end of the Mesoproterozoic and during the whole Neoproterozoic when rather strong orogenetic phases favoured the amalgamation of the Gondwana supercontinent. No evidence has been found of a glacial record in the deposits what means that possibly the relief of that time did not favour the development of ice caps. The youngest molassic deposits reflect the transition stage after the Brasiliano orogeny towards the stabilization of the South American platform.

7.5.5.3 Parnaíba Basin

Generalities. The Parnaíba Basin has been implanted as a syneclise upon the crystalline basement of the Borborema Province. In this basement, a few molassic rifts are found in the basin subsurface and at its borders. The proper Parnaíba Basin shows a spectrum of sedimentary environments ranging from marine to continental in various stages which correspond to the cyclic tectonic-sedimentary episodes. These environments are represented in the stratigraphic record by a number of formations composed chiefly of clastic sedimentary rocks.

The basin has been studied successively by various authors, among which may be mentioned Kegel (1957), Mesner and Wooldridge (1964), Aguiar (1971), Mabesoone (1977), and more recently by Góes and Feijó (1994), from which most of the available data have been collected.

Stratigraphical sequences. In Table 7.7, the stratigraphical sequences as proposed by Góes and Feijó (1994) have been mentioned. They show neatly the tectonic-sedimentary episodes of the Phanerozoic since Late Ordovician.

The Serra Grande Group represents the Late Ordovician-Silurian oscillatory-emergent episode. It lies unconformably on top of the crystalline basement and

Table 7.7: Lithostratigraphic sequences in Parnaíba Basin.

Cenozoic oscillatory-emergent episode		
		relief correlated deposits
Jurassic-Cretaceous submergent episode		
	Cretaceous formations	Itapecuru Formation
		Codó Formation
		Grajau Formation
	Mearim Group	Sardinha basalts
		Corda Formation
		Pastos Bons Formation
		Mosquito basalts
Late Carboniferous-Triassic oscillatory-emergent episode		
	Balsas Group	Sambaiba Formation
		Motuca Formation
		Pedra de Fogo Formation
		Piauí Formation
		Poti Formation (upper lithosome)
Devonian-Early Carboniferous submergent episode		
	Canindé Group	Poti Formation (lower lithosome)
		Longá Formation
		Cabeças Formation
		Pimenteira Formation
		Itaim Formation
Late Ordovician-Silurian oscillatory-emergent episode		
	Serra Grande Group	Jaicós Formation
		Tianguá Formation
		Ipu Formation
Cambrian-Early Ordovician submergent episode		
		molasse deposits

has been subdivided into three lithostratigraphic units, in a geocratic phase with continental deposits.

The next submergent episode is recorded by the Canindé Group, composed of five formations, with ages ranging between Devonian and Early Carboniferous. The sequence is separated from the underlying Serra Grande Group by an erosional unconformity. The lithostratigraphic units indicate a thalassocratic phase with marine and littoral deposits.

Follows unconformably on top of the foregoing sequence the Balsas Group, of the next oscillatory-emergent episode of Late Carboniferous to Triassic age. As a geocratic phase, it is represented chiefly by continental sediment successions, subdivided into five formations.

The Jurassic-Cretaceous submergent episode is present in the basin recorded by the basalts and intercalated sediments of the Mearim Group and some three more lithostratigraphic formations deposited in marine and continental realms.

During the last oscillatory-emergent episode, various sequences of Tertiary and Quaternary ages have been accumulated, mainly as deposits correlated with the relief development.

All episode-related sediment groups are separated between them by regional basin-wide erosional unconformities. The continental geocratic phase deposits show also many small, often erosional hiatuses between them. In contrast, the marine and littoral deposits of the thalassocratic phases pass gradually one into the other. The ages attributed to the different sequences are based on macro-and microfossil associations, mentioned by Mabesoone (1994). The oldest sediment sequences crop out at the borders of the basin, with exception of its northern limit. Towards the basin center, the outcropping sequences become ever more younger (Fig. 7.10). In the subsurface, the entire sediment succession is almost completely present.

Sediments and their depositional environments. Within the crystalline basement developed molassic rifts have been filled in with coarse to medium-grained clastic sediments accumulated in the foreland of the mountainous area formed by the Brasiliano orogeny in the Borborema Province.

The Serra Grande Group has been deposited during an oscillatory-emergent episode, with the area in a rather elevated topographical position. The Late Ordovician (Ashgillian) glaciation, with its centre S of the Sahara, reached also the Parnaíba Basin, located at that time near the South Pole (Caputo and Lima, 1984). The lower Ipu Formation, composed of diamictites (tillites) and conglomeratic sandstones with a current direction towards SW-SE, that is out of the basin, suggest then a deposition by glaciers and glacial tongues for the diamictites, and in periglacial fans and outwash plains for the conglomeratic sandstones (Fig. 7.11). The middle Tianguá Formation is composed of shales, siltstones and fine sandstones interpreted as of shallow marine origin, conse-

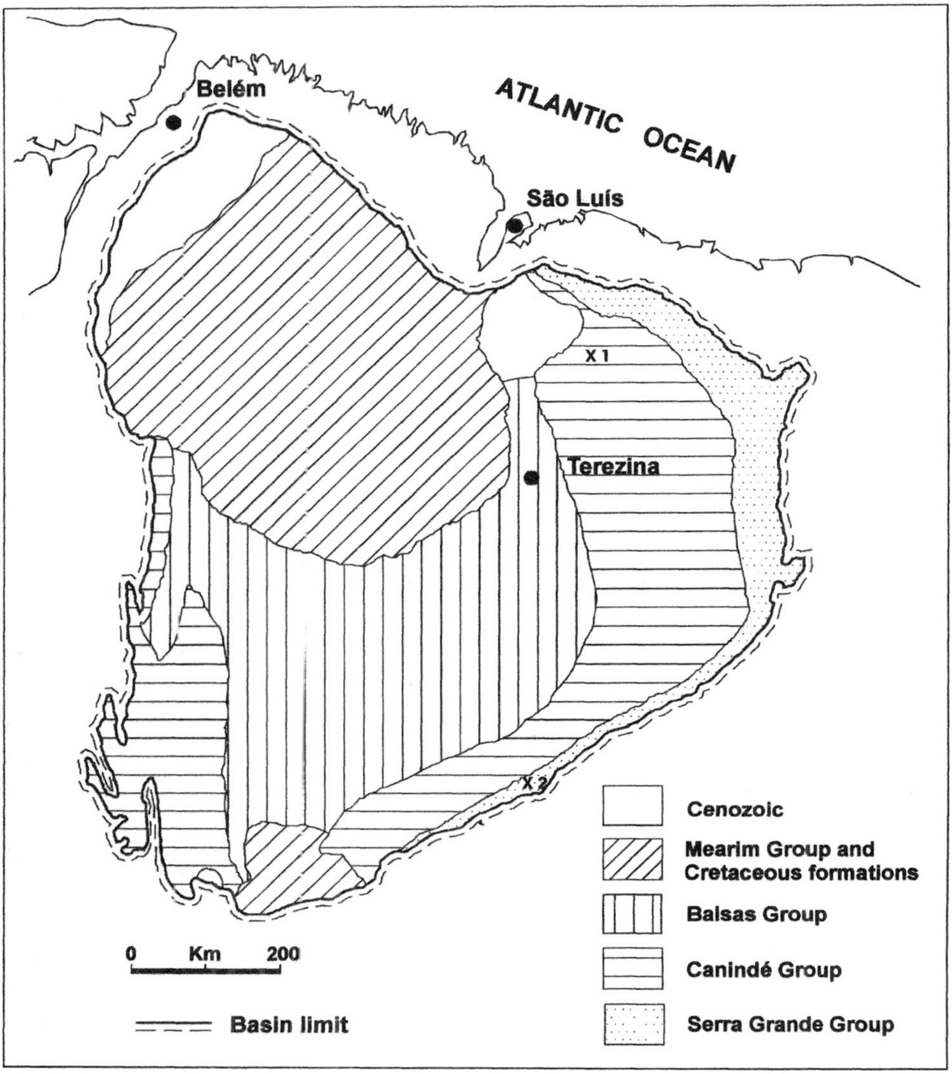

Fig. 7.10: Lithostratographic sequences of Parnaíba intracontinental basin.

quence of the transgression taken place after the melting of the ice in the Early Silurian (Llandoverian-Wenlockian). In addition, the upper Jaicós Formation represents the regression of the sea (Ludlovian-Pridolian) that left behind medium-grained to conglomeratic sandstones and conglomerates composed of small quartz pebbles in a lithic matrix. These deposits accumulated in alluvial fans, fan deltas, and braided river plains, partly reworking the Ipu Formation sediments. In this latter case, the current directions are towards the basin depocenter.

Fig. 7.11: Parnaíba Basin: Serra Grande Group fluvioglacial coarse sandstones.

The erosional and parallel unconformity that separates the Serra Grande Group from the overlying Canindé Group is related to the cratonic emergence at the end of the oscillatory episode, and lasted until the beginning of the Devonian, when the episode turned to be submergent. The Parnaíba Basin suffered then a continuous slow subsidence so that the sea could encroach upon the area, establishing one single long-term transgression-regression cycle (Mabesoone, 1977), with small oscillations. The lower Itaim Formation (Eifelian?) is composed of fine-grained white to pink-coloured sandstones with intercalated grey shales, deposited near the basin borders in river facies of coastal plains and more towards the basin centre by coastal facies (nearshore, beach and bar) in rather high-energy environments. The overlying Pimenteira Formation (Gedinnian-Famennian; Daemon, 1976) represents a shallow sea in which mostly red-coloured shales, claystones and siltstones became deposited, with some intercalations of fine sandstones; geochemical studies effectuated in the microclastic deposits confirm the suggested environment (Mabesoone et al., 1985). At the end of this epoch, a small regression took place, recorded by a rather thick bed of fine to medium-grained sandstones in a nearshore facies. With respect to the high energy and stormy seas, as concluded from the aspect of the sediments, this is due to the geographical position of the area, at that time still at a rather high southern latitude (Mabesoone, 1975), with cold temperate climates, also confirmed by the boreal character of the fauna. Certain equilibrium between subsidence and sedimentation established gradually, resulting in coalescent deltas along a more or less rectilinear coastline, with the deposition of the fine to medium-grained sandstones of the Cabeças Formation (Frasnian-Strunian). In continuation, the realm became more restricted with the development of coastal lagoons in which the water circulation decreased considerably; in this situation the black to grey shales and siltstones of the Longá Formation accumulated under reducing circumstances as is confirmed by geochemical analyses (Famennian-Tournaisian). These lagoonal depressions were nevertheless local occurrences so that the black shale unit does not appear uninterruptedly through the whole basin. Slight tectonic movements during the Longá time caused small advances of the sea resulting in the deposition of the beach and nearshore sands of the lower lithosome of the Poti Formation (Strunian-Visean, Early Carboniferous), the upper unit of the Canindé Group.

After a while, in the beginning of the Late Carboniferous, the sea withdrew definitively from the area, when a new oscillatory-emergent episode started. In a short time the sea disappeared, the coastline receded and a marine coastal plain came to surface over which the rivers coming from the adjacent continent extended their courses, changing it into an alluvial plain. The upper lithosome of the Poti Formation was the consequence, representing typical alluvial plain deposits, of medium to fine sandstones with microclastic intercalations. Herewith began the Late Carboniferous-Triassic Balsas Group sediment sequence, in a geocratic phase, and in which five lithostratigraphic formations are recognized. The respective depositional environments have generally been separated based upon climatic differences, as the paleoclimatic indicators are often strongly represented in the many sedimentary responses. After the sedimentation of the Poti Formation upper lithosome, the next deposited unit is the Piauí Formation. This sequence consists of three members: (1) lower – alternated section of shales, siltstones and fine sandstones, partly in a rhythmic succession, deposited in lacustrine and alluvial plain facies, under temperate and rather humid climatic circumstances, as is recorded by a few intercalated coal lenses; the geochemical indicators point to a fresh to brackish water hydrofacies (Mabesoone et al., 1985); (2) middle – a micritic limestone with a rich faunal assemblage of Late Carboniferous age, being the consequence of a short-lived marine invasion; the sequence was laid down in nearshore shallow water to offshore shallow shelf realms, of low to moderate energy, under an already quite warm climate; (3) upper – due to regression of the sea and return of continental circumstances, a section of sandstones deposited in river and eolian environments, under warm subhumid to semiarid climate conditions. The overlying Pedra de Fogo Formation is composed of siltstones with a pisolithic chert layer at the base. The chert appears to be a silicified algal limestone from a lagoonal to shallow marine environment, as has also been confirmed by geochemical analyses. The fossil content suggests an Early Permian age; a labirintodont amphibian species has been found in this formation. The next Motuca Formation is recognized by a rapid change in lithology towards brick-red coloured fluvial sandstones, near-coastal to marine sabkha deposited microclastics with some limestones, dolomites and anhydrite, and upward again a sandstone sequence accumulated in intermittent river realms, all deposited under dry climatic circumstances. The age of the formation is supposed to be Late Permian. The geocratic sedimentary sequences of the Balsas Group finish with the Sambaiba sandstone formation. Its lithological aspect is that of well to medium-sorted light-coloured sandstones, deposited by intermittent rivers and by wind action. The suggested environment was that of dune fields in wide river valleys under dry, although not completely desertic climatic conditions. The unit is non-fossiliferous and its age is inferred as Early Triassic. After the sedimentation of these Sambaiba sandstones the Parnaíba syneclise passed through a period of non-deposition or a deposition so reduced that possible remainders of it should have been totally removed by later erosion.

In the beginning of the following Jurassic-Cretaceous submergent episode, the basin showed a flat denudation relief (Gondwana Surface) upon which various continental sediment sequences were laid down (Petri and Campanha, 1981). During this episode, the Gondwana supercontinent started to break up, causing a tectonic reaction in the basin. The thermal arching caused basalt outflows in the southern part of the basin, with peaks in the Jurassic (180–150 Ma) and Early Cretaceous (120 $\pm$ 10 Ma). These volcanics and their intercalated thin sediment complexes have been taken together in the Mearim Group (Aguiar, 1971), with the following formations: Mosquito – lower basalts; Pastos Bons – chiefly microclastics and some sandstones (age Callovian-Oxfordian, Middle Jurassic); Corda – chiefly sandstones (age Kimmeridgian-Portlandian, Late Jurassic); Sardinha – upper basalts. The respective depositional realms were lacustrine in paleo-depressions and fluvial in semiarid climate. The basin subsidence continued with in its centre the deposition of the Grajau sandstone formation, accumulated in braided river channels and floodplains; the Codó Formation, with dark bituminous shales, limestones, and anhydrite, deposited in a restricted sea under dry climatic circumstances; Itapecuru Formation, of sandstones and some shales, under regressive conditions in river floodplains (Petri and Campanha, 1981). The respective ages determined by fossils, range between Late Barremian and Cenomanian. The few Late Cretaceous lithostratigraphic units have been deposited as a consequence of the opening Equatorial Atlantic rift, and occur at the northern border of the basin.

The current Cenozoic oscillatory-emergent episode is characterized by a continuous slow basin centre subsidence and an uplift of its borders, forming the cuestas, which limit the basin chiefly at its eastern and southern sides. A few relief-correlated sequences have been deposited in local depressions as microclastic sediments and in river valleys as sands and sandstones.

Basin history. The Parnaíba Basin history may serve as an example of the cyclic succession of tectonic-sedimentary episodes in an intracontinental setting. The time span ranges from about Late Ordovician to Holocene, during which five different episodes have been recognized together with their corresponding sediment sequences. The drift of the South American continent through the different climate zones, from near-pole glacial to actual tropical and equatorial has been reflected in the character of the periodically appearing sediment successions, from glacial to continental and marine clastic and carbonate deposits. During the Late Ordovician-Silurian geocratic phase the accumulated sediments are chiefly of glacial origin and its corresponding fluvio-glacial, fluvial, and marine facies. The Devonian-Early Carboniferous thalassocratic phase is recorded by marine and nearshore sediment sequences. The Late Carboniferous-Triassic geocratic phase shows continental deposits and some invasions of the sea, finishing with an extensive denudation surface. The next Jurassic-Cretaceous thalassocratic phase is recorded by continental clastic sediments and a few marine limestones, because the sea could almost not invade the area then in full uplift as a tectonic reaction on the separation of South America and Africa. The present Cenozoic geocratic phase is determined by the basin relief development and the deposition of its correlated sediments.

Special flavors. Two special features, for which the Parnaíba Basin is also known, may still be presented here.

The first feature is that of the Sete Cidades National Park, in the north of Piauí State (area 1 in Fig. 7.10). Here the weathering of the Cabeças Formation sandstones resulted in beautiful ruinous erosional features, preserved as a national park (Fig. 7.12).

In the second place there is the Serra da Capivara National Park in the south of Piauí State (area 2 in Fig. 7.10), where under the almost vertical to overhanging scarps of small and narrow creeks cutting through the basin border cuestas, habitation existed since at least 50,000 years ago. Rupestral pictures of animals and hunting scenes have been painted on these walls, made between 10,000 and 3000 years ago (Fig. 7.13).

7.5.5.4 Afrobrazilian Depression

Generalities. The occurrence of Middle and Late Paleozoic and of Early Mesozoic sediments in the coastal areas along both sides of the South Atlantic Ocean,

Fig. 7.12: Parnaíba Basin: ruinous erosion of Cabeças Formation sandstones, Sete Cidades National Park (Piauí State).

suggests a sedimentary depression to have existed at this site since the beginning of the Phanerozoic. This "basin" has been called Afrobrazilian Depression by Ponte et al. (1972) and was redefined by Dias (1991). It should have had more or less the same character as the Parnaíba Basin, thus evidently an intracontinental sag basin upon a rifted basement within the Gondwana supercontinent.

The preserved deposits of marine as well as continental origin are actually found in Brazil, in Bahia, Sergipe, Alagoas, and Pernambuco States. The most important data have been collected from the papers of Barretto (1968), Schaller (1969), Viana et al. (1971), and Ojeda and Fugita (1974). In Africa,

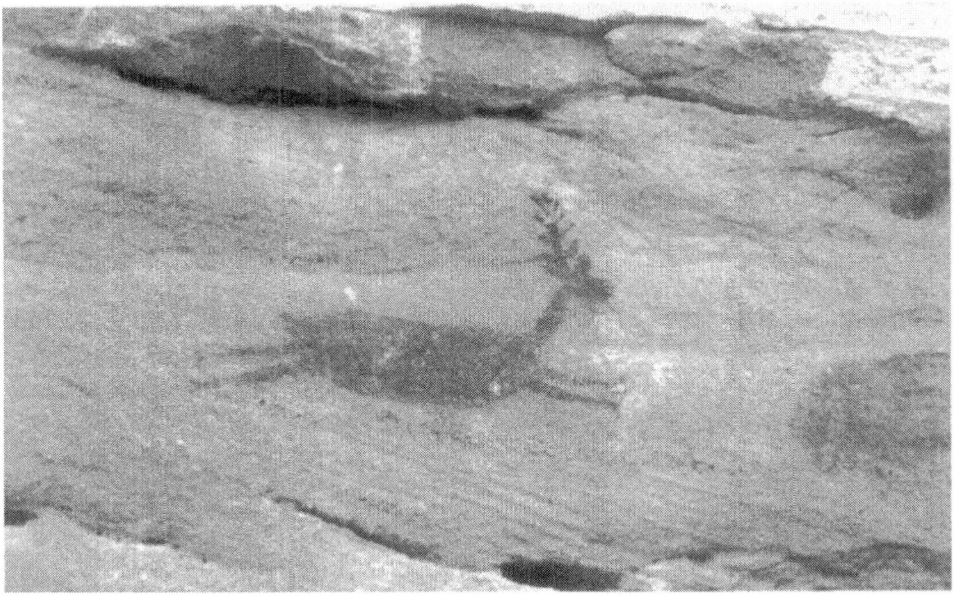

Fig. 7.13: Parnaíba Basin, Serra da Capivara National Park: pre-historic painting on Serra Grande Group coarse sandstones.

correspondent sequences have been found in the Gabon, Congo-Cabinda, and offshore Congo basins.

In the depression, the South Atlantic rift developed since Early Cretaceous. An aborted rift arm with N–S trend is also present as a real aulacogen.

Stratigraphical sequences. In Table 7.8, have been presented the stratigraphical sequences found in the basin, and ordered after their tectonic-sedimentary episodes to which they belong. Also mentioned are the lithostratigraphic units of the South Atlantic successor basins, in the sense of Klemme (1975), formed during the latest submergent episode, and which may be divided into three groups: (1) Pernambuco Basin, (2) Sergipe-Alagoas Basin, and (3) South Bahia basins: Camamu, Almada, Jequitinhonha, and Cumuruxatiba. Of these last basins, most of the deposited sedimentary sequences are found only in the subsurface of the continental shelf; the Camamu and Almada basins show also small occurrences cropping out on the continent.

The Devonian-Early Carboniferous submergent episode is only recorded in the Jatobá Basin and the northern part of the Tucano basin, preserved in tilted fault blocks. The complex lies unconformably upon the crystalline basement, and has been divided into four lithostratigraphic formations: Tacaratu, Inajá, Ibimirim, and Moxotó, with the last two present only in subsurface, as proven by wells.

Table 7.8: Lithostratigraphic sequences in Afrobrazilian depression and its successor basins.

Cenozoic oscillatory-emergent episode, possibly started in Late Campanian		
		relief correlated deposits on land and on the continental shelf, in phases caused by sea level oscillations
	alongside the whole coast	Barreiras Formation (Plio-Pleistocene)
	Sergipe-Alagoas basin	Piaçabuçu Formation
	South Bahia basins	Rio Doce Formation Caravelas Formation Urucutuca Formation
Jurassic-Cretaceous submergent episode		
		most complete sedimentary sequences, deposited in four phases
– open-ocean drift phase (Albian-Santonian)		
	Pernambuco basin	Estiva Formation (upper part)
	Sergipe-Alagoas basin	Cotinguiba Formation Riachuelo Formation
	Camamu and Almada basins	Algodões Formation
	Jequitinhonha and Cumuruxatiba basins	Barra Nova Group São Mateus Formation Regência Formation
– proto-oceanic gulf phase (Aptian-Albian)		
	Pernambuco basin	Estiva Formation (lower part)
	Sergipe-Alagoas basin	Muribeca Formation
	Recôncavo-Tucano-Jatobá basins	Marizal Formation
	Camamu and Almada basins	Taipus-Mirim Formation
	Jequitinhonha and Cumuruxatiba basins	Mariricu Formation, Itaúnas Member
– syn-rift phase (Berriasian-Aptian)		
	Pernambuco basin	Cabo Formation (Aptian)
	Sergipe-Alagoas basin	Coruripe Group, with six formations

Table 7.8: Cont'd. Lithostratigraphic sequences in Afrobrazilian depression and its successor basins.

	Recôncavo-Tucano-Jatobá basins	São Sebastião Formation
		Ilhas Group, with two formations
		Salvador Formation
		Candeias Formation
		Itaparica Formation
	South Bahia basins	Mariricu Formation, Mucuri Member (Aptian)
– pre-rift phase (Callovian-Tithonian: Middle-Late Jurassic)		
	Sergipe-Alagoas basin	Serraria Formation
		Bananeiras Formation
		Candeeiro Formation
	Jatobá, Tucano, Recôncavo and South Bahia basins	Brotas Group, with two formations
Late Carboniferous-Triassic oscillatory-emergent episode		
Afrobrazilian depression		
	Sergipe-Alagoas basin	Aracaré Formation (Permian)
		Batinga Formation (Late Carboniferous)
	Recôncavo-Tucano basins	Santa Brígida Formation (Permian)
		Curituba Formation (Late Carboniferous)
Devonian-Early Carboniferous submergent episode		
Afrobrazilian depression		
	Jatobá-north Tucano basins	Moxotó Formation (Early Carboniferous)
		Ibimirim Formation (Late Devonian)
		Inajá Formation (Early-Middle Devonian)
		Tacaratu Formation (Early Devonian)

The respective ages, partly defined by mollusks and spores, range between Early Devonian and Early Carboniferous. The accumulated sediments are of marine and continental origin.

From the following oscillatory-emergent episode, Late Carboniferous and Permian sequences have been determined in the Sergipe-Alagoas and Recôncavo-Tucano basins, as well as probably also in the subsurface of the Camamu and Almada basins. Every basin shows two lithostratigraphic formations, which rest unconformably on top of the foregoing sections: submergent episode sediments or crystalline basement rocks. The sediment sequences have essentially been deposited in continental realms, possibly even with glacial influence (Batinga Formation).

A long hiatus happened between the Late Permian and Middle Jurassic. During this latter period, already under submergent tectonic-sedimentary circumstances, a doming or thermal uplift (Matos, 1992) followed by early stretching stages of the lithosphere (Ribeiro, 1992) took place, as a prelude for the separation of South America and Africa. A so-called pre-rift phase started, recorded in the South Bahia and Recôncavo-Tucano-Jatobá basins by two formations, and in the Sergipe-Alagoas Basin by three lithostratigraphic units, all accumulated in river and lake environments.

Follows a period of intense rifting that started in the beginning of the Cretaceous and lasted until about the Aptian. The Afrobrazilian Depression became definitively subdivided into a number of more or less individual basins: South Bahia, Sergipe-Alagoas, Pernambuco, and the Recôncavo-Tucano-Jatobá aulacogen, every basin with its own lithostratigraphic sequence and with various different formations, often determined through lithic aspect and depositional environment. These environments remain still continental, with fault scarp, river and lake sediment accumulations.

At last, the rift became so deepened that the sea could invade the area, thus beginning the very South Atlantic Ocean opening (Aptian-Albian). In the ocean marginal basins clastic, carbonate, and evaporitic sediments became deposited, stratigraphically taken together into one lithostratigraphic formation with different names for the individual basins, and subdivided into members determined by rock character. In the intracontinental Recôncavo-Tucano-Jatobá aulacogen a relief-correlated sandstone became accumulated.

With the open ocean drift phase that lasted from Albian to Santonian, finishes the Jurassic-Cretaceous submergent episode. With exception of the Pernambuco Basin, in the other marginal basins deposited a fan-delta, carbonate shelf, clastic slope sequence, united into two lithostratigraphic formations with their respective members, and with different names for the Sergipe-Alagoas and South Bahia basins.

During the Cenozoic oscillatory-emergent episode, that started actually already in the Late Campanian, diverse successions of relief-correlated sediments have been accumulated, chiefly along the present coast and on the adjoining continental shelf in various sequences caused by the oscillatory movements of the sea level. From the land seaward there appears the common fan-delta – carbonate shelf – clastic slope trend. In different basins, several formation names exist, although the well-known Barreiras Formation of Plio-Pleistocene age occurs along the whole Atlantic coastline.

As is common, the episode related sediment groups are separated between them by erosional unconformities, which can be regionally traced. Moreover, the small local erosional breaks in the geocratic continental sequences are present. In Fig. 7.14, the occurrences of the on the continent outcropping sequences have been presented.

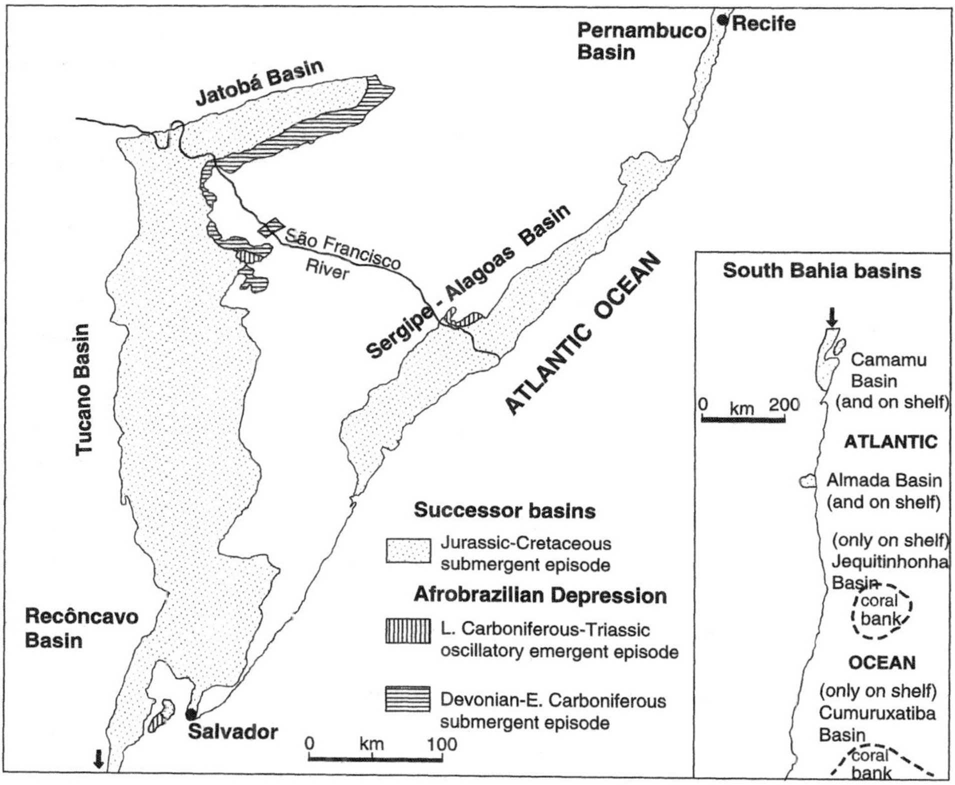

Fig. 7.14: Lithostratigraphic sequences of Afrobrazilian Depression.

Sediments and their depositional environments. The first deposited sediments in the Afrobrazilian Depression have been preserved in a fault block that comprises the Jatobá and the northern part of the Tucano basins. Due to the submergent character of the episode, most of the accumulated sequences have a marine to littoral origin. The lowermost Tacaratu Formation is represented by a succession of sandstones and pebble conglomerates, sometimes with lenses of kaolin. Their depositional environment ranges from river inundation plains to littoral realms. The next succession of clastic deposits, known as the Inajá Formation, covers conformably the underlying Tacaratu sequence, with excellent exposures in the Jatobá Basin. The unit is composed of fine sandstones, siltstones, and shales, reddish-coloured at the surface. It is rather rich in fossils of Middle-Late Devonian age (Muniz, 1979), with bivalves, brachiopods, gastropods, and also a palynoflora. The suggested depositional environment was near-coastal marine, with shallow water and a not so high salinity. The two following lithostratigraphic units, Ibimirim and Moxotó Formations, occur only in the subsurface, with a record too scarce for a definitive confirmation of their depositional environments. Both units have been found in wells drilled by Petrobrás. The underlying non-fossiliferous section is composed chiefly of medium-coarse grained sandstones with a few intercalations of shales. The overlying sequence is represented by a succession of white clayey sandstones with intercalated siltstones and shales as well as a few micritic limestones. Spores have been determined in the sequence pointing to a Late Devonian-Early Carboniferous age. The depositional environments of both units are supposed to show a transition from shallow marine to lagoonal, estuarine, and costal plain in regressional facies.

Then, from the Late Carboniferous onwards, the environments turned continental again during the following oscillatory-emergent episode. The sediment succession determined in the present Tucano and Recôncavo basins seems to represent the western border of the Afrobrazilian Depression. The lower Curituba Formation is composed of green shales with thin layers of calciferous fine sandstones; the unit is non-fossiliferous and supposed to be of Late Carboniferous age. Its depositional environment remains doubtful, with suggestion for a quiet lake realm. Conformably on top of it, the next unit occurs more frequently; this Santa Brígida Formation is composed of sandstones, calciferous siltstones and dolomites, in a positive cyclic recurrence. The siltstones and dolomites are fossiliferous, including bone breccias; through spores, the unit has been dated as Late Carboniferous-Permian. The depositional environment was fluvial of dry climate to lacustrine in its lower part, and lagoonal to marine in its upper part. The few Late Paleozoic formations still present in the Sergipe-Alagoas Basin give the impression of deposition in a more central part of the Afrobrazilian Depression. The succession starts with the Late Carboniferous Batinga Formation, subdivided into three members representing different depositional systems.

The lower member is composed of coarse diamictites, sometimes with striated pebbles. The middle member shows polymictic, matrix-supported conglomerates, and the upper member is represented by rhythmites, laminated and variegated silts and clays, having the aspect of varvites. The inferred environment of deposition is of cold climate, possibly proglacial (Mabesoone and Rolim, 1981). Unconformably on top of this formation, there occurs still the quite homogeneous succession of medium-coarse sandstones of the Aracaré Formation, with intercalated green shales, nodular limestones, and chert concretions. The fossils found in this unit are spores pointing to a Permian age. The depositional environment was evidently fluvial, from braided to meandering in wide alluvial plains.

After the Triassic-Early Jurassic period of tectonic quiescence during which an extensive denudation surface developed, the thermal arching that preceded rifting that is testified by the alkaline volcanism within the Borborema Province, caused a removal of the soil material from this denudation plain. The common subsidence of the beginning of the next Jurassic-Cretaceous submergent episode was irregular, and various peripheral basins formed. This pre-rift phase saw the deposition of the eroded soil material as the lower fine clastic sequence known under various formation names in the different basins. Once totally removed the fine clastic matter, erosion of the underlying bedrock produced the accumulation of a sandy sequence. Because the deposited sediments are almost the same in all basins, it is suggested that there has been one extensive shallow depression in the whole Afrobrazilian area, in which the sedimentation took place in coalescent alluvial fans, passing into alluvial plains and lakes (Netto, 1978).

The subsidence of the area continued and due to the movement of the continents an intense rifting started in the Early Cretaceous. Consequently, the sedimentary fill in the basins became essentially terrigenous clastic and alike in all trends. The sediment section is composed of conglomerates deposited in high-constructive fan-deltas, passing into sandstones and microclastics with some limestone intercalations, dispersed into fluvial and lacustrine systems. The paleo-environmental interpretation is as follows: Almost continuous coarse alluvial fans along the fault scarps of the rift valley borders grade distally into braided and meandering river plains which border lakes, quite deep in their central parts, with finer-grained lake deltas and clayey to calcareous sediments in the proper lake. Occasional slides caused the deposition of turbidites in various places (Bruhn and Moraes, 1988).

The following proto-oceanic gulf phase lasted only during the Aptian, and is characterized by periodic invasions of the sea into the widening and deepening rift, with its consequence for the lithic infilling. In the intracratonic Recôncavo, Tucano, and Jatobá basins, sedimentation became scarce and remained chiefly clastic, as a correlate relief deposit. In the opening ocean marginal basins a few conglomerates still accumulated along the borders, distally grading into calciferous sandstones. Upward there appear gradually medium to fine-grained clayey sandstones, dolomicrites, gypsum, and anhydrite. In the deepest parts of the basins, a thick section of halite, sylvinite, carnallite, and tachydrite is found, covered again by dolomites, gypsum, anhydrite, and barite.

The episode finishes with an open-ocean drift phase, in which the Albian and Cenomanian-Coniacian transgression epochs may be distinguished. Because the ocean was still narrow, the prevailing environment was that of a shallow marine shelf. The first sediment succession is the record of the Albian transgression, showing along the borders of the basins terrigenous conglomerates and sandstones in alluvial fan and river systems deposited during the sea level rise. In the deeper parts of the ocean, shales and biomicritic fossiliferous limestones accumulated. The Late Albian carbonate microfacies have been studied in detail by Turbay et al. (2004) as did Berthou and Bengtson (1988) for the Cenomanian-Coniaician sequences. During the following regression, the marine environment became shallower at some places, turning into warm-climate tidal flats and the consequent deposition of a variety of limestones, dolomites, calcareous shales, and finely laminated algal-biomicrites. The sequence finishes with an erosional unconformity. At the end of the Cenomanian and chiefly during the Turonian, another transgression affected the area, with the accumulation of clastic sediments along the coast, shallow-water micritic limestones on the shelf, and fine calciferous pelites and thin-bedded biomicritic limestones on the slope, of a widening Atlantic ocean. The Coniacian regression record seems to have been eroded later.

The current Cenozoic oscillatory-emergent episode seems to have started as early as in the Late Campanian. The South Atlantic Ocean had already widened but still not enough to cause the deposition of open-ocean limestones. However, in the from now on accumulated lithic facies types, the common ocean border trend of fan-delta, carbonate shelf and clastic slope deposits became evident. The fan-delta complex along the continental border, with its proximal, medium, distal, and pro-fan facies, shows clastic material derived from the erosional surface in formation on the continent at that time. The carbonate shelf complex is composed of two facies: bedded limestones and dolomites of the proper shelf, and biolithites of reef-like mounds at the edge. The clastic slope system is represented by a thick shale sequence, with intercalations of turbidites. In the Sergipe-Alagoas Basin, the unit is called Piaçabuçu Formation that shows excellent exposures. This episode's section started in the Late Campanian and lasted until the Early Eocene. Braga and Della Fávera (1978) recognized during the episode three more such successions (Middle Eocene-Early Oligocene, Late Oligocene-Early Miocene, Middle Miocene-Holocene), separated by depositional hiatuses, and representing the sea level oscillations common for the episode.

Basin history. The geological history of the Afrobrazilian depression and its successor basins alongside the South Atlantic margin of Brazil presents another example of the cyclic succession of tectonic-sedimentary episodes since Devonian. However, due to its proper characteristics the preserved record is not always complete for every episode in question. The depression developed probably already in the Precambrian upon a basement that then suffered an initial rifting (Ebert, *apud* Beurlen, 1961). The reactivation of this proto-Atlantic rift must have occurred since the Late Triassic whence the beginning of the dismemberment of the Gondwana supercontinent, and when extensive basalt outpours took place chiefly at the Brazilian side of the shear zone. The Devonian-Early Carboniferous submergent thalassocratic episode shows marine and littoral deposits. The Late Carboniferous-Triassic continental geocratic phase is recorded by terrigenous sediments, possibly with some glacial influence in Late Carboniferous times.

The South Atlantic rift opening manifested itself in the Early Cretaceous during the Jurassic-Cretaceous submergent episode. The gradual propagation of the rift occurred essentially from south to north. Where this propagation was locally impeded by basement anomalies, it led to a complex deformation near the tip of the propagating rift. The deviation to the NE that opened the Sergipe-Alagoas Basin resulted from the differential clockwise rotation of South America relative to Africa (Szatmari et al., 1984). The NE direction follows the general trend of the extensional fractures in the Borborema Province. In addition, this caused also the abortion of the Recôncavo-Tucano-Jatobá S–N trending rift, transforming it into an aulacogen. The episode is recorded by the common accumulation sequences of the pre-rift, syn-rift, proto-oceanic gulf, and open-ocean drift phases. However, only in the still ongoing Late Cretaceous-Cenozoic oscillatory-emergent episode, the wide-open ocean lithic sequences have been accumulated along the South Atlantic margin, together with the often-occurring relief-correlated deposits on the continent. Not earlier than the Late Campanian, the last link of continental crust between South America and Africa became severed at the northernmost limit of the depression in Pernambuco State (Mabesoone, 2000b).

7.6 FINAL REMARKS

From the foregoing paragraphs, it becomes evident that sedimentary basins form cyclically, in particular during submergent tectonic-sedimentary episodes. There appear even alternating, and cyclically returning phases during which rift basins are preferentially developed, or interior sag basins by simple intracratonic subsidence. Moreover, this may be due to alternating phases of stronger and weaker tectonic activity. The sedimentary sequences deposited in these basins appear to be surprisingly uniform, with nearly always the same type of sediment succession.

Two other types of sedimentary sequences, of secondary importance, have still to be mentioned: glacial deposits and relief correlated sediments.

Generally, glaciations occur during oscillatory-emergent tectonic-sedimentary episodes (Table 7.1). This has been proved also for the study area at least since the Late Mesoproterozoic. Evidence of glacial deposits, mostly in the form of tillites which remain recognizable even after strong metamorphism, has been recorded from various basins through the whole area. As examples serve the Neoproterozoic sequences of the North and Central Brasília belts (Marini et al., 1984), those within the São Francisco craton and the Ribeira Basin (Kaufman, 1998), and in the Silurian of the Parnaíba Syneclise within the Borborema Province (Góes and Feijó, 1994), mentioned above.

Relief-correlated sediments, deposited under relief denudation and erosional conditions, are only recognizable as such since the Cretaceous, in the whole study area (Mabesoone, 2003b). They are often present as thin sheets of sands and sandstones over flattened surfaces, or as fluvial deposits in river valleys and coastal cliffs, where they appear during the relief-forming oscillatory-emergent episodes. The last deposited relief correlated sequence in Late Pliocene to Pleistocene times, is known as the Barreiras Formation that is present along almost the whole coast of northeastern and eastern Brazil. Its variegated clastic sediments which range among sands, silts, and clays provide a colourful aspect to the coastal cliffs of the region.

With respect to the opening of the Atlantic Ocean and the development of the Atlantic marginal basins, it appears that the rifting that opened the Atlantic already started in the Proterozoic (Beurlen, 1961). Maybe even the Ribeira Basin has been affected by it, and certainly the development of the Afrobrazilian Depression. This seems to have later enabled the opening of the ocean and the break-up of Gondwana, separating South America from Africa at that place.

Finally, it may thus be concluded that the cyclic behaviour of earth's geological phenomena is also reflected in the sedimentary basin formation and evolution, and consequently in the sediment type accumulation. This can be demonstrated since the beginning of the Mesoproterozoic until the very Holocene, as has been shown here for the Brazilian Borborema and São Francisco tectonic provinces.

8. RUSSIAN ARCTIC SHELF SEDIMENTARY BASINS

O.I. SUPRUNENKO AND M.K. KOS'KO

8.1 INTRODUCTION

General structural features and trends of the evolution of sedimentary basins are known mainly from ancient platforms. Recent achievements in marine geology on the continental margin provided new data, which have not yet been fully considered in global structural and historical models. Recent geological and geophysical data on the Russian Arctic shelf were synthesized and sedimentary basin models were presented by scientists from VNIIOkeangeologia in a recently published monograph (Gramberg et al., 2004). The present review is focused to distinguish and correlate basic tectonic and sedimentary events and settings, with examples of better studied shelf sedimentary basins to perform correlations with global geodynamic cyclicity on the basis of data provided from the monograph and other publications.

Russia possesses a vast share of the Eurasia Arctic shelf. Peculiarities of the Eurasian Arctic Ocean passive continental margin are extensively described in detail in the following publications: Gramberg and Pogrebitsky (1984); Grantz et al. (1990); Gramberg et al. (2000); Dodin and Surkov (2002). They are, in many respects, determined by the young age of the Arctic Ocean. This young age and the moderate area of the deep ocean floor correlate with the unusually great width of the shelf and the enormous thick sedimentary cover in the shelf basins.

Peculiarities of the Amerasian Basin are not adequately considered in most of the general geodynamic constructions. The western part of the Amerasian Basin including the Lomonosov Ridge and Alfa-Mendeleev Rise appears to be an assemblage of marginal plateaus common for Eurasia and North America. Typical oceanic features of regular linear anomaly magnetic fields are not clearly established in the eastern part of the Amerasia-Canada Basin, and attempts to distinguish traces of the symmetric spreading lead to extremely complicated results.

Cyclic Development of Sedimentary Basins
J.M. Mabesoone and V.H. Neumann, editors.

8.2 BARENTS-NORTH KARA BASIN

8.2.1 GENERALITIES

The Barents-North Kara Basin comprises the Barents Sea and Pechora Sea, the Bolshezemelskaya Tundra areas, and the north of the Kara Sea (Fig. 8.1). The

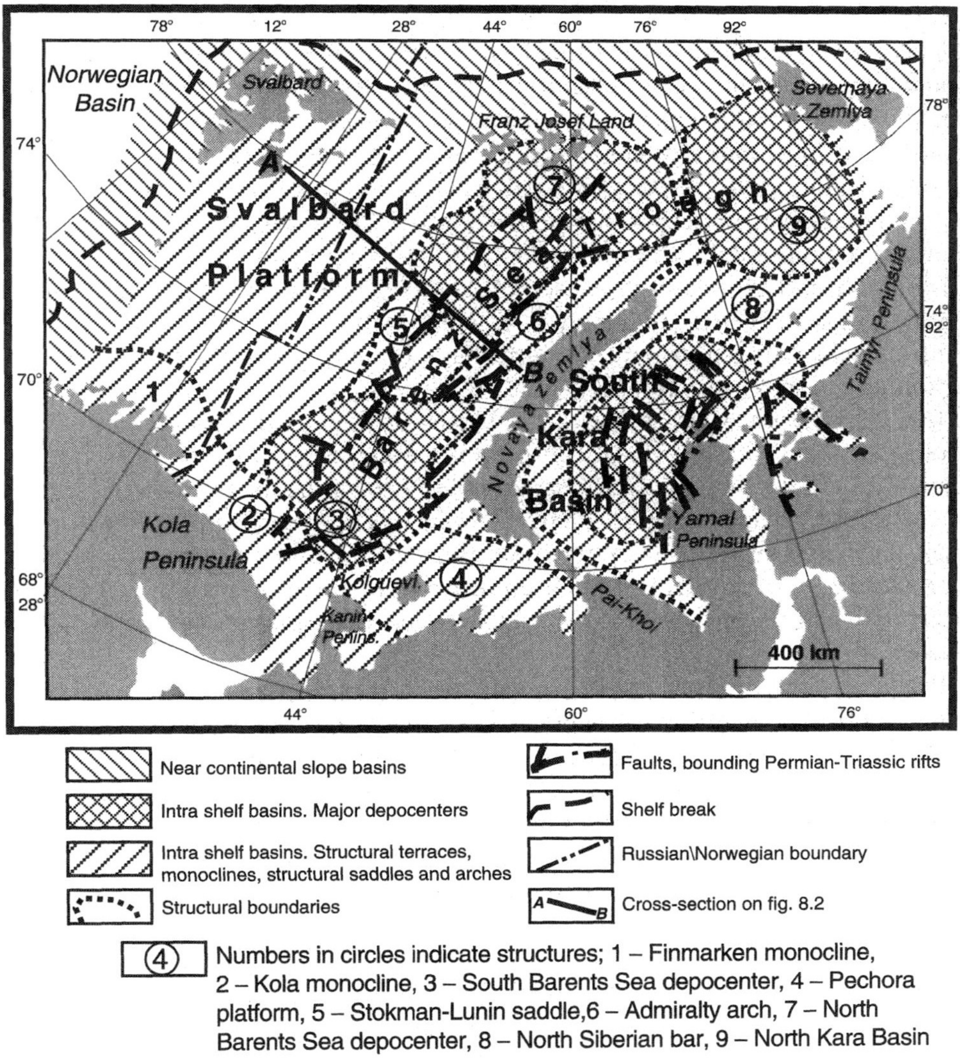

Fig. 8.1: Sedimentary basins on the West Russian Arctic shelf.

total area of the basin is 2,000,000 km^2. It is separated from the West Siberian inland sedimentary basin by structural and topographic highs, such as the Polar Urals, Pai-Khoi, Vaigach Islands, Novaya Zemlya Archipelago, and the buried North-Siberian sill with the Izvestia CIK Islands and North Taimyr. The boundary with the Norwegian-Greenland and Eurasia oceanic basins is marked by a chain of uplifted basement domes on the outer shelf crowned with the Archipelagos Svalbard, Franz Josef Land, and Severnaya Zemlya. These archipelagos are separated from each other by troughs transverse to the continental slope. The Baikalian fold structures including ancient crystalline blocks are widespread in the basement of the basin. In the north-west the basement is Grenvillian in age, whereas in the far west, mainly in the Norwegian sector, the basement is Caledonian in age. Archean complexes stretch to the south from the Kola Peninsula. The age of the basement in the west of the Kara Sea is most likely to be Proterozoic-Pre-Vendian.

The axial zone of the Barents-North Kara sedimentary basin is formed by the Barents Sea Trough with the North Kara Basin in the east. Structural saddles divide the trough into a series of depocenters, being third-order structures in regard to the Barents-North Kara Basin.

The area between the Barents Sea Trough and surrounding uplifts comprises the Svalbard plate in the west, the Kola monocline and Pechora plate in the south near Novaya Zemlya, and near the Severnaya Zemlya regions in the east and south-east.

8.2.2 SEDIMENTARY COVER

8.2.2.1 Generalities

The sediment cover is up to 18–20 km thick in the Barents-North Kara Basin. This cover is exposed and can be examined mainly on the periphery of the basin on the Arctic islands and the coast, and has been cored in a few marine-prospecting boreholes. The majority of the marine boreholes are in the south of the Barents Sea and Pechora Sea. Northward from 77°N, marine boreholes are absent. The density of the seismic coverage is quite irregular. The best coverage is in the Pechora Sea and in some parts in the north of the Barents Sea. The lowest density of seismic lines is in the Barents Sea northward from 78°N, and in the north Kara Sea. In the greater part of the Barents-North Kara Basin the sedimentary cover can be divided into three large lithostratigraphic complexes: Proterozoic (predominantly Vendian)-Lower Permian terrigenous-carbonate, Upper Permian-Triassic terrigenous, and Jurassic-Cretaceous terrigenous.

8.2.2.2 Late Proterozoic-Lower Permian Terrigenous-Carbonate complex

This complex occurs throughout the entire area of the Barents-North Kara Basin. Its thickness varies from 1 to 2 up to 5 km, and reaches 12 km in some places. Depth to the top of the complex ranges from a few meters below sea bottom to as much as 10–13 km. The base of the complex was distinguished as an acoustic basement. The top of the complex is the intra-Permian unconformity with an age of about 260 Ma. The lowest Upper Proterozoic (?)-Early Paleozoic part of the complex cannot be divided by seismic data.

Platform-type Vendian and Cambrian deposits are known in the Barents Sea on Bear Island and on North-East Land (Nordaustlandet) Island of the Svalbard Archipelago. Their distribution is limited, and they are mainly terrigenous sediments and dolomite. Their thickness does not exceed a few hundred meters. The strata occur apparently only in deep basement depressions offshore.

The Ordovician, Silurian and Early Devonian deposits rest on underlying sediments with erosion. Hiatuses in the succession are common. The deposits are carbonaceous and terrigenous-carbonate with vertical and lateral facies variability. Conglomerate occurs frequently within the sequence. The thickness varies widely and may reach 5000 m. The Ordovician deposits dominate within the section. Evaporites are perhaps present in the east. The Silurian section is not complete and often totally eroded. From the Middle Devonian onwards the succession is predominantly organogenic carbonate with occasional reefs and widespread evaporites. The strata are similar to the deposits of the same age on the Russian Platform. It can be traced along the entire margin of the Barents-North Kara Basin, in the Tyman-Pechora province, on the Kolguev Islands (Preobrazenskaya et al., 1995), Bear Island, West Svalbard, Alexandra Land Islands, and in the South Island of Novaya Zemlya.

Under favorable seismogeological conditions in structurally uplifted areas in the Pechora Sea in particular, the lower part of the sediment cover could be divided as follows:

- Ordovician, mainly terrigenous strata with a carbonate unit in the upper part. An unconformity was recognized at the base of the carbonate unit.
- Silurian carbonate-terrigenous strata.
- Lower-Middle Devonian predominantly terrigenous and sulfate-terrigenous-carbonate strata, which are greatly varying in thickness and areal distribution. Lower Devonian sulfate-terrigenous-carbonate units tend to be confined in grabens. Their thickness reaches 1–2 km. The top of the Lower Devonian is a regional disconformity. Middle Devonian deposits are sandstone and siltstone with beds of volcanic ash in the upper part. An erosional unconformity is observed in the upper part of the Middle Devonian deposits.

- Upper Devonian carbonate strata.
- Carboniferous carbonate-evaporite strata.
- Permian carbonate strata occur throughout the southern part of the Barents Sea shelf. The age of the top of the Permian carbonates varies from Early Permian in the marine continuation of the Tyman-Pechora province to Late Permian in the western regions of the Barents Sea shelf. In the Norwegian sector the Late Permian age of the strata is proved by drilling on the Finnmarken (Ivanova, 1997).

The upper strata starting from Upper Devonian can be traced in the northern part of the shelf due to the presence of Early Carboniferous and Early Permian regional reflectors related to the World Ocean regression 360 and 250 Ma ago (Baturin et al., 1991).

8.2.2.3 Permian-Triassic Terrigenous complex

The Barents Sea rift originated in the Middle Permian and developed through the Triassic in the eastern part of the Barents Sea. The rift stretched for 1000–1200 km from south to north. The thickness of the granite-metamorphic layer is greatly reduced, M break is abnormally elevated, sedimentation rate three to four times higher than that of the underlying and overlying sediments within the rift. The rift boundaries can be most reliably detected by Permian and Triassic high sediment thickness gradients.

The Permian-Triassic complex is the main fill of the rift. According to geophysical data the thickness of the complex reaches here 10–12 km. Outside the rift Permian-Triassic strata are widespread, with but a lower thickness of 2–6 km. The sediments are terrigenous varying from continental and lagoonal in the south to shallow-marine in the north. The Permian-Triassic deposits have avalanche sedimentation typical features such as clinoforms and large cross stratification. The strata have been studied in outcrops on islands and drillholes on islands and offshore.

The Permian, mainly Upper Permian units represent the regressive part of the large Upper Paleozoic cycle. Its lower part is formed by marine, predominantly clayey-calcareous and cherty deposits and the upper part consists of coastal and paralic coal-bearing argillaceous-silt-sandy strata. Basaltic tuffaceous material is present locally in the upper part. The transgressive–regressive cyclicity is clearly observed in the Upper Permian strata in the east and south-east. West- and north-westward the sedimentary structure is dominated by large clinoforms seen on seismic profiles. It is particularly well marked at the beginning of the Late Permian. In the west and north-west of the region the Upper Permian deposits were quite often completely eroded in Pre-Triassic time. The thickness of the Permian deposits is up to 3–3.5 km.

The deep-water environment changed to accumulation of thick mud deposits related to an increase in clastic supply from the south and south-east in the South Barents Sea depocenter.

A significant portion of the Barents Sea region sediment cover was formed by Triassic sediments. The thickness of these deposits in depressions reaches 7.5 km. They have been studied in outcrops on Bear Island, West Svalbard, Edge Island, and the Novaya Zemlya Archipelago, and from drillholes on Kolguev Island, Franz Josef Land on the Admiralty Rise, on the Svalbard Plate, and offshore in the Barents Sea depression. The deposits are terrigenous with interbedded carbonates and concretions. Tuffaceous sediments are present locally at the base. In the north of the region, the Triassic is mainly marine, and in the south lagoonal and continental sediments are widespread.

The top of the complex occurs as an erosional unconformity: Pre-Rhaetian erosion in connection with the global sea level lowstand at 215 Ma ago.

The seismic facies analysis enables the detection of some peculiarities of the Triassic sedimentation. Marginal zones of large structural depressions are bounded by continental sand and clay deposits with typical clinoforms. Towards the middle of the depression they are replaced by rapidly accumulated thick, mainly clayey strata with poorly defined lamination. This environment changed in the Upper Triassic when the rate of deposition reduced and shallower water and laminated sediments extended across the whole area of even the most stable depocenters. The marine horizon present in the west was a response to the eustatic sea-level maximum at the end of the Ladinian age (224 Ma).

Sheets of basic rocks occur in the Permian-Triassic deposits. These rocks in the borehole in the central part of the Barents Sea are Jurassic in age (131–159 Ma; Komarnitsky and Shipilov, 1991).

8.2.2.4 Jurassic-Cretaceous Terrigenous complex

The Jurassic-Cretaceous deposits of the Barents-North Kara Basin and West Siberia have much in common: the same major cyclicity, predominance of marine deposits in the Jurassic and the Early Cretaceous, occurrence of shallow-marine and lagoonal sediments in the second half of the Early Cretaceous, and return of a marine regime in the Late Cretaceous. In both basins, rich with organic matter specific marine clay deposits are restricted to the end of the Jurassic (Kimmeridgian and Volgian/Tithonian).

The Lower Jurassic deposits, 700–750 m thick, overlie unconformably the Upper Triassic strata. They are mainly non-marine sands with rare and thin interbeds of mudstone, siltstone, conglomerate, and sometimes coal. Sandstone is fine- to coarse-grained, quartzose, and rarely polymictic. The sandstones are mainly alluvial, rarely deltaic, and coastal-marine; the clayey units are usually lacustrine-alluvial.

The Aalenian-Bathonian strata show an irregular alternation of sandstone, siltstone, and clay sediments, up to 400 m thick. The Callovian strata, 150–200 m thick, are marine clays with separate, sometimes quite thick interbeds of sandstone and siltstone; the sandstones represent bars and shallow-marine sandbanks. The Middle Jurassic sandstones are variably mature. Kaolinite is predominant in the cement. Coarse-grained sandstone together with gravelstone and sedimentary breccia are restricted to the lower part of the Jurassic sequence.

The Upper Jurassic strata overlap the Middle Jurassic ones in the east of the Barents Sea. High organic-content clays, with II and III types TOC of 4–23%, dominate in the Upper Jurassic section. It is assumed that the Upper Jurassic series overlay the Callovian clay with an erosional disconformity. Drilling revealed Oxfordian sandstone and siltstone, sometimes with glauconite, at the base. Black clay accumulated in the shallow waters.

The thickness of the Jurassic deposits reaches 2.2 km in the South Barents Sea depocenter. The deposits are sometimes completely eroded on the inner and bordering highs. The Jurassic sedimentation was controlled by an expanding transgression. At the beginning of the period, the marine regime existed only in the South Barents Sea depocenter. It was surrounded by palustrine alluvial plains rarely flooded by the sea. By the beginning of the Middle Jurassic, the limits of marine sedimentation expanded, and by the Bathonian-Callovian the marine regime advanced over the North Barents Sea depocenter. At the end of the Jurassic, the sea covered most of the Barents Sea shelf. Isolated zones of Late Jurassic stable specific marine sedimentation were located in the South and North Barents Sea depocenters, where black clays accumulated.

The Neocomian part of the Jurassic-Cretaceous terrigenous deposits is clay, sometimes carbonate clay, rarely with interbeds of siltstone. The Neocomian deposits formed in an open-marine shelf environment. Clinoform bedding is typical for these strata. Their thickness reaches 500 m.

The Aptian-Albian deposits are best studied in offshore boreholes on the Shtokman-Lunin sill. They are an irregular alternation of sandstone, siltstone, and clay deposited in lacustrine-alluvial and shallow-marine environments. The Aptian is dominated by sandstone, often coal-bearing. Maximum regression of the sea occurred during the mid-Aptian. The Albian deposits are predominantly marine clays, especially in the upper part. The Aptian-Albian sediments are up to 1000 m thick on the Shtokman-Lunin sill.

Jurassic and Early Cretaceous traps are well developed in the north of the Barents-North Kara Basin (Gramberg, 1988; Makar'eva, 1997; Stolbov, 1997, 2000; Dibner, 1998).

8.2.2.5 Upper Cretaceous-Cenozoic

The Upper Cretaceous strata in the Barents Sea region are highly eroded. On the Shtokman-Lunin sill they are Cenomanian sandstone, siltstone, and clay, less than 100 m thick. A thin Cenozoic sedimentary cover overlaps units that are all more ancient.

The Barents-North Kara Basin was intracontinental prior to the formation of the Arctic oceanic basin, in the same way as the present-day West Siberian Basin. Provenance areas were located not only in the south-west and south-east of the basin, but also in the north, where the Eurasia deep-sea basin formed in the Cenozoic. Simultaneously, the Novaya Zemlya orogen originated; it separated the Barents Sea and West Siberian regions, which developed in a similar way in the Jurassic and Cretaceous.

The Cenozoic sedimentary basins are known only on continental slopes and near slope shelf zones. The West Svalbard graben along the shelf break of the Norwegian Basin and the Cenozoic sedimentary basin on the continental slope of the deep-sea Nansen Basin can serve as an example.

During the Paleogene and the major part of the Neogene, the Barents-North Kara Basin was an area of uplift and erosion of the Cretaceous deposits, as well as a bypass zone of clastic material. Sedimentation recommenced at the end of the Neogene and continued through the Quaternary. However, the thickness of the sediments is not great, and therefore the Barents-North Kara Basin should be considered as Paleozoic-Mesozoic.

8.2.3 REGIONAL STRUCTURES

The Barents Sea Trough is located in the east of the Barents Sea and continues into the north of the Kara Sea. It is about 2000 km long and up to 600 km wide. The continental basement in the trough is greatly reworked and apparently replaced by suboceanic-type crust in its axial part. The boundaries of the trough are controlled by flexure and fault zones (Fig. 8.2). The trough is divided into the South Barents Sea depocenter, North Barents Sea depocenter and North Kara Basin, by structural saddles and sills. The North and South Barents Sea depocenters are separated by the Shtokman-Lunin sill.

The thickness of the sedimentary cover of the Barents Sea Trough is up to 19–20 km. Pre-Upper Permian, mainly carbonaceous deposits in the major part of the trough are buried to depths of over 7 km, and are more than 5 km thick. The terrigenous Permian and Triassic series are characterized by an abrupt thickness increase from 6 km on the flanks to 12 km towards the axis. Permo-Triassic rifting plays a major role in the formation of the modern structure of the Barents Sea Trough.

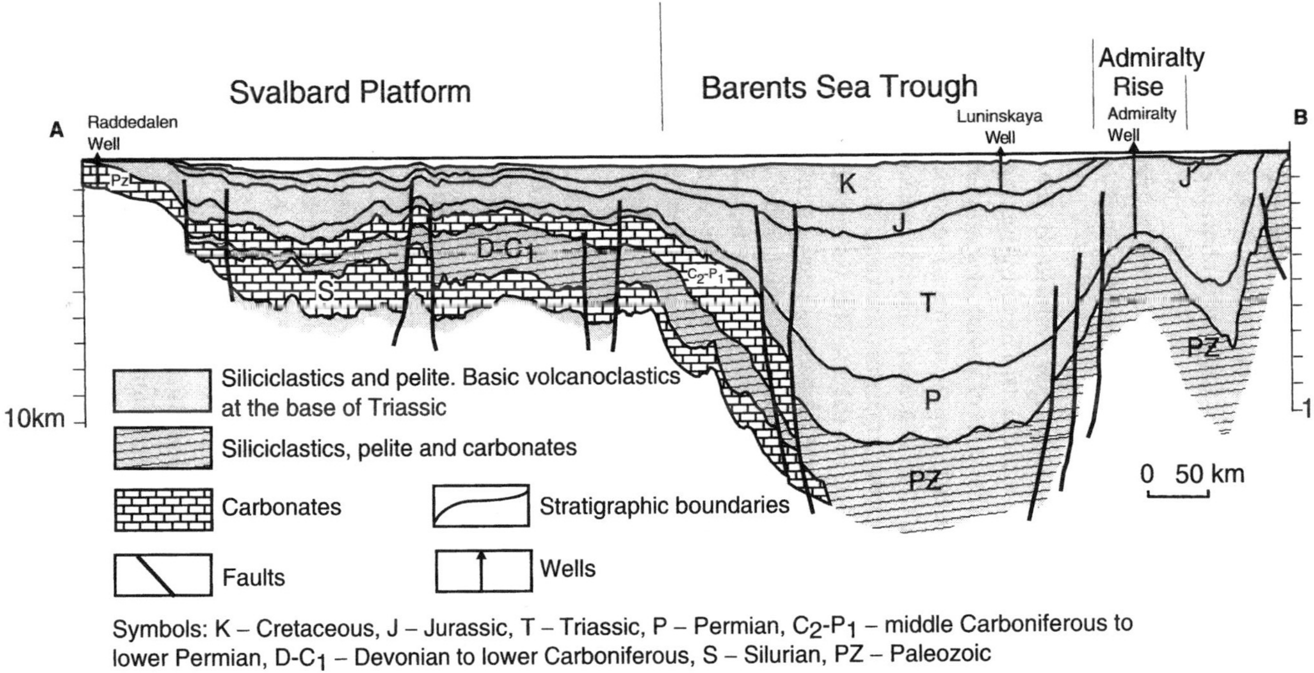

Fig. 8.2: Geological cross-section of the Barents-North Kara Basin; line AB in Fig. 8.1.

The *South Barents Sea depocenter* (500 × 400 km) is distinguished by the distribution of the Jurassic-Cretaceous deposits. The depth to the top of the Carboniferous-Lower Permian carbonates exceeds 13 km, and the depth to the top of the Jurassic sediments reaches here 2500–2800 m.

The *North Barents Sea depocenter* (375 × 150 km) is characterized by a significant thickness of the sedimentary cover (up to 17 km). The thickness of the Upper Permian-Triassic deposits reaches 12–13 km, and the Jurassic-Cretaceous thickness is about 2.5 km. The presence of intrusive sheets in the Mesozoic succession is indicated by geophysical data.

The *Shtokman-Lunin sill* (475 × 175 km) is manifested by the top Jurassic deposits occurring at depths from 23 to 12 km. The amplitude of some individual structures on the sill at the top Jurassic level sometimes may exceed 500 km.

The *North Kara Basin* is subdivided into a western and eastern part. In the east, the sediment cover consists mainly of Paleozoic deposits, which are affected by considerable faulting. The fault border grabens of various orientations with the basement are buried to a depth of 14–16 km. The vertical amplitude of the individual structures could reach a few kilometers. The bedding steepness quite often exceeds normal platform values, reaching here 600 m/km. The western part varies due to the dominance of Mesozoic deposits in the known part of the cover. The maximum depth to the topmost Upper Jurassic deposits is 2.5–2.9 km where it is overlapped by the Jurassic and Cretaceous deposits.

The *Svalbard Plate* occupies the north-western part of the Barents Sea. The plate is a structural terrace between elevated blocks with an uncovered basement on Spitsbergen Island and the Barents Sea Trough. Contrasting platform-type folds of lower rank are typical for the Svalbard Plate; usually they do not exceed 300 km in diameter. Horsts and grabens bounded by faults with their amplitude up to 1.5 km frequently occur in the lower horizons of the sedimentary cover. The largest graben is the Nordkapp rift with a reduced continental crust and a length of 500 km. Diapiric structures, including stocks, up to 4 km high, occur in this rift, and in other troughs containing evaporites.

In the south the Barents Sea Trough is bounded by monoclines connecting the trough with the Baltic shield in the west, and the Pechora Basin in the east. Normal faults are parallel with the monoclines. The fault block structure displays a stepwise northward burial towards the depocenter.

The *Pechora Basin* overlies the Baikalian basement. It is contoured by the distribution of weakly deformed Middle Jurassic-Cenozoic strata. The lower stages of the sediment cover form an assemblage of troughs, ramparts, structural terraces, and monoclines accompanied by faults. The general structural trend is to the north-west.

The *Near Novaya Zemlya structural area* embraces the Barents Sea Trough from the east and south-east. It is a transitional zone between the Novaya Zemlya fold belt and the trough. The area is separated from the Novaya Zemlya orogen by large-scale thrusts and overthrusts, and from the Barents Sea Trough by normal faults. From the orogen to the trough, the near Novaya Zemlya area consists of structural terraces, a trough, and a structural high, called the Admiralty Rise. The sedimentary cover comprises Paleozoic and Triassic strata. Jurassic-Cretaceous deposits are absent within the major part of the area. The sedimentary cover on the Admiralty Rise is 5 km thick, and on the structural terraces up to 12 km.

The eastern closure of the Barents Sea Trough, the North Kara Basin, is bordered in the east and south-east by structural terraces and monoclines with exposed lower horizons of the sedimentary cover and, of the basement, on the islands and on the seafloor surface.

8.3 SOUTH KARA BASIN

The South Kara Basin appears to be the northern, most subsided structural depression of the West-Siberian midland plate (Fig. 8.1). It is separated from the North Kara Basin by the North-Siberian (Novaya Zemlya-Taimyr) sill. In the west it is bounded by the Pai-Khoi-Novaya Zemlya and in the south-east by the Taimyr fold systems.

The basin is filled with strata from Permian to Quaternary age, and more than 10 km thick. The following parts are distinguished in the sequence: rift related Permian-Lower Triassic; Lower-Upper Triassic, traditionally assigned to the rift type, but representing in fact, a transitional regime from rifting to general subsidence; a Jurassic-Quaternary part, responding to the stage of total areal downwarping of the basin. Rock samples from the Early Cretaceous have been studied from wells and outcrops. The properties of the Permian and Triassic formations were extrapolated from conjugated regions, developed in similar geodynamic settings.

Syn-rift volcanic-terrigenous strata are admittedly dated as Permian-Early Triassic, and are present only in central parts of the Kara Sea basin where they are up to 4 km thick. Volcanogenic deposits are widely distributed throughout the Lower Triassic formations in the north of West Siberia, southward from the Yamal peninsula (Bochkarev, 1985). In Tyumen well UD-6 the Induan (Lower Scythian, Early Triassic) deposits are represented by a non-marine sedimentary-volcanic unit. On the left bank of the Enysei river estuary the Induan interval of the section is built of variegated non-marine deposits of rhythmically alternating claystone, siltstone, tuff, and tuffaceous sandstone, with lava flows (Zaporozhtseva, 1958).

According to data from the Enysei-Khatanga trough the lower-Upper Triassic deposits are rhythmically alternating sandstones, siltstones and claystones. The terrigenous unit is presumably Olenekian (Upper Scythian)-Late Triassic in age. It constitutes a continuous cover in the South Kara Basin, gradually pinching out towards the margins. The Olenekian sedimentation took place in separate lacustrine depressions. Maximal Triassic transgression happened in the Middle Triassic. Towards the end of the Middle and beginning of the Late Triassic, the arid climate changed to humid, and formation of gray-colored strata with abundant plant detritus, and coal interlayers commenced. Sedimentation in the Barents-Kara region ceased in the Middle Norian, and Late Norian-Rhaetian deposits are missing. The thickness of the Olenekian-Upper Triassic strata is 0.5–2 km after seismic data.

Transgression of the sea started in the Early Jurassic, and the Jurassic-Cretaceous episode was generally marked by a consistent spread of the sedimentation across the region. In Hettangian-Sinemurian time shallow marine, predominantly sandy deposits accumulated in the South Kara Basin, alternating with alluvial sands toward the northern and south-western flanks. Later, in the Bajocian, the marine basin widened with the maximum transgression in the Early Toarcian when a regionally consistent pelitic unit became deposited (Devyatov et al., 1994).

The Jurassic deposits are spread throughout the basin except its margins. Closest to the South Kara Basin area, in the north of West Siberia, the Jurassic deposits have been studied in the western part of the Enysei-Khatanga trough. The Lower and Middle Jurassic deposits here are strata of siltstone and sandstone, or siltstone and pelite. According to the well and seismic data from the Yamal and Gydan peninsulas, these deposits spread into the South Kara Basin. The thickness of these deposits in this basin is up to 2.8 km.

Marine clayey siltstones up to 1 km thick, with rare sandy beds are predominant in the Upper Jurassic. Formation of a siliceous-calcareous "black clay", with TOC 5–17%, was the most prominent feature of the Late Jurassic sedimentation in both the Barents-North Kara and South Kara basins. In this latter area, the sediments accumulated in the Volgian and beginning of the Berriasian. The thickness of the sequence does not exceed 100 m. The TOC content decreases towards the flanks of the depression.

The Cretaceous deposits of the Lower and Upper series occur throughout the basin. They were penetrated by offshore wells and wells on the Belyi and Sverdrup islands. A complete Cretaceous sequence has been studied in wells on the Yamal peninsula. The sequences found in the South Kara Sea and on the Yamal peninsula are practically identical.

The Lower Cretaceous sequence begins with a marine Berriasian-Valanginian clay unit. Sandstone and siltstone beds occur upward in the section, increasing in number towards the east.

The Neocomian is characterized by tectonic activation such as the growth of local structural highs, evidenced by the erosion of the "black clay" (Fomichev and Volkova, 1989). Clinoforms of a prograding shelf are typical for the Neocomian deposits. They are evidence of infilling of the basin from the east and, to a lesser extent, from the south-west and north-west. The clinoforms consist mainly of clay with irregularly intercalated sandstone and siltstone beds.

In the Hauterivian-Barremian the marine clay deposits were replaced by more sandy coal-bearing coastal-marine and continental sediments.

The Aptian was marked by widespread accumulation of sandy coal-bearing deposits of coastal-continental and coastal-marine origin. In the South Kara Basin the sea retreated to the south, however intermittent transgressions repeatedly spread out across the area. In Aptian time, the relief was almost peneplanated in the whole western sector of the Russian Arctic shelf, excluding Taimyr. The Early Albian transgression resulted in the formation of a regional marine substantially clay unit in the South Kara Basin. Apparently, differences in development of the Barents-North Kara and South Kara basins started from the Albian onwards.

Maximum Early Cretaceous sediment thickness reaches 2 km.

Marine conditions continued into the Late Cretaceous. The sequence is composed of a mainly silt and clay shallow-water, 1000 m thick unit. Rather deep-water conditions were typical for the Turonian and Late Santonian. Opoka and opoka-like clay accumulated in the Late Santonian in the South Kara Basin.

Marine clay, silt, and sand were deposited in the Lower Paleogene. Beds of diatomaceous clay, diatomites, and opoka occur as well in the upper part of the Paleogene sequence. The strata are up to 260 m thick.

The Neogene-Quaternary deposits are predominantly marine loam, loamy sand, and infrequent gravels, having a thickness of 250 m.

The depocenter in the central part of the South Kara Basin is termed the South Kara Depression. The peripheral zones of the basin are formed by monoclines and structural terraces. Thickness of the sediment cover in the South Kara Depression reaches 12 km, and from the south northward it decreases from 6–12 km to 4–7 km. This reduction in thickness is mainly due to thinning of the Triassic and part of the Jurassic. The depression is characterized by the presence of specific structures of various ranks, isometric in the south and elongated in the north. According to peculiarities of potential field anomalies, intersected rift-related grabens trending NW and NE are confined to the base of the sediment cover. The width of the largest specific structures reaches a few hundreds of kilometers. Vertical amplitudes on the base of the sedimentary cover usually do not exceed 1.5 km, but reach 5 km in exceptional cases. Upward in the section the amplitudes decrease and in the Upper Cretaceous beds they do not exceed 100 m. Steepness of the slopes ranges from 120 to 450 m/km.

In the peripheral zone the sediment cover subsides towards the depocenter, from exposure of the base to the surface down to a depth of 5.5 km. Inclination of the layers in the base is estimated as 25–60 m/km, while near the Upper

Cretaceous level, it becomes reduced to 10 m/km. Within the peripheral zone, individual Triassic and Jurassic units pinch out divergently. Structural terraces are complicated by platform-type folds of lower rank, similar to folds in the inner depression of the basin.

8.4 LAPTEV SEA BASIN

The Laptev Sea Basin is a topographically and structurally defined centrocline open to the Eurasia Basin (Fig. 8.3). This centrocline is located on the closure of the vast area of subsidence and extension of the lithosphere surface which projects from the oceanic area to the continental shelf. In the west, the Laptev

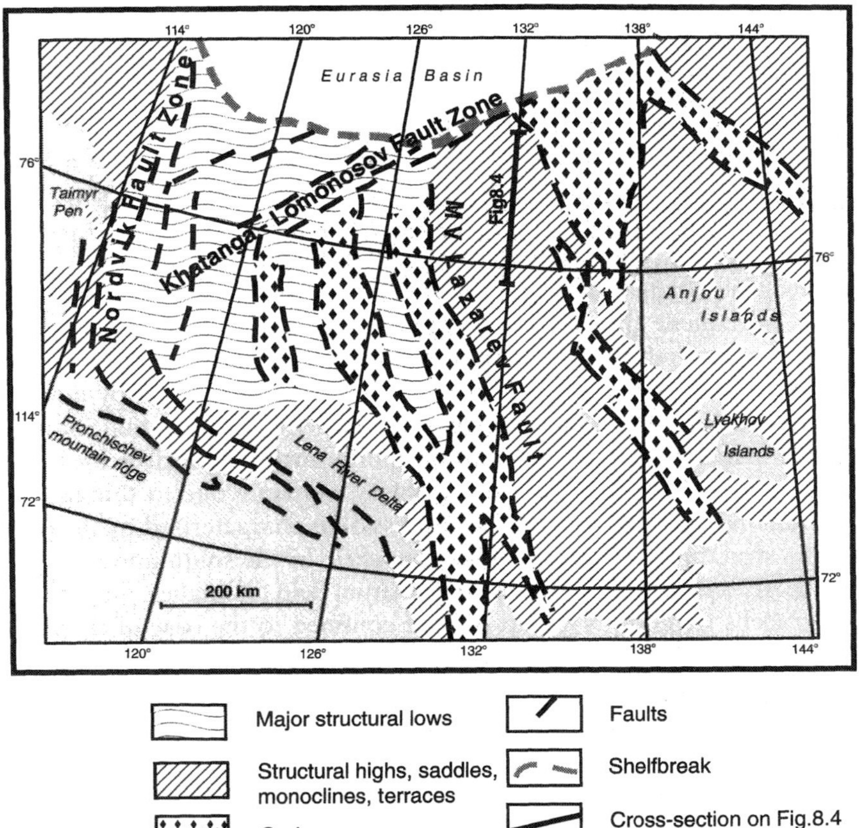

Fig. 8.3: Laptev Sea sedimentary basin.

Sea Basin is bounded by fold belts of Severnaya Zemlya and Taimyr; and in the south, it is separated from the Lena-Anabar trough of the Siberian Platform by fold structures of the Pronchischev Ridge. In the south-east, the basin continues into the Primorskaya lowland up to the ridges of the Late Kimmerian Verkhoyansk fold belt. In the east, on Novosibirskie (New Siberian) islands, the boundary of the basin is defined by structural uplifts with exposed basement and lower horizons of the sedimentary cover. In the north, the near-slope sedimentary basins are located between the Laptev Sea Basin and the Eurasia Basin.

Structural zones traced from the mainland intersect in the Laptev Sea Basin. The Khatanga-Lomonosov zone continues the Late Paleozoic-Triassic continental rift trend, and stretches from the Khatanga gulf in a north-easterly direction towards the shelf, and continues further parallel to the continental slope towards the termination of the Lomonosov Ridge. The Nordvik fracture zone appears as an extension of the Udzha fault zone of the Siberian craton and can be traced close to the Taimyr coastline. Tectono-magmatic activity in the Udzha zone was manifested in the Proterozoic and was repeatedly reactivated throughout the Phanerozoic.

The initial basement of the basin is Precambrian, most likely Proterozoic. In the west, this basement was rejuvenated in the Early Kimmerian orogenic phase, in a narrow zone proximal to Taimyr and Severnaya Zemlya. The rest of the basement in the basin was rejuvenated in the Late Kimmerian. The areas of these rejuvenations of the basement are divided by the Nordvik and Khatanga-Lomonosov zones. In the far south-east, close to the Lyakhov Islands, the age of the rejuvenated basement is Early Kimmerian.

8.4.1 SEDIMENTARY COVER

The sedimentary cover is subdivided into two structural stages. The lower one consists of complexes formed during a long time interval between the Late Proterozoic and the Late Mesozoic (Figs. 8.4 and 8.5). The age of the upper structural stage ranges from Late Mesozoic to Recent. A structural unconformity between the stages reflects tectonic events which preceded the formation of the modern Eurasia Basin as well as the Early and Late Kimmerian orogeny in the Mesozoic fold belts on Taimyr and in the North-east of Russia. The thickness of the sedimentary cover in the west of the Laptev Sea Basin is significantly greater than 10 km, while in the east it usually does not exceed 3–5 km.

The sedimentary cover has been studied in outcrops on the islands and on continental rises bordering the Laptev Sea Basin. On the basis of a dense coverage of CDP seismic lines offshore together with refraction and high-frequency seismic data, the principal features of the shelf structure were proved

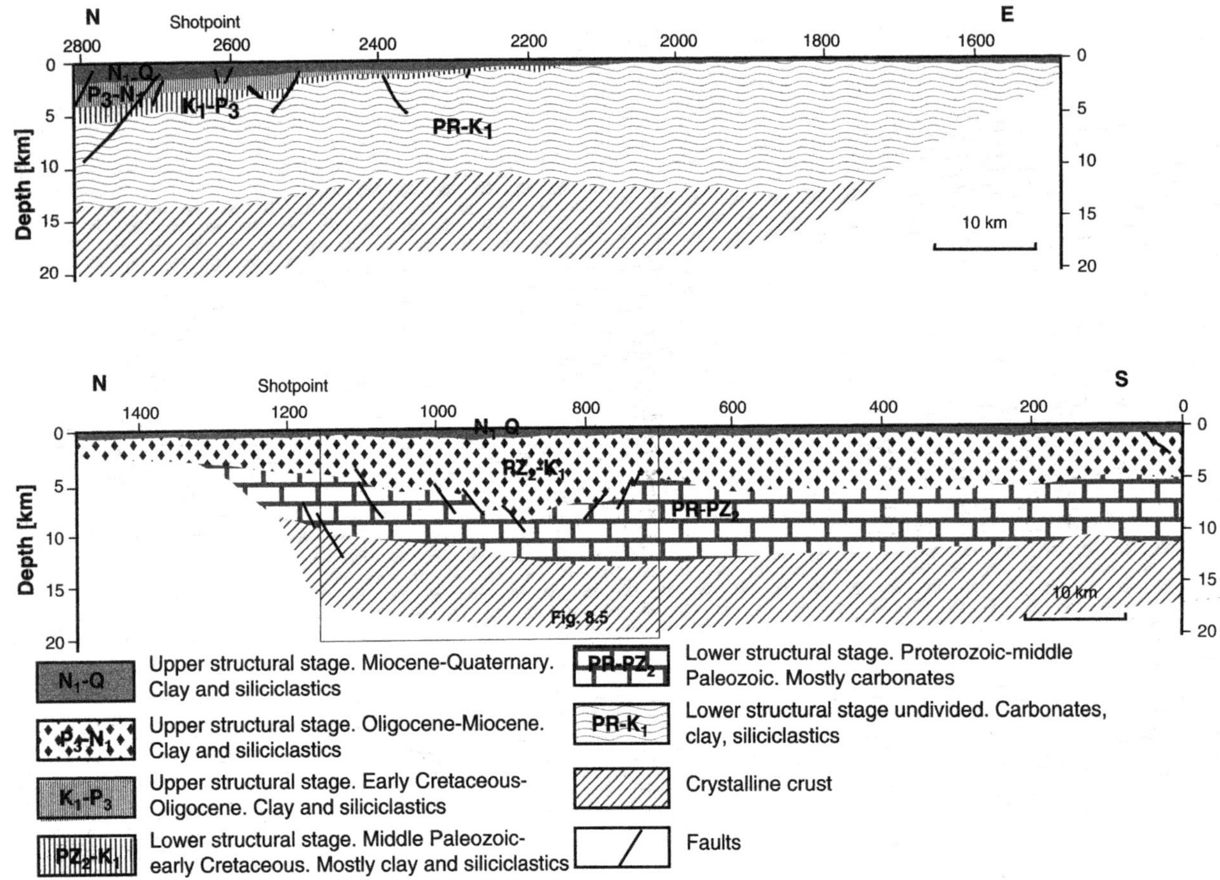

Fig. 8.4: Geological cross-section in the northeast Laptev Sea Basin; after interpretation of seismic line BGR97-05 (Franke and Hinz, 1999; Franke et al., 2001). For location, see Fig. 8.3.

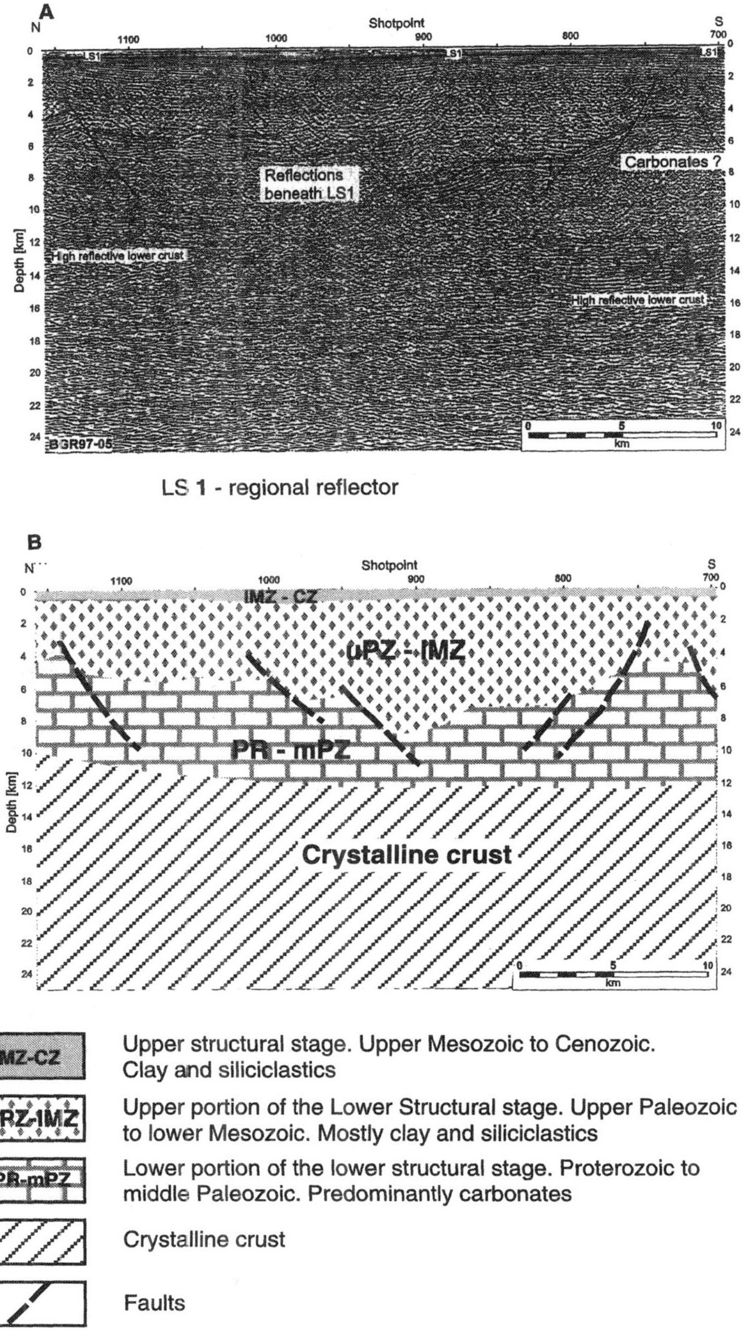

Fig. 8.5: Interpreted seismic section, showing constituent parts of the sedimentary fill of the Laptev Sea Basin. The line segment shown in Fig. 8.4: (A) interpretation by Franke and Hinz (1999) and Franke et al. (2001), (B) interpretation in this chapter.

to be concordant with the tectonic pattern earlier revealed by potential field survey data (Ivanova and Secretov, 1989; Kim and Ivanova, 2000). In the south-west and south-east of the Laptev Sea the sediments of the upper structural stage were penetrated by test boreholes to a depth of usually less than 200 m. Offshore deep drilling has not been carried out. Some parametric boreholes with a depth up to 3.6 km were drilled on the coastal part of the Siberian Platform. Seismological observations on the Novosibirskie Islands and in the Lena river delta, provided data on modern geodynamics and deep structure of the Earth's crust (Avetisov, 1982, 1991; Avetisov and Guseva, 1991).

8.4.1.1 Lower structural stage

In our understanding of the seismic-stratigraphic model compiled by Franke et al. (2001), the lower structural stage is placed between the top of the acoustic basement (horizon LS 1) and the high reflectivity zone topping the lower crust. Between these limits, discontinuous reflectors are identifiable, interpreted as lithostratigraphic boundaries between carbonate and terrigenous strata. Bounding the sedimentary cover from the bottom by the high reflectivity zone is accepted as an unverifiable working hypothesis.

The lower structural stage comprises mainly relics of the ancient cover of the Siberian craton. Outside the craton proximal to the Taimyr, the basement is Baikalian, and in the north-east, in De Long Islands area, it is Caledonian. The lower structural stage was formed in different geodynamic and sedimentary settings, and suffered a repeated reworking associated with constructive and destructive tectonism. During constructive tectonism, the sedimentary cover together with the basement underwent horizontal compression. The initial stage of compression resulted in minor warping of the strata, which remained a constituent part of the sedimentary cover. More advanced stages of constructive tectonism resulted in intense folding, metamorphism, and calc-alkaline magmatism, and the transformation of the sedimentary cover partly or completely into basement. Localities of the secondary young basement are apparently present in the Laptev Sea Basin, but they cannot be mapped using the present-day database.

The composition of the lower structural stage can be extrapolated from the Siberian Platform, Taimyr, Severnaya Zemlya, and Novosibirskie Islands. Riphean, Vendian, and Cambrian strata can be clastics, carbonates, pelites, and perhaps evaporites. Ordovician to Middle Paleozoic deposits are predominantly carbonates. Starting from the Eifelian and ascending in the sequence, the carbonate series are replaced step by step by clay and terrigenous sediments, marine, non-marine and paralic. Terrigenous and clayey deposits dominate in the Mesozoic. Rift-related sequences are found together with typical platform sequences. Details of the intermediate structural stage are seen by correlation of sequences from Severnaya Zemlya, Anjou Island, and the north-east Siberian platform.

On the Severnaya Zemlya Archipelago, the lower boundary of the sedimentary cover is well defined at the base of the Ordovician, although it is possible that it may range into the Cambrian. The uncertainty is caused by the fact that the Cambrian deposits are highly deformed in a long-lived fault zone, but unlike the Vendian strata, they are practically unmetamorphosed. Cambrian strata are marine clay and clastics.

Ordovician deposits overlap the Cambrian strata most likely disconformably. At the base, the Ordovician is quartz sand with interbedded limestone, tuffaceous sandstone, tuffaceous siltstone, conglomerate with quartz, microquartzite, basalt, and andesite pebbles. From the Ordovician up to the Middle Devonian, various limestone, dolomite, marl, mature terrigenous and clay sediments dominate; gypsum-bearing series are also widespread. The sedimentary environment varied from shallow-marine to lagoonal, and coastal subaerial. Lagoonal settings are best developed in the Middle Ordovician, shallow-water marine realms in the Upper Ordovician and Early Silurian. Shallowing and subaerial environments are typical for the Silurian-Devonian boundary. This boundary is also marked by a hiatus. At the beginning of the Emsian, shallow-marine carbonate precipitation was widespread for a short time.

At the beginning of the Ordovician the N–S trending rift zone originated in the east of the Oktyabrskaya Revolutsia Island. Lavas and volcanoclastics of basic to acidic composition, usually highly alkaline, are widespread within the rift (Gramberg and Ushakov, 2000).

In the Eifelian, the carbonate deposits were replaced by clastics. These are red beds and variegated quartzitic and arkosic sandstone, siltstone, and argillaceous rocks with interbeds of marl, sandy and detrital limestones, and gritstone. Sedimentation took place in coastal to continental, lagoonal, deltaic, and coastal-marine environments.

The latest Devonian to Middle Carboniferous deposits are absent.

Late Paleozoic strata overlap unconformably different-aged deposits, occurring locally. On the islands Bolshevik and Oktyabrskaya Revolutsia, Late Carboniferous to Early Permian continental sandstone, siltstone, rare claystone, and conglomerate with plant detritus were mapped. In the north of Komsomolets Island, sandstone, claystone, conglomerate with lenses of shelly limestone with Late Permian marine fossils, were observed.

Mesozoic sediments are rare. Early to Middle Jurassic conglomerate with interbeds of silt and sand, of Early-Middle Jurassic age, and sandstones and clay with fragments of pumice, of Volgian-Berriasian age, were penetrated by boreholes in topographic lows on Bolshevik Island. Their thickness is not more than a few tens of meters. Coastal-marine Upper Jurassic-Lower Cretaceous deposits in the south of Bolshevik Island and in the west of Oktyabrskaya Revolutsia Island are suggested to be present by analogy with the Cheluskin Cape.

A carbonate sequence of the Early Ordovician to Middle Devonian was mapped on the eastern margin of the Laptev Sea Basin on Kotel'ny Island (Kos'ko, 1977; Kos'ko et al., 1990). The sequence comprises limestone and dolomite with subordinate clastics deposited in variable settings. Silurian deposits occur in two facies zones trending north-west, which correspond with the modern structural grain. In the south-western zone, the Lower Silurian to the base of the Upper Silurian, are basinal graptolite-bearing pelitic-carbonate-cherty deposits. In the north-eastern zone, shallow-water carbonates with various benthos, organic buildups, and brachiopod banks predominate. The beds on the boundary between Silurian and Devonian are chiefly sedimentary dolomites. Upward in the sequence of Early Devonian strata, the number of normal marine layers increases. In this interval facies zoning is not clearly seen.

The Eifelian to Givetian part of the complex is separated from the underlying strata by washout. The deposits are algae bioherm layered limestones, dolomites, limestone and dolomite sedimentary breccias, and conglomeratic breccias. Clay, siltstone, claystone, and sandstone are also present. In the Middle Devonian, the distinctive facies zonation of a north-west trend occurred again.

In the Late Devonian, the carbonate deposits conformably or with insignificant washout, was overlain by clastics and carbonates. The section is complete in the Belkov-Nerpalakh aulacogen in the south-west of Kotel'ny Island, oriented in accordance with earlier facies zonation. The lower part is formed by siltstone, claystone, and rare sandstone. In the upper part appears cross-bedded quartz sandstone with beds of conglomerate and gritstone, with pebbles of dolomite, limestone, quartz, cherty rocks, and rare volcanics. Limestone and dolomite occur throughout the sequence. Red beds occur irregularly but quite often.

Carboniferous and Permian deposits are limestone, clayey limestone with chert, siltstone, claystone, and sandstone with beds containing a marine fossil fauna and plant detritus. Dolomite and marl, lenses and interbeds of gritstone and conglomerate, and sometimes andesite and dacite lava flows are also present. Contacts with the underlying sediments are conformable and disconformable.

All stages of the Triassic system have been identified. They overlie disconformably the Paleozoic strata. Clay and claystone with various concretions are interbedded with subordinate limestone and dolomite is predominant in the Triassic. Siltstones and sandstones are significant in the Upper Triassic. Basalt occurs close to the base of the Triassic, and tuffaceous admixture in the sediments is present here. The Triassic sediments accumulated in a normal marine and abnormal salinity basin.

Three stages of the Jurassic system are present in the area. Their contact with the Triassic was not observed. Silt-clay, silt, and sandstone with marine fossils are widespread in the Jurassic.

Lower Cretaceous deposits in marine facies were observed in a borehole at a depth of 100 m in the south of the Zemlya Bunge. These deposits are clayey siltstone and sandstone with Valaginian foraminifera.

In the north of the Siberian platform the sedimentary cover includes deposits from Riphean to Recent (Malich, 2002). In the Riphean-Early Carboniferous the sediments are mainly shallow-water normal marine and high-salinity lagoonal. Non-marine deposits occur infrequently. The sequence consists of silty and sandy pelite, carbonate, clayey-, and clastic-carbonate units. Alternation of these units constitutes a multi-order cyclicity. Basic and alkaline basic magmatic manifestations, weathering, and hiatuses tend to occur at the boundaries of the cycles. Starting from the Late Vendian, formation of gypsum-bearing units occurs frequently. The sedimentary environment changed abruptly in the Early Carboniferous, and from the Middle Carboniferous in a shallow-water low-salinity basin coal-bearing sandstone-siltstone and argillaceous siltstone strata accumulated. On the Permian-Triassic boundary, traps magmatism was intensely manifested. At the end of the Triassic, a general uplift of the Siberian platform took place. Sedimentation recommenced in the Jurassic. Since then and up to and including the Late Cretaceous, the Lena-Anabar trough existed in the north of the platform. It is separated from the Laptev sedimentary basin by the Pronchischev Ridge. The Lena-Anabar trough is filled by paralic clayey, terrigenous, coal-bearing strata. The deposition of these coal-bearing units was interrupted by short marine transgressions and ingressions in the Toarcian-Aalenian and Turonian.

8.4.1.2 Upper structural stage

This stage comprises deposits from Aptian to Holocene. According to the seismo-stratigraphic model compiled by Franke et al. (2001), the structural stage is placed between the seafloor and regional horizon LS 1, interpreted as an erosional unconformity. Unconformity LS 1 is the top of the acoustic basement. Above this within rift troughs, the LS 2 horizon is placed. It forms the top of a highly reflective series. Unconformity LS 3 is manifested by an erosional truncation and by a change in pattern of seismic reflectors from clearly defined parallel lamination to poorly defined lamination. Above occurs another poorly defined unconformity that cannot be easily traced due to its shallow depth and permafrost effects. Our estimation of the age of the seismic horizons is given below after the description of the upper structural stage series by on-land observations.

The upper structural stage of the sedimentary fill of the Laptev Sea Basin was studied in outcrops and shallow boreholes in the eastern margin of the basin on the Novosibirskie Islands.

On De Long Island, the Late Mesozoic and Cenozoic volcanics are widespread. On Benett Island sandstone and coaly claystone containing a pollen assemblage of the second half of the Early Cretaceous, 20 m thick, occur at the base of the Late Mesozoic sequence. The unit overlies unconformably Early Paleozoic strata. The unit is overlain by Cretaceous traps. The Zhokhov and Vil'kitsky Islands are relics of Cenozoic volcanoes (Vol'nov et al., 1970; Kos'ko et al., 1990; Silant'ev et al., 1991; Layer et al., 1993; Surnin et al., 1998).

The Aptian-Albian coal-bearing deposits represent Late Kimmerian molasse. They were mapped on Kotel'ny Island, Zemlya Bunge, and on Faddeev Island (Kos'ko and Trufanov, 2002). In the central part of Kotel'ny Island, these deposits overlap the underlying formations with an angular unconformity. The thickness of the molasse is up to 500 m. The deposits are clay, mudstone, sandstone, sand interbedded with conglomerate, acidic tuffs and tuffaceous sandstone, and beds of coal up to 25 m thick. The sequence is capped by a rhyolite flow of 60 m thick.

The Upper Cretaceous strata, with a thickness of up to 300 m, occur on Zemlya Bunge, on Faddeev Island, and on Novaya Sibir Island. The sediments are non-marine clay, silt, and siltstone with intercalations of sandstone, pebbly sandstone, tuffaceous sandstone, and brown coal. The pebbles are of rhyolite, basalt, hornfels, quartz, brown coal, and siliciclastics. Their Late Cretaceous age is supported by fossil plants, pollen, and spores.

The Cenozoic complex consists of the Paleocene-Eocene, Oligocene-Miocene, Pliocene, and Quaternary sequences. Cenozoic deposits transgressively overlay unconformably all older formations with a weathering profile at the base. Hiatuses and transgressive relationships were established at the boundaries of the sequences.

The Paleocene-Eocene deposits are distributed discontinuously on the Anjou Island. Paleocene layers were distinguished paleontologically in one borehole close to the south coast of Zemlya Bunge. They are clay, mudstone, silt, poorly sorted sand with pebbles, grit, allochthonous, and autochthonous brown coal interbeds. A high content of kaolinite was identified in the clay fraction, pointing to a reworking of chemical weathering products. Lacustrine-alluvial and alluvial deposits predominate as whole, with only a few intercalations containing marine fossils. The thickness of the Paleocene-Eocene series on the Anjou Island is thought to reach 90 m (Lopatin, 1999).

The Oligocene-Miocene sequence is distributed significantly wider than the Paleocene-Eocene one. It disconformably overlaps transgressively all older units. The sequence consists of coastal-marine and non-marine sand and sandstone, siltstone, gritstone, and clay, sometimes with interlayers of brown coal. Unlike the Paleocene-Eocene deposits, this sequence has a higher content of marine sand and less coal-bearing layers. Its maximum thickness is 190 m.

The Pliocene sequence is composed of non-marine and transitional to marine mudstone, a few tens of meters thick. In east, Novaya Sibir Island the Pliocene section includes the uppermost Miocene beds.

On Lyakhov Island the Late Cretaceous is absent. The Tertiary deposits are similar to those on the Anjou Island. Unlike on this latter island, marine deposits are absent here. The thickness of the Tertiary sediments in a composite section is about 120 m (Dorofeev et al., 1999).

On Severnaya Zemlya Archipelago, Aptian-Albian palynologically dated alluvial quartz sand and micaceous silt, 8 m thick, occur on the west Oktyabrskaya Revolutsia Island.

The Cenozoic sediments cover variably elevated topographic plains. There exists a Paleocene-Eocene weathering profile at the base of the Cenozoic strata. The Oligocene-Miocene sediments are sand, conglomerate, breccia, and clay with boulders, of non-marine and marine origin. The sediments were weathered in the Miocene. Alluvial grit and pebbles in present-day river valleys are believed to be Miocene in age. The Pliocene to Lower Pleistocene marine pebbles, sand, clay, silt with mollusks, and foraminifera occur on relics of a peneplain elevated above 260 m. Within the Tertiary, different series vary in thickness from a few meters to 30 m. Quaternary sediments are polygenetic from clay to cobble round stones (Gramberg and Ushakov, 2000).

Aptian sand crops out on the north-east Taimyr. Relics of a Tertiary peneplain underlain by variable in age rocks, weathered in the Paleogene, are widely spread in the region. Sediments of the Late Pliocene transgression occur at the base of the Late Pliocene-Holocene sequence.

The principal features of the upper structural stage sequences on the Laptev Sea south coast are a Pre-Albian unconformity in the Lena-Anabar trough, a weathering profile at the base of the Tertiary, hiatuses separating Oligocene strata from Eocene strata, in the Miocene and close to the Miocene-Pliocene boundary in grabens within the Verkhoyansk fold belt (Drachev et al., 1998).

It is assumed that the strata known on the islands and the mainland project offshore. A more widespread distribution of Lower Cretaceous, Oligocene-Miocene, Pliocene, and Quaternary strata thicker and with an increased portion of marine sediments was assumed.

Regional seismic reflectors mapped by Hinz et al. (1997) and Franke et al. (2001) in our understanding should be related to the base of more widely spread sequences. In this case, the LS 1 unconformity is Aptian, the LS 2 unconformity mid-Oligocene, the LS 3 unconformity Late Miocene-Pre-Pliocene in age that correlates with eustatic sea-level events. The above dating varies from the dating by the above-cited authors who put horizon LS 1 at the Cretaceous-Paleogene boundary.

8.4.1.3 Regional structures

The Laptev Sea shelf is a segment of the divergent boundary between the North American and Eurasian lithosphere plates. Scientists are focused on revealing the history of rifting in the area (Ivanova and Secretov, 1989; Fujita and Cook, 1990; Hinz et al., 1997; Drachev et al., 1998; Franke et al., 2001). The Laptev Sea Basin is divided in a western deeply buried (3–13 km) and a relatively elevated eastern area (1–3 km) as based on the upper structural stage. The latter is a structural terrace between the western area and the Kotel'ny-Svyatov Nos arch bounding the basin on the east. The boundary of the eastern structural terrace and the depression in the west is a listric fault. The basin as a whole in both buried and uplifted parts has a horst and graben structure of north-west or near longitudinal trend. The horst and graben structure is detectable on seismic profiles from the acoustic basement most often up to the base of the Neogene strata (LS 3 reflector). The upper portion of the sedimentary cover forms synclines over grabens. Vertical displacement on the faults flanking the grabens is up to 6 km decreasing upwards in the section.

8.5 SEDIMENTARY BASINS OF THE EAST-SIBERIAN AND CHUKCHI SEAS

The East-Arctic (Amerasia) sector of the Russian Arctic continental margin is a topographic and structural terrace, which divides the Amerasia Basin from the mountain ranges on the mainland. The terrace is traced from the Novosibirskie Islands in the west to Alaska in the east (Fig. 8.6). At present, the East-Siberian and Chukchi Seas shelf appears as a single sedimentary basin (Gramberg et al., 2004).

The oceanic structures neighboring the shelf are starting from the west Lomonosov Ridge, Podvodkinov (Toll's) Basin, Mendeleev Rise, Chukchi Abyssal Plain, Chukchi Plateau, and Northwind Ridge.

These structures represent different stages of continental crust transformation. The initial stages are manifested in the eastern part from the Mendeleev Rise to the Northwind Ridge. In the Podvodkinov Basin this process is most advanced. But even here the crust thickness, the presence of thinned "granite" bed, much shallower (2.5–3.0 km) than typical oceanic water depth and the absence of a zebra-pattern magnetic field prevents from supposition that this is a normal oceanic structure, and one must assume only far gone continental crust destruction.

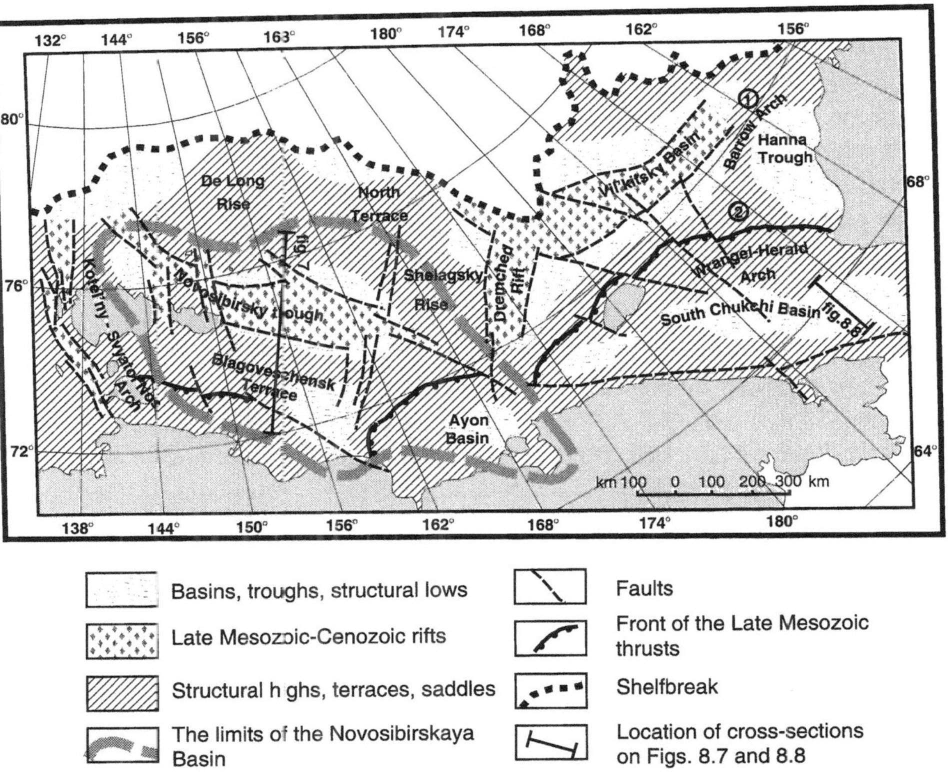

Numbers in circles: 1 – North Chukchi Rise, 2 – Chukchi Platform

Fig. 8.6: Major structural features of the East Siberian and Chukchi Seas shelf.

The shelf sedimentary basin is separated from the oceanic core by a strip of near-slope basins confined to the continental slope and rise, and filled by clinoform complexes. The shelf sedimentary sub-basins are bounded from the continent by neotectonic uplifts with exposed Late Kimmerian fold belt complexes, stretching from Verkhoyanie to the Bering Strait and further to Alaska and the Brooks Range. Structures of the belt spread offshore and from the Late Kimmerian tectonic basement in the southern part of the shelf. In the north, the Late Kimmerian basement is replaced by areas with Ellesmerian, Caledonian, and Pre-Cambrian basement (Kos'ko et al., 1998, 2000; Kos'ko et al., 2002a, b).

The geology of the islands and eastern seacoasts is known in enough detail to project the structural style offshore and to restore tectonic events. The upper part of the sedimentary cover offshore was irregularly and poorly sampled by dredges and tube samplers from vessels (Semenov, 1965; Lapina et al., 1968;

Pavlidis, 1982; Yashin and Kosheleva, 1994; Kosheleva and Yashin, 1999). Deep drilling was performed only in the American sector of the Chukchi Sea (Sherwood et al., 1998, 2002). One borehole of 600 m deep was drilled in the Russian sector on Ayon Island in the Chaunskaya Bay. The shelf was covered by airborn magnetic survey with spacing of 10–40 km. Some individual islands and coastal offshore areas were mapped in a detailed scale up to 1:50,000. Gravity maps at a scale of 1:1,000,000 were compiled for the major part of the area. Mapping at a scale of 1:200,000 and 1:500,000 was carried out in individual key areas.

Seismic investigations in the north-western part of the East Siberian Sea were carried out in 1985 and 1974–1980 by NIIGA and PMGRE NPO "Sevmorgeologia" by applying the method of individual sounding from the ice drifting polar stations "SP-13" and "SP-22". Results from CDP seismic profiling in the west of the East Siberian Sea, east from Novosibirsk Island and in the De Long Archipelago have been published (Drachev et al., 2001). A comprehensive geophysical survey along the longitudianl geotraverse in De Long Islands and the Podvodnikov Basin areas was carried out under the program "Transarctica" (Poselov et al., 2002). Reconnaissance CDP results from the eastern part of the East Siberian Sea carried out in 1993–1994 and 1997 by BGR scientists in association with "Sevmorneftegeophysika" have been published (Hinz et al., 1997).

Seismic research carried out on the Chukchi Sea shelf by Russian and USA geological surveys together with US oil companies is sufficient to build general geological models. The results were published by Thurston and Theiss (1987), Grantz et al. (1990), and Sherwood et al. (2002).

The sedimentary cover in the east of the Eurasia Arctic continental margin is subdivided into two structural stages: Paleozoic-Early Mesozoic and Early Cretaceous-Cenozoic. Extrapolations from islands and the continent enable to infer the structural pattern and the composition of the sedimentary cover offshore. The Paleozoic is represented by terrigenous, clayey and carbonate deposits. In the Late Devonian to Early Carboniferous, trap magmatism may occur. In Triassic-Neocomian times, clay and mudstone with insignificant amounts of terrigenous sand-size deposits predominate. In the Early Triassic, the presence of basic flows and sills is extremely likely. At the base of the upper structural stage, Early Cretaceous orogenic formations, marking the completion of the Late Kimmerian orogeny, are present. The presence of orogenic magmatic manifestations is highly probable. In the north of the East Siberian Sea the Early Cretaceous plateau basalts are distributed over a vast area. The Late Cretaceous to Cenozoic strata are alternating coastal-marine and fresh-water clay, silt and sand, often with brown coal beds. In the north of the East Siberian Sea, Upper Miocene-Quaternary alkaline to basic volcanics are present (Gramberg et al., 2004).

The Lower structural stage is subdivided in detail in the east of the Chukchi Sea. In the rest of the offshore area, it is included in the acoustic basement, where it is recognized by separate non-correlative reflectors. A detailed seismo-stratigraphic model for the American sector of the Chukchi Sea was created, based on drilling data and detailed seismic survey (Thurston and Theiss, 1987; Sherwood et al., 1998, 2002). The acoustic basement is represented by an Early Paleozoic-Devonian Fanklinian sequence. The following sequences are distinguished in the Lower structural stage: Carboniferous (Middle Devonian?)-Permian, Lower Ellesmerian; Permian-Upper Jurassic, Upper Ellesmerian; Late Jurassic-Neocomian, rift sequence.

The Upper structural stage is built of the Lower Brookian sequence, early Cretaceous in age, and the Upper Brookian sequence, Tertiary in age, separated by a middle Brookian unconformity. The Upper Cretaceous deposits are absent in the offshore and coastal drill holes. Tertiary deposits are present in three out of seven boreholes. In the CRACKERJACK #1 borehole (Sherwood et al., 1998, 2002), the Paleocene to Upper Eocene deposits are overlapped by Pliocene ones. Oligocene deposits are absent. In the POPCORN borehole, the Paleocene-Upper Eocene sediments are overlapped by the Upper Oligocene-Neogene strata. Consequently, hiatuses during Late Cretaceous and in the Early Oligocene were established. A seismo-stratigraphic model of the Upper structural stage for the western part of the East Siberian Sea along CDP-profile "Indigirska Bay – Jannetta Island" was published by Drachev et al. (2001).

Comparison of these models with stratigraphic sequences on the coast (Laukhin and Patyk-Kara, 2002), on islands (Kos'ko and Trufanov, 2002), and from boreholes in the east of the Chukchi Sea (Thurston and Theiss, 1987; Sherwood et al., 1998, 2002), together with seismo-stratigraphic models of the Laptev Sea sedimentary cover (Drachev et al., 1998; Kim and Yashin, 1999; Franke et al., 2001), and with global eustatic events (Vail et al., 1977; Haq et al., 1987) allow to hypothesize the following succession constituting the Upper structural stage ascending the section: Early Cretaceous (*Kolville*) sequence, Late Cretaceous-Paleogene (*Bungin*) sequence, Late Miocene-Pliocene (*Nerpichin*) sequence, Middle-Late Miocene (*Faddeev*) sequence, Late Miocene-Pliocene (*Kanarchak*) sequence, and Quaternary sequence seismic complexes. The Kolville and Nerpichin sequences are areally continuous, having the greatest thicknesses. The Kanarchak sequence is most likely the same as the areally continuous one, but is not that thick. The Bungon and Faddeev sequences are discontinuous, comprising units variable in age and composition, separated by unconformities.

Novosibirskaya Basin. The East Siberian and Chukchi Seas shelf appears to form one single sedimentary basin at the base of the Pliocene-Quaternary sequence. A complicated tectonic ensemble is revealed at the base of the sedimentary cover

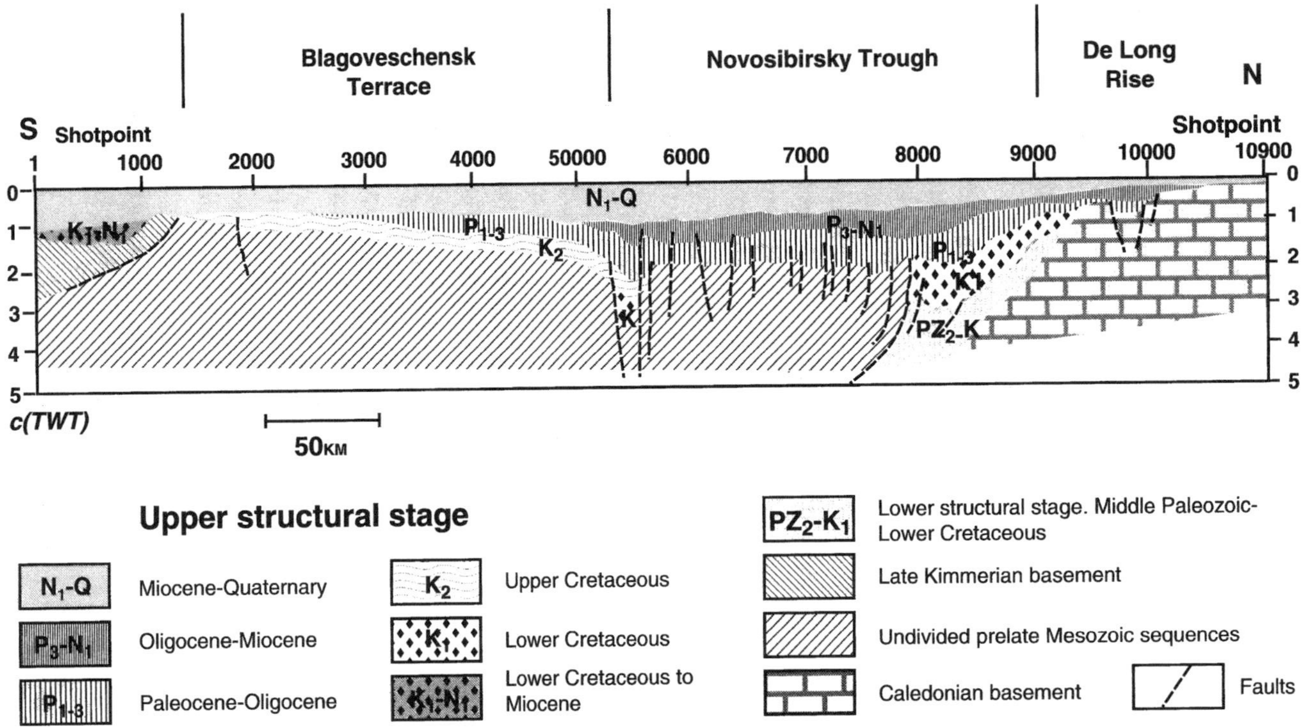

Fig. 8.7: Geological cross-section in the west East Siberian Sea. Interpretation in this chapter of seismic line LARGE-98001 (Drachev et al., 2001). For location, see Fig. 8.6.

(Fig. 8.7). The major part of the East Siberian Sea is occupied by this Novosibirskaya Basin (Vol'nov and Litinsky, 1976).

Novosibirsky Trough. The main depocenters of the megabasin are confined to the Novosibirsky trough (Gramberg and Progrebitsky, 1984; Drachev et al., 1998). Rifts with a reduced granite layer and filled with Cretaceous-Cenozoic strata, up to 12 km thick, occur along the axis of the trough. Northward from the rifts, the basement is Caledonian, to the south it is Late Mesozoic and/or Ellesmerian in age. The position of the Novosibirskaya trough along the border between blocks varying in age, of the basement and along the front of the Late Kimmerian orogeny suggest inheritance of positioning and combination of rifting mechanisms and formation of foredeeps at the initial stage.

The *Blagoveschensk structural terrace* spreads to the south of the Novosibirskaya trough, and is separated from it by a flexure and fault zone. In the south, the structural terrace is bounded by deformations over the thrust front in the Late Mesozoic basement. In the west of the structural terrace, a gently northward dipping monocline can be distinguished on the acoustic basement. The base of the Upper structural stage here is most likely Paleogene in age. A local basin can be traced in the east of the Blagoveschensk terrace.

The *Ayon Basin* locates farther on in the Chaunskaya Bay area. It closes the Novosibirskaya Basin in the south-east. The following sedimentary cover succession was revealed by drilling in the rim of the basin on Ayon Island: a 10 m thick weathering crust with debris of sandstone and siltstone; Paleocene-Eocene sand with silt and clay; Oligocene interbedding of sand, silt and clay with abundant brown coal; Early-Middle Miocene sand and silt; Middle-Late Miocene silt and clay; Pliocene sand; Quaternary mud. The thickness of the Paleogene is 490 m, of the Neogene 155 m, and of the Quaternary 30 m (Slobodin et al., 1990; Laukhin and Patyk-Kara, 2002; Gramberg et al., 2004).

The *De Long rise* occupies the north-west of the East Siberian Sea. In the north the rise is bordered by the sedimentary basins of the continental slope. In the south-west and south, it neighbors with the Novosibirsky trough. Towards the south-east it is replaced by the North structural terrace. Saddles connecting the Novosibirsky trough with the Vil'kitsky Basin are located between the North terrace and the Shelagsky rise. Within the borders of the De Long rise, the Caledonian basement is exposed on Henrietta, Jannetta, and Bennett islands. On Bennett Island, it is Middle Cambrian-Lower Ordovician distal, essentially clayey-silty turbidites. Volcanic proximal turbidites containing flows, sills and dykes of andesitic basalt, basalt, dolerite, and diorite porphyry of a calc-alkaline island arch series, with Ar–Ar age of 440 and 444 ± 2 Ma (Gramberg et al., 2004), were discovered on Henrietta Island. The presence of Paleozoic strata in the sedimentary cover is suggested by the collection of Carboniferous silicified limestone in debris on Zhokhov Island (Makeev et al., 1991). The Upper structural stage of the sedimentary cover is composed of Cretaceous and Cenozoic lava flow units and terrigenous sediments. The De Long rise is transected by faults. Grabens striking north-west and elongated north-eastward blocks, gradually descending the north-western boundary of the rise, are well defined.

The *Vil'kitsky Basin* is located in the north of the Chukchi Sea and the north-east of the East Siberian Sea (Gramberg and Pogrebitsky, 1984). The basin is up to 17 km deep. A rift forms the axial zone of the trough. Seismic reflections from the upper strata of the Lower structural stage were obtained on the southern flank of the trough. The Upper structural stage starts with Aptian-Albian molasse with a thickness up to 3.2 km in the southern flank of the trough close to the front of the Kimmerides (Grantz et al., 1990). The thickness decreases towards the trough axis. The Late Cretaceous sedimentation was controlled by faulting. The Tertiary series fill in a platform-type syncline, which is slightly faulted in the lower horizons. Longitudinal faults younger than those paralleling the southern flank of the trough are extensively manifested in the east. There exist Pre-Late Cretaceous and Pre-Eocene faults; tension faults predominate. North-vergent thrusts were mapped along the border with the Wrangel-Herald arch. In the southern flank between this arch and the axial rift, reflectors relating most likely to Mesozoic strata were observed beneath Cretaceous strata. Carbonate and terrigenous platform-type strata, known from the Northwind scarp, can continue under the northern flank (Grantz et al., 1998; Kos'ko et al., 2002a, b). In the axial rift the granite layer is absent or extremely thinned.

The *North Chukchi rise* of the Barrow arch separates the Vil'kitsky Basin from the near-slope basins of the Alaska Arctic margin. The rise was formed in the Late Cretaceous-Cenozoic. Previously a single sedimentary basin possibly stretched to the east to the Mackenzie River delta.

The *Wrangel-Herald arch* is traced from the south-east of the East Siberian Sea from about the longitude of Billings Point across the southern part of the Chukchi Sea up to Cape Lisburne in Alaska. The acoustic basement ascends up to the seafloor on the range crest. The most uplifted areas are represented by Wrangel and Herald Islands and also by basement exposure on the Lisburne Peninsula in Alaska.

Late Kimmerian north-vergent fold and overthrust structure was mapped on Wrangel Island. Late Precambrian metamorphosed volcanics, white mica and chlorite schist, phyllite, sills and dykes of granite-porphyry, metamorphosed gabbro, dolerite, and aplite, together with small intrusions of granite and aplite are exposed here. The most reliable age of granites is 633 ± 2 Ma (Cecil et al., 1991; Kos'ko, 1993; Kos'ko and Ushakov, 2003). Upper Silurian-Lower Devonian shallow-marine sediments, slates, and carbonate occur on top of the basement. These sediments were replaced by Devonian slate and sandstone turbidite with interbeds of conglomerate and chert. Lower Carboniferous strata comprise conglomerate, slate, siltstone with minor carbonates and gypsum. Acidic lava and basalt in the center of the island are admittedly assigned to the Lower Carboniferous. Middle Carboniferous deposits are mainly shallow-marine limestone in the north-west of the island and deeper water marine limestone and slate in the south-east. The Permian strata are slate and limestone with minor sandstone, conglomerate and chert. The Late Permian facies vary from shallow-water marine clastics and carbonate in the north-west to basinal slate with pyrite and rhodochrosite, and chert in the south-east. The Triassic is terrigenous flysch. The Paleozoic and Mesozoic rocks are metamorphosed to the lower greenschist facies.

The Wrangel-Herald arch is a north-eastward convex bend. The N-NE rim is a wide north-vergent fold and overthrust zone, the basement overthrusting the deformed sedimentary cover. In the east of the Chukchi Sea this zone was named the Herald overthrust. The dip angle of the overthrust is 8–10°, minimum horizontal displacement is 20 km. The Wrangel-Herald arch was segmented by longitudinal faults; the most significant ones are those that control the Hearld canyon along 175°W, east of the Herald island. The south-west limb of the arch in the east Chukchi Sea suffered normal faulting that resulted in a horst and graben structure. The grabens were filled by the Paleocene-Miocene, possibly Late Cretaceous-Miocene, syn-tectonic sediments. Upward in the section, the amplitude of the faults distinctly reduces. The upper Pliocene-Quaternary sequence is but slightly faulted. The sequence thickens gently towards the axis of the South Chukchi Basin. The Wrangel-Herald arch together with mainland sources supplied the clastic matter filling this basin.

The *Barrow arch* is traced as a single structure from the Beaufort Sea shelf in the east to the central part of the Chukchi Sea. In this way it was identified in early publications (Grantz et al., 1975) and it is accepted as such here, contrary to the modern schemes in which the name is used only for the northern section of the arch (Thurston and Theiss, 1987; Grantz et al., 1990). The Barrow arch separates the Hanna and Vil'kitsky basins; in the south, it contacts with the Wrangel-Herald arch.

The *Chukchi platform* on the southern extremity of the Barrow arch is a poorly structured regional monocline between the Wrangel-Herald arch and the Vil'kitsky Basin. The saddle, with a thickness of the Late Cretaceous-Cenozoic deposits up to 2500 m, is placed between the platform and the North-Chukchi rise of the Barrow arch.

The Chukchi platform existed from the Late Devonian as a tectonic rise. It served as the western limit of accumulation of Carboniferous-Early Permian deposits. On the more uplifted areas, the sedimentary cover is represented by thin Late Cenozoic sediments. The acoustic basement reaches the sea floor in some localities. A thick sedimentary cover developed in individual structures of second-order and on the rim of the platform. This platform is dissected by almost N–S trending faults. Some of these faults started in the Late Devonian-Early Carboniferous, and controlled the distribution and thickness of the Paleozoic-Early Mesozoic cover. The faults were reactivated in the Late Cretaceous-Cenozoic as strike-slip faults.

The *North Chukchi rise* is the northern segment of the Barrow arch. The arch as a single structure is distinguished by the base of the Late Mesozoic-Cenozoic strata. Structurally variable domains alternate below this surface. On the rise the Early Cretaceous strata form part of the basement. The age of the cover is not older than Late Cretaceous.

The *Hanna trough* occupies the eastern part of the Chukchi Sea. The Colville foredeep continues from the east into the southern part of the trough. It separates the Late Mesozoides of the Brooks Range in the south from the Arctic platform in the north. In the south the Hanna trough is limited by the fold and overthrust Wrangel-Herald arch. In the north and north-west it is bounded by the Barrow arch. The Hanna trough is connected with the Vil'kitsky Basin by a structural saddle between the Barrow arch segments. The trough developed from Middle Paleozoic to Recent under alternating geodynamic settings and orientations of tectonic stress. This resulted in an extremely complicated inner structure of the trough, in particular in the south where the Mesozoic Colville foredeep basin was superposed on the Paleozoic trough.

The Hanna trough overlies the Early-Middle Paleozoic folded basement. The lower portion of the cover is a Devonian to earliest Carboniferous clastic graben infilling. These deposits are overlapped by the Carboniferous shallow-marine limestone and clay, siliceous and carbonate deposits of an euxinic-type basin. The overlying sequence comprises Permian, Triassic, Jurassic, and partly Lower Cretaceous deposits. This distribution of the sequence is not limited to the Hanna trough. It overlaps with a reduction in thickness, parts of the Chukchi platform and the northern segment of the Barrow arch. Elevated blocks of this arch were a provenance area for the Hanna trough filling sediments in the Late Permian-Jurassic. The Permian-Early Cretaceous deposits in the trough are most likely pelite and mudstone with minor sandstone. The upper part of the sedimentary cover in the south Hanna trough is assigned to the sequence of the Colville foredeep. It is mostly mud, clay, and sand. The Early Cretaceous Wrangel-Herald arch and Brooks Range orogen served as provenance area.

Faults are the major structural feature of the Hanna trough. Longitudinal faults started in the Paleozoic. In disjunctives of the NW trending faults in the south-west were active in the Early Cretaceous. The Late Cretaceous-Cenozoic faults follow the Paleozoic ones to a large extent.

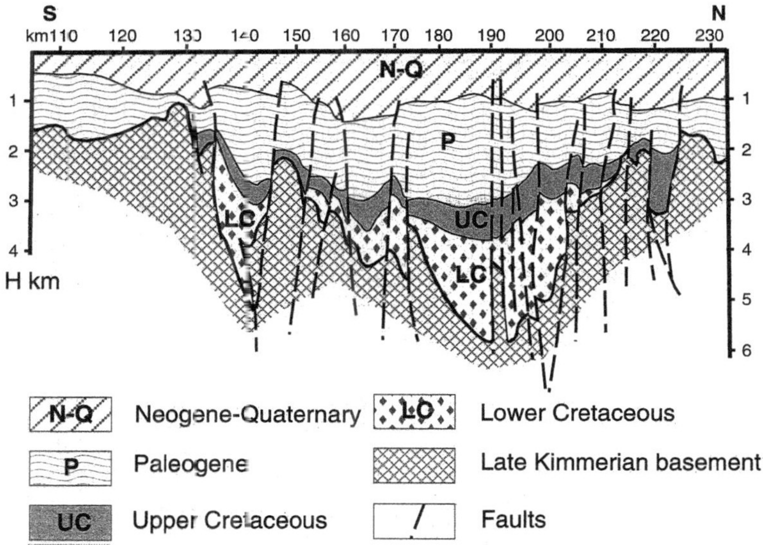

Fig. 8.8: Geological cross-section across the South Chukchi Basin. Interpreted fragment of seismic line DMNG-87004 (Kim et al., written communication). For location, see Fig. 8.6.

The *South Chukchi trough* occupies the southern part of the Chukchi Sea and a minor portion of the south-eastern East Siberian Sea (Figs. 8.6 and 8.8). In the north and north-east, the conjugated structure is the Wrangel-Herald arch. In the south, the trough is bounded by basement exposed on the Chukchi mainland. The trough is filled by Late Mesozoic and Cenozoic strata up to 8 km thick, overlapping the Late Kimmerian basement. The following units were distinguished in deep depocenters: Early Cretaceous strata correlated with onland Early Cretaceous molasses; Upper Cretaceous and Paleogene non-marine and coastal-marine sediments; Neogene-Quaternary coastal-marine and non-marine sediments. Eocene-Oligocene, Lower Miocene, and Middle Miocene-Pleistocene seismic sequences were distinguished in the east (Tolson, 1987). The Eocene-Oligocene deposits are coal-bearing and they contain basalt flows and volcaniclastics.

8.6 CONCLUSIONS

Due to the rather poor and irregular geological and geophysical data on the Russian Arctic shelf sedimentary basins, the study of the cyclic development of their respective sequences has only started now.

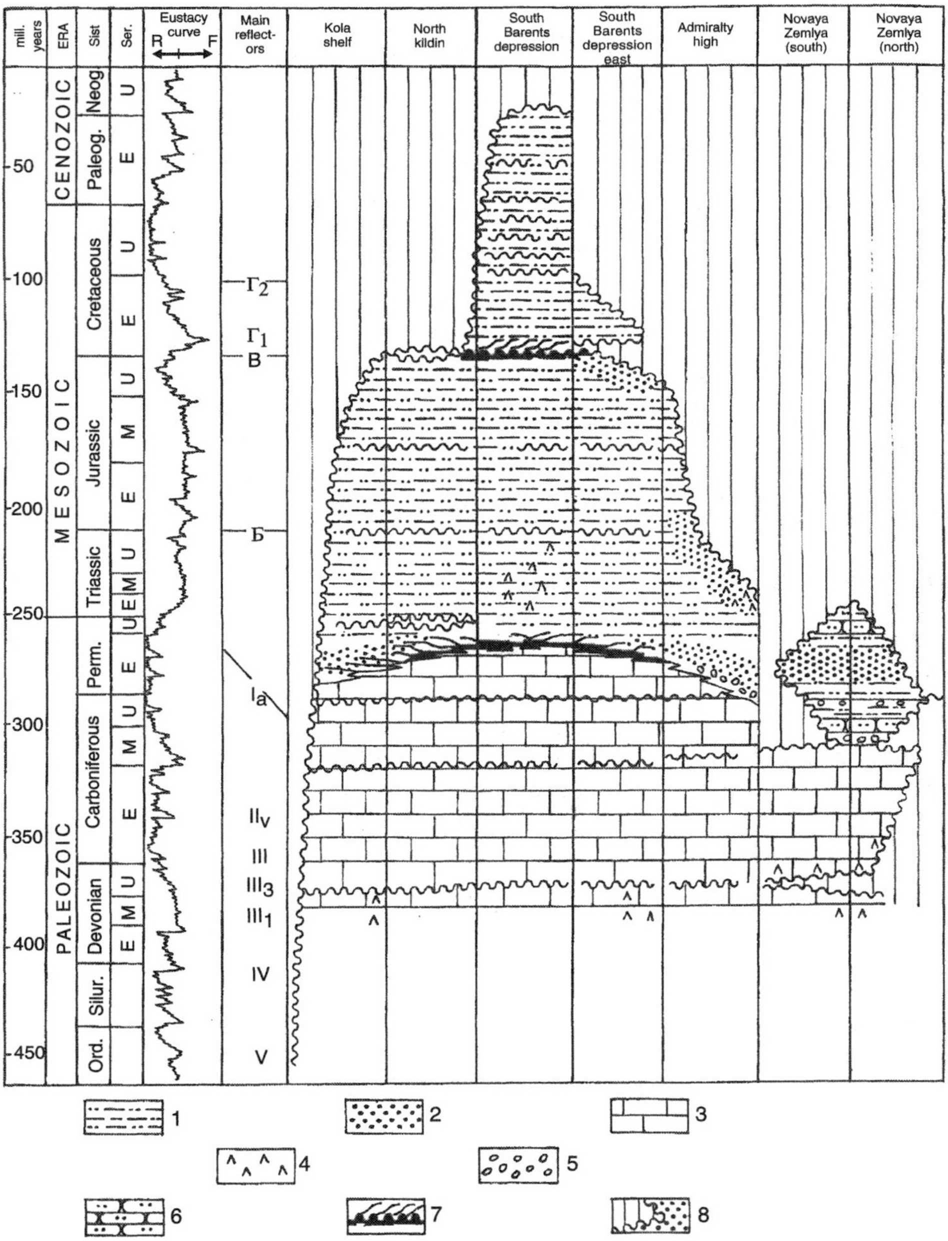

1 - clay, claystone; 2 - sandstone, 3 - limestone, 4 - volcanics (flows, dikes and sills), 5 - conglomerate, 6 - siltstone, 7 - clinoforms, 8 - unconformity and hiatus

Fig. 8.9: Correlation of the lithostratigraphy and seismic reflectors of the Barents Sea (simplified after Baturin et al., 1991).

At present it is assumed that the development of the basins was guided: (1) in accordance with established general peculiarities of the Earth and the World Ocean evolution; (2) under the influence of the initial stage of the Arctic Ocean, as it was for the first time noticed by Gramberg et al. (1983).

Within the Russian Arctic shelf sedimentary basin system, the processes of the platform development, rifting, compression-extension, and uplift-submergence migrated from the west eastward in a complicated way. The most prominent event in the history of the Phanerozoic sedimentation was the transition from carbonate-terrigenous sedimentation through the major part of the Paleozoic to terrigenous sedimentation from the Late Permian to Mesozoic and Cenozoic. The age of the sedimentary cover ranges from the west to the east from Precambrian to Cenozoic in the Barents Sea to Late Cretaceous-Cenozoic in the sedimentary basins of the southern East Siberian and Chukchi Seas.

An attempt to study cyclic development of the sequence, mainly of the Barents and South Kara Seas was undertaken. The best known theories were offered by Baturin et al., (1991; Fig. 8.9) and Suprunenko and Bro (1994). The first authors considered this phenomenon based on seismic and structural analysis in connection with fluctuation of the World Ocean level. The other authors examined it mainly with regard to the division of the sequence, distinguishing oil and gas producing, reservoir, and sealing series.

Consequently, the comprehensive study of the cyclic development of the Russian shelf Arctic sedimentary cover is now only in its initial stage.

ACKNOWLEDGMENTS

The authors wish to thank for the financial support for the research provided by the Russian Foundation for Basic Research, Projects 03-05-65174 and 02-05-65273.

9. CYCLICITY IN THE EVOLUTION OF THE NEOGENE NORTH CROATIAN BASIN (PANNONIAN BASIN SYSTEM)

D. PAVELIĆ

9.1 INTRODUCTION

The formation of the Pannonian Basin System started in the Early Miocene due to the continental collision and subduction of the European Plate under the African (Apulian) Plate. Subduction caused thermal perturbation of the upper mantle, resulting in attenuation and extension of the crust, and formation of a back-arc type basin (Royden, 1988; Horváth, 1993, 1995; Kováč et al., 1998).

The Pannonian Basin System is surrounded by the Alps, Carpathians and Dinarides, and belongs paleogeographically to the area of the Central Paratethys (Fig. 9.1). After the destruction of the Western Tethys into the Paratethys and Mediterranean in the Late Eocene, a large area of northern Croatia became land. The sedimentation continued until the Oligocene only in north-western Croatia (Hrvatsko Zagorje Basin, Mura depression and the north-western part of the Drava depression; Fig. 9.2). During the Miocene, sea-level oscillations strongly controlled sedimentation because a connection of the Central Paratethys with the Mediterranean and Indo-Pacific Ocean was several times established and interrupted (Steininger et al., 1988; Rögl, 1996). Marine transgressions, especially during the Early Miocene, did not flood the entire basin. Therefore, the basement was disconformably covered by deposits of different age ranging from Early to Late Miocene, formed in marine, brackish and freshwater environments, while some parts of the basin were characterized by temporary emersions. The final isolation of the Central Paratethys began some 10.5 Ma ago (review in Steininger et al., 1988; Rögl, 1996). The nature of the Central Paratethys evolution and the occurrences of endemic faunas have necessitated the establishment of local Miocene stages.

Two basins with different sedimentation evolved in the area of North Croatia during the Early Miocene. The first one named the Hrvatsko Zagorje Basin occupied a small area in the north-western part, while the second named North Croatian Basin covered almost the entire area of northern Croatia (Fig. 9.2).

Cyclic Development of Sedimentary Basins
J.M. Mabesoone and V.H. Neumann, editors.

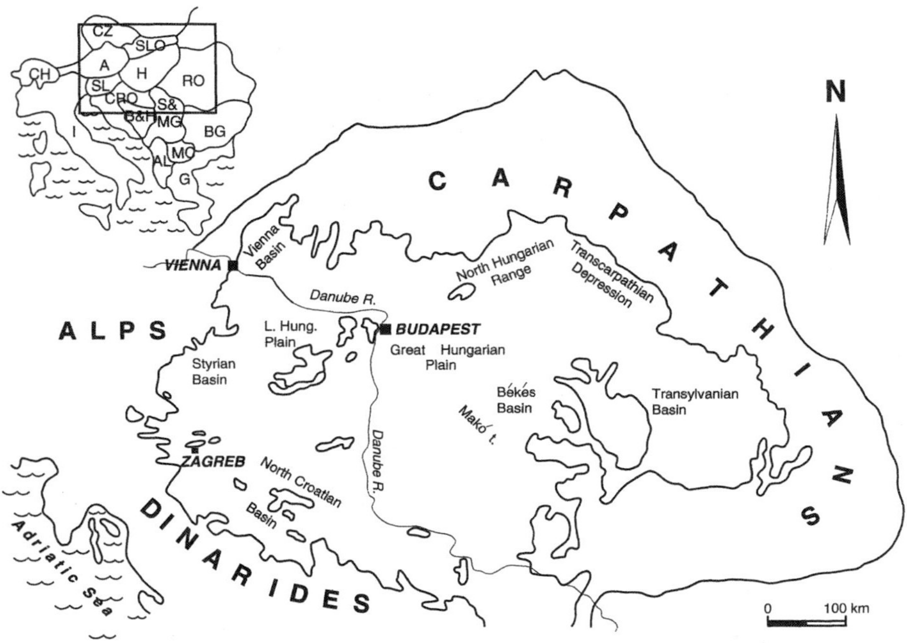

Fig. 9.1: Geographic position of Pannonian Basin system.

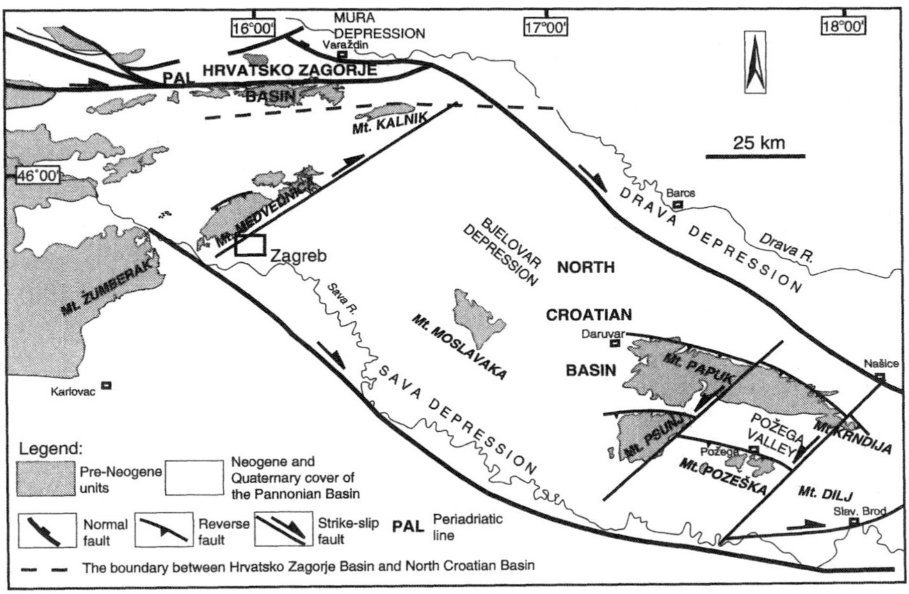

Fig. 9.3: Map showing main tectonic lines and depressions (after Marton et al., 2002).

The Miocene siliciclastic deposits of northern Croatia unconformably overlie a strongly tectonized basement. In geotectonic sense, the Paleozoic-Mesozoic-Paleogene basement belongs to the Supradinaricum, i.e. a part of the northern marginal zone of the Inner Dinarides, and to the Paradinaricum or Tisia. These geotectonic units were separate and became juxtaposed by subductional processes, including oceanic subduction from the Middle-Late Jurassic to the Middle Eocene (Herak, 1986; Pamić, 1998; Pamić et al., 2000).

The North Croatia Basin is a Neogene rift-type basin generated by continental passive rifting that began in Ottnangian time (Pavelić, 2001). The main tectonic lines are WNW-ESE trending elongated depressions, which represent relics of half-grabens. The maximum depth of the Pre-Neogene substrata in these depressions is about 6500 m. The thickness of the Mesozoic sediments on the surface is three to four times less than in the depressions (Fig. 9.3).

The syn-rift phase lasted until the Middle Badenian (Middle Miocene), and resulted in the formation of half-grabens characterized by a large sediment thickness strongly influenced by tectonics and gradually increasing volcanism (Fig. 9.3). Towards the end of the syn-rift phase sinistral strike-slip faulting took place, transverse to oblique to the master faults, which disintegrated the longitudinal structures contemporaneously with volcanic activity. The depositional environments gradually changed from alluvial and lacustrine to marine. The syn-rift–post-rift boundary was characterized by significant erosion of the uplifted fault-block footwalls. The post-rift phase extended from Middle Badenian to Quaternary. Tectonic influence drastically decreased, volcanism ceased, and subsidence of the basin was controlled predominantly by cooling of the lithosphere. Marine connections gradually decreased, resulting in a transition from marine to brackish, and finally fluvial-palustrine environments. By the end of the Miocene the basin was finally infilled. Upper Miocene deposits are conformably or unconformably overlain by Pliocene siliciclastic deposits accumulated in small fresh-water lakes, swamps, and rivers, characterized by local sources of the material. The Quaternary deposition was similar to that in the Pliocene, with exception of deposition of large quantities of eolian material (Fig. 9.3). The evolution of this type of the basin is usually interpreted as a first-order tectonic event (in the sense of Vail et al., 1991).

The basin evolution was additionally complicated by alternation of phases of extension and compression (Fig. 9.3). Extension generated subsidence during the syn-rift phase causing deepening of the basin. In the late syn-rift phase rotation of tilted blocks generated uplifting and shallowing. The early post-rift phase was characterized by thermal subsidence interrupted by two compressional events generated by intraplate stress that caused regional shallowing, and emergence of the Pre-Neogene basement.

The cyclic sedimentation in the Pannonian Basin System has been very well described in this system out of Croatia. There are several papers which present

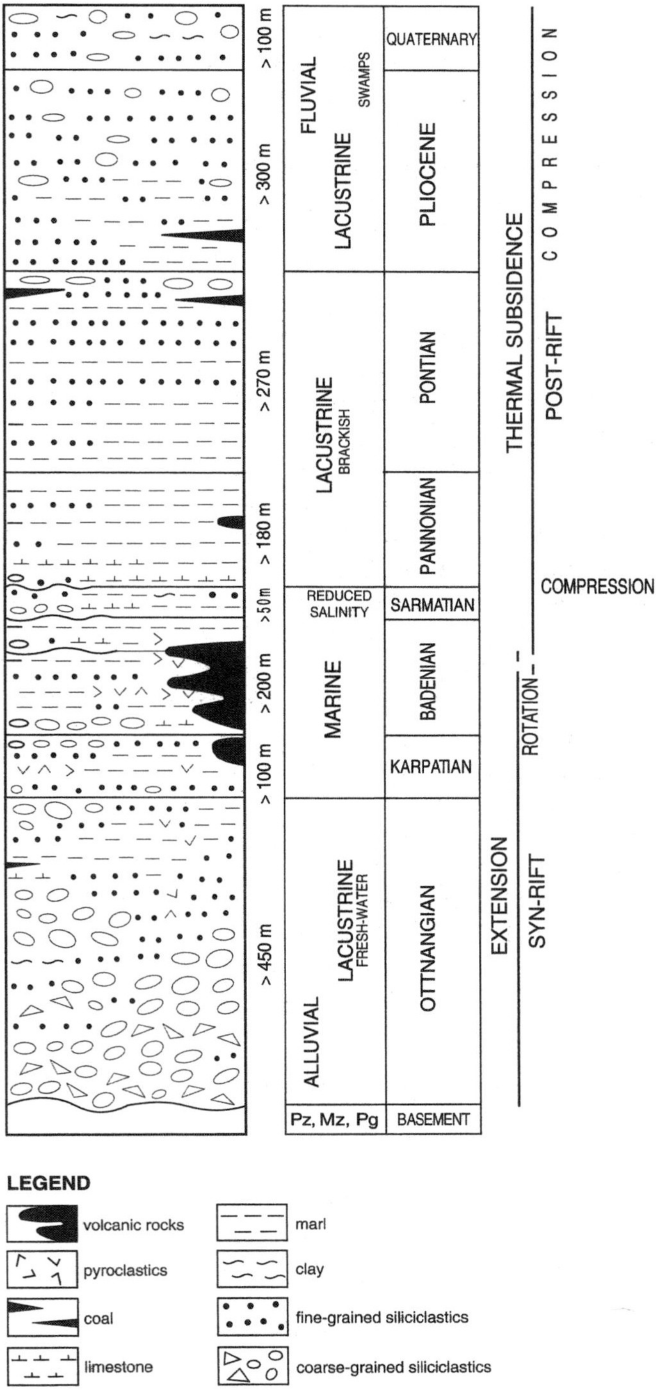

Fig. 9.2: Geological column of the Neogene succession of the North Croatian Basin.

interpretation of second-order, third-order, and higher order cycles (Tari et al., 1992; Csató, 1993; Vakarcs et al., 1994, 1998; Juhász et al., 1997; Kováč and Hudáčková, 1997; Kováč et al., 2001, 2004, and others). However, in the North Croatian Basin only second-order cycles have been interpreted (in the sense of Vail et al., 1991; Saftić et al., 2003), except high-resolution cycles within the uppermost Miocene deposits (Saftić, 1998). The Neogene sedimentary succession of the North Croatian Basin is composed of three depositional "megacycles", encompassing rocks deposited influenced mainly by structural evolution of the basin, which are separated by major unconformities (Saftić et al., 2003). The first megacycle consists of the Lower-Middle Miocene deposits, the second megacycle is composed of the Upper Miocene deposits, and the third megacycle consists of the Pliocene and Quaternary deposits. Except regional tectonics, these megacycles are the product of different controls on sedimentation, such as global and regional sea-level changes, change of climate, and delta progradation. Individual controls and their combination composed transgressive–regressive cycles which can be interpreted as third-order cycles (in the sense of Vail et al., 1991). Here they are named North Croatian Basin Cycles (NCBC).

9.2 DEPOSITION IN THE NORTH CROATIAN BASIN DURING THE NEOGENE AND CYCLICITY

In the description of the depositional sequences, data and interpretations from the general geological investigations on the outcrops is supplemented with the results of recent, mostly sedimentological studies. Based on the published data on the surface geology, a general geological column was compiled (Fig. 9.3). However, it has to be emphasized that the column is based on researches on many small or large scattered outcrops. An additional problem is the Early and Late Miocene endemicity that disables chronostratigraphic correlation in detail, and the interpretation of sequences in sense of their cyclicity should be considered only generally.

9.2.1 OTTNANGIAN – NCBC 1

The Ottnangian sedimentation is represented by fresh-water sediments generally showing a transgressive–regressive cycle which belongs to the lowermost part of the first megacycle (Fig. 9.4; in the sense of Saftić et al., 2003). The basin evolution in the early Ottnangian was characterized by predominance of coarse-grained deposits, occurrence of breccias along master faults, normal faulting, extension of the basin, and formation of half-grabens during the early stage of the syn-rift phase (Fig. 9.3). Elevated footwalls were intensely eroded

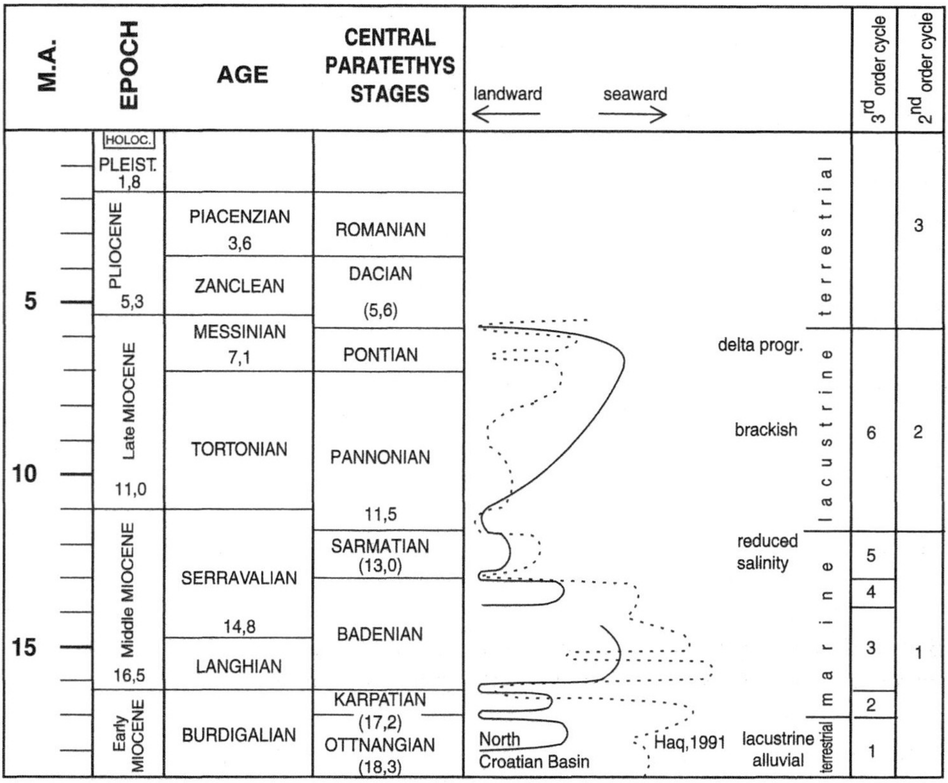

Fig. 9.4: Global and relative sea-level changes of depositional environments of the North Croatian Basin: third-order and second-order cycles after Saftić et al. (2003).

during short periods of high precipitation, in between long dry periods, and main depocenters were located in the braided alluvial fans in the deepest part of the hanging wall. The fine-grained fraction was transported over the alluvial material towards the north and north-east. After the end of alluvial deposition a hydrologically open deep lake was formed in the late Ottnangian, covering the whole area, primarily characterized by fine-grained deposits with high contents of organic matter and occurrence of pyroclastites (Fig. 9.3). Lacustrine deposition reflects rapid tectonic subsidence of the hanging-wall blocks resulting in an increase in basin accommodation space faster than its infilling with deposits. Shift from the semi-arid to a humid climate also influenced significantly the lake evolution. The lacustrine sedimentation ends with coarse-grained delta progradation as a consequence of a relative tectonic quiescence (Pavelić et al., 2000).

The possibility of indirect eustatic influence on the lake level might have played a role, since in the Ottnangian marine environments existed in the neighbouring Hrvatsko Zagorje Basin as well as in some areas of the Central Paratethys (Rögl, 1996; Pavelić, 2001). However, the late Ottnangian lake formation corresponds to the falling stage of the cycle TB 2.1 of global sea-level changes (Fig. 9.4). That means that the main controls of the lake evolution were regional tectonics as well as increased water mass due to a humid climate.

9.2.2 *KARPATIAN* – NCBC 2

The Karpatian marine sedimentation represents a short transgressive–regressive cycle and belongs to the lower part of the first megacycle (Fig. 9.4; in the sense of Saftić et al., 2003). Formation of a marine regime in the Karpatian is an event that is characteristic for almost the entire Pannonian Basin System. This event is explained by short-lasting marine flooding caused by opening of a Paratethyan seaway in the Mediterranean (Rögl, 1996; Kováč et al., 2003). This flooding is correlative with the global sea-level rise in the cycle TB 2.2 (Fig. 9.4; Haq, 1991). Continuous lacustrine to marine deposition and the prevalence of fine-grained offshore deposits indicate further subsidence of the basin (Bajraktarević and Pavelić, 2003), most likely due to the continuation of hanging wall tilting (Fig. 9.3). The occurrence of trachyandesites (shoshonites) indicates gradual increase of volcanic activity (Pamić et al., 1992/1993). The end of the Karpatian is characterized by regression and deposition of sands on the nearshore, being synchronous with the global sea-level fall at the base of the TB 2.3 global sea-level cycle (Fig. 9.4; Haq, 1991).

The sea-level fall at the end of the Karpatian has also been interpreted as the consequence of the complex tectono-sedimentary event (Pavelić, 2001). Generally, uplift in the rift system is explained as a consequence of rotation of the fault blocks. Fault block crests, thus uplifted above sea level, were strongly eroded, and large quantities of the mostly coarse-grained syn-rift deposits were resedimented particularly in marine shallows. Strong erosion reduced the load at the uplifted fault block crests, what could have resulted in additional uplift caused by isostatic adjustment to the removed load and repeated occurrence of the rocks of the Pre-Miocene basement at the surface. The uplift was contemporaneous with sinistral NE-SW strike-slip faulting transverse-to-oblique to the master WNW-ESE elongated structures (Prelogović et al., 1995; Jamičić, 1995). These faults disintegrated the elongate half-graben structures, and in this way reduced the effects of the uplift in some part of the blocks, a common case in the Pannonian Basin System (Horváth, 1993), and resulted in continuous Karpatian to Badenian sedimentation. The consequence of this was non homogeneous extension and complex basin morphology. In that sense, regression at the end of the Karpatian can be explained as a consequence of both by global sea-level fall and regional tectonics.

9.2.3 EARLY BADENIAN – NCBC 3

The Lower Badenian deposition occupies the middle part of the first megacycle (Fig. 9.4; in the sense of Saftić et al., 2003). Due to a relative sea-level fall at the end of the Karpatian, uplifted blocks were exposed and affected by strong erosion, which in some places resulted in a complete removal of Lower Miocene deposits by erosion and in the occurrence of basement rocks at the surface. Eroded coarse-grained siliciclastic material was transported into high-energy shallow marine environments, where it was reworked by waves and currents, and also into the rather deeper marine environments (Pavelić et al., 1998; Velić et al., 2000). The deepening of the environment in the Early Badenian caused deposition of marls and gravelly calcarenites in the offshore area (Pavelić et al., 1998). During the Badenian, the volcanic activity reached its peak (Fig. 9.3).

Tectonic controls of the deposition were very important at the end of the syn-rift phase. The activation of sinistral strike-slip NE-SW faults during the latest Karpathian caused formation of a short-lived active basin margin and evolution of the stacked Gilbert-type fan deltas in the coastal area during the earliest Badenian. Succeeding deepening caused overlaying of the shallow-marine deposits by offshore, mostly fine-grained material (Pavelić et al., 1998).

The Early Badenian sea-level rise that followed the uplift of the blocks near the end of the syn-rift phase correlates well with the global sea-level rise at the beginning of the Middle Miocene cycle TB 2.3 (Fig. 9.4; Haq, 1991). The rapid deepening from the newly formed shallow-water to offshore environment during this relatively short period indicates the dominance of the tectonic uplift by eustatic sea-level rise during the Early Badenian.

The evolution of the basin at the end of the Early Badenian is not clear because of erosion. There are only some indications of the sea-level fall in the western part of the North Croatian Basin (Avanić, 1997). This regression, which produced a regional unconformity between syn- and post-rift deposits, can be a consequence of closure of the presumed south-eastern seaway to the Mediterranean in the "middle" Badenian, and is correlative with the global sea-level fall at the end of the TB 2.4 cycle (Fig. 9.4).

9.2.4 LATE BADENIAN – NCBC 4

The Late Badenian transgressive–regressive cycle belongs to the upper part of the first megacycle (Fig. 13.4; in the sense of Saftić et al., 2003). During this epoch, which represents the earliest post-rift phase, the last Miocene marine transgression definitely flooded even the peaks of the elevated blocks that formed isolated islands in the Early Badenian. The first deposits were gravels, overlain by algal banks (Fig. 9.3). Further deepening resulted in predominant deposition of the offshore marls. The end of the Badenian was characterized by general shallowing.

The Late Badenian sea-level rise which resulted in the final flooding of all uplifted blocks and deposition of offshore mostly fine-grained material, influenced the entire Central Paratethys realm, and it is probably connected with the broad re-opening of the Indo-Pacific seaway (Rögl, 1996). The timing of this event coincides with a short-lasting sea-level rise in the beginning of the TB 2.5 cycle (Haq, 1991). The regression is correlative with sea-level fall at the end of this cycle (Fig. 9.4).

9.2.5 SARMATIAN – NCBC 5

The Sarmatian transgressive–regressive cycle represents the end of the first megacycle (Fig. 9.4; in the sense of Saftić et al., 2003). In the Early Sarmatian the salinity of the sea decreased. Due to the beginning of the isolation of the basin, which caused sea-level fall during the latest Badenian, morphologically prominent peaks were affected by weak erosion, resulting in deposition of shallow-water gravel, biocalcarenites and limestones, and large quantities of resedimented Badenian faunas. Subsequent deepening caused widening of the basin, and general dominance of fine-grained deposits. The horizontally laminated marls (similar to varves) and massive marl dominated in this period and episodical input of sands into the basin took place by sediment gravity flows (Fig. 9.3). The excellent preservation of the horizontal lamination probably reflects unfavourable hypoxic or even anoxic conditions for benthic life. During the latest Sarmatian, there was a general shallowing trend.

A reduced salinity and evolution of the Early Sarmatian brachyhaline fauna reflect a reduced connection with the Mediterranean and the opening of a link towards the Caspian domain of the eastern Paratethys (Rögl, 1996). One of the results of this event is an ecological change, reflected by the appearance of endemic species. The predominance of marls and carbonates can be interpreted by decreased influence of the tectonic activity.

A relative rise in water level resulted in the formation of a basin up to 70 m deep, which might be correlative with the beginning of the TB 2.6 cycle (Fig. 9.4; Haq, 1991). Onset of intraplate stress took place at the end of the Sarmatian and could have initiated the uplift of blocks, resulting in the water-level fall at the end of the Sarmatian and partial inversion of the basin (Horváth, 1995; Kováč et al.,

1998). This uplifting elevated some blocks above the base level producing a regional unconformity.

9.2.6 PANNONIAN-PONTIAN – NCBC 6

The Pannonian-Pontian transgressive–regressive cycle corresponds to the second megacycle (Fig. 9.4; in the sense of Saftić et al., 2003). In the early Pannonian, the deposition was influenced by changed ecological conditions due to definitive isolation of the Central Paratethys (Rögl, 1996). Salinity decreased further, and the environment became brackish (oligohaline), locally even fresh-water, followed by the expansion of endemic species of mollusks and ostracods of no stratigraphic value. The lower Pannonian deposits overlie the Sarmatian deposits mostly conformably, with general dominance of carbonates (Fig. 9.3). The deposits consist of lacustrine platy and thin-bedded limestones of littoral origin, with a few coarse-grained siliciclastic intercalations (Vrsaljko, 1999). Elevated blocks are characterized by deposition of terrestrial coarse-grained material which was also transported into shallow parts of the lake. Subsequent lake-level rise resulted in deposition of marls in the deep lake, which dominate the succession punctuated by occasional influx of terrigenous siliciclastic material. The infilling of the lake by sandy delta progradation was diachronous causing regression. In the western part of the North Croatian Basin progradation started in the Pannonian, while in the eastern part it started in the late Pontian (Kovačić, 2004). The delta sediments are overlain by fluvial sands, and silts and clays deposited in small lakes and swamps.

The correlation with the global sea-level changes is very speculative because of the basin isolation and formation of the lake. However, the time of deepening of the lake can be correlative with the beginning of the TB 3 cycles (Fig. 9.4; Haq, 1991). Regression due to progradation of siliciclastic deposits, and final infilling of the basin in the latest stage of the post-rift phase could be attributed to the reduced subsidence rates as thermal equilibrium was approached (Pavelić, 2001).

9.2.7 PLIOCENE-QUATERNARY

The Upper Miocene deposits are overlain by Pliocene siliciclastic deposits accumulated in small fresh-water lakes, swamps and rivers, characterized by more local sources of the material (Fig. 9.3). The Pliocene and mostly Quaternary is a new phase of basin evolution characterized by a transition towards overall compression and structural inversion in the entire Pannonian Basin System (Horváth and Cloetingh, 1996). In the North Croatian Basin, it is reflected by formation of many compressional structures, mostly by inversion of the preexisting listric normal faults, rotations, and uplift of basement blocks thus forming the present-day mountains in the area of North Croatia (Prelogović et al., 1998;

Tomljenović and Csontos, 2001; Márton et al., 2002). In these sediments, cyclicity has not been recognized.

9.3 CONCLUSIONS

The evolution of the North Croatian Basin represents a first-order cycle formed during the Neogene due to a continental rifting process. The sedimentary succession of the basin is composed of three second-order cycles, which are separated by major unconformities.

In the Neogene sediments of the North Croatian Basin, six transgressive–regressive sequences, which may represent third-order cycles, are interpreted. These are result of very complex interaction of tectonic uplifting and subsidence, global and regional sea-level changes, thermal subsidence, change of climate, and delta progradation.

The first cycles belong to the Ottnangian (TB 2.1) comprising alluvial, lacustrine, and coarse-grained delta deposits. The deposition was generated by tectonic subsidence, change of climate from semi-arid to humid, and delta progradation as a consequence of a relative tectonic quiescence. The second cycle developed during the Karpatian in marine environment due to opening of a new seaway to the Mediterranean. Deposition was on the marine offshore and later nearshore as a result of tectonic subsidence and global sea-level rise (TB 2.2), and global sea-level fall at the end of the Early Miocene, at the same time with uplift as a consequence of rotation of the fault blocks. The third cycle is of Early Badenian age. It is characterized by transgression and deepening which caused offshore deposition due to tectonic subsidence and global sea-level rise (TB 2.3) as well as shallowing at the end of the cycle producing a regional unconformity between syn-rift and post-rift deposits. The fourth cycle belongs to the Badenian, and is represented by transgressive deposits overlain with offshore sediments, ending with shallow-water sedimentation. The cycle was controlled by a short-lived global sea-level rise and fall (TB 2.5). A fifth cycle developed during the Sarmatian when deposition of reduced salinity occurred due to the onset of basin isolation. A relatively shallow basin formed as a consequence of a global sea-level rise (TB 2.6). A regression at the end of the cycle was a consequence of a tectonic uplifting of some blocks generated by intraplate stress. This uplifting produced a regional unconformity. The sixth cycle developed in an isolated brackish lake during the Pannonian and Pontian, that became closed by diachronous delta progradation. Deepening of the lake can be a consequence of thermal subsidence but is also synchronous with a global sea-level rise (TB 3). The interpretation of the whole cycle is not possible due to faunal endemicity and lack of chronostratigraphic data.

10. LONG-PERIOD CYCLES: A CASE STUDY FROM THE ARABIAN-NUBIAN CRATON

A.M. ABED

10.1 INTRODUCTION

The Arabian-Nubian Shield occupies an area of 3 million km^2 with outcrops in the following countries: Saudi Arabia, Yemen, Jordan, Palestine, Egypt, Sudan, Eritrea, Ethiopia, and Somalia. In the first three countries, it is called the Arabian Shield with extensive outcrops in Saudi Arabia, while in the latter five countries it is known as the Nubian Shield. Both shields are now separated by the Red Sea, which is a Miocene event, but prior to that both shields formed one continuous unit: the Arabian-Nubian Shield (Fig. 10.1). On the other hand, the Arabian-Nubian Craton involves more countries: Oman, United Arab Emirates, Qatar, Bahrein, Kuwait, Iran, Iraq, Syria, Lebanon, and Turkey. Figure 10.2 shows the countries involved in this chapter.

The Arabian-Nubian Shield is a late Upper Proterozoic feature. It was not present as a shield prior to about 650 Ma BP (Stoeser and Camp, 1984; Al-Shanti, 1993). Consequently, it is rather a young feature compared with the African cratons and other Archean cratons. Most workers agree that there are no rocks older than 1000–1200 Ma within the Arabian-Nubian Shield (Bentor, 1985; Al-Shanti, 1993, amongst many others).

The period between 1000–650 Ma was spent in the cratonization of the area occupied now by the Arabian-Nubian Shield which involved three stages (Al-Shanti, 1993):

1st stage: the separation of the Mozambique Craton and the formation of an ocean in eastern Africa (1200–950 Ma; Kazmin et al., 1978);

2nd stage: the formation of island arcs, subduction, accretion, erosion, sedimentation, magmatism and metamorphism (950–710 Ma; Jackson and Ramsey, 1980; Kemp et al., 1982);

3rd stage: the formation of the newly born Arabian-Nubian Shield by lateral accretion, suturing its segments, and continental collision as well (710–640 Ma; Stacey and Hedge, 1983; Vail, 1984; Davies, 1984).

The period from 640 Ma onwards to the onset of the Phanerozoic witnessed extensive faulting, graben and horst formation, intermontane basins, etc., asso-

Cyclic Development of Sedimentary Basins
J.M. Mabesoone and V.H. Neumann, editors.

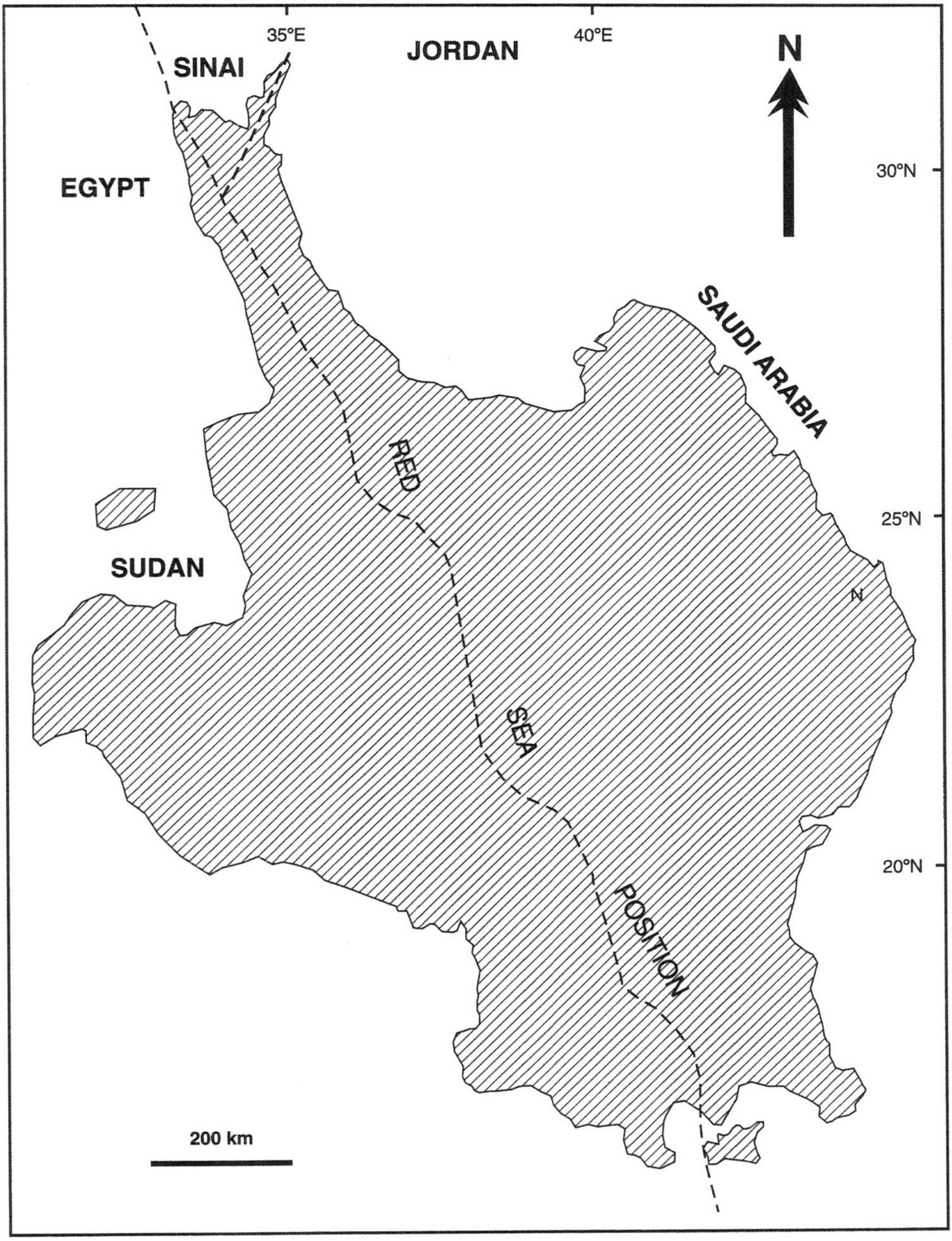

Fig. 10.1: The Precambrian Arabian-Nubian Shield. The dashed line indicates the position of the Red Sea, which came into existence at a much later time: the Miocene (modified from Al-Shanti, 1993; Powers et al., 1966).

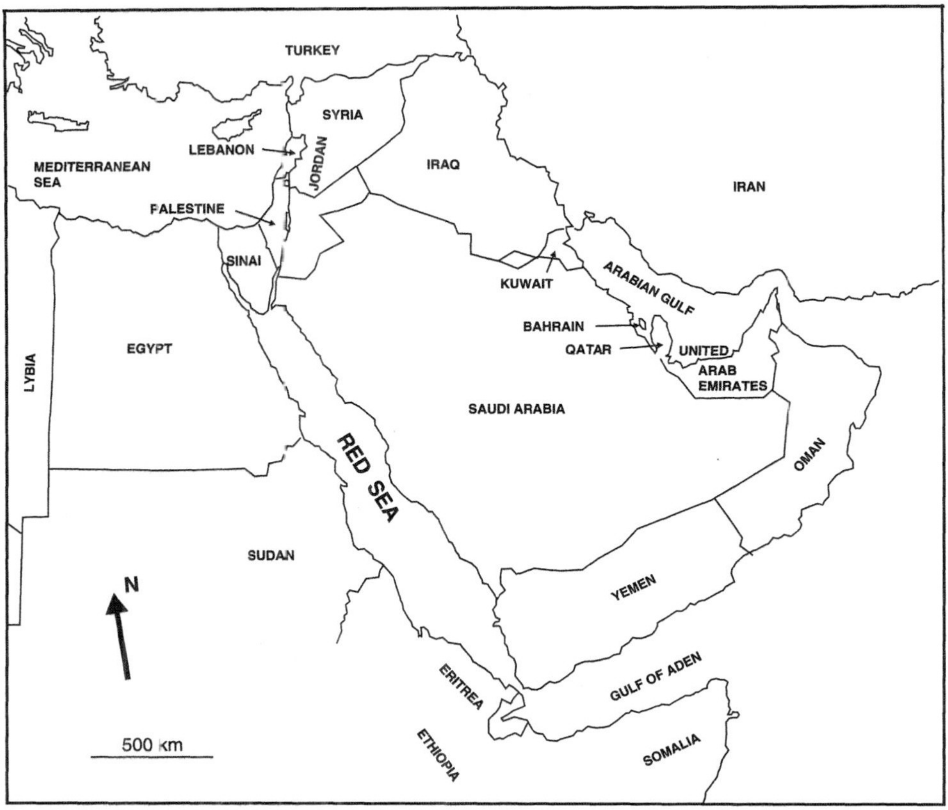

Fig. 10.2: Location map of the Middle Eastern countries involved in this chapter.

ciated with the Najd Fault system. The Phanerozoic sediments accumulated from all sides of the Arabian-Nubian Shield to create the Arabian-Nubian Craton. This craton includes the shield as well as the sediments on its flanks.

One of the major problems for this work is that the Middle East is politically highly fragmented. Each small or large country has its own set of formation and group names as well as literature. Correlation and generalization is thus extremely difficult and it is practically impossible to mention all references for the whole Middle Eastern countries throughout this long period of geological history.

The aim of this chapter is to discuss the long-period cyclicity and its causes within the Arabian-Nubian Craton since 640 Ma up till the recent. The names of the North American cycles of Sloss and Speed (1974), e.g. Sauk, Tippecanoe, etc., are not appropriate for use in the Arabian-Nubian Craton, because they do not fully coincide with the long-term cyclicity of the craton. Furthermore, no names for such cycles are present in the Middle Eastern geology. Consequently, the names of the cycles used in this chapter are proposed by the present author who hopes them to be accepted by others.

10.2 NAJD FAULT CYCLE

This is the first tectonic cycle involving the whole, newly formed Arabian-Nubian Shield. It started around 640 Ma and ended around the beginning of the Cambrian, with a duration of about 95 Ma. It is characterized by the activities of the Najd Fault system. This system is very conspicuous in Saudi Arabia, Yemen, Jordan, Egypt, and the Sinai, and is less clear in other parts of the shield. It is a sinistral, en echelon, strike-slip fault system extending in a NW-SW trend for about 2000 km (Yemen-Jordan) and several hundreds of kilometres wide. The total slip along the system is about 250 km with a slip on individual faults of up to 40 km (Moore, 1979; Brown et al., 1989). The Najd Fault activity continued from 640 to 540 Ma (Al-Shanti, 1993). The formation of this system of faults is due to the stresses that accumulated in the newly formed Arabian-Nubian Shield after the suturing and collision of its various segments (Stoeser and Camp, 1984; Davies, 1984).

Vertical tectonics along the Najd Fault was not less important than the horizontal slip. Normal faulting along the various segments of the system created rift basins, grabens, horsts, intermontane basins, and the like (Brown et al., 1963; Delfour, 1970; Jarrar et al., 1991). Uplift and erosion dominated the elevated blocks while subsidence and sedimentation continued in the lows. Due to these processes of uplift, erosion, and sedimentation, many rather thick sediment groups were deposited within the Najd Fault tectonic cycle as documented by absolute dating. No attempt will be made to give a full description of these groups because of space limitations. However, few of them are outlined below (Fig. 10.3).

The Jubaylah Group (600–570 Ma; Binda, 1981) in the northern Arabian Shield has a maximum thickness of 3300 m remaining after erosion (Delfour, 1977). It consists of immature, unmetamorphosed conglomerates, sandstones, mud rocks, and a thick sequence of limestone in its middle. The limestone contains stromatolites and the clastic rocks exhibit cross bedding, grading, and ripple marks. These structures are taken to indicate deposition in shallow marine environments.

The Murdama Group (620–570 Ma; Al-Shanti, 1993) in the eastern parts of the shield has essentially similar lithologies except that it is slightly older and metamorphosed. It attains a maximum thickness of 10,000 m deposited in a basin 500 km long and 80 km wide. Subsidence in both groups range from 100 to 200 m/million year.

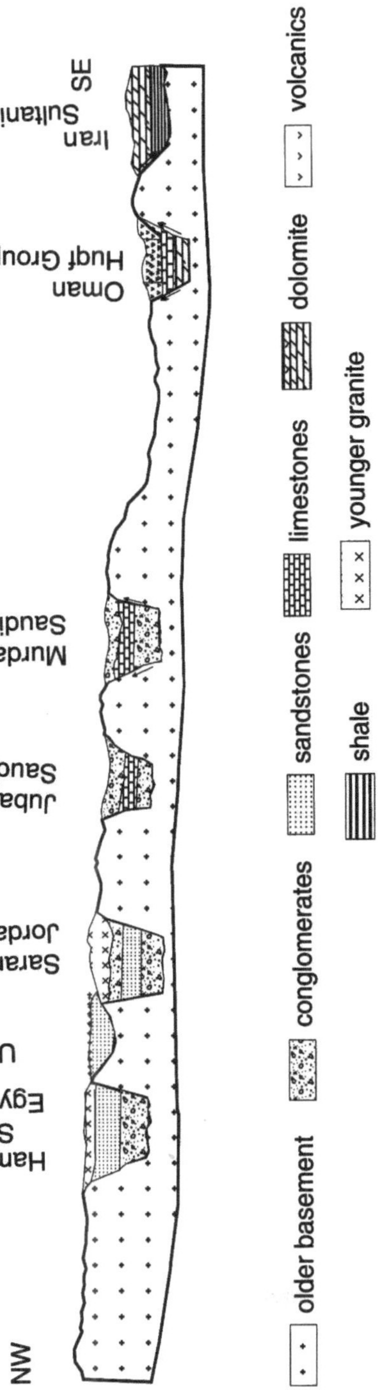

Fig. 10.3: Cross section from SW Iran and Oman in the southeastern study area, to Egypt in the northwestern part of the study area during the Najd Fault oscillatory cycle (640–545 Ma) (after data from Al-Shanti, 1993, and other authors).

Other examples from other countries, which have essentially the same age range, are the Saramouj Formation in Jordan (600–595 Ma; Jarrar et al., 1991; Fig. 10.4), the Hammamat Group in the Eastern Desert and Sinai in Egypt, the Homogar and Awat Groups in Sudan, and the Shiraro and Mathos Formations in Ethiopia and Eritrea.

Added to the above groups are the Infra-Cambrian sediments at the margin of the craton. Although not precisely dated, it is believed that they belong to this Precambrian cycle and not to the Cambrian. This is because they were deposited in a similar tectonic setting as the basins described above, i.e. rifted basins, grabens or pull-apart basins (see also Husseini, 1988). These Infra-Cambrian deposits are of two types. (1) The immature lithic sandstones with minor conglomerates deposited in intermontane basins; examples are the Umm Gadah Formation and the Sienaf (Zinfiem) Formation in Jordan and the Negev, respectively, with thickness up to 2000 m (Weissbrod, 1970; Amireh and Abed, 2000). (2) The thick sequence of evaporites and related rocks of Oman and the Arabian (Persian) Gulf; e.g. the Ara Formation and the Hormoz salts deposited in similar tectonic setting but seeming to be distal deposits of the craton.

An inter-regional unconformity overlies these sediments throughout the craton and marks the end of the Najd Faulting Cycle. The unconformity surface is a

Fig. 10.4: The Saramouj conglomerates near the SE tip of the Dead Sea: a typical example of similar deposits like the Hammamat of Sinai and the SE desert of Egypt, the Jubayal of Saudi Arábia, etc. These conglomerates were deposited in extensional basins produced by the Najd Fault tectonics at around 600 Ma.

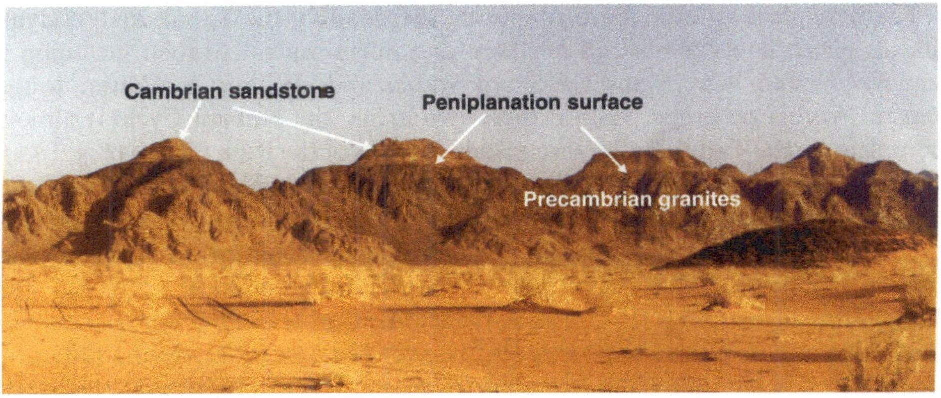

Fig. 10.5: The peniplanation surface at the end of the Najd Fault cycle in S Jordan. This unconformity is inter-regional and can be easily seen in Arabia and North Africa (from the collection of Khaled Al-Momani, NRA, Amman).

planation surface indicating the continuous erosion of the basement highs and the thick deposition of the above mentioned examples of molasses (Fig. 10.5).

10.3 NUBIAN CYCLE I

A conspicuous inter-regional unconformity separates the preceding Najd Fault Cycle from the overlying Nubian Cycle I. The highs of the former cycle were planated to produce this conspicuous surface that can be traced from North Africa, through Jordan and the Sinai, into central Arabia. This unconformity is considered through field studies, to separate the latest Precambrian from the Early Paleozoic (Bender, 1974; Al-Laboun, 1986, among many others).

Although the term "Nubian" is a loose term that involves sediments of variable age duration ranging from the Early Cambrian to the Lower Cretaceous, the lowermost part of the Cambrian-Lower Ordovician sandstones are the typical sandstones for this term. That is why the term "Nubian" was selected to represent this cycle.

The Nubian Cycle extended from the Early Cambrian to the Early Ordovician (Arenigian): a duration of about 70 Ma. The Arabian-Nubian Craton was a stable area with epeirogenic movements affecting it. Its margins were passive continental margins where mostly terrestrial mature sandstones were deposited, with minor marine ingressions. Consequently, this is an emergent-type episode where most of the craton was uplifted and erosion was rather active as to produce the thick (more than 1000 m) sediment sequence.

The best place to describe this episode are possibly the Tabuk and Widyan Basins which cover almost all northern and north-central Arabia, including S and SW Jordan, NW, N and NE Saudi Arabia, and SW Iraq; in addition to the central Arabia outcrops. Within these two areas, the Nubian Cycle is almost complete. In the Tabuk Basin, the episode consists of the Ram Group (Fig. 10.6), which is exactly equivalent to the Saq Formation in NW Saudi Arabia. Both units are made up essentially of braided-river, mature sandstones, with various types of pronounced cross bedding, channeling, fining-upward cycles, exotic quartz pebbles, etc. The heavy mineral suite indicates multiple recycling. The first marine ingression is that of the early Middle Cambrian: the Burj Formation of Jordan and its equivalents. It is totally siliciclastic near the shield, and in its middle part, carbonates appear away from the shield in Jordan and Syria. A second very short shaly marine horizon is found in the lowermost Ordovician within the Disi Formation and its equivalents. Furthermore, the Umm Sahm Formation at the top of the Ram and Saq units is terrestrial but with some marine affinity: transitional.

In central Saudi Arabia and further NW, the Nubian Cycle is represented by the Dibisiyah Formation, which consists almost completely of continental mature sandstones. The upper surface of this Dibisiyah Formation is an uncon-

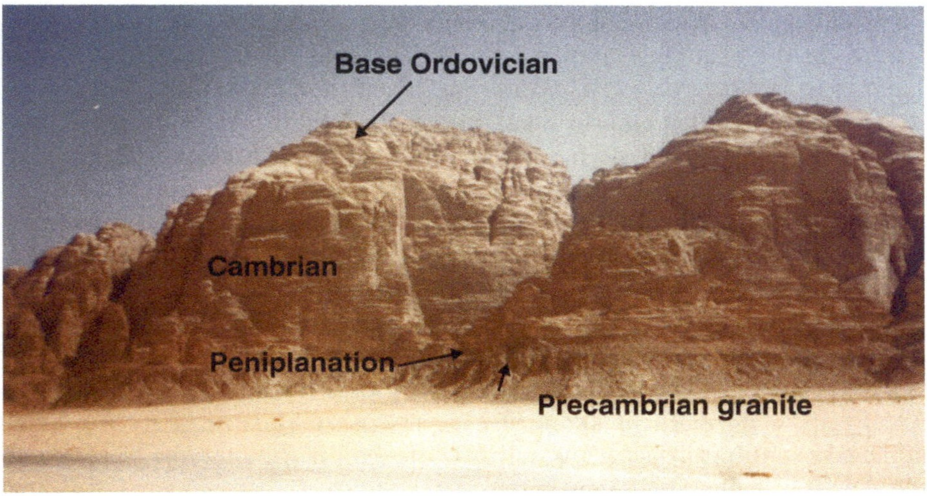

Fig. 10.6: Braided river quartzarenites at Wadi Ram, S Jordan: typical examples of similar sandstones of the Saq Group in NW Arábia, SW Egypt, and farther west (from the collection of Khaled Al-Momani, NRA, Amman).

formity, which is persistent in central Arabia and can be traced farther NW towards the Tabuk Basin (Vaslet, 1990) and into the Arab Emirates farther east. Although the beds in the Ram and Saq units of the Tabuk Basin are conformable with the overlying next episode deposits, the author believes the contact to be a disconformity. This is because of the abrupt change in depositional environment, from massive, mature, and onshore sandstones making the top

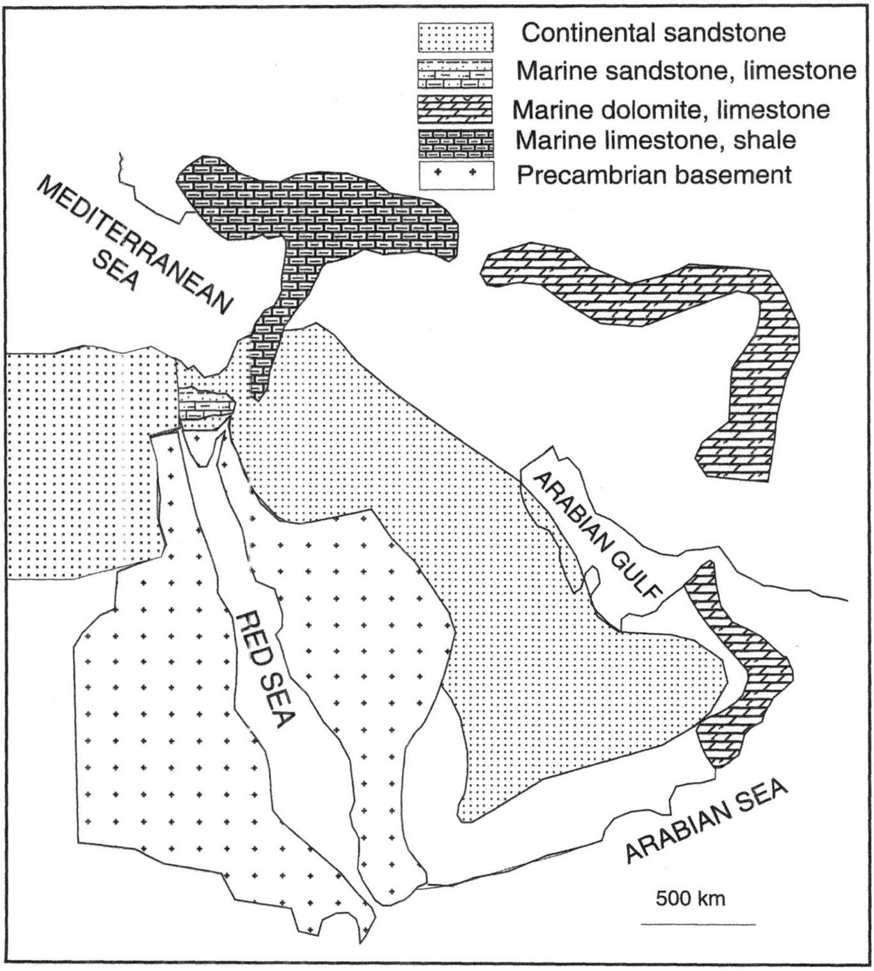

Fig. 10.7: The Arabian-Nubian Craton during the Nubian I emergent cycle, extending from the Early Cambrian to the Early Ordovician. Note that most of the craton is covered by fluvial sandstones (modified from Klitzsch, 1990; Alsharhan and Nairn, 1997; Abed, 2000).

of Nubian Cycle I, to pelagic shales with graptolites of the Hiswa (Hanadir) Formation of the overlying cycle.

In central Oman and farther NE, the Nubian Cycle is present as the lower Haima Supergroup (Hughes-Clarke, 1988; Boserio et al., 1995). The lower Haima is Cambrian to possibly lowermost Ordovician in age. It consists of mature quartz sandstones of braided rivers to eolian in its lower part, while the upper part consists more of claystones, marl with fossiliferous limestones, and ending with quartzarenites. The whole sequence is continental in the SW, with some marine influence towards the upper Cambrian in the NE. A clear unconformity separates the lower Haima from its upper part. In Yemen and southern Oman, the whole Nubian Cycle is completely eroded while in Egypt and Sudan, the outcrops of this cycle are rather fragmentary and isolated. It seems that most of Egypt was a high during this cycle (Klitzsch, 1990; Issawi, 2002). Figure 10.7 is a map showing the facies distribution within the study area. The fluvial sandstones predominate while the marine carbonates, shales, and sandstones are restricted to the craton margins in the north and east.

10.4 NUBIAN CYCLE II

This is a submergent episode where the sea had transgressed over a wide area of the craton. It is defined by the inter-regional unconformity at the top of the previous cycle and ends at the inter-regional unconformity marked by the widespread glaciations of the latest Ordovician (Caradocian/Ashgillian): a duration of about 20 Ma. Again, the Tabuk/Widyan basins in northern Arabia and southern Jordan show possibly the best continuous exposures for this cycle. It includes most of the Khreim Group of Jordan (Fig. 10.8) and its equivalent Tabuk Formation of Saudi Arabia (Al-Laboun, 1986; Abed, 2000). It consists of pelagic shales with graptolites, mid-shelf sandstone tempestites, and inner-shelf clastics with *Skolithos* and *Lingula*. A similar regime is reported in central Arabia (Vaslet, 1990).

The Tabuk Formation is penetrated in Kuwait and Qatar in the Arabian (Persian) Gulf. Equivalent marine deposits are also present in the Emirates and Oman mountains as well as central Oman (Ra'na, Amdeh, Ghudun, and Safiq formations, respectively). This cycle is missing in SW Arabia (southern Oman, Yemen, and SW Saudi Arabia) as younger sediments are overlying directly the Precambrian basement.

The marine sediments of this cycle have also penetrated into western Iraq, south and central Syria, and possibly SE Turkey. Towards the thrust zone of NE Iraq, the not-well dated Khabour Formation shows marine horizons of equivalent ages. Egypt seems to have remained a high during the Ordovician and only

Fig. 10.8: Three horizons of Skolithos from the Upper Ordovician of S Jordan and equivalent formations (from the collection of Khaled Al-Momani, NRA, Amman).

few tens of meters of fluvial and marine sediments are present in W and SW Egypt (Klitzsch, 1990; Issawi, 2002; Fig. 10.9).

This marine regime with a net deposition far exceeding erosion is followed by the inter-regional unconformity throughout the whole Middle East and North Africa, representing the latest Ordovician glaciation described in the next cycle.

10.5 HOGGAR EMERGENT CYCLE

Although short (Caradocian/Ashgillian or part of them, around 10 Ma), the continental to marine glacial deposits of this cycle are reported throughout the Middle East and North Africa, indicating an inter-regional exposure at the latest Ordovician. That is why this event is taken as a new episode/cycle. The type sediments are present in the Hoggar Mountains in southern Algeria (and hence the name) near where the South Pole at the latest Ordovician was possibly located. Consequently, continental glacial deposits are well displayed in North Africa (Rust, 1981), Egypt (Gabgaba Formation; Issawi, 2002), central and northwest Arabia (Sarah and Zarqa formations; Vaslet, 1990) and

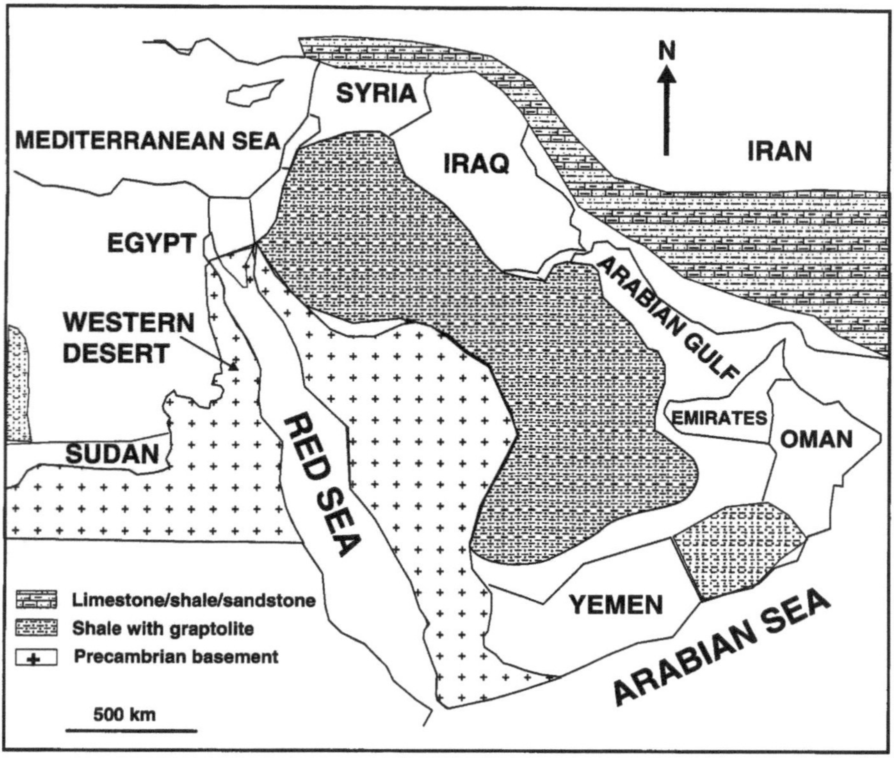

Fig. 10.9: The Arabian-Nubian Craton during the Nubian II submergent cycles, extending from the late Early Ordovician to the Late Orodivican, with a duration of about 20 Ma (modified after Klitzsch, 1990; Alsharhan and Nairn, 1997; Abed, 2000).

southern Jordan (Ammar Formation; Abed et al., 1993). Glacial sediments of this event are also reported from Turkey, Iran, and Oman. In areas where such sediments are not reported, a major unconformity exists with continental deposits, as e.g. the Afandi Formation of NE Syria.

10.6 NUBIAN CYCLE III

This a submergent episode that includes the Silurian-Devonian times, possibly also the Early Carboniferous, with a duration of about 120 Ma. It is identified by the glacial unconformity of the previous cycle at its base and by a pronounced inter-regional unconformity associated with the Hercynian Orogeny around the middle Carboniferous. It starts with a major transgression with deep marine facies rich in organic matter (hot shale) and graptolites, a deposition possibly caused by the melting of the ice sheet of the previous emergent Hoggar Cycle. These facies are typified by the Quseiba Formation of central, SW, and N Saudi Arabia; the Batra Member of the Mudawwara Formation of S, SE, and E Jordan (Fig. 10.10), and their equivalent formations in western Iraq; the Tunf Formation of Syria, southern Turkey, and eastern and southern Iran (Fig. 10.11). The relative sea level starts to fall through the Late Silurian, and nearshore deposits of the Sharwara Sandstone Formation (or its equivalents) were deposited throughout much of Arabia, as e.g. Qatar, the Emirates, eastern Iran, and SE Turkey. This cycle continues through the Early Devonian by the deposition of the interbedded siliciclastics and carbonates of the Jauf Formation in a shallow marine to bay environment. The cycle ends with the Sakaka Sandstone Formation (and its equivalents) of the Late Devonian, being fluvial to alluvial. The episode is a typical shallowing-upwards cycle involving the Silurian and Devonian (Al-Laboun, 1986; Hamam and Nasrulla, 1989; Vaslet,

Fig. 10.10: A horizon rich in graptolites from the Batra mudstone member, S Jordan, at the base of the Silurian. The horizon correlates with the Kuseuba Formation of Saudi Arábia and equivalent formations. It indicates the onset of a major ingression on the whole craton (from the collection of Khaled Al-Momani, NRA, Amman).

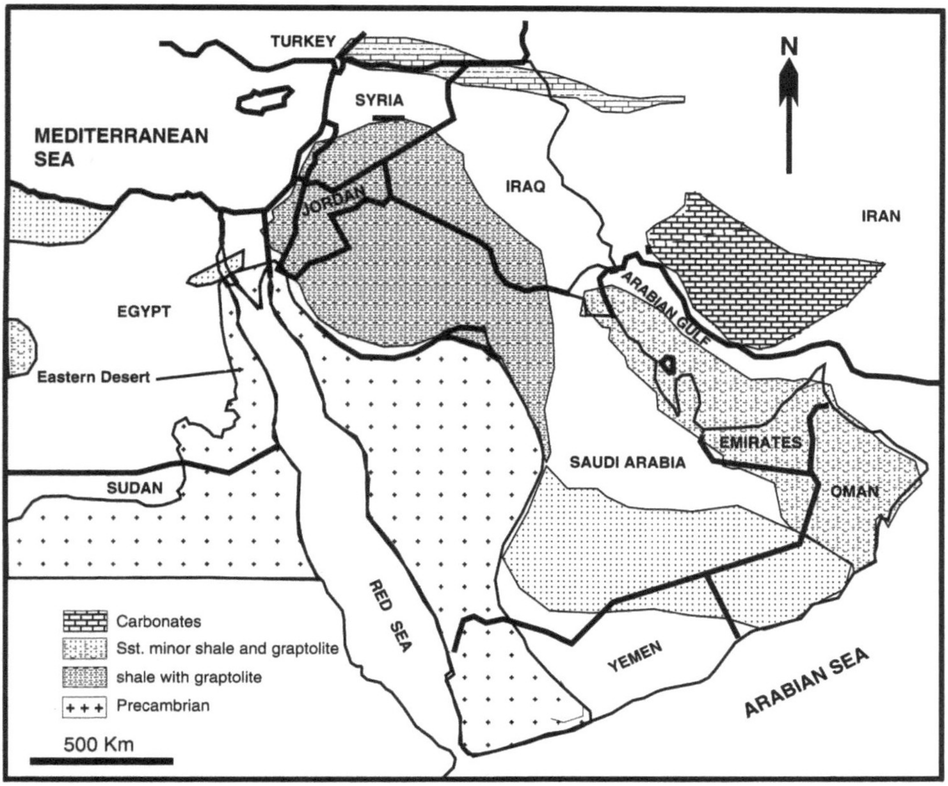

Fig. 10.11: The Arabian-Nubian Craton at the time of the Early Silurian Quseiba Formation. This is the Nubian III submergent cycle involving the Silurian-Devonian and possibly the earliest Carboniferous (modified from Klitzsch, 1990; Alsharhan and Nairn, 1997; Abed, 2000).

1990; Abed, 2000, among many others). A similar picture is also present in North Africa till western Egypt. The Silurian and Devonian in the rest of Egypt are again sporadic, with fragmentary outcrops (Issawi, 2002). However, the above-described complete shallowing cycle is not present throughout the Middle East. The whole Devonian, for instance, is missing in Yemen, Oman, Jordan, Syria, and Palestine, possibly due to erosion associated with the following Hercynian orogenetic cycle (Powell, 1989; Andrews, 1991).

10.7 HERCYNIAN CYCLE

This cycle is an emergent episode involving the late Early Carboniferous till the early Late Permian with its maximum in the middle Carboniferous: a duration of about 80 Ma. This cycle is characterized by regional uplift throughout the Middle East, having caused the erosion of the previously deposited cycles. No true deformation appears associated with this cycle, except for tilting and uplift. It is underlain and overlain by pronounced inter-regional unconformities and loss of strata. Although the erosion caused by this orogeny had removed the strata down to the Precambrian in certain parts of the Middle East like western Jordan, Palestine, and southern Syria (Figs. 10.12 and 10.13), the maximum effect of this orogeny seems to have taken place in the early Late Carboniferous or simply around the mid-Carboniferous (Gvirtzman and Weissbrod, 1984; Andrews, 1991; Abed, 2000). This conclusion is drawn from the fact that the mid-Carboniferous is missing throughout the Middle East, except in the eastern Syria-western Iraq basin, the Alborz area in southern Iran, and the Gulf of Suez and SW Egypt.

It could be said with confidence that the Carboniferous-Early Permian deposits, when present, consist of sandstone and sandy shale of continental origin (fluvial, alluvial, deltaic). This is typified by the Berwath Formation (Pre-Unayzah) of the Carboniferous and the Unayzah Formation of the latest Carboniferous-Early Permian and their equivalents throughout Arabia (e.g. Al-Laboun, 1986; Vaslet, 1990). Shallow marine deposits are present in the eastern Syria-western Iraq basins, Alborz area in southern Iran, the Gulf of Suez, and SW Egypt. From southern Oman (Alkhlata and Asfar formations) and North Yemen (Akbara Formation, Early Permian), glacial deposits have been reported.

10.8 JURHOM CYCLE

This cycle is also defined by inter-regional unconformities at its base with the previous cycle and at its top at the end of the Triassic/Early Jurassic. It involves the Late Permian-Triassic-early Jurassic, with a duration of about 60 Ma. The name Jurhom is taken from the name of an extinct Arabian tribe in central Arabia. This cycle is interpreted as an oscillatory episode according to Sloss and Speed (1974) as is shown later.

The best place to describe the full cycle is possibly N Arabia, that is the inner platform. Here the cycle starts with the carbonate Khuff Formation of Late Permian age, representing a transgression covering the terrestrial sandstones of the previous cycle (Hughes-Clarke, 1988). Figure 10.14 shows in a map, the facies distribution at the maximum Khuff (Late Permian) throughout the Middle East. It is overlain by the regressional deltaic-intertidal clastics of the Sudair Formation of the Early Triassic. The Middle Triassic is represented by the transgressional Jilh Formation sandstones, shales, and minor carbonates. The Late Triassic is represented by the Minjur Formation continental sandstone,

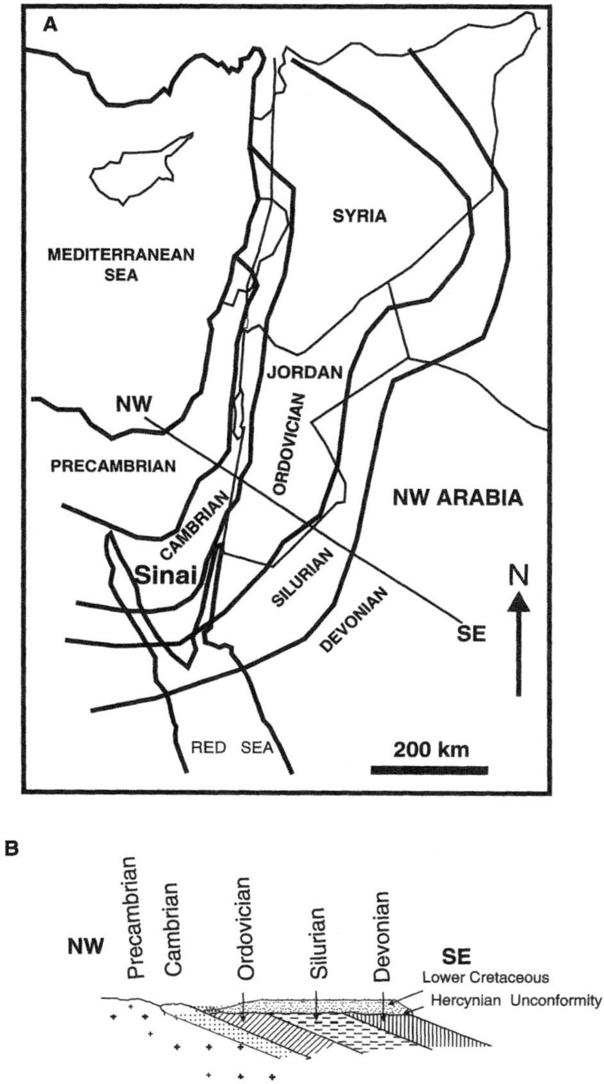

Fig. 10.12: The Hercynian emergent cycle: (A) Map showing the removal of the early Paleozoic sediments progressively towards the NE Sinai, where the whole pre-Carboniferous Paleozoic has been removed. (B) Cross section from N Sinai to NW Arabia (Tabuk Basin) showing the preservation of the early Paleozoic sediment away from the center of uplift somewhere near NE Sinai (modified from Gvirtzman and Weissbrod, 1984; Andrews, 1991; Abed, 2000; cross section by the author).

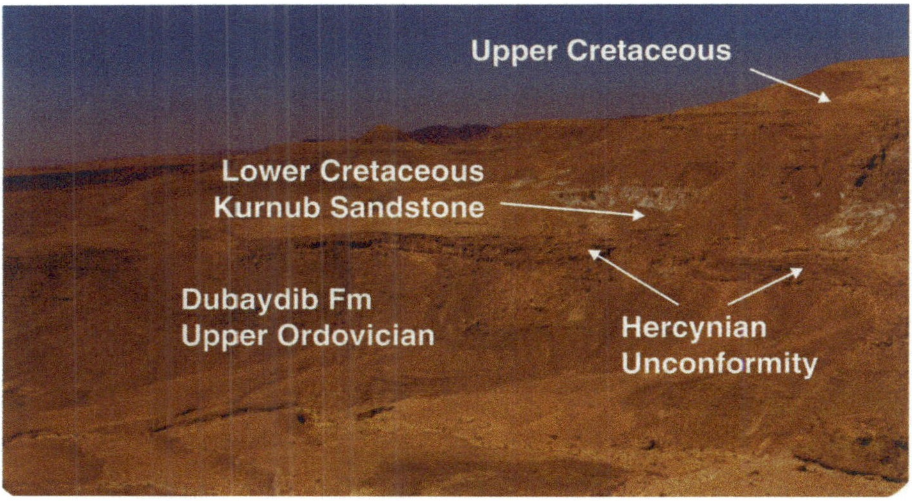

Fig. 10.13: The Hercynian unconformity shown from S Jordan, separating the Upper Ordovician from the lower Cretaceous (from the collection of Khaled Al-Momani, NRA, Amman).

which clearly indicates a retreat of the sea from the area (Al-Laboun, 1986; Al-Jallal, 1995).

The above-mentioned description of the cycle is essentially the same in Kuwait, Bahrein, and Qatar even in the formations names (Khan, 1989). The basin depocenter was situated in SW Iran and the southern Gulf where marine conditions continued throughout the whole cycle. However, sandstones are present in the south as well as immediately on the shield in central Arabia. In the extreme S and SW of Arabia, southern Oman, and Yemen, the Khuff is not present, possibly due to later erosion.

In the northern Arabian platform of Jordan, Palestine, Syria, Lebanon, Iraq, SE Turkey, and the eastern North African coasts, that is the margin of Gondwana at that time, fault-related rift basins developed as a response to the extensional tectonic activities associated with the separation of Eurasia from Gondwana in the Early-Middle Triassic, the opening of the Neotethys along the Zagros zone, and the closure of the Paleotethys. These tectonic activities led to the stretching of the Gondwana margins, the formation of horsts and grabens or basins separated by arches, and volcanic activities. Examples are the Mardin High in SE Turkey, Khleissa High in NE Iraq, Ha'il-Rutba Arch from northern Arabia into Iraq. Troughs are the large Mesopotamian Basin in central Iraq, Palmyra-Sinjar Trough in eastern Syria that seems to have reached the eastern Mediterranean, and the Euphrate-Anah Trough between Syria and Iraq. Relatively deeper

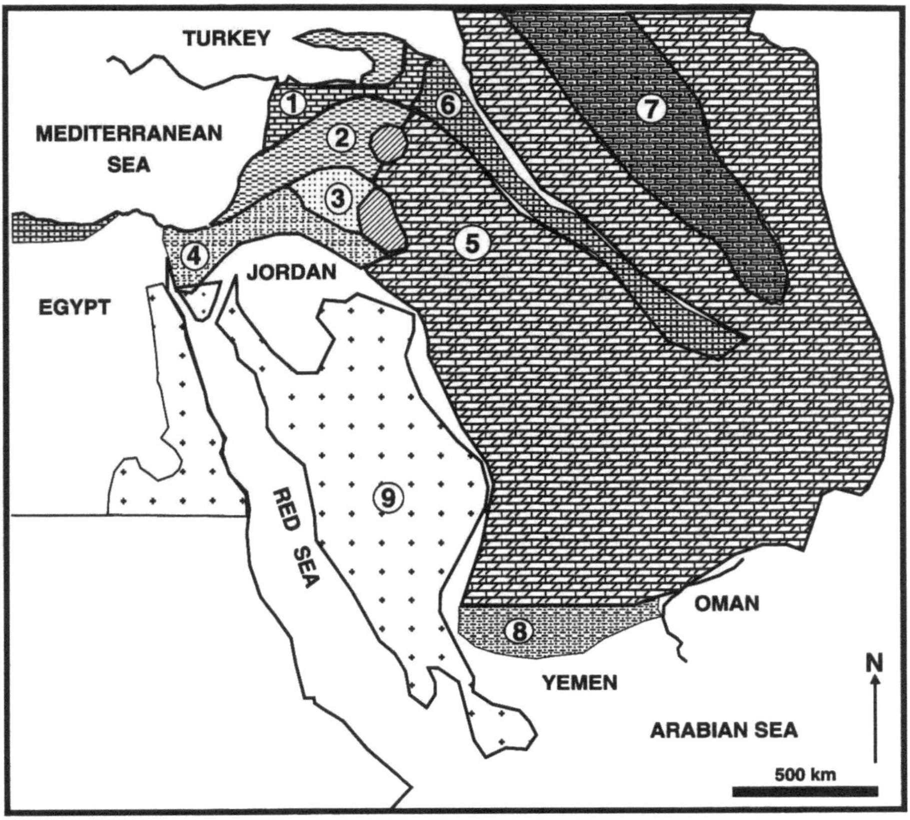

Fig. 10.14: Depositional setting of the Late Permian. Designations: (1) shallow carbonate shelf, (2) deep-water mixed shelf, (3) alluvial plain, (4) shallow mixed shelf, (5) restricted inter-subtidal carbonate shelf, (6) shallow carbonate build-up, (7) deep-water clayey limestone, (8) non-marine clastic sediments, (9) Precambrian basement (modified after Klitzsch, 1990; Alsharhan and Nairn, 1997).

carbonates were deposited in the trough with shallower carbonates and evaporites or fluvial deposits on the highs. Some of these highs became emerged, with erosion and clastic sedimentation in the adjacent trough (Sharief, 1983; Lababidi and Hamdan, 1985; Al-Laboun, 1986, among many others).

Furthermore, DeLaune-Myers (1984) reported rifting in the Baër-Bassit in NW Syria with rather deeper carbonate facies in the trough, becoming shallower with sandstone deposition farther east. Volcanic intrusions have also been recorded (Ponikarov et al., 1967). Garfunkel and Derin (1984), and Drukman

(1984) refer to rifting in the coastal plain of Palestine during the Early-Middle Triassic to Late Triassic-Early Jurassic. Changing rates of subsidence, erosion and sedimentation were reported throughout this cycle. The eastern North African coast (W Sinai, N Egypt and Libya) formed unstable margins of Gondwana. Fault-controlled depocenters and highs are characteristic, with shallow-marine, deltaic, and continental deposits (Sistini, 1984). Volcanic activities have also been recorded in the Triassic of Egypt (Issawi, 2002), the Triassic of Jordan (Bender, 1974; Bandel and Khoury, 1981), and the Oman Mountains.

10.9 NEOTETHYS CYCLE

This cycle is a submergent episode associated with the opening and closure of the Neotethys Ocean. It started in the Early Jurassic (Sinemurian, at about 170 Ma) and ended at about the Middle Miocene (15–10 Ma), a duration of about 185 Ma. Figure 10.15 shows the thicknesses and lithofacies of the sediments in both periods, which clearly indicates continuous deposition from the Neotethys. There was an inclination to end this cycle by the end of the Eocene, when a major uplift took place as an introduction to the formation of the Gulf of Aden-Red Sea-Dead Sea Transform and the subsequent separation of the Arabian Plate from the African Plate. Moreover, major parts of the study area became emerged by the end of the Eocene. However, some considerable parts of the study area remained still submergent like northern Syria, most of Iraq except its western part, the Arabian Gulf and its surroundings, northern Egypt and Libya, and southern Yemen, and Somalia (Alsharhan and Nairn, 1995; Issawi, 2002). That is why the end of the Eocene is not a suitable time to end this episode (Fig. 10.16). The Neotethys was not yet fully closed till the Middle-Late Miocene. At that time, the whole study area turned emergent and very similar to the present-day configuration.

The Neotethys started at about the earliest Jurassic with the separation of Eurasia from Gondwana after which it deepened and widened to from a rather E–W elongated seaway with oceanic crust at its bottom. These deep conditions continued from the Middle Jurassic till the Late Cretaceous. By this time, subduction of the African Plate, including Arabia, started, thus indicating the beginning of a compressional regime in the area. Ophiolites were emplaced at this time in several positions like Cyprus, NW Syria, Oman, and SW Iran (Delaloye and Wagner, 1984; Alsharhan and Nairn, 1997). Also formed at this time, the Syrian Arc Fold system that extends from Libya, through northern Egypt, into the Sinai, Palestine, Jordan, and SE Syria. Africa continued to migrate northward until Arabia collided with Eurasia, thus creating the Tauros-Zagros mountains that fully closed the Neotethys by the Middle

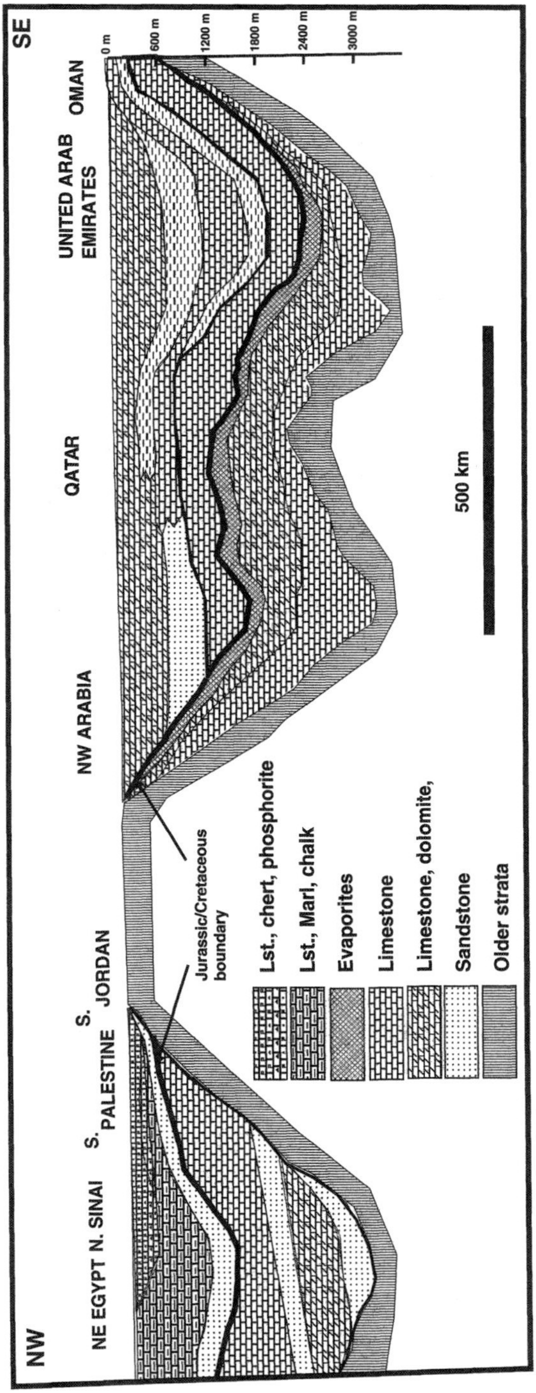

Fig. 10.15: Generalized cross section across the Arabian Craton from Oman in the SE to NE Egypt in the NW, showing the thicknesses and lithofacies of the Jurassic and Cretaceous deposits. Note the predominance of the marine carbonates throughout this period. The Jurassic and Cretaceous sediments form most of the Neotethys submergent cycle (drawn by the author from data in the various references mentioned in the text).

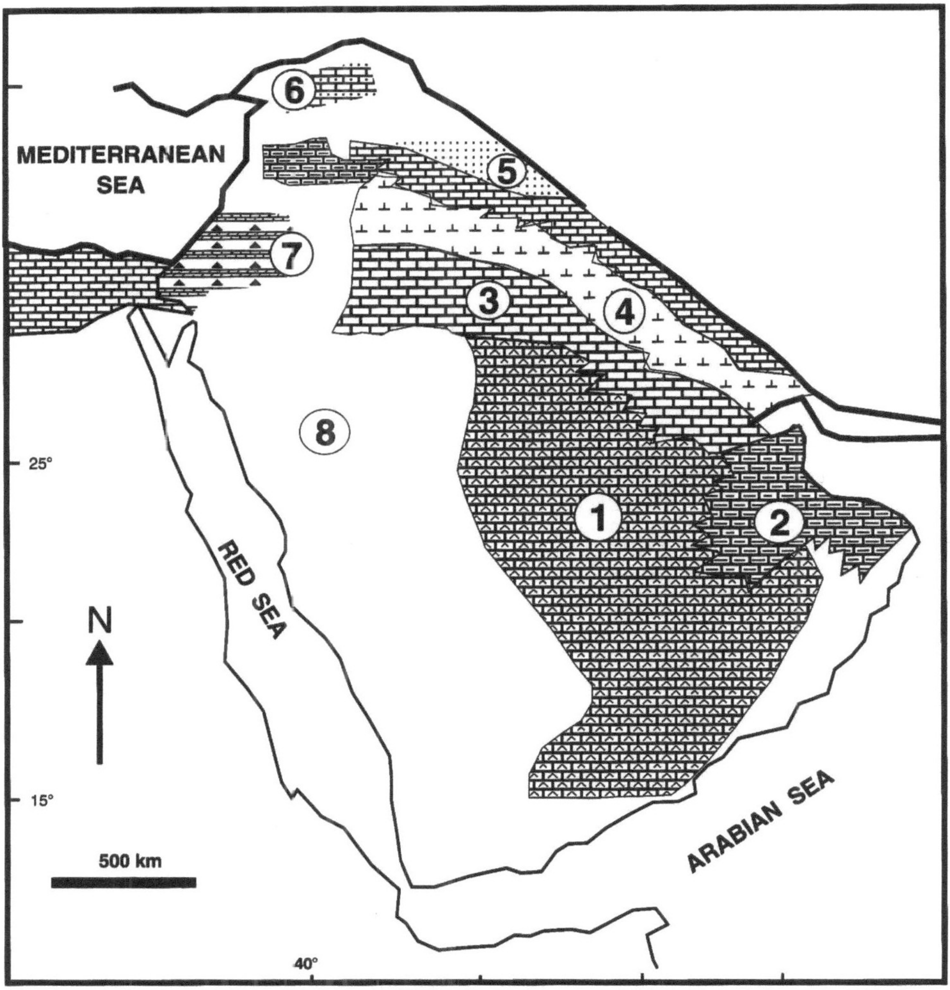

Fig. 10.16: Facies distribution of the Middle-Late Eocene in the Middle East. Designations: (1) evaporitic carbonate platform, (2) clayey limestone, (3) platform carbonate, (4) pelagic carbonate, (5) coarse clastics, (6) mixed deposits, (7) limestone-chert facies, (8) emerged area (modified after Jones and Racey, 1994; Abed, 2000; Issawi, 2002).

Miocene (Coleman, 1984; Hempton, 1987). To all that, other important tectonic events may be added, as e.g. the beginning of the separation of India from Africa very early in the cycle, and of South America from Africa in the Late Jurassic (Seyfert and Sirkin, 1979).

With this continuous tectonic regime in mind, net sedimentation was much greater than erosion. Thousands of meters of sediment, especially carbonates (Fig. 10.17), were deposited in the northern and central Arabian Platform as well as in the Arabian Gulf, Iran and the eastern North African coast. Several second-order cycles due to local tectonic can be identified in this thick sequence, as well as many third-order cycles due to fluctuations in the sea level (Murris, 1980).

Fig. 10.17: The Mujib canyon in central Jordan, showing the dominantly carbonate regime characterizing the Neotethys cycle. Upper Cretaceous strata from the top Turonian to the base Maastrichtian with a thickness of 250 m (from the work of Ahmad Al-Masri, NRA, Amman).

With the presence of the Neotethys north and east of the Arabian Craton, three major transgressions took place in the Early, Middle, and Late Jurassic, respectively. Consequently, a widespread carbonate regime was established throughout the area. A large carbonate ramp developed dipping E to SE, that is towards the basin along the Zagros deep. The Late Jurassic transgression advanced westward to cover parts of the Precambrian Arabian Shield. During this period, two intra-shelf basins developed: the first is found in eastern Iraq, while the second occupied parts of the Arabian Gulf. Euxinic conditions prevailed in these basins, where organic matter, source rock for oil, was deposited and preserved (Murris, 1980; Al Silwadi et al., 1996). By the end of the Jurassic, arid conditions coupled with regression, led to the deposition of an extensive evaporate regime, covering most of the area except for the two intra-shelf basins mentioned above (Powers et al., 1966; Alsharhan and Kendall, 1994).

During the Cretaceous, the carbonate platform environment was reestablished. However, this period can be divided into three unconformities in the Arabian Platform (Harris et al., 1984; Scott, 1990). The Early Cretaceous (Berriasian-Aptian) is represented by the Thamama Group consisting of carbonates in most parts of the platform. The two deep basins developed in the Jurassic in eastern Iraq and the Arabian Gulf, were reestablished, and continued throughout most of the Cretaceous. The Thamama Group is separated by an unconformity from the Wasia Group representing the middle Cretaceous (Albian-Turonian). Fluvial-deltaic sandstones are widespread especially in the western part of the Arabian Platform as well as in the southern part of the northern platform (Abed, 1982). The carbonate regime continued in the east (the Gulf, SW Iran, and NE Iraq). The clastic-dominated regime is unconformably overlain by the Aruma carbonates of Late Cretaceous age (Coniacian-Maastrichtian).

In the north Arabian Platform as well as northern Egypt and Libya, shallow-marine carbonates were deposited especially from the Cenomanian onwards (Bender, 1974; Abed, 2000; Issawi, 2002). Relatively deeper facies were deposited in the troughs while shallower facies were associated with the arches. There seems to have been a continuous shallow carbonate deposition in the north Arabian Platform across the Cretaceous/Tertiary boundary. A major transgression took place in the Late Maastrichtian-Paleocene, which accumulated open-shelf carbonates throughout the northern platform, Egypt, and Libya. This continued into the Early Eocene. While in the central Arabian Platform the Late Cretaceous Aruma Group is overlain unconformably by the Paleocene Umm Er Rhaduma Formation, further southwestward clastic sediments dominate. It could be said that most of the platform had been reestablished as a shallow carbonate regime during the Paleocene-Early Eocene.

With the uplift during the Eocene associated with the preparations for the opening of the Red Sea, a regression started to affect the whole carbonate platform. By the Late Eocene, sedimentation continued in SE Arabia, the Gulf, and SW Iran. The facies were pelagic carbonates along the Zagros trend in the east grading into shallow-water carbonate facies in and around the Gulf and carbonate-evaporite facies in SE Arabia. In the north Arabian Platform and Egypt, limestone, chalk, marl, chert, and phosphorite became deposited, as e.g. in Jordan, Palestine, and northern North Africa (Abed, 1994; Jones and Racey, 1994; Issawi, 2002). A major uplift pulse took place in the Oligocene, and the sea was pushed farther north and east. Marine sedimentation was restricted to the peripheral parts of the platform. The Arabian Gulf remained still connected to the Mediterranean through central-north Syria via Iraq (except its west). Carbonate deposits both shallow- and deep-marine are present chiefly in eastern-central Iraq, northern and central Syria, and along the present-day coast of the Mediterranean. This trend of sedimentation continued through the Early and lower Middle Miocene where it was restricted to the Zagros-Tauros seaway. By about the Middle Miocene, the Arabian Plate became separated from Africa, moved NNE and docked or collided with Eurasia in Iran and Turkey, completely closing the Neotethys (Fig. 10.18). No connection existed after this date between the Arabian Gulf and the Mediterranean. The whole study area was emerged and had become very similar to the present-day situation.

10.10 ARABIAN PLATE CYCLE

This cycle is the last episode in the evolution of the Arabian-Nubian Craton (Fig. 10.19). It started by the latest Middle Miocene or the earliest Late Miocene,

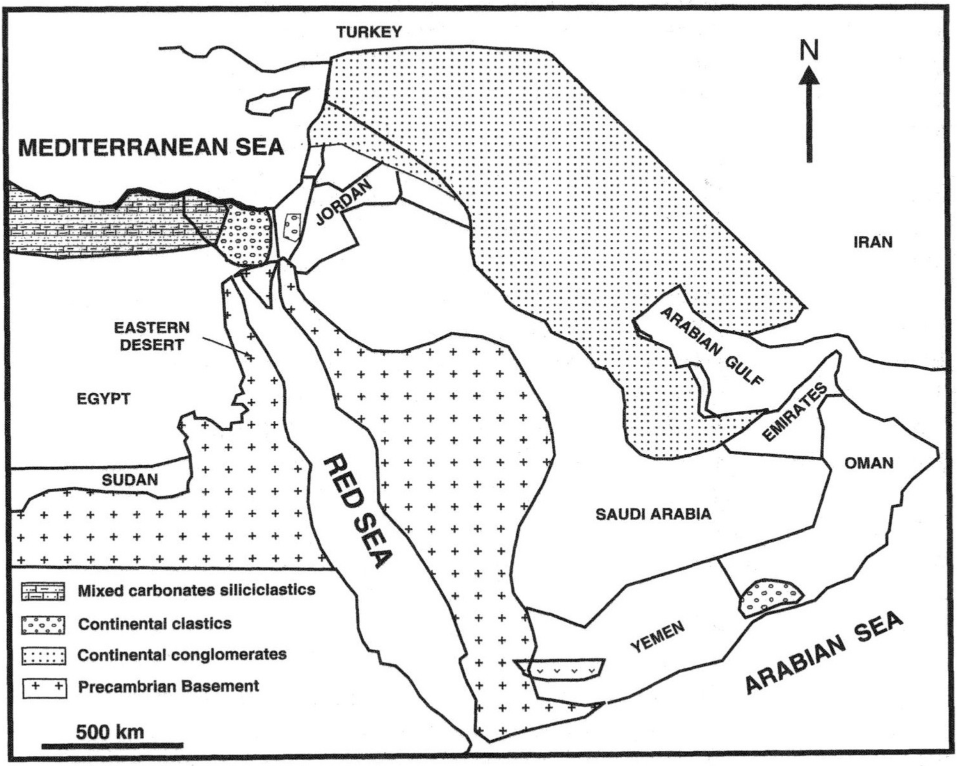

Fig. 10.18: Facies distribution during the latest Middle Miocene. Note the complete closure of the Neotethys by this time. Marine deposition is restricted to the Mediterranean coast as the whole craton has become emerged (modified from Jones and Racey, 1994; Issawi, 2002).

Ma		age	Mabesoone (1988)	Sloss and Speed (1974)	this work	orogeny	
0–65	PHANEROZOIC	Quaternary-Tertiary	Oscillatory-emergent	Oscillatory (Tejas)	Emer.-oscil. (Arabian Plate)	ALPINE	T, K
65–203	PHANEROZOIC	Cretaceous-Jurassic	submergent	Submergent (Zuni)	Submergent (Neotethys)	ALPINE	J
203–320	PHANEROZOIC	Triassic-Late Carbon.	Oscillatory-emergent	Oscillatory (Absaroka)	Oscillatory (Jurhom); Emergent-oscillatory (Hercynian)	VARISCAN	TR, P
320–410	PHANEROZOIC	Early Carbon.-Devonian	submergent	Submergent (Kaskaskia); Emergent	Submergent (Nubian III)	VARISCAN	C, D
410–455	PHANEROZOIC	Silurian-Late Ordovic.	Oscillatory-emergent	Submergent (Tippecanoe); Emergent	Emer.-oscil.(Hoggar); Submergent (Nubian II)	CALEDONIAN	S
455–540	PHANEROZOIC	Early Ordovic.-Cambrian	submergent	Submergent (Sauk); Emergent	Emergent-Oscillatory (Nubian I)	CALEDONIAN	O, C
540–650	L. PROTERO.	Latest Proterozoic	Oscillatory-emergent	Submergent? "Windermere"	Oscillatory (Najd Faulting)	NAJD	

Fig. 10.19: Comparative table of tectonic-sedimentary episodes presented in this chapter with those of Sloss and Speed (1974) and Mabesoone (1988). Summary of the long-period cyclicity throughout the history of the Arabian-Nubian Shield starting from around 640 Ma ago until present.

around 15 Ma ago, and is still ongoing. Around the Middle Miocene, the Arabian plate had separated from the African Plate and moved independently in an NNE direction. The Red Sea and the Gulf of Aden were flooded with water and became the site for thick sedimentation. Both gulfs are widening since then, thus pushing the Arabian Plate to collide with the Eurasian Plate along the Tauros-Zagros suture zones in Turkey and Iran, respectively. The divergent movement in the Red Sea in the south was transformed into collision in the north through the Dead Sea Transform, a 1100 km strike-slip fault starting from the southern tip of the Gulf of Aqaba to the East Anatolian Fault in the north.

An epeirogenic uplift of the whole plate started some time during the latest Eocene and is still ongoing especially along the Red Sea-Dead Sea Transform. This led to the emergence of almost the whole Arabian Plate. Two long mountain ranges have formed on both shoulders of the Red Sea-Dead Sea Transform, due to uplift of the shoulders, spreading of the Red Sea floor, and subsidence of the floor of the Dead Sea Transform. Erosion from these mountains and the deposition at their centers has led to the accumulation of a thick sediment sequence, as e.g. more than 7000 m in the Dead Sea Basin since the Middle Miocene. In general, sedimentation is actually restricted to the margins of the Mediterranean (mixed), the Arabian Gulf (chemical and biological), and the Gulf of Aden-Red Sea-Dead Sea Transform system (mixed). Consequently, this cycle is an emergent episode.

10.11 CONCLUSIONS

By about 1200 Ma, the Pan-African thermal event started to affect the Mozambique craton and led to its separation into an ocean in the area occupied now by the Arabian-Nubian shield. Island arc formation, subduction, collision, obduction, and suturing continued until the cratonization of the shield at about 640 Ma. Intrusive, volcanic, and sedimentary rocks were forming throughout this period in accordance with the geological setting.

The fully cratonized Arabian-Nubian Shield came into existence at about 640 Ma ago. During the following 100 Ma, the shield was unstable due to the dissipation of stresses accumulated throughout the cratonization period. An oscillatory cycle is assigned to this episode: the Najd Fault Cycle. The stable craton then underwent seven cycles of emergent and submergent, and one oscillatory cycle, shown in Fig. 10.19. The length of the cycles is variable lasting the submergent cycles longer.

The main reason for this cyclicity is tectonics on the craton and/or global scale with minor modifications due to eustasy. The duration of the cycles is variable, but the submergent cycles are usually longer than the emergent cycles. These cycles are mostly first-order cycles and partly second-order cycles.

The names of the cycles are coined by the present author and it is hoped that they will find acceptance by workers on the geology of the Middle East.

ACKNOWLEDGMENTS

The author would like to thank Mr. Khaled Al-Momani and Mr. Ahmad Al-Masri of the Natural Resources Authority of Jordan, and Prof. Galeb Jarrar of the University of Jordan for providing the field photos from their own collections.

11. CYCLICITY IN PALEOPROTEROZOIC TO NEOPROTEROZOIC CUDDAPAH SUPERGROUP AND ITS SIGNIFICANCE IN BASINAL EVOLUTION

P.K. DASGUPTA, A. BISWAS, AND R. MUKHERJEE

11.1 INTRODUCTION

Attention has recently been focused in the Indian subcontinent on the development of cyclicities in Proterzoic basins. Various aspects of cyclic sedimentation including tectonic controls, role of source area and different types of autocyclic and allocyclic processes supported by facies assemblages have put forward an enormous work load that is to be addressed (Chakraborty and Bhattacharyya, 1996; Dasgupta and Sengupta, 1997). Moreover, geochronological dating of the different scales of cyclicities is a feedback that has just started (Crawford and Compton, 1973; Bhattacharji and Singh, 1984; Jachariah et al., 1999).

Sedimentary cycles and cyclic sequences in basins are composed of at least three different beds and interbeds, which form a repetitive succession (Einsele, 1992). The term "cyclothem" is widely used in North America in a purely descriptive way especially for coal-bearing sequences (Heckel, 1986; Klein, 1989; Riegel, 1991). In seismic stratigraphy, the term is replaced by depositional sequence, which is defined by its lower and upper boundaries, as well as by lowstand, transgressive, highstand deposits, and parasequences (Van Wagoner et al., 1987; Vail, 1987; Vail et al., 1991; Coe, 2003). In the present chapter the terms *depostional sequence* or *depositional cycle* and for a succession of cycles, *cyclic sequences* have been used.

11.2 EVOLUTIONARY TRENDS OF THE PROTEROZOIC BASINS

In India, Proterozoic sedimentary sequences occur in the mobile belts of Singhbhum, Rajasthan, and Eastern Ghats as well as in epicontinental/rift basins in the northern fringe of Peninsular India bordering the Indo-Gangetic alluvial trough (Vindhyan Supergroup, Bijawar Group and Chattisgarh basins), along

Cyclic Development of Sedimentary Basins
J.M. Mabesoone and V.H. Neumann, editors.

the Pranhita-Godavari valley near the east coast, and further south in the Bhima, Kaladgi, and Chuddapah basins (Fig. 11.1).

While addressing cyclicities in Proterozoic basins in India, the evolution of such basins deserves mention. The geometry of the Proterozoic basins in India should have been essentially shaped either by their deposition around the Archean accretions or nuclei, and the related crustal upwarps and downwarps, or, curving of the "highs" and "lows," or, depressions along lineaments, or,

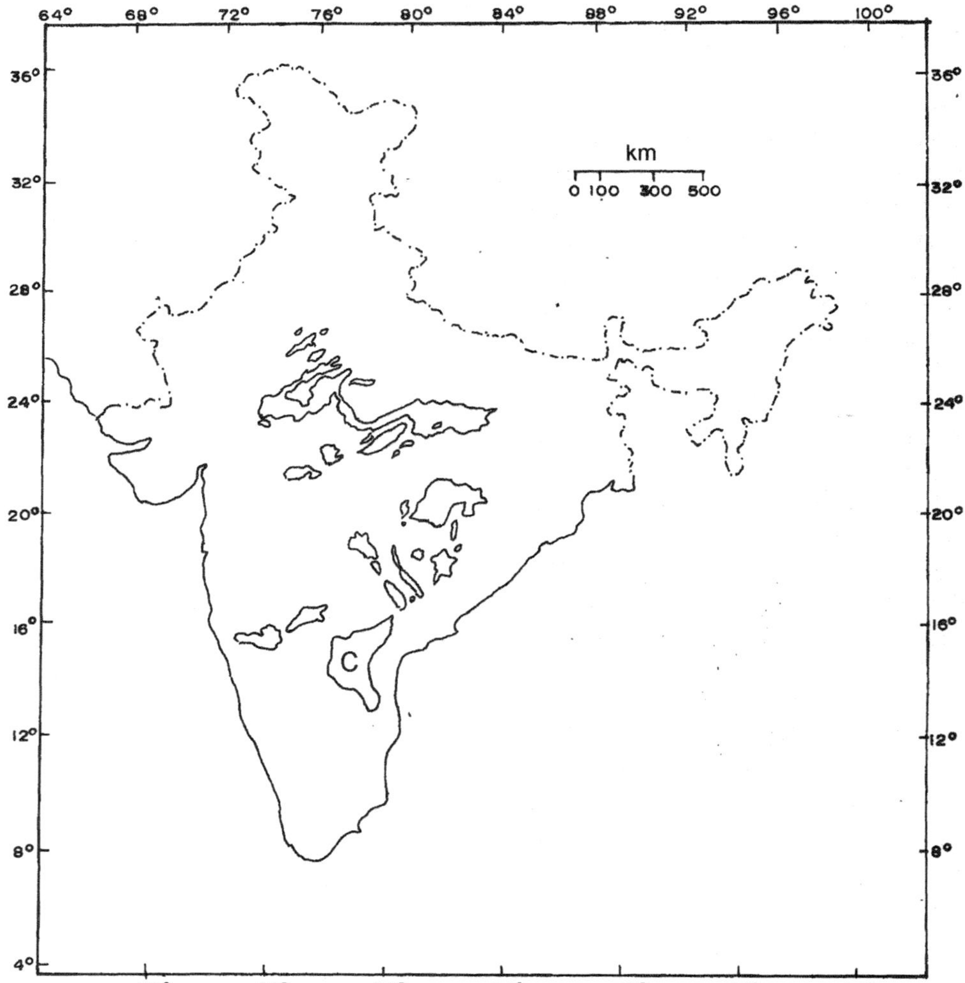

Fig. 11.1: Distribution of Proterozoic sedimentary basins in India. C – Cuddapah Basin.

combination of the two processes. The distribution of such major Proterozoic basins has been documented (Fig. 11.1). Ideas of specific basin models are far from being universally acceptable (Bhattacharji and Singh, 1984; Dhoundial, 1987; Narain, 1987; Radhakrishna, 1987).

To what extent the present geometry of the exposed Proterozoic basins is a reflection of original depositional geometry is by itself a subject of enquiry. The role of boundary faults and the attendant thickening of sedimentary prisms and structural complications along such selected zones have received considerable attention in deciphering the geometry of the basins (Bhattacharji and Singh, 1984; Narain, 1987). The characteristics of foreland basins, continental or epi-continental basin-like grabens or half-grabens have been attributed to some of the earlier stages of the large complex Proterozoic basins. These, testify strong phases of deformational history before the simpler platform cover deposits were laid down in wider and narrower basins (Kaila et al., 1979; Dhoundial, 1987). Out of the major Proterozoic basins, the 44,400 km^2, the Paleoproterozoic to Neoproterozoic Cuddapah Basin in Peninsular India (Fig. 11.2) housing sediments shed over approximately 1 Ga comprises autocyclic as well as allocyclic sequences. Such cyclicities in the stratal assemblages of the Chuddapah provide a case history for interpretation.

11.3 THE CUDDAPAH BASIN

11.3.1 GEOLOGIC SETTING

The Cuddapah Basin is bordered on the west by the Peninsular Gneiss showing remnants of greenstone belts. The Nellore Schist Belt and the Eastern Ghats Granulite terrain appear along the eastern margin. The basal formations of the Cuddapah Supergroup lie unconformably on the Peninsular Gneiss along the eastern margin (Fig. 11.2). The flat lying strata of the basal units have low easterly dips. In contrast, the sedimentary sequences in the eastern part of the basin are steeply dipping, isoclinally folded, and slightly metamorphosed. Along the eastern margin of the basin, the Nellore Schist Belt is thrust over the sedimentary sequences. Deep seismic soundings indicate that the thrust and the related imbrications are steeply inclined and penetrate the base of the crust (Drury and Holt, 1983; Kaila and Tewari, 1985; Kaila et al., 1987).

Active rifting, in which impingement of a thermal plume on the base of the lithosphere causes convective thinning (dilatation), domal uplift, crustal extension, sagging, and extensive episodes of igneous activities. These geodynamic controls were operative in the protobasinal setups through time (Sen and Narasima Rao, 1967; Bhattacharji and Singh, 1984; Bhattacharji, 1986). Passive

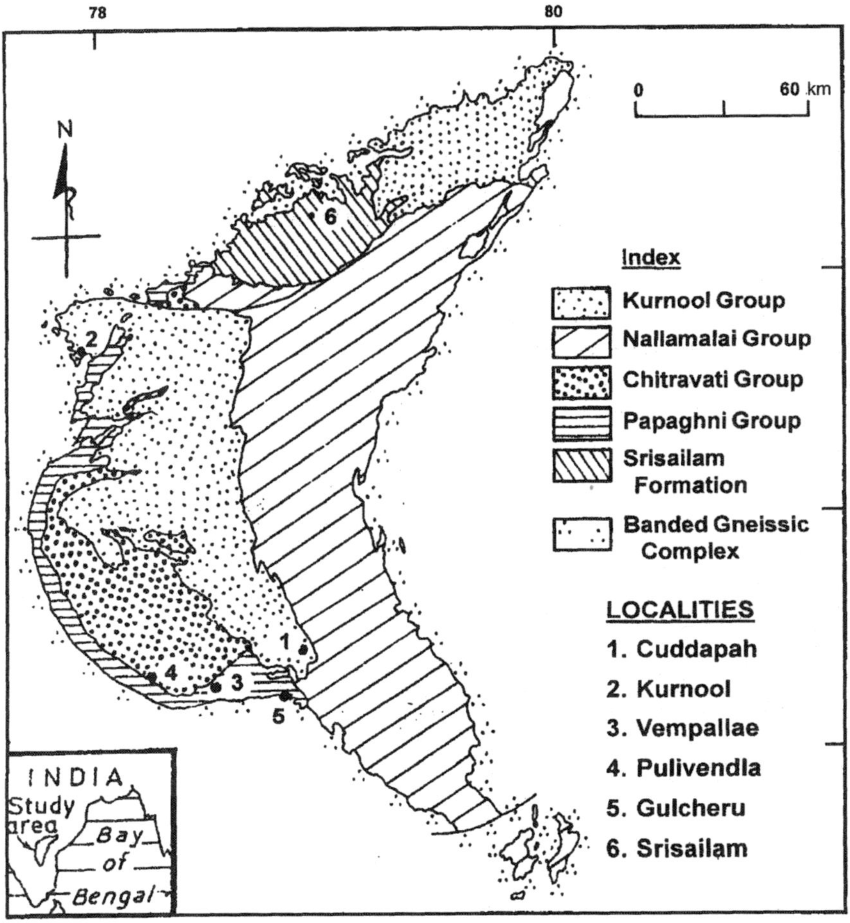

Fig. 11.2: Generalized geological map of the Cuddapah Basin, Peninsular India. For location, see Fig. 11.1.

rifting, in which extensional stresses in the continental lithosphere caused thinning and episodic upwelling of hot astenosphere, appears to be a widely acceptable model for the origin of sedimentary basins. Crustal doming and volcanic activity are only secondary processes (Turcotte and Oxburgh, 1973; McKenzie, 1978). Cyclic sedimentation in intracratonic basins with its tectonic controls is of relevance to test the role of passive rifting in such basin modeling. Cyclicity in the Cuddapah Supergroup, in this context, is an important case in hand.

11.3.2 STRATIGRAPHY

Four major subdivisions of the Cuddapah Supergroup, arranged in ascending order, were recognized by King (1872). Each subdivision was taken to be laterally continuous throughout the basin. Nearly a century later, an attempt was made to develop a scheme of chronostratigraphic classification for the Cudddapah succession (Sen and Narasima Rao, 1967). This was immediately replaced by a lithostratigraphic classification, which appears to be an improvement on the scheme given by King (Rajurkar and Ramalingaswamy, 1975, 1978; Nagaraja Rao and Ramalingaswamy, 1976; Murthy, 1979). Extensive mapping coupled with lateral tracing of the lithostratigraphic units for more than two decades resulted in a lithostratigraphic classification, which accounts for most of the observed stratigraphic relations (Murthy, 1979; Table 11.1).

11.3.3 LITHOLOGY

The Cuddapah Supergroup is dominantly arenaceous and argillaceous. Calcareous sediments constitute a small proportion of the total thickness. Volcaniclastics

Table 11.1: Stratigraphy of the Cuddapah Supergroup.

		thickness [m]
	Kurnool Group unconformity	500
	Engalapenta Member	440
Srisailam Formation	Tapasipenta Member	210
	Kaklet Vagu Member angular unconformity	320
	Kolamnala Shale Formation	950
Nallamalai Group	Irlakonda Quartzite Formation	1050
	Bairenkonda (Nagari) Formation angular unconformity	1500
	Gandikota Formation	1200
Chitravati (Cheyair) Group	Tadpatri Formation	4600
	Pulivendla Formation disconformity	1–75
	Vempallae Formation	1500
Papaghni Group		
	Gulcheru Formation nonconformity	28–250
	Archean and Dharwar	

and volcanics make up a considerable proportion of the Cuddapah sequence at different stratigraphic levels, the thickness of the individual units ranging up to a maximum of 3.5 km. The Cuddapah sediments mostly belong to fluvial, eolian, and lagoonal facies, though intercalations of marine sediments are rather common (King, 1872; Banerjee, 1974; Rajurkar, 1979; Singh, 1980; Dasgupta, 1996).

11.3.4 DEPOSITIONAL SUB-BASINS, FACIES MODELS, AND CYCLICITY FROM ROCK-RECORD SEQUENCES

Earlier workers recorded a progressive shifting of sub-basins with their sub-parallel axes trending NE-SW (Sen and Narasima Rao, 1967; Murthy, 1981). The following sequences of development of sub-basins has been recognized: Papaghni sub-basin (Papaghni Group), Chitravati sub-basin (Chitravati Group), Nallamalai sub-basin (Nallamalai Group) and Srisailam sub-basin (Srisailam Formation).

11.3.4.1 Papaghni sub-basin

Sediments of the Papaghni sub-basin (Gulcheru-Vempallae Formation) form a clastic-carbonate couple represented by bouldery conglomerate – impure sandstone facies and quartz-arenite facies overlain by cyclic carbonates and fine-grained clastics. Genetic controls of cycle hierarchy of the group have been interpreted (Schwarzacher and Fischer, 1982; Goodwin and Anderson, 1985; Einsele et al., 1992).

Cyclicity in Gulcheru Formation. The Gulcheru strata develop on a broadly undulatory erosion surfaces on the Precambrian gneisses basement. The sequence, extending along a strike length over 48 km, is characterized by a dominance of boulder-conglomerate beds (Fig. 11.3). Large incised channels with widths of 0.4–0.85 km having channel walls as high as 22 m consist of channel fills (Fig. 11.4). The channel facies mainly comprises thick beds of boulder deposits with or without interbeds. The lower Gulcheru coarse clastics interfinger with an onlap over the incised channel facies and show cyclicity in both large and small scales. Cycles (6–30 m) of stratified to well-stratified discontinuous matrix-poor upward coarsening boulder/conglomerate facies are discernible (Fig. 11.5). In addition, their vertical stacking patterns indicate an overall coarsening and thickening upward trend. Intermittent minor cycles, which are 2–4 m thick also, develop upward coarsening trends both in grain size and thickness. Stacks of strata within the minor cycles, however, are punctuated by fining upward and thinning layers of very coarse to coarse sandstones.

Fig. 11.3: Regional exposure of the Gulcheru Formation near Chitravati Gorge, Cuddapah Basin, India.

Fig. 11.4: Stacked channel facies of Gulcheru Formation. Scale: height of the shrub in the right hand corner = 1.2m.

Fig. 11.5: Stacked cyclic channels near basal Gulcheru Formation over the granitic basement. Couplet facies with cyclicity is discernible. Scale: length of hammer = 40 cm.

Interfingering of fine-grained deposits and coarse facies dictate the formation of lensoidal bed geometry with slight imbrication in detached outcrops. The facies assemblage is haphazardly arranged and grades laterally and vertically into beds of pebbly, coarse to granule-rich sandstones and granulestones.

Facies analysis of the Gulchery clastics from the best developed stratigraphic sections (Fig. 11.6) near Chitravati Gorge (ca. 520 m thick) reveals a spatio-vertical development of eight facies which have been classified following Miall (1978) (Blair and Mcpherson, 1994; Table 11.2).

Interplay of genetic factors and cyclicities in the Gulcheru. Sedimentary cycles including coarsening and fining upward sequences as well as complete and incomplete cycles are discernible from the development of different Gulcheru facies. Alternations between beds and interbeds of cyclic sequences were characterized by quasi-cyclic and dicyclic aspects. The thickness of corresponding bed types did change owing to variations in sedimentation rates in their cyclic deposition.

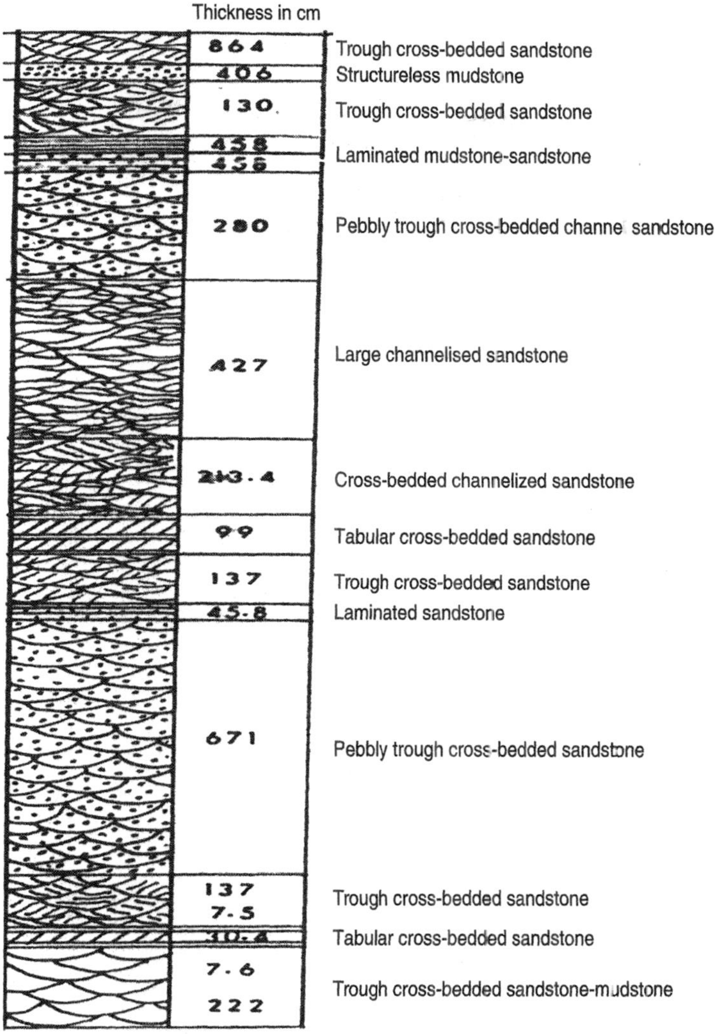

Fig. 11.6: Representative facies log of Gulcheru Formation near Chitravati Gorge, Cuddapah Basin.

Sheet-flood couplet facies (Fig. 11.7) in association with channel-fill deposits (Fig. 11.8) characterize the Gulcheru sequence with distinct facies attributes of alluvial fans.

The suite of facies common in both lower and upper Gulcheru sequences indicates deposition by channelized and non-channelized alluvial systems.

Table 11.2: Major facies attributes of the Gulcheru Formation.

name	brief description	interpretation
F1: Pebbly conglomerate	Organised erosively based, pebble conglomerate, lenticular and wedge-shape in cross-section, pebble conglomerate cross-stratification, normal grading, clast supported, fining upward trends within individual beds with 0.3–0.6 m thickness, similar to Gm, Gp facies of Miall (1978)	waning stage minor channel-fills and bar flank accretions
F2 : Moderately sorted pebble-cobble conglomerate	Organised, moderately sorted pebble-cobble conglomerate, clast imbrication, horizontal bedding, cross-stratification, clast supported, a(t) b(i) imbrication, bed thickness within the range of 0.8 and 5+ m, lateral extension upto tens of metres, occasional crude upward fining trends, similar to Gm, Gt, Gp facies of Miall (1978)	gravelly longitudinal bars and side bars, channel-fills
F3 : Disorganised cobble-boulder conglomerate	Disorganised, poorly sorted, structureless, cobble-boulder conglomerate, matrix and clast supported, clasts 0.5–0.6 m long, beds upto 6 m thick and of very large lateral extension, similar to Gm facies of Miall (1978)	hyperconcentrated flows
F4 : Medium to coarse sandstone	Cross-stratified, rippled and horizontally laminated, medium to coarse sandstone, one or more of following types of sedimentary structures: trough cross stratification, ripple cross lamination, horizontal to subhorizontal lamination, low-angle cross stratification showing broad (1 m) upward convex and upward concave laminae, mostly as lenticular bodies in upper parts of stratigraphic sections, tabular beds (0.1–0.8 m thick) as capping of conglomerate or 1–3.5 m wide, wedge-shaped bodies within conglomeratic units, individual laminae few mm thick, packages of concordant laminae (0.1–0.4 m thick) with truncation at low angle (5–18°) by adjacent and overlying laminae	waning stage channel and bar-top deposits
F5 : Massive, very coarse to granular sandstone	Massive, very coarse to granular sandstone, floating sand pebble size grains, tabular, lenticular or wedge-shaped beds, wedge shaped units most common in conglomerate lithosomes, tabular beds 1 m to several meters thick between beds of conglomerate	sandy mud flows
F6 : Very fine sandstone	Laminated or cross laminated very fine sandstone, siltstone and/or mudstone	drape deposits formed in pools of standing water
F7 : Couplet facies	Bouldery pebble beds interstratified with laminated pebbly granulestone and/or granular coarse sandstones. Facies variants: discontinuous parallel laminated pebbly sandstone, lensoidal conglomerate and low angle sandy conglomerate, occasionally found associated with the dominant couplet facies. Similar to couplet facies of Blair and Mcpherson (1994)	sandy alluvial sheet flood deposits
F8 : Lensoidal cross-stratified sandstone	Lensoidal bodies with high angle (20–32°) tabular and wedge planar cross strata with supercritical and subcritical climb, alternate grain fall and grain flow layers, upper and lower contacts in association with small channelised sandbodies	aeolian dunes in association with interdune deposits

Overall upward-coarsening and thickening cycles within fan bodies are punctuated by thinning and fining upward cycles. The most likely explanations for such small-scale cyclicity appears to be owing to fan retreat followed by flood

Fig. 11.7: Cyclicity within couplet beds, Gulcheru Formation. Scale: length of hammer = 40 cm.

basin aggradation and then by graded fan progradation (cf. Selley, 1978; Reading, 1996; Blair, 1999). In other words, autocyclic episodes are either represented by responses to periods of base level lowering against the marginal fault or they are partly due to lateral fan lobe shifting. The latter was important as it clearly happened within a context of basin floor subsidence easily triggered by the tectonic movements. Further support for direct and indirect tectonic controls on cyclicicty in the Gulcheru comes from the lateral persistence of boundaries of the fan facies and axially transported alluvium comprising braided and non-channelized fluvial systems. Similar case histories have also been recorded from other rock records (Steel and Gloppen, 1980).

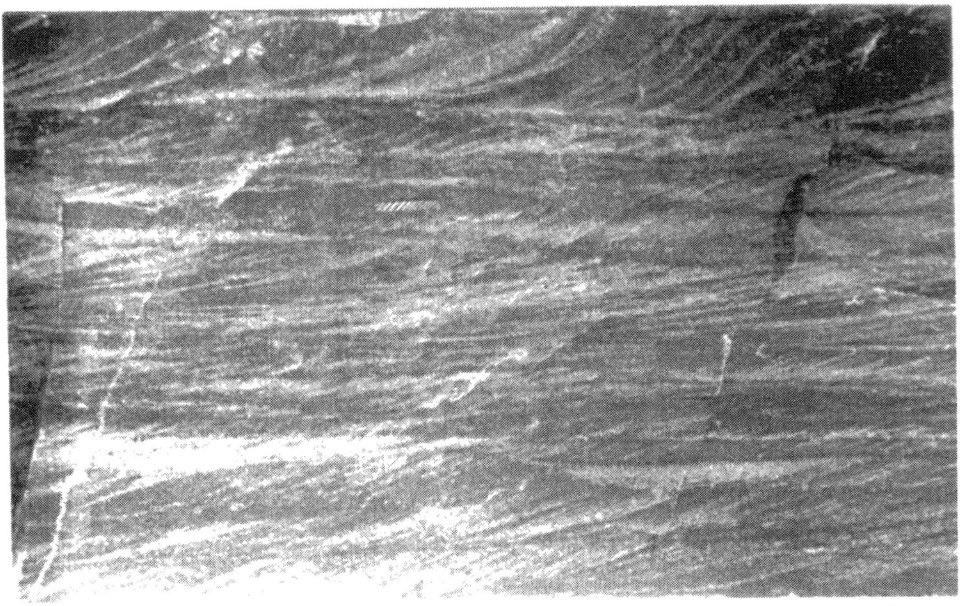

Fig. 11.8: Torrential cross-beds with epsilon cross stratification at the top, Gulcheru fluvial clastics. Length of the bar scale = 10 cm.

A vertically stacked cyclic sedimentation of alluvial fan and fluvial or eolian deposits in the setup occur in the tectonically most active side of the half-graben extensional basin as a result of periodically changing rates of basin subsidence, which is overprinted by allocyclic influence (cf. Denny, 1965; Heward, 1978; Blair, 1987).

In addition, the Gulcheru alluvial fan and braided river system record sequences of fluvial sediments alternating with deposits of sheet floods and debris flows.

Alternating phases of transportation and deposition of coarser gravel and fine gravel ± sand give rise to facies VIII. This is deposited under distinctive hydraulic conditions (high Froud number, high flow attenuation rate and high deposition rate) related to flow expansion and decreasing slope as well as intrinsic variations in depth and velocity of supercritical flow (Allen, 1982, 1983; Blair, 1985a, b; Blair and Mcpherson, 1994). Such couplet facies in association with other three facies variants owing to deposition by fluid gravity flows have been described as characteristics of sandy alluvial sheet-flood deposits and have been reported from modern fans in California (Van de Kamp, 1973; Blair and Mcpherson, 1994), Spain (Harvey, 1984), and England (Wells and Harvey, 1987). Thick (100+ m) sequences of sheet flood stratification due to sediment gravity processes generated by the failure of bed-rock cliffs have been documented from stratigraphic records. Analogues of the Triassic Mount Toby conglomerate of Massachusetts (Hand et al., 1969), the Jurassic Todo Santos Formation of southeastern Mexico (Blair, 1985a, 1987), and the Oligocene-Miocene Pantano Formation of southern Arizona (Balcer, 1984) can be cited.

More specifically sheet flood deposition is a product of migration and washout of submerged antidune bedforms present on the fan surface below the wave base which shoot down-slope and then dissipate by mingling with the rest of the sheet flood (McGee, 1897; Zielinski, 1982; Blair, 1985a, b). Spatio-vertically developed such major sheet flood deposits during a geologically significant span of time forms a kind of episodic succession with irregularly spaced thin and thick event beds. On the other hand, moderately thick fluvial deposits in sequences display some sort of cyclicity that is discernible in areas of substantial subsidence, which are continuously filled. Deep-seated fault-dominated terrains under an overall control of extensional regime of the Papaghni basinal setup frequently induced reactivation along fault surfaces producing piedmonts which were subsequently modified through weathering and erosion that supplied a huge amount of detritus shed in deposition of terrestrial fan complexes and braided fluvial deposits. Deposition of major floods with periodicities of tens to hundreds of years is not transmitted in the rock record (Dott, 1988; Clifton, 1988). The recurrent interval of large Gulcheru flood episodes preserved in thick flood deposits including those of flash floods, however, is recorded in the Chitravati stratigraphic section. This, when compared with similar analogues, appears to be in ranges of hundreds to thousands of years. As these cycles of flood deposits are not strictly periodic, they have been designated in Gulcheru scenario as dicyclic in their development.

The upper Gulcheru in continuity with the lower sequence comprises cycles of tabular and overlapping channel facies of sandstones. Trough cross beds in profusion with infillings of ferricrete in few cases, are pervasive in the channelized deposits. Red colored lensoidal sandstone facies discernible in lower Gulcheru stratal assemblages becomes repetitive in the upper Gulcheru sections. Spatio-vertical facies variations frequently reveal a cyclic arrangement of red colored tabular sand bodies alternating with trough-dominated channel facies. The red colored lensoidal sand bodies document high angle (20–32°) tabular- and wedge-planar cross-strata with supercritical and subcritical climb, alternation of grain fall and grain flow layers.

Spatio-vertical assemblage of sedimentary structures suggests that braided fluvial deposition was quite common during the Gulcheru with vertical decrease of flash flood signatures. Cyclically developed red colored cross-stratified lensoidal bodies with other characteristic facies attributes can be ascribed to eolian dune-interdune deposition (Hunter, 1977a, b; Rubin and Hunter, 1983; Kochurek and Nielsen, 1986; Kochurek, 1988). The unique repetitive eolian and fluvial deposition in the Chitravati section is a case in hand, which substantiates records of allocyclic changes.

The terrestrial fan complex of the lower Gulcheru sequence appears to have been generated in response to successive marginal faulting and subsequently got modified owing to interference by haphazardly arranged facies of different fluvial processes which were controlled by variations in subsidence rates through time. The upper Gulcheru is however dominated by cyclic channelized fluvial facies, often interfered by non-channelized coarse fluvial clastic deposits, cyclically arranged with eolianites.

Repetitive interference of a high energy eolian environment in the fluvial depositional systems of the Gulcheru stratal assemblages appears to be the result of periodic changes in a deep-seated fault-controlled rejuvenated topography and paleoclimatic variations that affected precipitation, temperature and wind velocities promoting generation of cyclicity between eolian and fluvial facies.

Switching between arid to humid climatic influence is allocyclic and orbitally forced with a resultant cyclic arrangement of eolian and braided fluvial Gulcheru sediments (cf. Einsele, 1992; Frederiksen et al., 1998; Ulicny, 1999). In addition, such frequencies of climatic variations depicting quasi-periodic signals were recorded in cyclic changes from relatively coarse-grained channel-fill dominated Gulcheru braided systems to transition in the finer-grained flood plain dominated sequences as documented in the Chitravati Gorge section.

The Transition Facies. Under the influence of oscillatory emergence of the Papaghni sub-basin, a 72 m thick Transition Facies was deposited over the Gulcheru sediments. Episodic reactivation of faults, repetitive redistribution of "highs" and "lows" in physiography and voluminous supply of terrigenous detritus were no longer there. A source area with subdued topography approaching peneplanation and intermittent felsic volcanism fed the basin with a feeble supply of intermixed fine to very fine sands and mud or interlayered carbonates and volcaniclastics. Passage from extrabasinal Gulcheru to intrabasinal Vempallae sedimentation is represented by lithosomes of mixed lithology of the Transition Facies with some evidences of cyclicity.

Interrelations of limestone and dolomite vary from standpoint of bedforms, their spatio-vertical relationships, continuity of bedding surfaces and internal structural details. The limestone are mainly erosively based, lensoidal and flat-bottomed layers. The lenses are internally cross-bedded. Foresets are asymmetric at their bases. The lensoidal limestones show a slight overriding and relatively abrupt decrease in wavelength and amplitude. Profused ripple-drift laminations, small ripples, interference ripples, tabular cross beds are salient facies attributes. Micro-ripple-drifts in graded laminites with admixture of glauconitic fine clastics, carbonates, and volcaniclastics have been identified. Thinly laminated felsic ash with pinch and swelling varies laterally and fades out. Ash beds are finely laminated or structureless and possess grading in the coarse tails. Volcaniclastics are discernible from microscopic to exposure levels. Fine-grained volcaniclastic-dominated rhythmites with slumps are common in the basal part of the Transition Facies. Typical eolian bimodality has also been documented from the clastics at four different stratigraphic levels of the formation. Thin-bedded mauve-colored sandstones-siltstones have been recorded. Shales and glauconitic fine sandstones are thinly laminated, laterally continuous with mild wavy parallelism and horizontal disposition. Chert or cherty bands are intermixed with laminated dolomitic layers showing an overall decrease in the fine clastics. Dolostones with algal laminites interbedded with mudstones and sandstones form a distinct cyclic arrangement. The dolostones fill the wavy depressions and ultimately cover the limestone dunes and ripples with horizontal laminae or thin beds. The flat-bottomed limestone lensoidal bodies are found to develop on this substrate. Cyclicity within the limestone-dolostone beds within the range between "ripple laminae" and "dune thin-bed" associations is discernible. Each cyclic arrangement of limestone-dolostone begins where chert-dominated interlayering with dolomitic carbonate ends. The association of cherty lithounits and limestone-dolostone couples also show cyclicity. Cyclicities on the basis of lithic character, their interrelations, gradations from a thin bed to thin lamina, transition from thin-bedded tabular asymptotic foresets and ripple-drifts are identifiable throughout the Transition Facies. The lithic units commonly are 3–15 cm thick. Their lateral extent ranges from few meters to several hundreds of meters. The beds are separated from one another by thinner or thicker, usually weaker intercalations. Intercalated lithic units form distinct lithosomal assemblages.

Intercalations are designated as interbeds with differing composition or structure (cf. Campbell, 1967; Reineck and Singh, 1980; Collinson and Thompson, 1982). Beds with intercalations forming bedding couplets in repetition form bundles of several bedding couplets. Amalgamated thin beds of ash and volcaniclastics define event layers in the Transition Facies. Simultaneous supply of sand, mud, and volcaniclastics often got mixed up with intrabasinal carbonates and deposited lithosomal assemblages in the facies. Depositional rhythms and cycles in the Transition Facies resulted from abrupt changes in sedimentation due to depositional events or episodes at random to quasi-periodic time intervals forming episodic and dicyclic bedding. The most prominent of this group are rhythmites with slumps, laterally extended over hundreds of meters and subaqueously settled repetitive pyroclastic ash falls into basins receiving normal background sedimentation (cf. Einsele, 1992; Reading, 1996; Middleton, 2003). Bedding variations in the Transition Facies were caused by slow gradual changes in sedimentation generating cyclic bedding even involving formation of three-bed cycles (cf. Einsele, 1992). Some of the laterally continuous thin ash layers serve as marker event beds and their episodic occurrence in stratigraphic record of the Transition Facies attest to intermittent volcanism in the source area. Interlayering of the ash layers with clastics and carbonates is ascribed to subaqueous settling of the ash. Glauconitic fine sandstones in the sequence are suggestive of deposition in a sea-marginal setup and their repetition in the stratigraphic column is indicative of oscillatory marine incursions probably owing to eustatic sea-level changes. Dolostones with algal laminites, interbedded with mudstones and sandstones, lend support to repeated lagoonal impress through time.

Attendant cyclicities in the Transition Facies suggest different phases of oscillation in its depositional setups. Two extremes are fine clastic dominated lagoonal condition and shallow-marine environment. Marine incursion only punctuates the cyclic trend in Transition Facies. Sands with typical eolian bimodality at different stratigraphic levels attest to episodic reworking of eolian sands in shallow-marine and lagoonal environments.

The Vempallae Formation. The Transtion Facies grades into a sequence (ca. 1500 m) of well-defined carbonate beds of the Vempallae Formation with sheet-bedded limestones, lenticular dolostone comprising stromatolitic bedding, subordinate fine-grained sandtones, shales, mudstones, and volcaniclastics, traceable laterally over tens of kilometers. Impress of the Transition Facies, however, is discernible in a portion of the lower part of the sequence. The Vempallae Formation, in its extents of northern and southern outcrops, depicts variations in depositional style and stromatolitic growth (Fig. 11.9).

Facies analysis of the Vempallae carbonates reveals twenty one facies which include: intraclastic dolomitic packstone, intraclastic dolomitic packstone with sparsely developed stromatolitic laminae, laminated dolosiltite/dololutite with ash pockets, ash beds, laminated mudstone with ash layers, laminated mudstone, tufa, stromatolitic dolomite, laminated dolosiltite/dololutite, oolitic chert, oolitic dolomite, massive dolomite, rippled dolostone, trough cross-stratified turbiditic dololutite, and cherty dolostone.

A representative vertical facies log shows some major facies associations of the Vempallae (Fig. 11.10). The dolostone facies consisting of lenticular-bedded dolostones associated with subordinate wavy and lenticular-bedded limestones is predominant in the middle part of formation. Medium- and small-sized channels in dolostones are characteristic of the upper part of Vempallae carbonate facies. Different types of facies attributes discernible are dunes, ripples, ripple-drifts, and occasional mud-cracked surfaces within the middle dolostone facies. Large-sized cabbage-shaped, ovoid to domal stromatolites, and highly elongated stromatolitic forms have also been recorded. However, gradational size and shapes of the stromatolites appear and fade out frequently. Impertinent occurrence of different types of stromatolitic bedding is a characteristic feature of the facies.

Fig. 11.9: Prolific stromatolitic growth from Vempallae Formation, Cuddapah Basin, India. Bar scale = 10 cm.

Dolomite element of "primary" origin develop in the carbonate sequence on very low gradient tidal flats and coastal sabkhas which might be either marginal to shallow-marine or a lagoonal setup under extreme arid conditions dominated by circulation of hot supersaline brine (Reading, 1996). Transportation, breaking down and deposition of intraclasts signify wave action and frequently generated platformal slope-controlled deposition of debris flows (Fig. 11.11). Muddy limestone-dominated facies with sheet geometry in the lower part characterizes a low-energy, shallow-water, lacustrine to lagoonal origin (cf. Kukal, 1970; Selley, 1978; Davies Jr., 1983). Frequent presence of glauconitic sandstones in it lends support to the origin in a low-energy, shallow-marine condition below wave base (cf. Irwin, 1965; Selley, 1978; Reading, 1996). The morphometry of the stromatolites is an important indicator of the energy condition during its growth in depositional setup (cf. Krylov, 1976; Walther in Reading, 1996). Large colonies of stromatolites could only grow in quiet low-energy conditions. Owing to a possible presence of a paleotopographic high towards the north, the wave action might have increased promoting a development of highly elongated stromatolitic forms parallel to the current directions. On the other hand, two

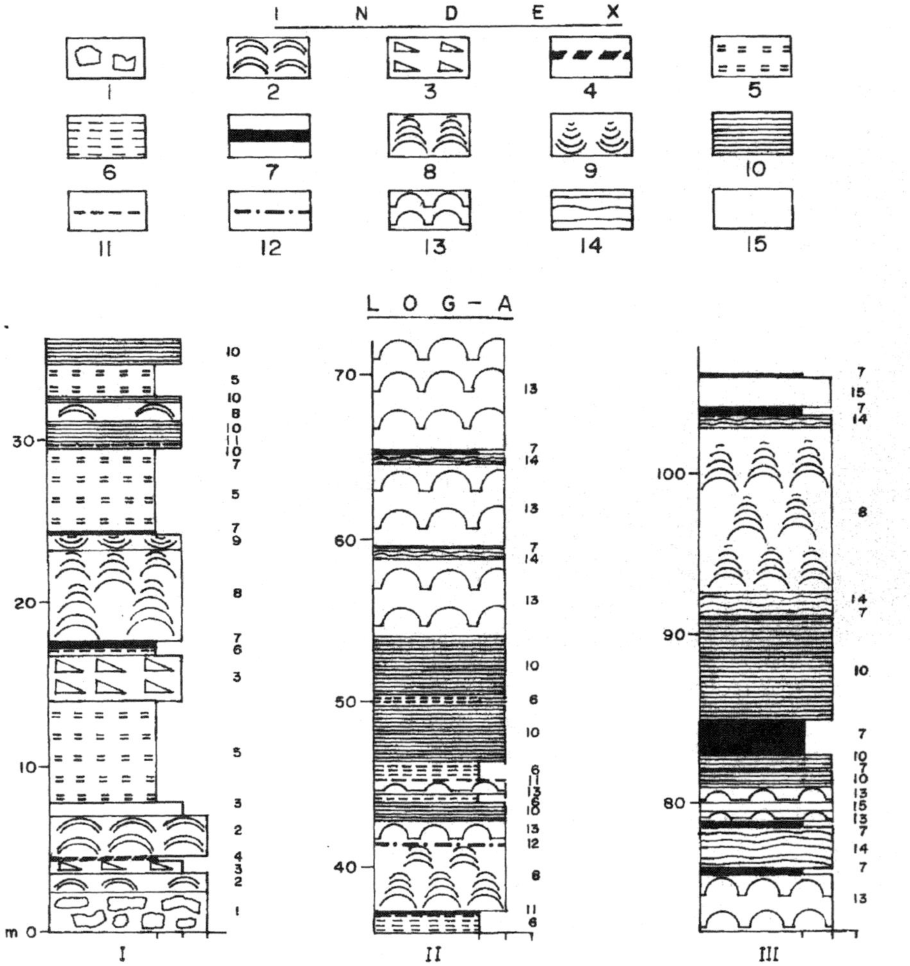

1. Intraclastic dolomitic packstone
2. Intraclastic dolomitic packstone with sparsely developed stromatolitic lamina
3. Laminated dolosiltite / dololutite with ash
4. Ash beds
5. Laminated carbonate mudrock with intervening thin ash layers / pockets
6. Laminated carbonate mudrock
7. Tufa with sporadic calcretes
8. Pseudocolumnar stromatolites associated with columnar varieties
9. Columnar stromatolites
10. Laminated dolosiltite / dololutite
11. Oolitic chert
12. Oolitic dolomite
13. Pseudocolumnar stromatolites
14. Cryptalgalamintes
15. Massive dolomite

Fig. 11.10: Representative facies log from Vempallae Formation.

Fig. 11.11: Debris flow deposits within the Vempallae carbonates. Scale: diameter of the coin = 2 cm.

levels of large cabbage-shaped, ovoid and domal stromatolites in the middle order dolostone facies corroborates a low-energy lagoonal-marginal environment where medium- and small-sized fluvial channels redistirbuted penecontemporaneous dolomitic mud, derived from low-gradient, arid subaqueous flats that got trapped in stromatolitic laminae.

The upper part of the formation is a cyclic carbonate facies with shales, rhythmites, tufas, oolitic cherts/dolostones, volcaniclastic ash, and chert layers (Fig. 11.12). The underlying dolostone facies is, however, gradational to the upper calcareous shaly facies. The shaly facies shows a gradual vertical increase of terrigenous clastics.

Successive increment of the terrigenous clastics is a response to the redistribution of topographic highs in the source area. Topographic highs can be added to a peneplained surface either due to oscillatory movements or tectonism related to intracratonic tectonic blocks (Beloussov, 1962; Leeder, 1982, 1999). The most intriguing feature of this basinal development is the associated volcanism. Cryptalgal and algal laminites with a slight tendency of pseudo-columnar growth float within carbonates intermixed with volcaniclastics, which indicates that mixed lithology favored only the growth of single forms of algal mats.

Fig. 11.12: Rhythmites of volcaniclastic ash beds, impure clastics below the lensoidal carbonates, Vempallae Formation. Scale: height of human figure = 170 cm.

Twenty-one lithofacies with variable proportions of volcaniclcastic elements in a platformal setup, often formed asymmetric to fining upward cycles, which was punctuated by monocyclic phases both in northern and southern sectors. The deposition, therefore, attests to an overall pulsating basin dynamics.

Cyclic and non-cyclic sequences in the Vempallae Formation. The stratigraphic record of geologic history is chacraterized by intervals of pronounced rhythmic or cyclic arrangement of facies such that repetitive groups of rock units have component facies, which tend to occur in a certain order (Grotzinger, 1986a). In the natural rock record throughout geologic time, such cyclic arrangement of facies is rarely complete, due to changes in environmental conditions as well as due to occurrence of non-cyclic units. Autocyclicities are responses of internal feedback mechanisms within the sedimentary systems. On the other hand, allocyclicity arises within the sedimentary piles of the formation in response to episodic subsidence, volcanism, eustatic sea-level changes, with are the external controls operative from domains outside the basinal configuration (Grotzinger, 1986a, b). Milankovich-forced controls on many cyclic systems of the formation can also be deciphered (cf. Grotzinger, 1986a, b). Upward shallowing platformal carbonate cycles of the Vempallae under variable paleoenvironmental dynamics, in this context, appear to have been formed in response to low-amplitude (Milankovich-band) sea-level oscillations. Cyclic carbonate sedimentation of Rocknest Platform, NW Canada, can be cited as a suitable paleoanalogue (Grotzinger, 1986a, b) for the Vempallae.

Cyclic and non-cyclic units. Regional distribution of cyclicities in the Vempallae outcrops highlights characteristic variations. Asymmetric shallowing upward cycles are usually incomplete with intervening non-cyclic intervals. The cyclic lithofacies contain tufa-capped sequences, dominated by impure carbonate mud rocks and laminated dolomites with frequent stromatolites towards the southern fringe of the Papaghni Basin. This passes northwards into cycles of stromatolitic boundstones and trough-stratified sequences.

The southern cyclic lithofacies is characterized by intraclastic dolomitic packstone, laminated dolosiltite/dololutite, impure carbonate mudstone associated with stromatolitic dolomites and also capped by thin tufa layers. The northern cyclic lithofacies consists of infrequent tufa-capped sequences of large stromatolitic growth with overlying trough-stratified beds and laminated dolosiltite/dololutite.

The non-cyclic units of the southern cycle comprise carbonates with volcaniclastics and chert as two major dominant elements. Volcaniclastic layers are however thin. The middle non-cyclic intervals include thin beds (10–20 cm thick) of oolitic chert/oolitic dolomites. The oolites show radial structures within either a fine-grained chalcedony, chert, or dololutite, locally layered with discontinuous chert beds. The oolitic horizons indicate a high-energy environment, prevalent for short duration. The non-cyclic units in the upper part of the southern cycles are massive structureless dololutite beds, indicative of deposition in low-energy condition.

The non-cyclic units of the northern cycle of the basin are more pervasive and thicker and comparatively more complete than its southern counterpart. The non-cyclic interval consists of stromatolitic beds with intervening milky-white chert horizons having interfingering, vertical gradation as well as sharp contacts with the dolomites. This is suggestive of a syn-sedimentary origin with the dolostones. These chert beds and those associated with the oolites appear to be genetically related to volcaniclastics.

11.3.4.2 Chitravati sub-basin

The Pulivendla Formation of the Chitravati Group unconformably overlies the Vempallae Formation. The Pulivendla in its lower part records cyclic development of trappean–intertrappean sequence (Fig. 11.13), followed up stratigraphically by boulders, bouldery conglomerates, and sandstone beds in a cyclic manner.

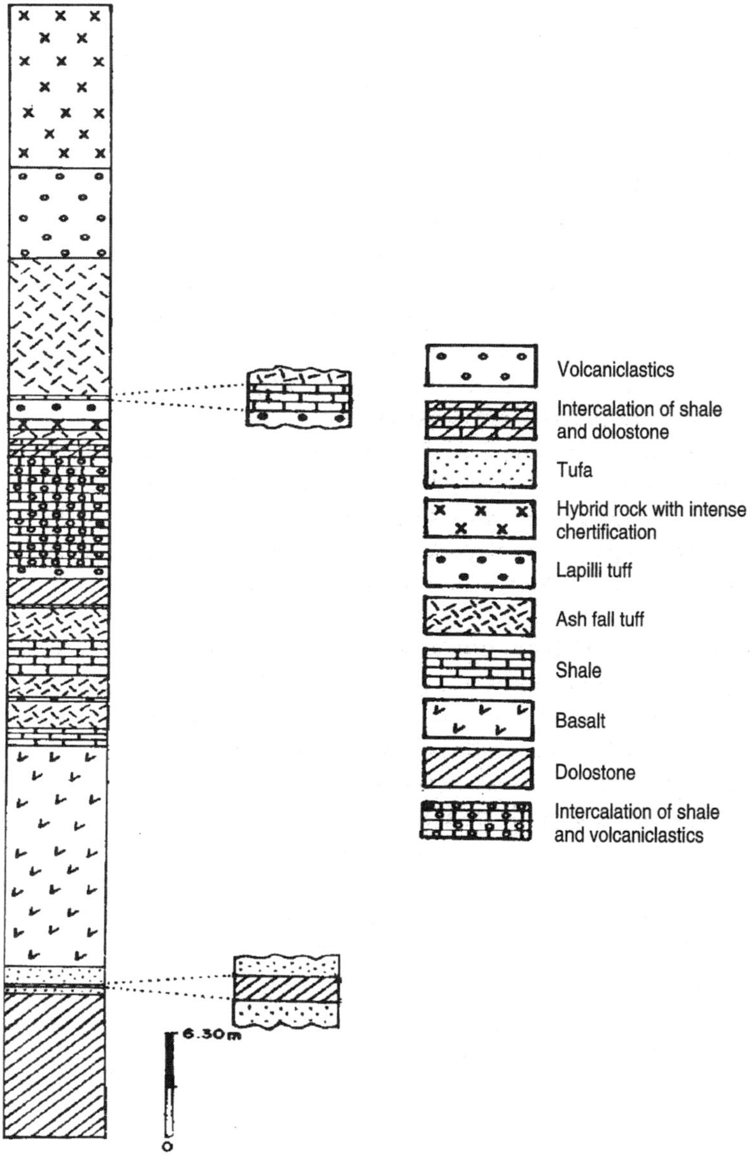

Fig. 11.13: Representative facies log showing trappean–intertrappean sequence from basal part of Pulivendla Formation, Cuddapah Basin, India. Scale: length of hammer = 40 cm.

Trappean–intertrappean sequence. The trap rocks record mafic and felsic volcanism. Subaerially basaltic lava was emplaced with intermittent pause over the unconformity surface of the Vempallae. During the quiescence, formation of red beds owing to weathering and erosion of basic lava took place. Basic lava flows with inclusions of unmodified mantle xenoliths (Dasgupta, 1986) constitute a 80 m thick sequence which is associated with intervening event layers of red beds indicative of episodic weathering and quiescence.

Associated intertrappeans comprise carbonates, terrigenous clastics, chert, hybridized rocks, impure doloarenites, and dolosiltites. The terrigenous clastics comprising shales, siltstones, fine-grained sandstones with sharp-crested current ripples, tabular cross beds, mud-cracked surfaces, wind ripples, and subcritically climbing translatent strata are some of the major facies attributes. Facies attributes of the intertrappeans can be ascribed to subaqueous deposition in isolated pools and ponds accompanied by formation of eolianites.

Stratigraphically basaltic eruption was followed in a later phase by felsic volcanism, which might be expected in the Paleoproterozoic scenario (Jachariah et al., 1999). The presence of lapilli tuff, lithic tuff, and ash fall tuffs indicate eruptive phases of felsic volcanism. Subaerially emplaced flows were basaltic in the beginning and intermittent quiescence heralded subaqueous sedimentation of limited basinal extent. Sedimentation was either intrabasinal or subaerial. Touch of eolian reworking of the subaerially exposed sediments was intermittent. The association of mafic and felsic volcanics in the Pulivendla stands for Paleoproterozoic bimodal volcanism. Such intracratonic bimodal volcanism of limited extent has also been recorded from other extensional basins of the Precambrian (Goodwin, 1981; Windley, 1984; Condie, 1992).

The Pulivendla clastics. The Pulivendla outcrops are laterally continuous over hundreds of meters. The Pulivendla clastics are bouldery and rudaceous in its lowermost part and followed up by rippled sandstones. Spatio-vertically trough-cross-stratified channelized sandstones are incised over the rippled sandstones and are capped by repetitive development of couplet beds, which are in turn, overlain by flat-bottomed conglomeratic channels grading to trough-cross-stratified channelized sandstones (Figs. 11.14 and 11.15). Couplet beds again dominate over the channelized facies overlain by planar and trough-cross-stratified channelized sandstone. Reappearance of couplet beds in broad open channels over the sandstones marks the closing phase of Pulivendla sedimentation (Fig. 11.16).

Boulder beds and repetition of couplet facies testify episodic interference by extremely high-energy depositional conditions in the Pulivendla Formation (cf. Blair and Mcpherson, 1994; Blair, 1999). Flash floods and sheet floods operated repetitively in the basin in response to high-energy discharges dictated by frequent changes in physiography of the source area and the basin itself under an

Fig. 11.14: Channelized fluvial clastics overlying conglomerate facies, Pulivendla Formation. Length of hammer = 40 cm.

overall control of extensional regime. Significant variations in depth and velocities of supercritical flows and high-energy gravity flows resulted in an intimate association of bouldery flash flood deposits and sandy alluvial sheet flood deposits characterized by a cyclic disposition of couplet facies (cf. Allen, 1983; Blair, 1985a, b; Blair and Mcpherson, 1994). Intervening channelized tabular and trough cross-stratified sandstones signify change in depositional style as represented by braided fluvial deposits.

The Pulivendla, in its upper part, therefore, attests to cyclic formation of alluvial fans and braided fluvial setups, as promoted by frequent changes in basin dynamics.

Tadpatri and Gandikota Formation. The cyclic facies assemblage of Pulivendla is overlain unconformably by Tadpatri Formation. It comprises another cyclic association of repetitive thick ignimbrites depicting phreatic volcanism (Figs. 11.17 and 11.18) with intercalations of carbonate and terrigenous clastics bearing signatures of fluvial-marine, lagoonal, and shallow-marine facies. The uppermost Gandikota Formation overlying the Tadpatri attests a development of cyclicity within sandstone-mudstone sequences bearing impress of braided fluvial setups.

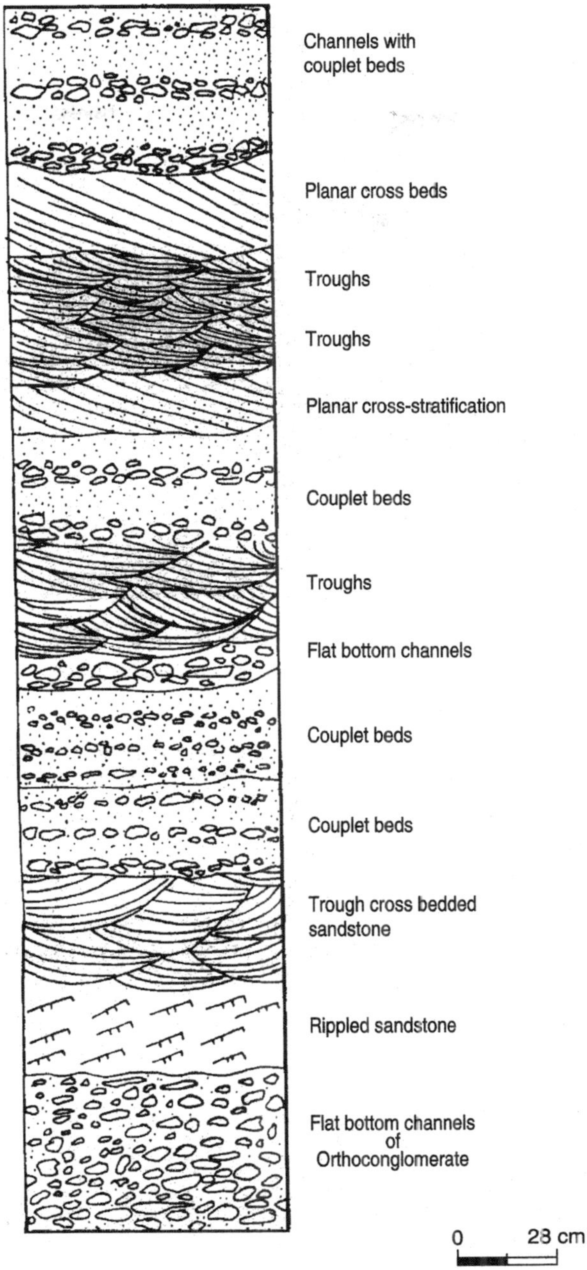

Fig. 11.15: Representative facies log showing cyclicity in Pulivendla Formation.

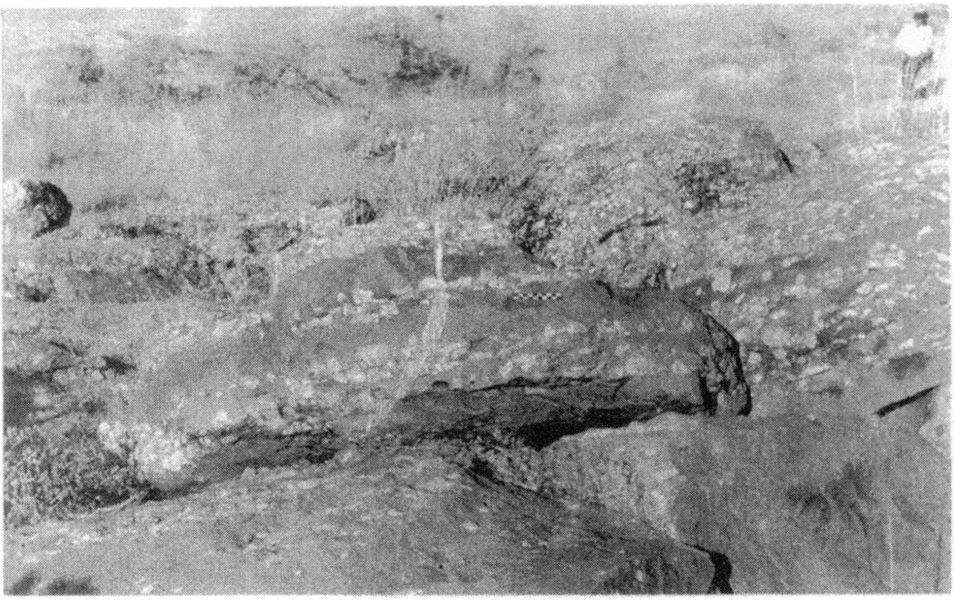

Fig. 11.16: Plano-convex channelized cyclic couplet beds, Pulivendla Formation. Bar scale = 10 cm.

Fig. 11.17: Repeated ignimbrite facies marked by color difference, Tadpatri Formation, Cuddapah Basin, India. Bar scale = 5 cm.

Fig. 11.18: Welded tuff with bedded pumice, Tadpatri Formation. Diameter of lens cap = 6 cm.

11.3.4.3 Nallamalai sub-basin

The arenaceous Gandikota underlies the Nallamalai Group. The basal arenaceous and argillaceous Bairenkonda of the Nallamalai Group is of fluvial and/or shallow-marine origin (Nagaraja Rao et al., 1987).

The facies assemblage of the overlying Cumbum Formation of the Nallamalai sub-basin includes conglomerate-sand facies (now conglomerate-quartzite association), sand-mud alternation facies (now quartzite-phyllite/schist alternation), mud facies (phyllite/schist), and carbonate-mud facies (metamorphosed into crystalline limestone/dolomite-phyllite/schist) at different stratigraphic levels. Mafic and/or felsic to silicic volcanics are often associated with the facies assemblage.

Identifiable facies attributes of the formation are: channelized coarse- to fine-grained sandstones with massive bedding and thin muddy intercalations, subparallel laminations containing Ta, Tb, and Tbc turbidites developed frequently with contourites in spatio-vertical continuity (Fig. 11.19). Dish structures, infrequent sole marks, abundant erosional channels, and low-angle wedge-shaped cross laminae are also discernible.

Thin interbedded sandstone-mudstone sequences are traceable for hundreds of meters. Laminated sandy beds (6–20 cm thick) are often replaced by muddy units. However, occasional laterally continuous thick massive sandstone

Fig. 11.19: Chertified interbedded sandstone-mudstone turbidites, Cumbum Formation, Cuddapah Basin, India. Scale: height of human figure on the right = 182 cm.

lithosomes with muddy contourites (70–90 cm thick) are noticeable. Tbe, Tde, and dominant Tce turbidites are other prominent facies attributes.

Thin, irregular, discontinuous, sandy-muddy beds showing wedging and lenticularity are common. Fine sandstone beds are frequently overlain by mudstones with sharp basal contacts. Tde and occasional Te turbidites interbedded with in situ chert characterize the facies.

Sediment deformation such as slumps, mudflows, intraformational folding, contorted strata, breccia, and isolated dispersed clasts are identifiable in chaotic facies (Fig. 11.20). In some stratigraphic sections, the chaotic facies is repeated with cyclic Tbc turbidites (Fig. 11.20). Strong facies elements such as disorganized to slightly organized lensoidal bodies with basal laminae containing mudstone clasts in muddy matrix, pebbly mudstones, slump folds capped by convolute laminations and clast-rich zone are noticeable.

Facies attributes point to repetitive development of chaotic, debris flow deposits in association with turbidite-contourite stratal assemblage in the Cumbum stratigraphic record (cf. Bouma, 1996; Stow et al., 1998). Disorganized to slightly organized lensoidal bodies with basal laminae containing mudstone

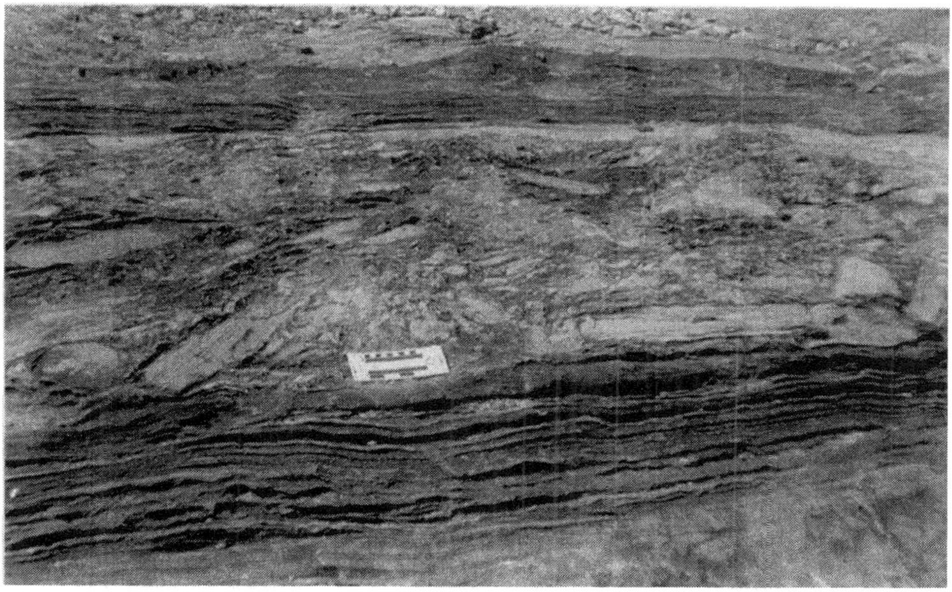

Fig. 11.20: A chaotic horizon from cyclic turbidite facies of Cumbum Formation. Repetition of Tbc turbidite is discernible. Length of the scale = 15 cm.

clasts in muddy matrix, pebbly mudstones, slump fold capped by convolute laminations and clast-rich zones, which characterize the debrites (cf. Shanmugan and Moiola, 1995; Bouma, 1996). A cyclic development of seismites, seismically affected (reworked) turbidites and tsunamiites at different levels of Nallamalai sections is attested by attributes comprising graded microfaults, fault-bounded graben-like down-sagging structures, disturbed bedding in restricted horizons (Fig. 11.21), hydroplastic deformation structures, extensional structures including pich-and-swells, "boudinage" dislocation of individual competent beds, brecciation of brittle layers, sand dikes, mudstones with thixotropic deformation and diastasis cracks (cf. Seilacher, 1969, 1984; Sims, 1975; Guiraud and Plaziat, 1993; Shiki and Yamazaki, 1996; Plaziat and Purser, 1998). Volcaniclastic detritus mixed up with terrigenous clastics include chertified pumices and lapilli, clasts of chertified welded tuffs and volcanic bombs. Association of in situ chert interbedded with Tde and/or Te turbidites points towards deep-water bathymetry for the Cumbum turbidites of the Nallamalai.

Presence of volumetrically significant felsic to silicic volcaniclastic elements and repetitive occurrence of pyroclastics enclosing volcanic bombs have been recorded (Figs. 11.22 and 11.23). Association of the attributes suggests that the earthquake waves repeated in the wake of explosive volcanism.

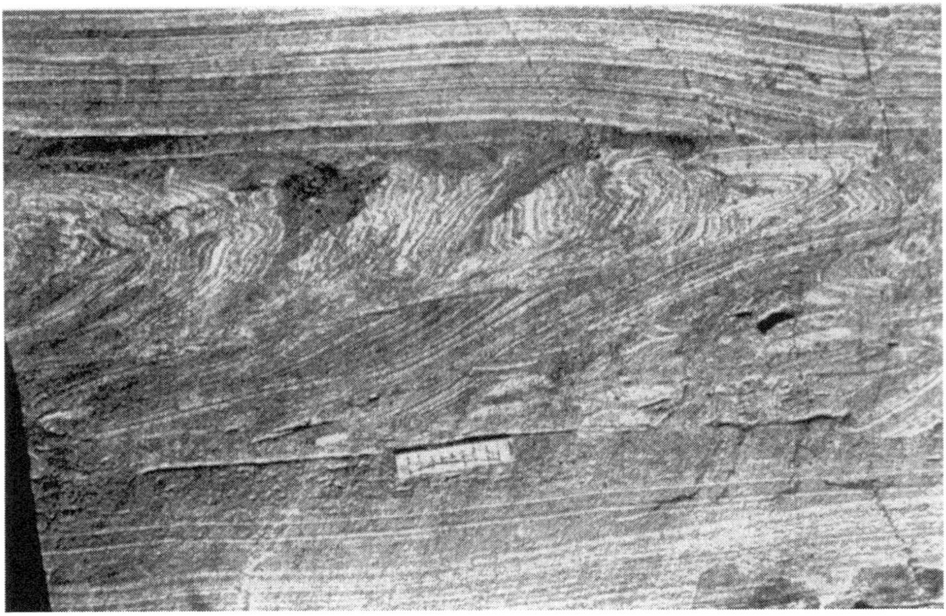

Fig. 11.21: A discrete seismite level from Cumbum Formation. Bar scale = 10 cm.

Fig. 11.22: Repeated chertified welded tuffs with volcanic bombs with long axis parallel to the bedding indicating subaqueous deposition, Cumbum Formation, Nallamalai Group. Scale: long axis of the volcanic bomb, centrally placed = 35 cm.

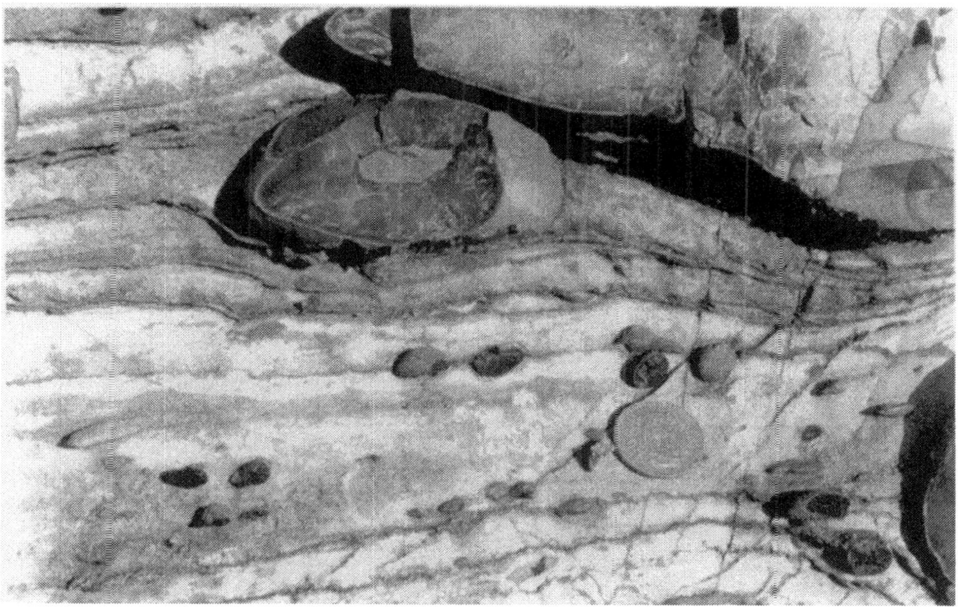

Fig. 11.23: Chertified layered welded tuff with lapilli and volcanic bombs, Cumbum Formation. Scale: diameter of the coin = 2.2 cm.

Shallow-marine storm deposits are marked by well-preserved laminites, wave ripples, hummocky and swaley cross-stratified beds arranged in several thickening- and thinning-upward sequences. Severe seismic activity in the syn-rift Cumbum scenario is indicated by seismites and seismo-turbidites developed also among the storm-induced sediments. These are deposited in transitional offshore (upper bathyal) setting in association with tsunamiites.

In addition, faulting, syn-depositional as well as basin-marginal might have been accompanied by ground tremors, tsunamis that induced the generation of seismites, seismoturbidites, and tsunamiites. Haphazardly arranged seismic signatures within discrete layers throughout the succession have relevance to the cyclic development of facies, which is common in Proterozoic sedimentation (Dasgupta, 1998).

The Cumbum turbidites with chaotic levels and debris flows in the stratigraphic record (Fig. 11.24) supported by recent and ancient analogues seems to have been resedimented owing to variable genetic controls such as overloading, volcanism, diapiric uplift and/or seismicity, and slope failure common in an extensional regime (cf. Shanmugan and Moiola, 1995; Sixsmith et al., 2004).

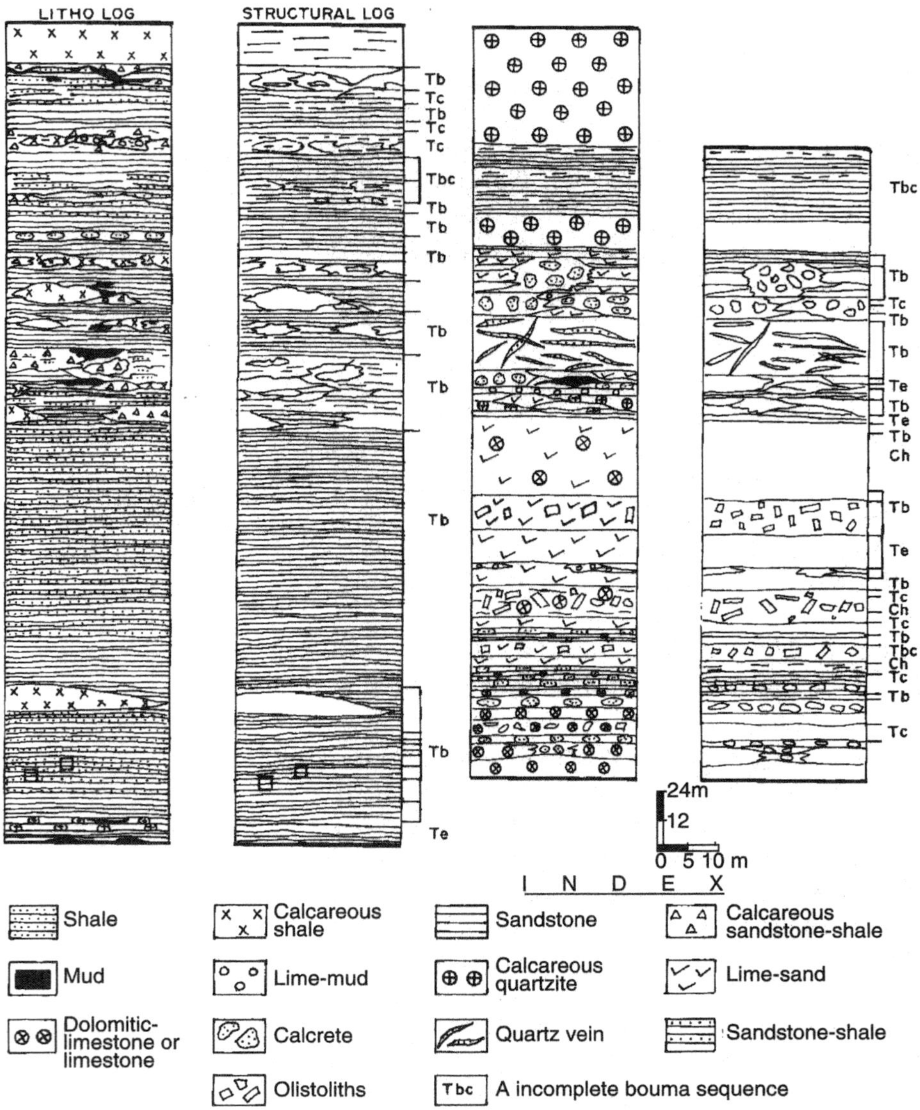

Fig. 11.24: Representative vertical facies log from Cumbum Formation; cyclicity between turbidite and chaotic facies is discernible.

11.3.4.4 Srisailam sub-basin

In the uppermost part of the Srisailam sub-basin, the Egalapenta Member records different types of cyclicities in stratal assemblages. Three different cyclic associations have been documented from the lower, middle, and upper parts of the Egalapenta Member.

In the lower part of the member, glauconitic sandstone-mudstones with attributes such as heterolithic sandy-muddy interlaminations, bidirectional cross-beds, rhythmically cross-stratified sandstones comprising sigmoidal cross-bedding, shallow troughs, symmetrical ripples are repetitive with channelized reddish-brown, medium- to coarse sandstone facies (Fig. 11.25) bearing signatures of profuse trough-cross-stratified beds, ferricrete clasts and clast-bearing beds, tabular cross-stratified sandstones, pebble and granule bearing disorganized beds and laminated muddy-sandy beds. The glauconitic sandstones and associated lithic units are of shallow-marine origin (Biswas, 2000). A braided fluvial environment can be ascribed to the channelized sandstones (Biswas, 2000). The lower part of the Egalapenta depicts cyclicity developed owing to repetitive sedimentation of shallow-water glauconitic sandstones, tidalites, and fluvial clastics. Onlaps by glauconitic sandstones and offlaps by fluvial sandstones indicate shifts between offshore and near-shore environments. Cyclicity between shallow-marine and fluvial strata is frequently cited (cf. Smith, 1974; Cant and Walker, 1978; Long, 1978; Selley, 1978; Bluck, 1981; Muir and

Fig. 11.25: Long profile of a channel with channel fills from Egalapenta Member, Srisailam Formation, Cuddapah Basin, India. Length of the hammer at the base of the channel = 40 cm.

Rust, 1982; Allen, 1983; Simpson and Eriksson, 1993; Miall, 1996; Reading, 1996; Middleton, 2003).

In the middle part of the member, cycles defined by impure carbonate-sandstone-mudstones and reddish-brown sandstone sheets are documented (Fig. 11.26). Attributes of the impure carbonate-sandstone-mudstones include evaporitc signatures such as anhydrite, enterolithic bedding, chicken wire and

Fig. 11.26: Flat-bedded impure sandstone-mudstone beds, interbedded with reddish-brown sandstone sheets showing cyclicity between coastal sabkha and eolian sandsheets, Egalapenta Member. Length of the hammer at lower right hand part = 40 cm.

bird's eye structures, mud cracked surfaces, shallow isolated troughs. Reddish-brown sandstone sheets are characterized by the development of profuse wind ripples, small cross-stratified dunes, horizontal- to low-angle cross-stratified beds, subcritically climbing translatent strata and thin pinches of laminated mudstone beds.

The cyclic occurrence of impure carbonate-sandstone-mudstone with signatures of coastal sabkha and low-angle stratified sandstones representing eolian sandsheets have been interpreted as episodic marine incursions that promoted growth of such cyclicities in arid climatic zones under eustatic control (Biswas, 2000). Similar cyclic associations, as reported and interpreted, point towards Milankovich forcing (cf. Wood and Wolfe, 1969; Hardie and Eugster, 1971; Fairbridge and Bourgeois, 1978; Friedman and Sanders, 1978; Hunter, 1981; Kochurek, 1981; Driese and Dott, 1984; Grotzinger 1986b; Chan, 1989; McLane, 1995).

The upper part of the formation records another cyclic association that developed between stacks of climbing lensoidal cross-bedded medium to fine sandstones and thin beds of laminated, wind-ripple-bearing low-angle cross-stratified sandstone-mudstone beds (Fig. 11.27). Attributes of large climbing cross-stratified lensoidal sandstones (Fig. 11.28) stand for a token representation of the migrating dunes forming draas (Fig. 11.29; Biswas, 2000). Wind-ripple-bearing, low-angle cross-stratified sandstones (Fig. 11.30) and thin mudstones bear the impress of interdune deposits (Biswas, 2000). Such cyclicities between climbing dunes (draa) and interdume deposits on a regional scale deposit sand sea or ergs. A portion of an erg system between cyclic draa and interdune deposits crops out in the upper Egalapenta near the Krishna gorge (Fig. 11.31). Sands derived from the granitic-gneissic basement underwent episodic reworking in high-energy environments including an eolian environment with distinct imprint of eolian bimodality (Fig. 11.32). Cyclicities in eolian and associated facies have been frequently addressed (Inman et al., 1966; McKee, 1966; Kochurek and Dott, 1981; Kochurek and Fielder, 1982; Fryberger et al., 1983).

The Egalapenta displays a complex interplay of fluvial, shallow-marine, arid sea-marginal sabkha and eolian processes. The Egalapenta complex represents a spectrum of cyclic eolian system types in which end members are dry, wet and stabilized systems with records of interference by fluvial-eolian interaction, sabkha phases and allocyclic shallow-marine milieus. Gradual change from wet fluvial condition to dry arid climate might point to mechanisms such as fluctuations in groundwater volume in continents or thermal expansion and contraction of seawater (Ulicny, 1999), which could have controlled to some extent the eustatic fluctuations. Association of fluvial and cyclic eolian dune-interdune and coastal sabkha facies corroborates climatic fluctuations from humid to arid and possible controls of the climatic variation on the dynamics

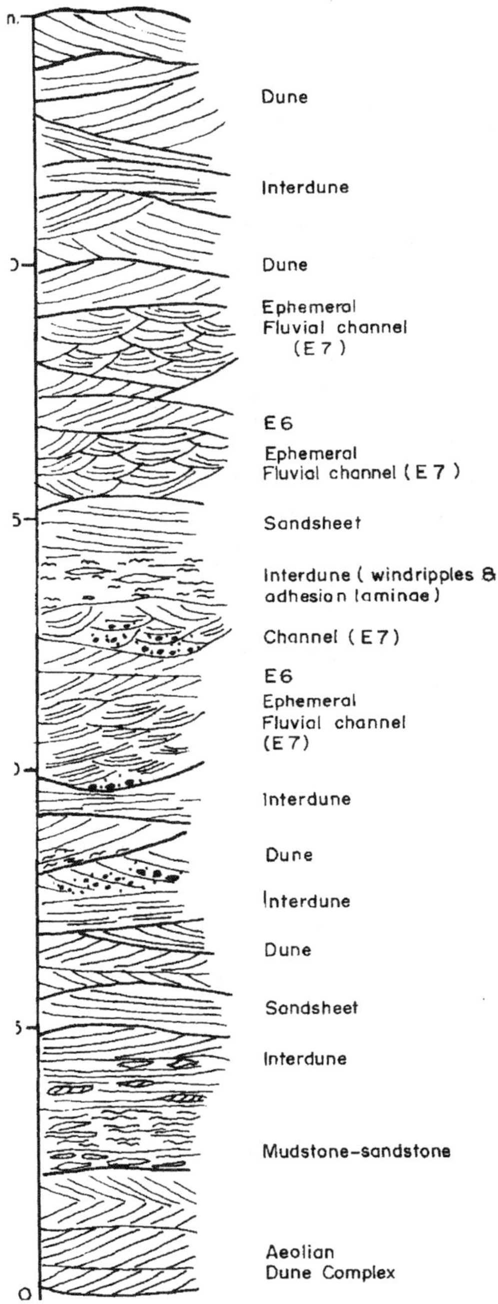

Fig. 11.27: Representative vertical log showing cyclicity between dune-interdune deposits, Egalapenta Member.

Fig. 11.28: A dune showing high-angle foresets overlying the mudstone-sandstone interdune deposits, Egalapenta Member. Scale: length of hammer = 40 cm.

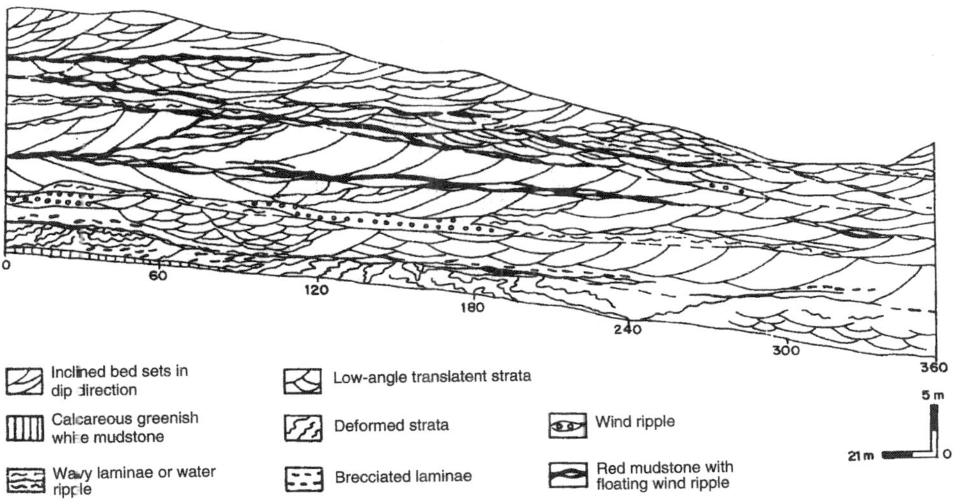

Fig. 11.29: A 360 m panel diagram showing cyclic association of climbing dune (draas) with intervening interdune deposits, forming a part of an erg, Egalapenta Member.

Fig. 11.30: Low-angle cross-stratified reddish brown sandstones characterizing the eolian sandsheets, Egalapenta Member. Scale: height of the shrub in the middle = 80 cm.

Fig. 11.31: Superimposed sets of dunes (draas) separated by thin interdune deposits together representing part of an erg, Egalapenta Member. Scale: lenght of the tree on the top = 5 m.

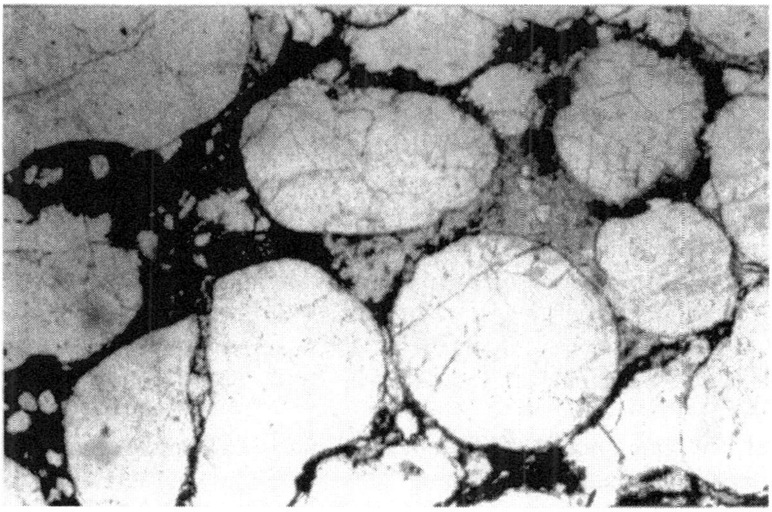

Fig. 11.32: Typical eolian bimodality among quartz sand and silt from Egalapenta Member. The diameter of largest quartz grains = 0.74 mm.

of the desert systems. The cyclicity also appears to be the response of climatic fluctuations that might have been orbitally controlled (Frederiksen et al., 1998).

11.4 DISCUSSION

Unlike the Phanerozoic, cyclicity and its interpretation from the Precambrian is limited owing mostly to overprinting by subsequent geologic events. Because cycles of several periods are superimposed, the resulting pattern of cyclicity is very complicated and rather random-appearing. However, seismicity of the depositional system to an external process may vary geographically and temporally, and long stratigraphic sections commonly drift from one kind of cyclicity to another. The basic problem is how to distinguish between the "forced" cyclicities (allocyclicity) and cyclicity generated internally (autocyclicity) (Einsele, 1992; McLane, 1995).

The excellent preservation of stratal assemblages in Paleoproterozoic to Neoproterozoic Cuddapah Supergroup is discernible in large outcrops. The influence of deformation and metamorphism is absent or feebly recorded. The transition from a highly mobile and widespread tectonic regime, reducing atmosphere and acidic ocean of the Archean to a state of lithosperic rigidity and near-modern air and water provided a geodynamic setup for the deposition of the Cuddapah sediments over 1 Ga, only punctuated by minor unconformities in between the sub-basinal stages. Preservation of cyclicities in the supergroup attest to least interference by latter geological changes. Periodicities of cycles is not possible to establish owing to limited radiometric age data available from stratal assemblages (Bhattacharji and Singh, 1994; Jachariah et al., 1999). Some types of cyclic bedding and certain cyclic sequences though, reflect allogenetically controlled strictly cyclic or periodic sequences with a regular time record; it is however, very difficult to prove whether or not a given cyclic sequence is really caused by a mechanism with constant time period (Einsele, 1992). Based on regional depositional controls and global tectonics, sea-level fluctuations, changes in climate, variations in water volume of lakes, sedimentary cycles are caused by various processes and discernible from a stratigraphic record depicting their relative stratigraphic values (Einsele, 1992; McLane, 1995; Middleton, 2003).

When cyclic phenomena of different frequencies and natures are superposed, the resulting sedimentary records may show a complex multicyclic pattern. This reflects added effects of two or more cyclic processes. Cyclic hierarchy includes such interplay between various factors controlling the actual stratigraphic record. In this context, sedimentary cycles may appear either symmetric or asymmetric (coarsening- or fining-upward sequences) as well as complete or incomplete, depending on subsidence rates. "Punctuated" aggradation cycles (Goodwin and Anderson, 1985) represent truncated shallowing-upward sequences deposited in shallow-water environments under the influence of fluctuating sea level. Depositional sequences or depositional cycles and cyclic sequences from stratigraphic records of the Cuddapah have been discussed in a framework of cyclic hierarchy (Goodwin and Anderson, 1985; Einsele, 1992).

Cyclic bedding, dicyclic bedding, rhythmic bedding, platformal cycles, and orbitally forced cycles have been identified and interpreted. The role of cyclicity in basin evolution of the Cuddapah has been highlighted by depositional styles of the different sub-basins.

Fault-controlled cyclicity in a terrestrial fan complex attested by sheet flood couplet facies in association with channel-fill deposits marks the initiation of the Gulcheru sedimentation of the Cuddapah Supergroup and demonstrates a significant role of extensional tectonics in basin evolution. Fining-upward cycles in alluvial fans that punctuate coarsening-upward trends indicate recession of piedmont-like sources from an earlier site of sedimentation as a subsequent shift of proximal setup away from the basin. Thickness of the corresponding bed types did change owing to variations in sedimentation rates in their cyclic patterns including quasi-cyclic and dicyclic developments. Switching between arid to humid climatic influence is allocyclic and orbitally forced with resultant cyclic arrangements of eolian deposition and braided fluvial deposition. Other minor depositional characteristics appear to be the background noise. In addition, such climatic variations attest quasi-periodic signals as recorded in cyclic changes from relatively coarse-grained braided fluvial clastics to transition in fine-grained flood-plain deposits.

The role of the Transition Facies is important as it records a span of time needed for changing depositional styles from extrabasinal Gulcheru to intrabasinal Vempallae sedimentation. Oscillatory emergence induced by the pulsatory nature of basin dynamics generated this transition facies. Spatial and vertical facies associations in the transition facies attest to formation of cycles involving lagoonal, shallow-marine, and terrestrial sedimentation. Terrestrial volcanism in the source area appears to be intermittent as subaqueously settled volcaniclastic elements were found intimately mixed in the sedimentary column of the transition facies.

The carbonate-dominated Vempallae with variable proportions of volcaniclastic elements in a platformal setup frequently formed asymmetric- to fining-upward cycles, which was punctuated by non-cyclic phases both in its intimately related northern and southern outcrops. Such stratal assemblages also attest to an overall pulsating basin dynamics. Upward shallowing platformal carbonate cycles under variable paleoenvironmental dynamics suggest that sedimentation occurred in response to low-amplitude (Milankovich-band) sea-level oscillations.

The cyclic trappean–intertrappean sequence in the lower part of the Chitravati Group bears signatures of bimodal volcanism, which is rare even in the Paleoproterozoic. Quiescence among subaerially emplaced flows is attested by the occurrence of intermittent red boles. Presence of unmodified mantle xenoliths in the basic flows points to deep-seated derivation of the mafics. Boulder beds and repetition of couplet facies again testify episodic interference by extremely high-energy depositional conditions in the Pulivendla clastics. Flash floods and sheet floods, repetitively operative in the basin was dictated by frequent changes in topography of the provenance area and the basin itself under an overall control of an extensional regime.

The cyclic association of ignimbrites and fluvial sandstone-mudstone sequence of the Tadpatri is overlain by the fluvial sandstone-mudstone cycles of the Gandikota Formation. The acid emplacement to silicic emplacement of the ignimbrites was phreatic in nature and was probably related to episodic generation of acid magma to silicic magma at shallower depths.

The Nallamalai Formation demonstrates a cyclic association of turbidites with chaotic facies, debrites, felsic volcaniclastics to silicic volcaniclastics, seismites, seismoturbidites, and tsunamiites owing to a variable genetic control under the influence of extensional tectonics.

The Srisailam Formation, in its uppermost member, records three discrete cycles of fluvial to shallow-marine in the lower part, coastal sabkha-eolian sandsheet in the middle and climbing dunes (draa)-interdune forming a part of an erg in the upper part. Such cyclicities, marking the closing phase of the Cuddapah sedimentation owes its origin to eustatic fluctuations and/or orbitally controlled climatic variations.

ACKNOWLEDGMENTS

Funds were available from University Grants Commission and department of Science and Technology in the project ES/23/157/92. Logistic support was extended by Y.S. Prakash Reddy and Y.S. Vivekanada Reddy. Authors are indebted to F.J. Pettijohn and K.S. Valdhiya for suggesting the problem of Cuddapah basin evolution.

12. CYCLICITY OF THE MESOZOIC SEDIMENTATION ON THE EASTERN MARGIN OF THE CHINESE CRATON AS A RESPONSE TO THE MAIN GEODYNAMIC EVENTS IN THE ADJACENT ACTIVE OCEANIC AREA

P.V. MARKEVICH, V.V. GOLOZUBOV, I.V. KEMKIN, A.I. KHANCHUK, Y.D. ZAKHAROV, A.N. PHILIPPOV, AND S.A. SHOROKOVA

12.1 INTRODUCTION

Mesozoic volcano-sedimentary deposits of the Sikhote-Alin have been sufficiently well studied, but they were considered separately for the eastern margin of the Chinese Craton (directly to the east of its supposed fragments: Khanka and Bureya ancient massifs; Markevich et al., 2000; Markevich and Zakharov, 2004), and for the intensively dislocated mountain Sikhote-Alin, situated more eastward. As a result, synchronous Mesozoic events have not been compared and correlated between them. In this chapter, such an attempt has been undertaken.

In Fig. 12.1 (Khanchuk and Kemkin, 2003), the main structural elements of the Sea of Japan region, transition zone between the Chinese ancient craton on the west, and the marginal seas of the western Pacific on the east, are shown.

It has been tried to examine, how the development of these two enormous geological structures have acted reciprocally during the Mesozoic, more exactly how the events in the tectonically active, mobile eastern area influenced the epiplatform sedimentary processes on the eastern extremity of the stable Chinese Craton.

Before the Early Jurassic, the region represented a passive continental margin. In the Early Jurassic an accretion of the oceanic plateau occurred (see Fig. 12.13). In the Tithonian, probably, the direction of the oceanic plate movement changed from NNW to N, and the subduction of the plate turned into strike-slip zone sliding northward in relation to the latter. Thus, during the Late Jurassic-Hauterivian, in this region two geodynamic settings combined: the transform (strike-slip) border and the border of Andean type. Since that moment along the eastern margin of the South Asian continent thick deposits of the Zhuravlyovka-Amur terrane accumulated. Such regime continued till the end of the Albian.

In the Chinese and Russian ancient continental part of the Sea of Japan region, the fragments of the ancient continental massifs are represented by the Amur

Cyclic Development of Sedimentary Basins
J.M. Mabesoone and V.H. Neumann, editors.

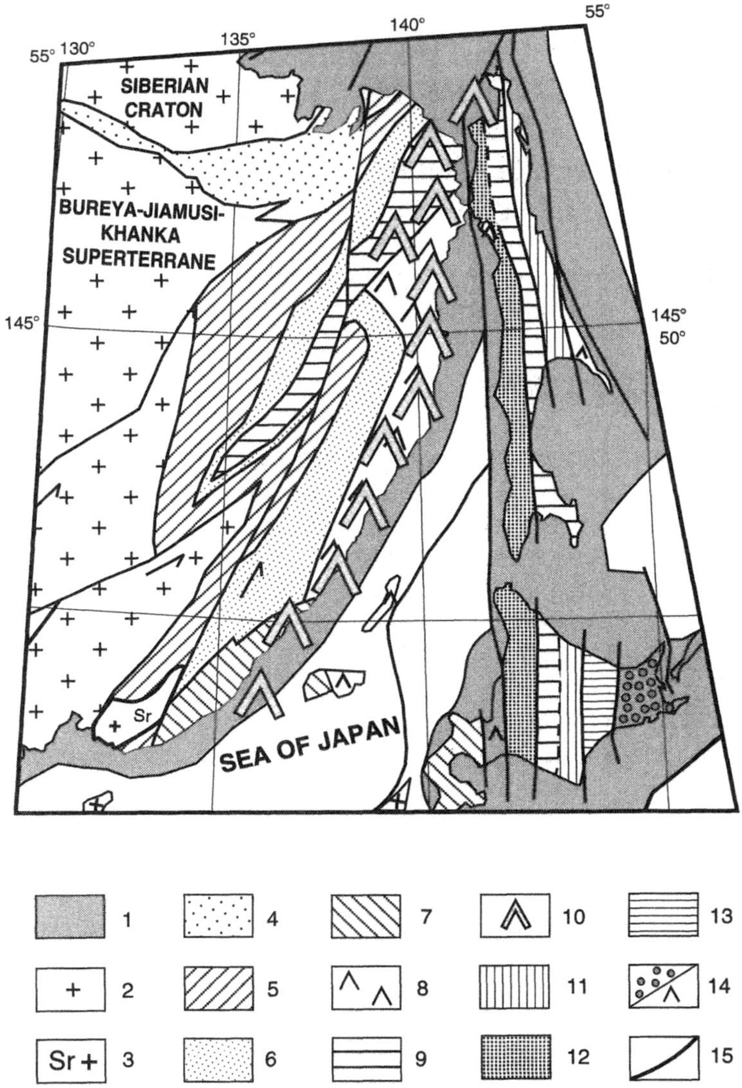

Fig. 12.1: Main structural elements of East Asia (after Khanchuk and Kemkin, 2003). Designations: 1 – recent continental and island shelf; 2, 3 – ancient crystalline massif with continental development regime (3 – Sergeevka continental block); 4 – Jurassic turbidite basin (Ulban and Unya-Boma terranes); 5 – Jurassic accretionary prism (Samarka, Nadankhada-Bikin, Khabarovsk, and Bajdal terranes); 6 – Early Cretaceous turbidite basin (Zhuravlyovka-Amur terrane); 7 – Tithonian-Hauterivian accretionary prism (Taukha, Oshima, North Kitakami,

Paleocontinent and the Sergeevka Terrane, which in the current structure of the Sikhote-Alin form a tectonic cover above the accretionary prism (Fig. 12.1).

Among the Mesozoic-Cenozoic folded terranes situated east of the Chinese Craton, several fragments can be recognized as follows:

- Early-Late Jurassic (Hettangian-Tithonian) accretionary prism – Samarka, Nadankhada-Bikin, Khabarovsk, and Badgal terranes;
- Late Jurassic-Early Cretaceous (Tithonian-Hauterivian) accretionary prism – Taukha terrane;
- Early-Middle Cretaceous (Hauterivian-Albian) accretionary prism – Kiselyovka-Manoma terrane;
- turbidite basin of the Early-Middle Cretaceous transform margin – Zhuravlyovka-Amur terrane;
- Early-Middle Cretaceous (Hauterivian-Albian) Kema volcanic island arc;
- Late Cretaceous-Paleocene – East Sikhote-Alin volcanic belt.

In the Mesozoic geodynamic evolution of the Sea of Japan region several stages have been distinguished, corresponding to the environment of the (1) passive margin (pre-Jurassic stage), (2) active margin of Andean type in combination with transform margin (Early-Late Jurassic stage), (3) transform margin in combination with active Andean-type margin (Tithonian-Early Cretaceous stage), (4) transform margin in combination with the active Japanese-type margin (Hauterivian-Albian stage), and (5) active Andean-type margin (Cenomanian-Paleocene stage). On that changing background of geodynamic regimes, the principal structure-forming processes in the Sea of Japan region were accretion of the paleo-oceanic formations to the margin of the eastern part of the Eurasia continent during the oblique subduction of the oceanic plate along the lithospheric plate convergent boundaries and subsequent deformation of the accretionary prisms under transform margin conditions.

Below the principal Mesozoic, geodynamic regimes and sedimentary peculiarities are enumerated: in the tectonically passive eastern margin of the Chinese Craton and in the adjacent active area of the Paleopacific.

South Chichibu, and Ryukyu terranes); 8 – Hauterivian-Albian volcanic island arc (Kema, Kamyshov, Shmidt, Moneron and Rebun-Kabato terranes); 9 – Hauterivian-Albian accretionary prism (Kiselyovka-Manoma, Goniva-Amon, and West-Hidaka terranes); 10 – Late Cretaceous volcanic arc (East-Sikhote-Alin volcanic belt). Fault and fault zones: 11 – Late Cretaceous accretionary prism (Nabil, East-Hidaka, and Shimanto terranes); 12 – Late Cretaceous fore-arc basin (West-Sakhalin and Sorachi-Edzo terranes); 13, 14 – subduction-accretionary assemblages of the Paleo-Okhotic subduction zone: Late Cretaceous accretionary prism (Tokoro Terrane, 13), Late Cretaceous fore-arc basin (Nemuro terrane, 14a), Late Cretaceous volcanic island arc (Terpeniya terrane, 14b); 15 – fault.

12.2 MESOZOIC SEDIMENTATION AND GEODYNAMIC SETTINGS ON THE EASTERN MARGIN OF THE CHINESE CRATON

During the Mesozoic the sedimentation was strikingly epiplatformal on the drowned eastern margin of the Chinese Craton, represented by several fragments. It is characterized by different shallow-water shelf sediments accumulated not deeper than in the sublittoral realm. In this area only marine (shelf seas) and freshwater (lakes, lake-swamps and rivers) basins of the continental margin existed.

Due to interruptions in sedimentation and the alternation of marine, continental-marine and continental environments, it is possible to mark the connection between the epiplatformal sedimentation and the geodynamic events in the, during the Mesozoic, tectonically active Paleopacific oceanic area.

The marine sediments contain shallow-water marine fossils. These are known only for short intervals of the Berriasian, Middle Valanginian, and the beginning of the Middle Albian, when the sea invaded the continent from the east and south-east.

Sediments of the Early Cretaceous continental margin basins of the Sikhote-Alin region differ sharply from the deposits in other basins. They are strongly laterally changeable, characterized by a rather little thickness, often sedimentary interruptions, and traces of terrestrial volcanism. Besides, they contain commercial reserves of coal.

In freshwater depressions gravel-pebble and sand-clay sediments accumulated. These have been recognized mainly by the abundant, often coalified, fragments of terrestrial plants and remains, imprints and nuclei (inner imprints) of freshwater invertebrates, mainly mollusk shells. Freshwater deposits were closely connected with marine shelf environments and particularly, with marine coastal shallow-water, beach and sand bank, lagoon and bay, bar, coastal bank, and also emerged parts of river bed falling into the sea. These deposits include freshwater invertebrate and terrestrial plant remains.

All this demonstrates a peculiar epicontinetal style that happened in the Mesozoic in that area. At last, due to the position and development of these basins on the consolidated stable basement (Khanka crystalline massif), their sediments lay quite horizontal, and folding has been sporadical.

The eastern edge of the Chinese Craton is represented by the Khanka and Bureya massifs, their large fragments, directly joining the Sikhote-Alin fold system, situated in the east. In the depressions of the Mesozoic eastern edge at the background of the epiplatformal shelf sedimentation, transgressions and regressions took place, stipulating the alternation of marine, continental-marine and continental deposits, leading to regional stratigraphical interruptions, and sometimes to angular structural unconformities. They were caused by the formation at the very east of positive structures, marking the beginning of the Sikhote-Alin orogeny, and towards the west, negative pull-apart basins, accompanying movements along the Tan-Lu strike-slip system.

Peculiarities of the epiplatformal sedimentation were also determined by the position of the considered region during the Mesozoic, on the junction of the northern part of the warm Tethys Paleo-ocean and south-western part of the cool Paleopacific Boreal province. Such situation created a particular life condition for the mixed northern and southern marine faunas (Zakharov et al., 1996), which have usually been explained by Tethyan transgressions to the north.

12.3 TRIASSIC

In the Triassic and preceding Late Permian, in the Sikhote-Alin mobile belt, the South-Primorsky trench, and East-Primorsky fore-arc basin seas, divided by the Ussuri mountains, existed (Buriyi, 1997).

The inner sea, that appeared during the penetration of the Early Triassic transgression from the Tethys and Paleopacific oceans, occupied South-Western Primorye and extended to the west, reaching Maly Khingan. That sea was vast, with numerous bays and islands, and its coastline was greatly indented, reminding the recent one of the Pyotr Velikiy Gulf. Sediments accumulated in superimposed depressions. Another sea opened into the Paleopacific and there mainly limestones, cherts, volcanics, and little terrigenous material deposited.

Among other complexes in Primorye, the Triassic includes both marine and continental sequences (Buriyi, 1997). Its sedimentation, paleogeographical and geological history was considered by a number of researchers (Sarkisyan, 1958; Korzh, 1959; Kaplan, 1965, 1966; Miroshnikov, 1971; Buriyi, 1997; Zakharov, 1997). The paleogeographical environments of the epiplatformal terrigenous Triassic of the south-western Sikhote-Alin have been considered by various researchers, but all of them agree that at the beginning of the Triassic a strikingly expressed stage of transgression occurred here, then in the Middle Triassic a regression took place, and at last in the Late Triassic, the sea left this territory, where a continental regime set in.

12.3.1 SOUTHERN PRIMORYE

12.3.1.1 Early Triassic

The Early Triassic epiplatformal terrigenous assemblage (Fig. 12.2) lies everywhere unconformable on the Upper Paleozoic and more ancient marine

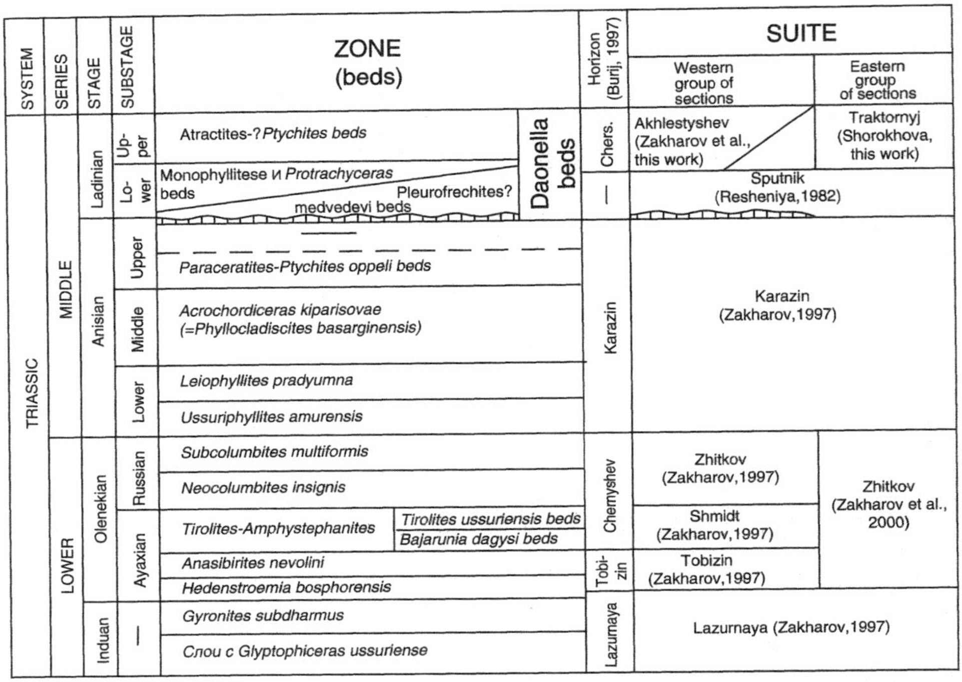

Fig. 12.2: Stratigraphical scheme and corresponding units of the South Primorye Lower and Middle Triassic.

and continental sedimentary and volcano-sedimentary deposits, volcanics and granitoids.

The problem of the Permian-Triassic boundary on the Sikhote-Alin is not finally solved. The study of the boundary strata shows that in all cases, conglomerates or sedimentary breccias are present at the base of the Induan (Early Scythian) stage. However, the most Induan *Otoceras* beds, known in the Verkhoyansk, Kolyma areas and Queen Elizabeth Island of the Arctic basin, have not been found in the Sikhote-Alin.

Induan. The Induan stage (Fig. 12.2) is represented as a whole by coarse clastic beds at the base, relating to two zones: *Glyptophiceras ussuriense* and *Gyronites subdharmus*, constituting the Lazurnaya complex (horizon). In the western and eastern sequence groups of this age, the levels differ (Fig. 12.3). In the west, the Induan stage is represented only by coarse-clastic sequences. In the east, it is represented equally: below – by thick coarse-grained clastics and thick sandstones with lenses of coquina shell rocks (mainly *Glyptophiceras ussuriense* beds),

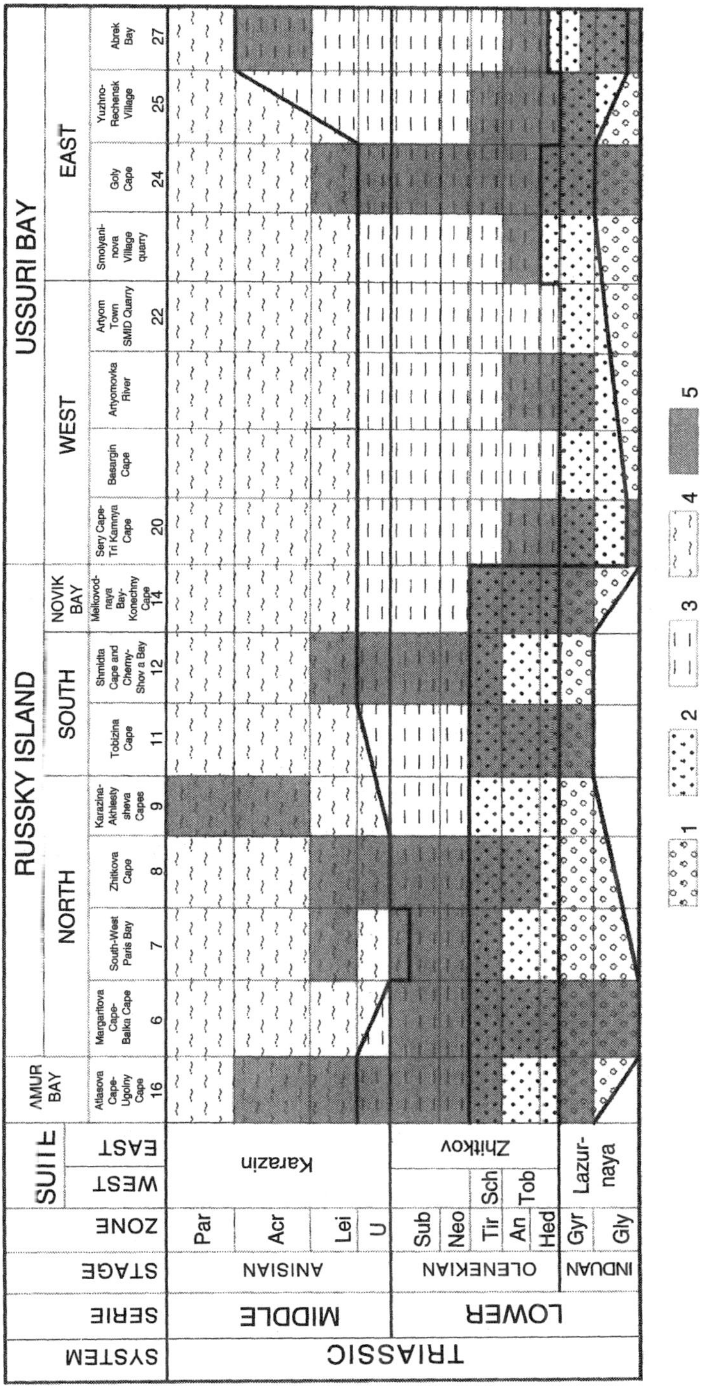

Fig. 12.3: Main Lower and Middle Triassic lithologic formations of the South-Western Primorye. Designations, units: 1 – fine-grained fucoid sandstones with nodules; 2 – siltstones and claystones with nodules; 3 – sandstones with coquina sandstone lenses; 4 – coarse clastics.

and above – only by thick sandstones with coquina shell rocks (mainly *Gyronites subdharmus* Zone).

Although in earth's history the Triassic is known as a geocratic stage, in the south-western Sikhote-Alin at the very beginning it was marked by a regional and rather fast transgression of the sea. The subsidence of this area after Lyu-Khung-Yung (Buriyi, 1997) permitted the waters of the Chinese Lower Yangtze basin to penetrate there. In that area the sedimentary interruption ceased, apparently, only in the earliest Induan, while latest Changhsingian fossils have been discovered in the Partizanskaya river basin. The basal beds are 150 m thick. In the north-eastern part of the area, the interruption lasted longer, until the Olenekian, and the thickness of the basal beds, represented by sandstones and siltstones, is no more than 30 m there.

During the whole Early Triassic, the sources of the material were situated directly near the mentioned marine basin. The absence of the *Otoceras* Zone, known as the lowermost Triassic beds, shows that in the south of the Sikhote Alin, the transgression began, though still in Induan time, slightly later than at the very beginning of the Triassic. There are no convincing arguments that in the Early and Middle Triassic and Jurassic the sea reached the East-Asian continent mainly from the south, from the Tethys, and only as bays, resulting in sinuous coastlines. Moreover, the Paleopacific marginal seas could have fully existed at that time, in eastern East Asia, so that the boreal basin could have joined the Tethys without essential physical obstacles.

The Late Induan sedimentation of sandstones with coquina shell lenses occurred only in the east. Absolutely not-rounded shells with well-preserved sculpture, ribbing, etc., accumulated in their habitat, perhaps in a rather deep environment, having been transported from the coastal area onto the shelf.

Heavy minerals of the Induan sandstones and siltstones are of identical near-sialic (granite-metamorphic) nature, and their non-essential variations are fully explained by the complex mosaic geology, the provenance area composition, and the existence of rivers which drained their different parts, especially because the Khanka massif had then the same framework as now, consisting of land area in the west and north-west of the sedimentary basin.

All other characteristics of the Induan sedimentation correspond to a marine shelf realm; the water temperature was determined as moderately warm, characteristic for the Tethys.

During the Early and Middle Triassic, excluding the Ladinian, on the eastern edge of the Chinese Craton, a regime of deepening to the east of the sea shelf existed, where the accumulation of sediments occurred, from coarse clastic in the earliest Triassic to essentially silty-pelitic during the rest of the time (Fig. 12.3).

Numerous desiccation and interformational erosion tracks testify the coastal shallow-water sedimentation during the Early and Middle Triassic of the south-western Sikhote-Alin, with a constant coastline migration (Geological map, 1969).

Olenekian. In the west, the coquina shell sandstones consist of broken, differently oriented shells and detritus, mixed with sand, indicating that these beds constitute sediments of a coastal marine environment from the surf zone coastal banks.

The Olenekian (Fig. 12.2) includes the uppermost part of the Lazurnaya, Tobizin, and the lower part of the Chernyshev horizons (uppermost Lazurnaya, Tobizin and Schmidt complexes). Here however, like in the Induan, the western and eastern sections of the Sikhote-Alin differ lithologically. In the west, the upper beds of the Lazurnaya, Tobizin, and the lower part of the Chernyshev horizon (*Hedenstroemia bosphorensis*, *Anasibirites nevolini*, and *Tirolites-Amphystephanites* Zones) are represented by thick sandstones, and the upper two thirds of the horizon (Zhitkov Sequence: *Neocolumbites insignis* and *Subcolumbites multiformis* Zones), by thick siltstones and mudstones with concretions. Only in the unique sequence on the Russky Island, the upper beds of the Chernyshev horizon show in its upper part thick fucoid sands with concretions.

In the east, the whole age interval is represented only by siltstones and mudstones, with concretion formations, excluding the lower beds of the *Hedenstroemia bosphorensis* Zone that consists of sands with coquina shell lenses.

In Olenekian time (Baklanova et al., 1971) the sea extended to the east of the Partizansk depression, where the Olenekian lies on the Upper Permian, penetrating also into the Arsenyev depression in Central Primorye. The sea was connected by a narrow strait with the Kukan depression in the Khabarovsk region. Eastward in the Sikhote-Alin, the micaceous sandstones could be Triassic in age, as they conformably lie upon Upper Permian deposits and are covered by Ladinian siltstones. Several positive structures were identified here, crowned by reefs, which also existed in the Triassic, up to the Carnian stage.

Judging from the fact that the Early Triassic deposits are not known to the north of the latitude of the Khanka Lake (Sarkisyan, 1958), the sea formed in this time only a small bay, covering the subsided southern and eastern parts of the Khanka massif.

The erosion was deep, so far as the upward conglomeratic clastic material is represented by more and more different ancient rocks, up to Proterozoic ones. Moreover, the basal strata lie unconformably upon the most different Upper Paleozoic formations. Judging from the great resemblance of the Olenekian fauna of South Primorye, with Japanese, North American, and Indian assemblages, the Ussuri Sea was linked with these seas as a part of the Tethys.

It is supposed that most likely the sedimentation area extended much farther than the surface where their sedimentary fragments are known now. This opinion is based on the fact that there are no sequences consisting only of lower

Triassic beds, and these, starting from the Induan coarse clastic shallow-water and coastal deposits, thicken the Triassic by many hundreds of metres, reaching deep-water environments. Moreover, sequences are not known where younger than Induan beds lie upon pre-Triassic deposits.

The silty-clayey formation that lies on top of the sand formation, indicates that the sea, from the end of the Induan stage, but mainly in the Early-Middle Anisian, turned deeper, especially in the east, and that accumulated in it, chiefly clayey sediments as well as chemical limestones and carbonate concretions. Maybe this means that during that time the provenance area was of a rather smooth relief.

12.3.1.2 Middle Triassic

Anisian. The Anisian (Figs. 12.2 and 12.3) is composed mainly by fucoid sandstones with septarian concretions, but in the east it gradually passes in its upper part into microclastics with concretions.

The Anisian deposits of the Far East (Birobidzhan region, South Primorye, and Kitakami) are a typical example of a sandy-clayey shale facies; noticeable higher phosphorous concentrations have been found in the Anisian sequences in comparison with the Lower Triassic ones (Zakharov and Shkolnik, 1994). In the *Acrochordiceras kiparisovae* Zone of the Basargin Peninsula in South Primorye, P_2O_5-content reaches 6.0% (in large septarian concretions) and 2.8% (in some lenses of sandy-clayey rocks), but the phosphorous concentrations in the Lower Triassic calcareous concretions with ammonoids of Russian Island are lower (less than 0.28–0.40% P_2O_5). No signs of phosphorite were found in adjacent Lower Triassic sandstones (sandy carbonate facies), with exception of the places where *Lingula* remains were expected.

The degree of phosphatization of global marine sediments as a process connected with the natural accumulation of phosphorous-containing organisms seems to be an indirect indicator for the paleoenvironments. Judging from data on phosphorous distribution in the Far East it is suggested that an arid climate existed during a significant part of the Early Triassic and a change to humid climate in the extensive territory (accompanied by the rise of sea level) during the Middle Anisian (Zakharov and Shkolnik, 1994). The upper continental slope conditions of the Anisian sandy-clayey shale facies in the Far East seem to be confirmed by the following important facts: (1) characterization by abundant fossils (ammonites, conodonts, radiolarian, and shark remains) with exception of shallow-water invertebrates; only a few brachiopod and bivalve shells were discovered in the Middle Anisian sandy-clayey shelf facies; (2) signs of underwater slump conditions have been recognized just before the Anisian sedimentation in the Primorye region.

Ladinian. Ladinian deposits (Fig. 12.2) rest unconformably on the erosion surface of the Anisian and Permian formations. They are represented mainly by siltstones, fine-, medium- and coarse-grained sandstones, including fucoid ones, and some intercalating mudstones.

During that geological time, the constantly migrating coastline and the changing of the sea level lead to the frequent alternation and laterally changing of continental and marine regimes and their corresponding sediment accumulation.

12.3.1.3 Late Triassic

Late Triassic deposits (Fig. 12.4) lie unconformably upon the Permian or Ladinian formations. The role of the Upper Triassic marine sediments is small in comparison with the freshwater ones.

SYSTEM	SERIE	STAGE	SUBSTAGE	ZONE	HORIZON
TRIASSIC	UPPER	RHAETIAN			PEREVOZNAYA
		NORIAN	UPPER	Monotis ochotica	
			MIDDLE	Eomonotis scutiformis	PESCHANKA
				Otapiria ussuriensis	
			LOWER	Pterosirenites tenuistriatus	
		CARNIAN	UPPER	Striatosirenites Arietoceltites	SADGOROD
			LOWER		KIPARISOVO

Fig. 12.4: Stratigraphical scheme of the Upper Triassic of the South Primorye.

Carnian. The Lower Carnian Kiparisovo complex (horizon) lies unconformably on top of the Permian and conformably on top of the Ladinian sequences.

They are represented by lagoonal sediments, conformably overlain by non-marine deposits. They are mainly thin intercalating siltstones, fine sandstones with wavy, lenticular, horizontal, and sometimes oblique-bedding, also with intercalations and lenses of medium- and coarse-grained, massive and cross-bedded sandstones containing fragments of gray sandstone, acid effusive rocks, and mudstone. In the basal horizon of the sequence fine and medium pebble conglomerates are found. Their pebbles consist of acid lava and tuffs.

The Upper Carnian Sadgorod complex, that in its turn, is conformably overlain by the Lower-Middle Norian Peschanka complex, is represented at the base by marine coastal (littoral) deposits, and upward by the sediments of two facies: in the west (Amba river) by marine, and in the east (Perevoznaya river) by coastal-marine deposits. The Sadgorod sequence consists of conglomerates, different-grained sandstones, silty sandstones, siltstones, mudstones, and coal, intercalated with coaly mudstones and mudstones forming layers of 10 cm up to 5.6 m. Acid and medium effusives are also known.

Norian. In the South Sikhote-Alin the Norian stage is represented by the conformably resting Peschanka, Amba, and Perevoznaya complexes. The Peschanka sequence consists of different-grained sandstones and in its upper part by tuffaceous sandstones. The Amba sequence contains siltstones, sandstones, conglomerates, and tuffaceous sandstones. The Upper Norian Perevoznaya sequence is represented by coastal marine and marine sediments.

12.3.1.4 Triassic Basin history

Before the Jurassic (see Fig. 12.13) a passive continental edge has been reconstructed on the basis that along the eastern margin of the Khanka massif and of the entire South-Asian part of the continent there is no accretionary prism. The paleocontinent and the joining plate moved together to north and north-west, drawing together with the Siberian Craton.

By this means, as a whole, two stages can be distinguished in the Triassic, as follows:

(1) Between the basal *Glyptophiceras ussuriense* Zone of the Induan stage, and the *Subcolumbites multiformis* Zone of the Olenekian stage, a transgression occurred. During that period the coarse clastic (1), the sandstone with coquina lenses (2), and the lower half of the mudstones with concretions (3), formations accumulated.

(2) From the end of the Anisian and Early Ladinian, a regression took place. During that phase, sandy formations were deposited.

These two stages constitute a complete transgressive-regressive cycle, accompanied by corresponding changes of the dimensions of the sea, its level, its depth, as well as the coastal configuration.

12.3.1.5 North-Western Primorye

In the North-Western Primorye, known as the Alchan tectonic zone, remains of a terrestrial flora in the lower part of the Late Triassic deposits, characteristic for swamped forests and flood-lands, the dominant sandstones with coarse clastic rock intercalations with well-rounded pebbles, show that these formations accumulated in the different areas of alluvial-lake valleys, sometimes reaching the sea coast. In the small swamps and lakes favourable conditions existed for coal formation.

As a result of the Late Norian transgression, the thick silty-sandy deposits with abundant *Monotis* remains formed here and there coquina shell rocks. These rocks also contain fossil brachiopods, ostracods, gastropods, foraminifers, sponges, and sea urchins. There, where the transgression reached the heights of the adjoining river valleys, a weathering crust developed on them, judging from the clayey sediments, probably kaolin.

In the Late Triassic, volcanic activity influenced the non-marine and shallow-water marine sedimentation. Thin fine-grained pyroclastic and volcano-sedimentary rocks show that the eruptions occurred far from the sedimentation area. The rocks are acidic, less basic volcanics and associated extrusive bodies, developed in the Chinese near-frontier regions (Geological Map, 1996; Zhao Chujing and Li Zhitong, 1988).

The clastic material of the Late Triassic coarse-grained rocks differs in different parts of the Alchan Zone, depending mainly on the composition of the underlying source formations.

On the left and right borders of the Bikin river valley, among the pebbles are many gabbroid serpentinites, cherts, with frequent sealing-watch jaspis and schists. The same rocks are known from the plates, overthrusted on the Upper Triassic deposits and composed of different Middle Paleozoic ophiolite associations and Permian volcano-cherty-terrigenous rocks. Middle Paleozoic and Permian deposits concern the accretionary assemblage of the Samarka terrane, and Norian ones are considered as overlying units of the Khanka massif (Khanchuk et al., 1995; Khanchuk, 2000).

In the southern part of the Alchan Zone, the Upper Triassic lies unconformably on the Late Permian dacites, and the clastic fragments in the conglomerates are represented mainly by these rocks.

In the upper Ulyanovka river, clastic material of the Norian coarse-grained rocks contains many quartz porphyrites, felsites, granites, quartz, andesites, sandstones, and siltstones derived from the Permian deposits.

The complex structure of the source provinces is confirmed by the rock-forming component and heavy mineral composition of the sandstones and siltstones. They have been studied in detail on three plots at the right borders of the Bikin river valley (the country between Vilyuika and Olon rivers), in the lower Chyornaya Rechka river and the Inchun settlement (Heilungkiang province of China). The sandstones of the country between the Vilyuika and Olon rivers are quartz-feldspar graywackes. They have up to 10% quartz and 45–67% feldspar. These minerals have usually irregular shapes, smooth, seemingly melted contours, and contain small inclusions of chloritizated volcanic glass, indicative of their effusive nature.

Two sandstone types have been identified. One of them contains more fragments of the acidic magmatic rocks with microfelsitic and microhipidiomorphic structures, as well as tuffs (?) or tuffites, probably of pyroclastic origin. Other sandstones contain many fragments of magmatic rocks, mainly diabases. The sedimentary rock grains are represented by cherts, schists, quartzites, and siltstones. By this means, the main sources for the sands of the country between the Vilyuika and Olon rivers were acid pyroclastic and basic magmatic rocks. Erosion products of these basic rocks are abundant among the silty clastic heavy minerals: epidote, hornblende, clinopyroxene, leucoxene, and ilmenite, consisting about 70% of the heavy fraction.

The sandstones of the near-mouth part of the Chyornaya Rechka river are quartz-feldspatic and feldspatic-quartz graywackes. The ratio of sandstone rock-forming components is about equal. Feldspars are represented by plagioclases and K-feldspars (up to 45%). Some quartz and plagioclase grains contain inclusions of the chloritizated hyaline mass, witnessing their direct pyroclastic nature or their origin from the acid tuff destruction.

Among the rock fragments in some sandstones, coal, clayey schists, clayey siltstones, sandstones, more rarely cherts and quartzites are present. In other sandstones, many acidic magmatic rock fragments with aplitic and microfelsitic structures have been determined. There is always present a small amount of the medium-basic volcanics with hyalopelitic or vitrophyric structures of the basic mass and granites. Therefore, the acidic volcanics, granites, and sedimentary rocks were widespread in the source area. This suggestion is confirmed by the high (up to 85%) zircon content among the heavy minerals.

In the neighbouring Chinese territory (Heilungkiang Province), the Upper Triassic deposits consist of sandstones, siltstones, and rare beds of poorly graded conglomerates with angular pebbles (up to 20 cm), made up of gabbro, andesites, felsites, and siltstones. The sandstones contain 40–50% of quartz and little (13–25%) feldspars (Fig. 12.3). The rock fragments are acidic volcanics, quartzites, and up to 75% of angular and rounded grains of the sericitizised rocks of a not clear nature. Among the sandstone clastic minerals, as a role, are zircon (60–78%), some leucoxene, apatite, magnetite, and hornblende.

12.4 JURASSIC

At the beginning of this part, a short description of each Jurassic unit has been presented, followed by comments on the paleogeography, sedimentary and geodynamic environments. The Jurassic units of South and North-Western Primorye have been described from ancient to the youngest (Fig. 12.5).

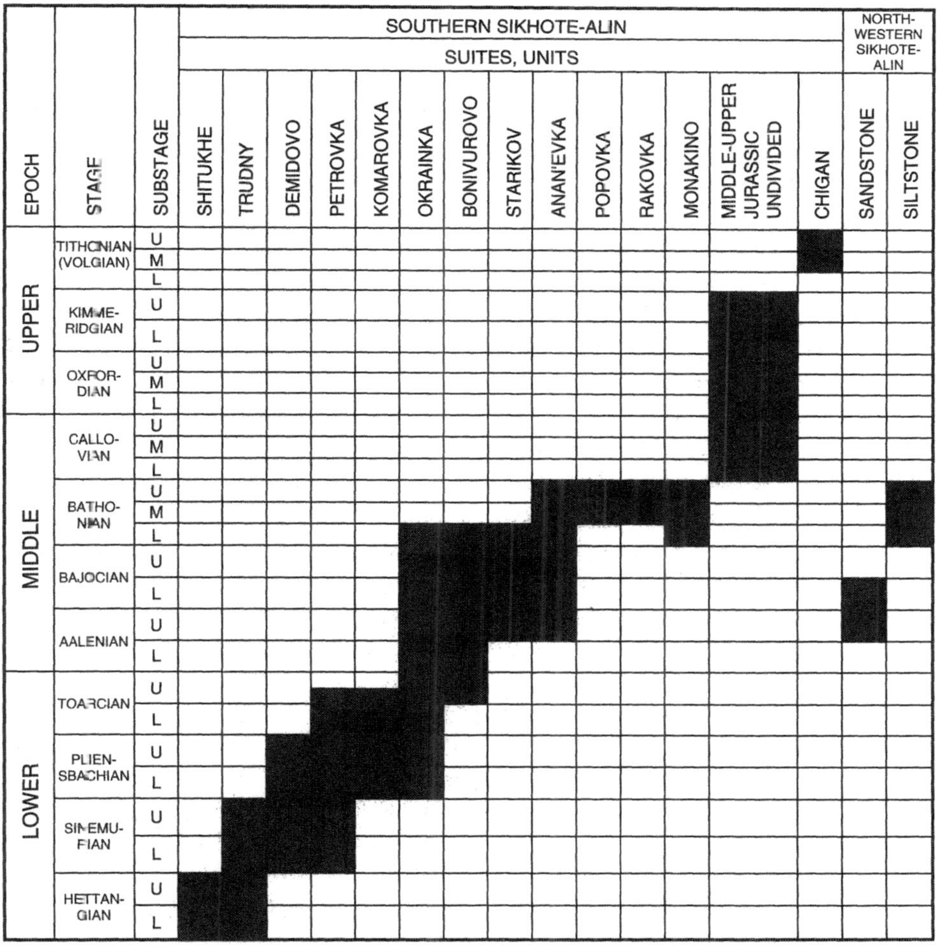

Fig. 12.5: Stratigraphical scheme and corresponding units of the South Primorye Jurassic.

12.4.1 SOUTHERN PRIMORYE

Shitukhe complex (Hettangian, 250–300 m). Nearshore marine and marine formations of this sequence occur with an erosional and little angular non-conformity on the Lower Triassic and are overlapped by the Petrovka complex with a little break in sedimentation.

The complex is represented by marine and non-marine deposits containing a fossil fauna and flora. It is composed mainly of siltstones and fine-grained sandstones with rare medium-grained ones, conglomerates, and coaly claystone intercalations. Their peculiarity consists in the presence of coquina limestone beds (10–15 cm), with myarian shells and nuclei.

The lower part of the complex begins with layered sandstones (6–10 m) with rare pebbles and fine-grained polymicitic sandstones with rudaceous breccia and siltstone interbeds (0.2–0.3 m). At the base of this sequence, the coquina limestone (0.5–2.5 m) consists of myarian shells.

The sandstones are mainly fine-grained, though coarse-grained ones are present, polymicitic and well sorted. Quartz constitutes about half of the grains, feldspars (mainly plagioclase) up to 30%, quartzite and cherty rocks up to 15%; heavy minerals are zircon and tourmaline. The siltstones are layered with a not clear to thin horizontal lamination, forming beds up to 15 m, intercalating with sandstones and containing coquina beds. Conglomerates form at the base of the complex interbeds of 0.5–0.6 m. Their pebbles (up to 4–5 cm) occur sometimes dispersed in the sandstones and consist of medium-grained sandstone, cherty siltstone, and rarely quartz. The coquina is composed of myarian shells, situated on the bedding surface, with the convex side oriented upwards. In most cases, the shells are open with united umbo. The lower section (300 m) of the marine sediments is constituted by fine- to medium-grained sandstones, sometimes banded, and siltstones with scarce marine organic remains of ammonites, crinoids, and myarians.

Trudny complex (Hettangian-Sinemurian, about 500 m). This complex lies upon the Upper Permian without a visible interruption or unconformity. It consists of equal quantities of siltstones, sandstones and conglomerates.

The siltstones show a weakly expressed horizontal bedding and lenticular interbeds, lenses, and nests of fine-, rarely coarse-grained, sometimes calcareous, cross-bedded sandstones. On the rock surfaces appear abundant mud-eating worm tracks. The sandstones are in one case, medium-grained quartz-feldspatic with rare interbeds of dark-gray fine-grained sandstones, siltstones and claystones. In other cases, they are fine- to medium-grained, micaceous and stratified (due to the alternation of light and dark-coloured laminae and of different composition) with thicknesses between 1 and 10 mm.

Approximately in the middle of the complex there occur two conglomeratic beds, separated by siltstone (1 m). The conglomeratic clastic components (up to 1 m) consist of siltstones and sandstones as well as well-rounded quartz and chert pebbles.

Fauna remains are rare and do not form dense accumulations, excluding marine lilies buried in lifetime position, of a myarian community reworked by swell.

Demidovo complex (Sinemurian-Pliensbachian, 480 m). This sequence has been divided into a lower and an upper member.

The lower member (Sinemurian, 300 m) is represented by siltstones and graywackes of different grain size, with tuff and tuffite interbeds. Most probably, the member occurs transgressively on the more ancient deposits. At its base, coarse-grained, gravelly sandstones consist of well-sorted and rounded grains (0.9–2.0 mm) of cherts, granites, liparites and their tuffs, schists, feldspars, and quartz. The graywackes (up to 30% of the member thickness) are fine- and medium-grained, and well sorted. Their clastic material is represented by fragments of acid effusive rocks, basalt porphyrites, chery-clayey and phyllite-like rocks (65%), and feldspars (20–25%) and quartz (15%). Argillaceous graywackes constitute no more than 3–5%. These are medium-grained with fragments (75%) of porphyrites and albite-clayey schists as well as feldspars (15%) and quartz (10%). Radiolarians are frequent. Volcanic rocks (about 40% of the member thickness) form interbeds of 1–10 m, within the sediment sequence; they are acid tuffs and tuffites.

The upper member (Pliensbachian, 180 m) occurs conformably upon the lower member and transgressive above the Lower-Middle Triassic. It consists of different-sized graywackes (70% of the thickness), more rarely quartz-feldspar graywackes, and tuffs, tuff breccias and porphyrites. The graywackes are medium-grained, well to medium sorted, sometimes with disturbed structure. The grains consist of quartz (20–50%), feldspars, mainly potassium (up to 55%), and acid effusive rocks, aplites, and porphyrites (up to 30%). Heavy minerals are represented by orthite and zircon. Silty claystones with disturbed structure constitute about 15% of the thickness. Their grains consist of quartz, feldspars, including K-feldspars, and micas. Claystones form beds up to 0.2 m thick within the sandstones. Marls form irregular concretions, containing silt-sized grains of plagioclase and quartz. The complex contains some myarians and brachiopod nuclei, a lot of shell fragments, sometimes forming accumulations, and wood fragments (0.3–5 cm). There appear also numerous ammonoid and ammonite fragments. The myarians appear probably in lifetime burial position, because their apical part is well preserved; the brachiopods occur as nuclei of the entire shells.

Petrovka complex (Sinemurian-Upper Toarcian excluding its upper part, 80–100 m). This sequence occurs with an erosional, but without angular disconformity above the Shitukhe complex, and is laterally constant. It is covered transgressively by the Upper Jurassic marine Chiganov complex or conformably by the Early-Middle Jurassic Bonivurovo complex. There are few organic remains in this complex, and these are mainly indeterminable. Determinable ones are extremely rare, and plant detritus is scarce. The complex consists essentially of sandstones; coarse-grained sandstones (in the upper part of the sequence cross-bedded and with disturbed structure) prevail. Like in the Shitukhe complex, there occur also siltstones, gravelstones and conglomerates, and overall it is more coarse-sized. The basal horizons of the Petrovka complex are represented by conglomerates, fine- and medium-grained sandstones with conglomerate interbeds and lenses, as well as by coarse-grained sandstones with conglomerate interbeds and lenses. The rest of the complex consists of different-grained, and at the base of the sequence, coarse-grained sandstones with conglomerate lenses.

The sandstones are mostly polymicitc, feldspathic-quartzose, and quartz-feldspathic. In these sandstones, quartz constitutes 25–35%, feldspars, mainly plagioclases, up to 40%, and rock fragments (cherts, siltstones, and micro-grained sandstones) 30–40%. Heavy minerals are zircon and apatite. The feldspathic-quartzose sandstones consist of quartz (up to 60%), acid plagioclases (20–25%) and fragments of hornfels, cherty and micro-grained quartzites (10–15%). Heavy minerals are zircon, leucoxene, and rutile. The quartz-feldspathic sandstones are fine- and medium-grained and well sorted. They consist of oligoclase-andesine (up to 60%), quartz (25–30%) and rock fragments: chert and cherty siltstones (less than 10%). The siltstones are of the same composition as the sandstones. The conglomerates consist of pebbles of sandstone and quartzite (up to 70%), quartz-porphyrites, cherty siltstones, and quartz.

Komarovka complex (Lower Pliensbachian-Upper Toarcian, excluding its upper part, 30–90 m). The complex occurs with an erosional contact, but without a regional disconformity on top of the Triassic, and is conformably overlapped by the Bonivurovo complex (Upper Toarcian-Lower Aalenian). The complex consists of feldspathic graywackes with rare thin interbeds of gravelstones and pelitic tuffs, and very rarely conglomerates (1–3 m).

Quartz-feldspathic graywackes constitute about 60% of the complex thickness. Among the grains about half is of acid effusive rocks, rare gneisses, cherts, and limestones, with quartz (up to 30%) and plagioclases (up to 20%). The prevailing heavy mineral type is zircon. The gravelstones form thin beds within the graywackes at the base of the complex. Their grains are represented by quartz, microcline, plagioclases, granite (sometimes greisenizated), liparite, albitite, cherts, and siltstones. Heavy minerals are zircon and apatite.

At this time in the Southern Sikhote-Alin a shallow-water marine bay existed, in which favourable life conditions appeared only for short periods and far from its whole surface. The majority of the oryctocoenoses are paleobiocoenoses. The presence in the Middle Liassic fauna associations of a great number of byssus-attached organisms shows that the basin was characterized by a normal to near-normal salinity and by near-bottom currents. The absence of typical stenohaline echinoderms and corals as well as a great number of oysters show some desalinization that judging from the poorly enduring salinity drop brachiopods, and from the prevalence of thick-valved myarians, was insignificant. The myarians *Vangonia*, *Ostrea*, and *Oxytoma* point to warm and well-aerated seawater.

Okrainka complex (Upper Pliensbachian-Lower Bathonian, up to 790 m). The relationship of this complex with the underlying Upper Triassic is not established; it is overlapped with an erosional surface by the Poga complex. The Okrainka complex includes tuffaceous siltstones and claystones, siltstones and claystones, rarer tuffaceous sandstones, limestones, and also basalt, dacite-andesite, diabasic, and andesitic porphyrites. The lower part of the complex is composed only of terrigenous deposits.

The quartz-feldspathic and feldspathic-quartzose graywackes are fine-grained and constitute the upper part of the Okrainka complex. They are quite friable rocks with poorly defined bedding. Quartz constitutes 17–58% of all grains, feldspars 22–30%, and rock fragments no more than 28%. The quartz is largely monogenic, but there is also quartz with wavy and polygonal extension?, and sometimes it is volcanic. The feldspars (up to 70%) are represented by plagioclase and rarely by K-feldspar. Among the rock fragments, acid lavas, tuffs and volcanic glass, also crystalline and coaly schists are predominant. Heavy minerals are zircon, tourmaline and garnet. The limestones (up to 7% of the sequence thickness) are recrystallized, with porous weathering surfaces, and contain fossil ammonite and myarian fragments, sponge spicules, and some radiolarian skeletons. The tuffaceous sandstones constitute about a fifth part of the complex. They are fine-grained, bed-sized, with a vague horizontal bedding. Their grains are represented by quartz and feldspars, acid lavas, tuffs, and pumice. The cement composition includes volcanic ash. The tuffaceous sandstones contain radiolarian skeletons and crinoid joints. The tuffaceous siltstones, forming about 30% of the complex thickness, are horizontally bedded, due to pelitic laminae, and their petrographic composition corresponds to that of the sandstones. They contain always radiolarian and crinoid remains.

Bonivurovo complex (Upper Toarcian-Lower Bathonian, 30–400 m). This complex occurs conformably upon the Komarovka or Petrovka complexes, or transgressively on the more ancient formations, including the Upper Triassic. It is divided into three members.

The *lower member* (upper part of the Upper Toarcian-Lower Aalenian, 30–120 m) consists mainly of feldspathic sandstones of different grain sizes without a clayey matrix. The member begins with gravelstones and sandstones (1–28 m). The boundary with the overlying member is drawn at the beginning of domination of the fucoid sandstones with abundant clayey matrix, and the change of diverse myarians of the lower member to the predominant *Mytiloceramus*. The member sometimes contains volcanic ash.

The *middle member* (Upper Aalenian-Bajocian, 270–400 m) is represented chiefly by polymictc sandstones of different grain sizes, containing much clayey matrix, small plant detritus, big and small fucoids, less medium-grained calcareous and feldspar sandstones, siltstones, and tuffites. The lower and middle parts of the member are evenly saturated by organic remains, among which the principal is of *Mytiloceramus*. The upper boundary of this member is drawn according to the appearance of the well-sorted sandstones of the upper member or at an erosion surface. Sometimes in the Bonivurovo complex appear tuffites.

The "face" of the complex's middle member is determined by characteristic "worm" (fucoid) rocks. This means an alternation of gray or dark-gray fine-grained silty sandstone and dark-gray sandy siltstone, with lenticular, striated, or irregular lamination and fucoids. The boundaries between the laminae are vague, changing one into another. In the principal rock background the subordinated rocks form dispersed isometric spots or bands of different length and width. In the outcrops, a sub-parallel orientation of all beds can be observed including very irregular ones. A regular and clear alternation resembles a rhythmical sequence with rhythm thicknesses between centimetres and a few decimetres. The fucoid are dark, worm-like, curved clayey thin bands with lens-shaped transversal or oblique sections of 4–5 cm length and 3–5 mm width. Usually fucoids occur parallel to the bedding, so that in the perpendicular sections only small lenses of different length and width can be seen, which change parallel or echelon-like one laterally into another, forming a striate lens-like lamination. In the siltstones and claystones, fucoids are regularly distributed in the rock or concentrated in separate laminae. In the fine-grained silty sandstones they become thicker, forming dark laminae and irregular silty and pelitic parts, or they disperse and disappear. The fucoid boundaries, depending on the granularity, are gradual or sharp. In plane, fucoids are S-, arc-, book-like, and similar shaped, but they do not intersect or branch in the unique flatness. It is not clear whether fucoids are worm grooves, filled with mud, or nereites, filled with clayey matter, reworked by worms.

The *upper member* (Upper Aalenian-Lower Bajocian, 61 m) consists of rather feldspathic calcareous sandstones of different grain sizes, with single thin siltstone interbeds.

The diverse fauna of the Late Toarcian-Early Aalenian in the middle and upper members of the Bonivurovo complex, changes into a depleted assemblage in the Late Aalenian-Early Bathonian, composed mainly of *Mytiloceramus*, a flourishing and almost unique myarian group with clearly expressed concentric ribbing.

Other Complexes and Series. The *Starikov series* (Upper Aalenian-Lower Bathonian, 50–170 m) consists of thin (1–5 mm) alternations of sandy siltstones, and silty, tuffaceous and graywacke-arkosic sandstones. Separate interbeds (0.2–0.3 m) of siltstones with carbonaceous concretions and of tuffaceous claystones occur. The terrigenous rocks contain a volcanic ash admixture, not characteristic for the Bonivurovo complex. The sandstones are silty, fine-grained, dark-gray with disturbed structures, lens-like, rarer horizontal lamination, conchoidal jointing, and siltstone interbeds up to 0.75 m thick. There are no normal stratigraphical contacts with the underlying and overlapping deposits.

The *Ananyevka series* (Upper Aalenian-Bathonian, 600 m) shows a lithologically diverse composition and is represented mainly by sandstones, and rarer siltstones, tuffs, coaly-clayey schists, and coal. Their contact with the underlying formation is not known. The sandstones are represented by feldspathic graywackes. Among their rock fragments, acid effusive rocks predominate, and they show clear indications of syn-sedimentary volcanic activity as suggested by the mixture of horn-like acid ash particles in all rocks, sometimes recrystallized into microfelsite, and pelitic tuffites (2–3 m) across almost the whole sequence. These sandstones contain sometimes myarian *Inoceramus*, as well as pelitic laminae with numerous fucoids and rarer biotite-feldspathic crystalloclastic tuffites. Sandstones with spots and "patterns" of volcanic material are also present.

The *Popovka series* (Middle-Upper Bathonian, 255–350 m) occurs conformably with a gradual transition above the upper member of the Bonivurovo complex. It is composed of essentially siltstone with silty sandstones at the base. At some places, the siltstones are pyritized and contain marcasite concretions, with 1–4 cm diameter.

The *Rakovka series* (Middle-Upper Bathonian, 150–190 m) consists only of feldspathic graywackes of different grain sizes, among the grains of which acid effusive rocks prevail. The series occurs unconformably on top of the Bonivurovo complex and is, probably conformably overlain by Upper Jurassic formations.

The *Monakino series* (Bathonian, more than 240 m) is non-marine and occurs disconformably with an angular unconformity above the more ancient formations. It is divided into a lower (rhyolitic) and upper (terrigenous-volcanic) member.

The *Chiganova complex* (Middle-Upper Tithonian, 252 m) consists mainly of fine- to medium-grained, often fucoid sandstones. Gravelstones and conglomerates, siltstones, silty claystones and silty sandstones also occur. The complex rests unconformably upon the Lower and Middle Jurassic formations.

12.4.2 NORTH-WESTERN PRIMORYE

In the North-Western Primorye (Alchan Zone) only the Middle Jurassic is known, represented by sandstone and siltstone complexes. The sediments, enriched by volcanic material, and the fossil fauna accumulated in a shallow-water fore-arc marginal sea.

12.4.3 JURASSIC BASIN HISTORY

The Jurassic sedimentary environments have been described after the manuscript of Konovalova, published in Markevich and Zakharov (2004) and after unpublished reports of Chernysh. Here new data on the stratigraphy and paleogeography, obtained since 1991, have been reflected. They allow refining the volume and paleontological description of a number of some "ancient" stratigraphic units and establishing of new ones.

In the *Southern Primorye* Jurassic deposits are represented mainly by marine sedimentary, less often by non-marine (continental) volcano-sedimentary and volcanic formations. In this region, three divisions of this system have been established containing a number of complexes. These are distinguished by paleontological and lithological criteria, their position within the sections and their relationship with underlying and overlapping deposits.

In this region (Fig. 12.5) Shitukhe, Trudny, Demidovo, Petrovka, and Komarovka complexes, also the Toarcian part of the Bonivurovo complex and the Pliensbachian-Toarcian part of the Okrainka complex are referred to the Lower Jurassic. The Middle Jurassic includes the greater part of the Okrainka and Bonivurovo complexes, the Starikovo, Ananyevka, Popovka, Rakovka, and Monakino series, as well as the lower, Callovian, part of the Middle-Upper undivided Jurassic, conditionally referred to the Callovian-Kimmeridgian. The Late Jurassic is represented by marine shallow-water deposits of the Middle and Late Tithonian Chiganova complex.

The Jurassic fauna was by and large poorer than the Triassic one, and is represented mainly by the remains of myarians, more rarely ammonites. The similarity of the number of forms with the Japanese ones, allows supposing that the basin was connected with the Boreal Province and the Tethys Ocean. The study of the Jurassic is quite complex because in the Sikhote-Alin there are no sequences saturated with fossil faunas, permitting to substantiate the stratigraphical boundaries. Relatively thick strata contain on the same or on different levels, fauna and flora remains that allow the determination of the stage or substage, but the boundaries between them are conditional.

Early Jurassic. At the beginning of the Jurassic (see Fig. 12.13), in connection with the closure of the Mongolo-Okhotsky Paleo-ocean and the collision of the Argun-Mamyn and South-Asian Paleocontinents, the further drifting of the latter northward became impossible. Along the major part of its eastern margin, the Jurassic oblique subduction zone formed, and the mode of passive continental margin changed into the setting of an active margin of Andean type, while the formation of the Jurassic accretionary prism started. In the recent structure of South-East Asia, fragments of this prism have been confirmed from the Udskoy Gulf coast in the north to the Palavan Island (Philippines) in the south. At that time the accretion of the oceanic plateau occurred, that, as a positive structure, could not be completely subducted. Along the steep fracture system it was broken into tectonical slices with a partial moving up of the lower ones under the upper, becoming accreted to the continental margin. Fragments of this plateau in the Sikhote-Alin are represented by the Kalinovka ophiolites, and in Japan by the Yakuno ophiolites.

The Lower Jurassic occurs always with washout above the more ancient formations. This indicates that before the Early Jurassic transgression, which advanced from the east and south, intensive denudation took place.

Monotonous fine-grained siltstones and sandstones of the Shitukhe complex, a good sorting of the sandstone grains and their clear horizontal bedding without the signs of cleavage and current structures, mean a rapid and quiet deposition. The rocks are dark-gray coloured, with limestone lenses and intercalations, pyrite microspheres and fossil animals and plants. There are also nearshore offshore bar sediments, interrupted by conglomerate and gravelstone beds, as well as limestones with many *Modiola* shells.

The composition and structure of the deposited marine sediments, in the beginning of the Early Jurassic, show that the environment was a shelf with bays and islands, sometimes subsiding below the sea-level surface.

The transgression started in the Hettangian. Then the sea invaded the Southern Sikhote-Alin land from the Sea of Japan, when the Shitukhe and the lower part of the Trudny complexes became accumulated.

In the Sinemurian the sea boundaries widened, and the upper part of the Trudny complex, as well as the lower parts of the Demidovo and Petrovka complexes became deposited. The shoreline of the central part of the bay coincides with that during the Hettangian. The basal beds of the Petrovka complex (conglomerates and coarse-grained sandstones with conglomerate lenses) show an erosional interruption between the Hettangian Shitukhe complex and the lower part of the Sinemurian Petrovka complex.

Fossil plant remains and abundant detritus at the base of the Sinemurian (in the lower part of the Petrovka complex) indicates the existence of a continental regime somewhere.

Interbedding of tuffs with the sandstones and siltstones, and also tuff intercalations in the lower part of the Demidovo complex, mean that there or nearby the volcanic centres existed. These volcanic deposits are the oldest, as in the Okrainka complex lava, tuffs, and tuffites are only known from the Pliensbachian. The volcanic glasses of the Komarovka complex are of the same age.

In the Pliensbachian the upper part of the Demidovo complex, the middle section of the Petrovka complex, and the lower part of the Okrainka complex accumulated. The lower part of the upper Demidovo complex contains abundant myarians and rare brachiopods, pointing to a sublittoral area of a normal saline sea. Moreover, in the Komarovka complex the abundance of fossil myarians, brachiopods, ammonites and belemnites, suggest a same type of environment.

During the Pliensbachian-Toarcian in a shallow-water normal saline bay, the upper parts of the Demidovo and Petrovka complexes, the whole Komarovka complex, as well as the lower parts of the Okrainka and Bonivurovo complexes deposited. Judging from the irregular disposition of the fossils and their confinement to certain stratigraphical levels, favourable life conditions lasted only for short time intervals and far from the surface of the bay.

Middle Jurassic. The Middle and Late Jurassic are characterized by subduction of the morphological weakly dissected part of the paleo-oceanic plate. Accreted paleo-oceanic fragments of the Jurassic accretionary prism are represented by the cherty-terrigenous sediment sequences, which differ only by the smooth rejuvenation of the transitional beds from the proper cherty part of the sequence to the terrigenous one. In the course of the accretion of the paleo-oceanic formations along the fractures, which appeared along the strike-slip fractures, dissecting the subduction part of the paleo-oceanic plate on its bends into the subduction zone, the astenospheric matter penetrated leading to the effusion of alkaline ultramafic and basic lava above the turbidites of the trench, and the formation of hyperbasites within the Jurassic prism.

In the Middle Jurassic the transgression reached its maximum, and the Bonivurovo complex deposits of that time in the Southern Primorye are more abundant than those of the Early Jurassic. The rocks of this Bonivurovo complex (Late Toarcian-Early Bathonian) are everywhere nearshore cross-laminated; they contain as well fucoids, plant detritus, and trunk and stem fragments. All these peculiarities suggest depths up to 30 m.

In the Toarcian-Early Aalenian (on the boundary between the Early and Middle Jurassic) the limits of the Pliensbachian-Toarcian bay broadened, and the genera composition of the inhabitant mollusks became renewed. They were represented by numerous *Trigonia*, *Vraiamussium*, *Meleagrinella*, *Oxytoma*, and rather rare *Ostrea*, *Grammatodon*, *Mytiloceramus*, *Pleuromya*, *Cardinia*, *Vaugonia*, *Modiola*, *Gastropoda*, and *Brachiopoda*. The lower Bonivurovo complex mainly corresponds to that epoch. The fossil fauna points to some community in their life conditions in the middle of the Early Jurassic and in the Late Toarcian-Early Aalenian. The abundant byssus-sedentary *Trigonia* and *Ostrea*, requiring for their life enough high temperatures, indicate a good warm mixing and aeration of the water with a normal or close to normal salinity.

The Late Aalenian-Bajocian corresponds to the middle parts of the Okrainka, Bonivurovo, Starikov, and Ananyevka complexes, as well as to the sandstone unit of the South-Western Primorye. At that time, the brachiopods and gastropods became completely extinct, and only widely occurred myarians, *Mytiloceramus* and a few *Pleuromya*, as well as traces of mud-eating worms. A similar picture is peculiar to the semi-closed basins, transitional from marine to fresh water. The disappearance of *Ostrea* and Pectinidae, typical for the Late Toarcian-Early Aalenian, and an enduring weak desalinization point to a strong basin freshening, and a regular organic remains distribution upwards to a stable ecological environment. The thin sandstone intercalations, full of belemnite rostra, indicate that the freshening was disturbed by a periodical invasion of saline seawater.

The fauna of the Bonivurovo sea consisted mainly of diverse *Inoceramus*, in their majority typical marine animals, forming banks in most of the shallow-water areas. The monotonous genus composition of the fauna was obviously connected with the abnormal seawater salinity. Among the *Inoceramus*, there are a lot of species common for the Boreal province and rarely close to the Japanese ones, indicating that the connection of the South-Primorsky basin with the Paleopacific occurred through the Boreal and Japanese seas. At the end of the Bajocian, the ammonites and belemnites appeared.

For the Bonivurovo complex pyrite microspheres and plant detritus are typical. Now and then among the deposits, volcanic glass of the syn-sedimentary eruptions is found.

The diverse fossil fauna of the Late Toarcian-Early Aalenian in the lower part of the Bonivurovo complex changed into the monotonous Late Aalenian-Early Bathonian fauna of the middle and upper parts of the complex, consisting mainly of *Mytiloceramus*, a flourishing and practically unique group. The basin limits remained as before, but the change of sediment-type composition and given the *Mytiloceramus* assemblages, as well as the disappearance of the *Pleuromya* remains and of the tracks of the mud-eating worms, show that the living conditions were another in comparison with the Late Aalenian-Early Bajocian. A general poorness of the organic remains and their mosaic distribution suggest that the favourable living conditions existed only in small areas.

Essentially reduced species and quantity composition of the *Mytiloceramus* assemblages, associated with a broad development of species with very big, coarse-ribbed flattened shells, and the complete disappearance of *Pleuromya, Belemnites*, mud-eating worms and plant detritus, became typical. The absence of sorting, the good remains preservation and orientation, prove that the fauna was undoubtedly a part of a life time community, buried in situ, and their unregulated position in the sediment, is an argument for the byssus-sedentary habitat.

In the Bathonian the marine basin became strongly reduced in size, and horizontal-bedded silty claystones accumulated. This means that the sea was deeper than in the beginning of the Middle Jurassic. The dark-coloured rocks and pyrite microspheres are evidence for the restoration of the environment existing in the basin at that time. Abundant *Inoceramus* changed into reasonably rare, more deep-sea, thin-valved *Posidonia* and rarer *Pecten.*

The Early Bathonian corresponds to the upper part of the Okrainka, Bonivurovo, and Starikov complexes as well as to the lower parts of the Ananyevka and Monakino complexes. During the process the species composition and quantity of *Mytiloceramus* essentially declined, appearing a wide expansion of species with coarse-ribbed flattened shells, and a complete disappearance of *Pleuromya, Belemnites*, mud-eating worms, and plant detritus. The generalized poorness in organic remains and the mosaic distribution of the paleobiocoenosis suggest that conditions favourable for life development existed only in small areas.

In the Late Bathonian on top of the Bonivurovo complex, the Popovka complex accumulated conformably with a gradual transition. On top, there occurs conformably the Late Jurassic.

Late Jurassic. At the beginning of the Late Jurassic, South Primorye turned into a nearshore plain with an unstable sea shoreline.

Still in the Callovian, in places where before marine sediments accumulated, a continental realm developed. In small depressions lake-swamp sediments deposited. These are now intercalating sandstones, siltstones, and claystones, including coal, pointing to a slightly arid climate, with thin seasonal banded bedding, abundant coalified plant detritus, and small hard coal lenses.

Until the second half of the Oxfordian, the South Primorye land was intensively eroded, so that the Oxfordian-Volgian deposits superpose with an erosional surface the other Mesozoic formations of different age. At the end of the Oxfordian a new transgression occurred. It penetrated only into the south-eastern part of the South Primorye, where the Chigan complex accumulated. It occurs with washout on the Middle Jurassic and more ancient deposits. Judging from the numerous pelecypods *Buchia*, at the beginning of the transgression the sea was closely connected with the Boreal basin. The remains of the ammonites *Virgatosphinctes* and *Partschiceras*, found together with *Buchia*, testify the connection of that sea with the Tethys Ocean with warm marine currents. Later the current impact increased, the fauna composition changed and its genus composition broadened. The Boreal *Buchia* became substituted by more thermophile myarians, as well as numerous ammonites.

The depths of the South Primorye Late Jurassic Sea were not great: at the base of the Chiganov complex, there occur conglomerates. Their pebbles are composed of quartz, cherts, sandstones, siltstones, and schistose claystones. The siltstones are analogous with the Triassic ones, and the claystones to the Late Permian ones, found nearby.

Upwards the conglomerates change into coarse- and medium-grained sandstones with wave-like cross lamination of nearshore type, more coarse-grained sandstone interbeds, fragments of plant trunks and stems, and small coal lenses. At the base of the sandstones interrupted beds of coquina limestones, disorderly overfilled by the *Tellina* sp. shells, are found. Much later, monotonous dark-gray fine-grained sandstones with fucoids, the tracks of mud-eating worms dominating among the bottom fauna, and with coarse plant trunks and stems and coal lenses, sometimes situated across the bedding, accumulated. The sandstones deposited in a shallow-water quiet basin, where temporary streams brought many plant remains.

During the fucoid sandstone accumulation, abundant *Buchia* lived and fossil ammonites and belemnites occur. Such diverse fauna shows that the Late Jurassic basin had a connection with the Boreal and Japanese seas. The coalified claystones with abundant plant detritus witness a regression of the sea, to lagoonal conditions and vicinity of the shore, covered by diverse vegetation.

The Late Jurassic transgression has been completed by the accumulation of different-grained sands, containing numerous rests of the diverse fauna and abundant coarse plant detritus. At that time, the sea was shallow with banks of pelecypods and brachiopods, settling in the mass of the sandy bottom. At the end of the Late Jurassic, the sea extensively regressed and then abandoned the South Primorye, where later in the Early Cretaceous, a continental regime developed, so that most of the lower Cretaceous horizons occur transgressively over an erosion surface.

In the Tithonian (Fig. 12.6), the mode of oceanic plate subduction changed to the mode of transform sliding of the latter northward, while along the eastern edge of the Eurasian continent the thick turbidites of the Zhuravlyovka terrane began to accumulate. Simultaneously in the southern part of the South Eurasian continent a new subduction zone developed, and the formation of the Late Jurassic-Early Cretaceous accretionary prism started, a fragment of which is represented in the structure of the Sikhote-Alin by the Taukha terrane. During the Tithonian-Early Hauterivian, fragments were consecutively accreted of the Late Devonian-Jurassic paleogyots and the sedimentary cover of the adjacent areas of the oceanic plate.

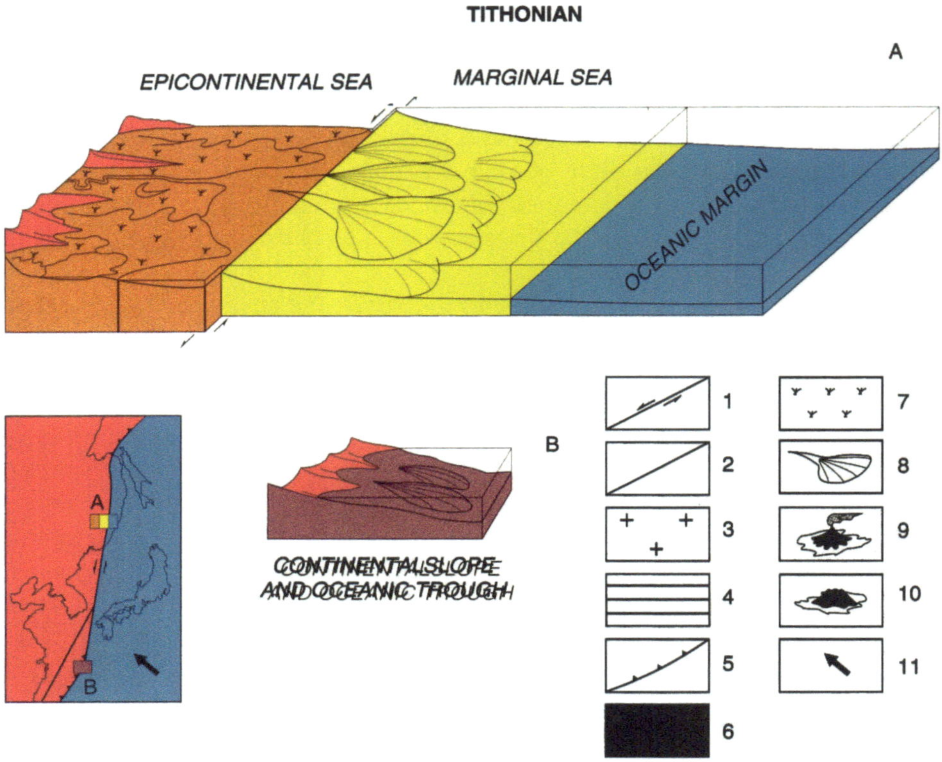

Fig. 12.6: Block diagrams of the sedimentary conditions and geodynamic scheme for the Tithonian. Designations: 1 – left-side (sinistral) strike-slip; 2 – other ruptures; 3 – continental crust; 4 – oceanic crust; 5 – erosion area; 6 – epicontinental sea; 7 – seaside sedimentary plain; 8 – submerged fan; 9 – active volcanic island arc system; 10 – extinct volcanic island arc system joint to continent; 11 – direction of plate movement.

12.5 CRETACEOUS

During the Early Cretaceous, the Izanagi oceanic plate moved to the north and north-west in relation to the not mobile Eurasia with a considerable rate (20 cm/year; Figs. 12.7, 12.8, and 12.9; Engebretson et al., 1985). In connection with this, at a significant distance of the East Asia continent between the latitudes 30°N and 55°N, the transform-sliding environment dominated. During the process, the Tau-Lu continental margin left-side strike-slip system developed up to 800 km deep into the continental margin, was formed (Xu, 1993;

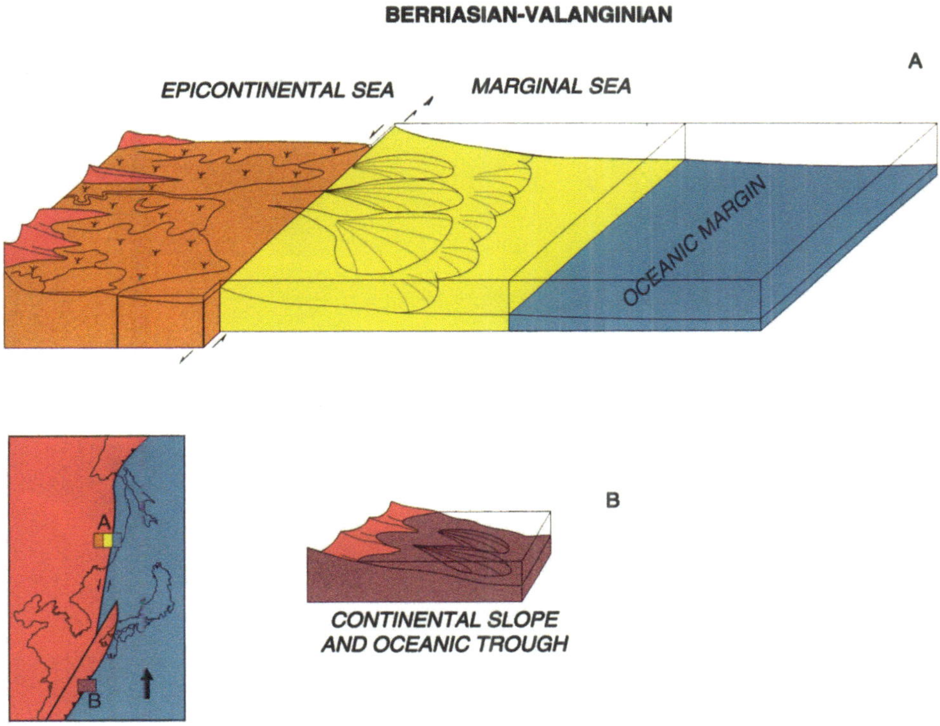

Fig. 12.7: Block diagrams of the sedimentary conditions and geodynamic scheme for the Berriasian-Valanginian. Designations: see Fig. 12.6.

Fig. 12.10), with strong movement, along which essentially broke the primary structure of the sedimentary basin, the active margin and the cratonic fragment distribution. On the background and as a result of these movements the continental margin syn-strike-slip and pull-apart basins were formed in the continental part of this margin. The movements were accompanied by intensive deformation of the teared-off continental margin fragments, their approaching to each other, with melting of big volumes of granitic magmas and as a result, the Sikhote-Alin orogenic belt was formed (Golozubov, 2004; Fig. 12.11).

The southern part of this orogenic belt includes fragments of the Middle-Late Jurassic and Late-Jurassic-Cretaceous accretionary prisms (Samarka and Taukha terranes) of the Early Cretaceous marginal syn-strike-slip turbidity basin (Zhuravlyovka terrane) and the Barremian back-arc basin (Kema terrane).

The *Taukha terrane* (Fig. 12.1) consists mainly of Berriasian-Valanginian terrigenous rocks: arkosic sandstones, siltstones, and mixed deposits with a silty-clayey matrix, containing Upper Paleozoic and Early Mesozoic fragments,

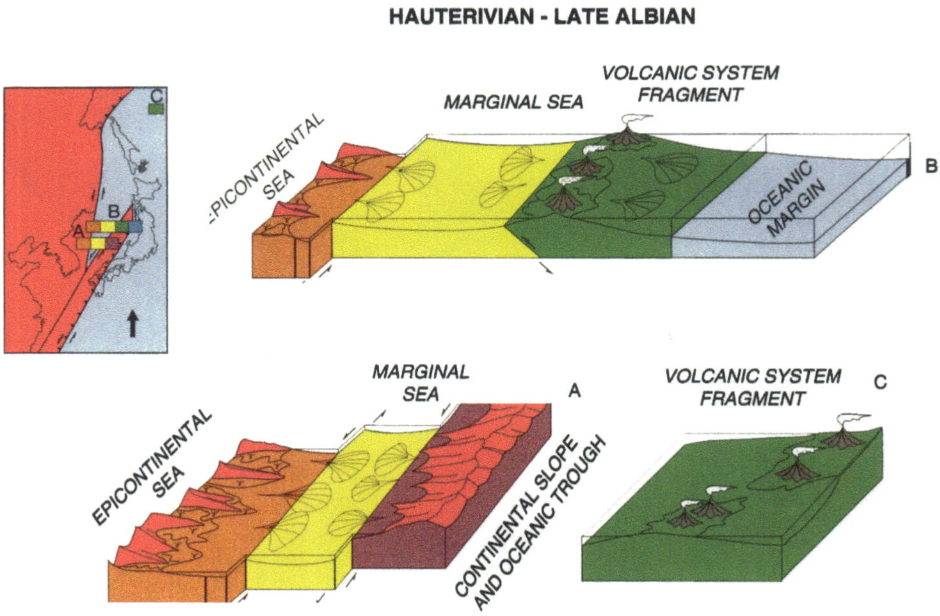

Fig. 12.8: Block diagrams of the sedimentary conditions and geodynamic scheme for the Hauterivian-Late Albian. Designations: see Fig. 12.6.

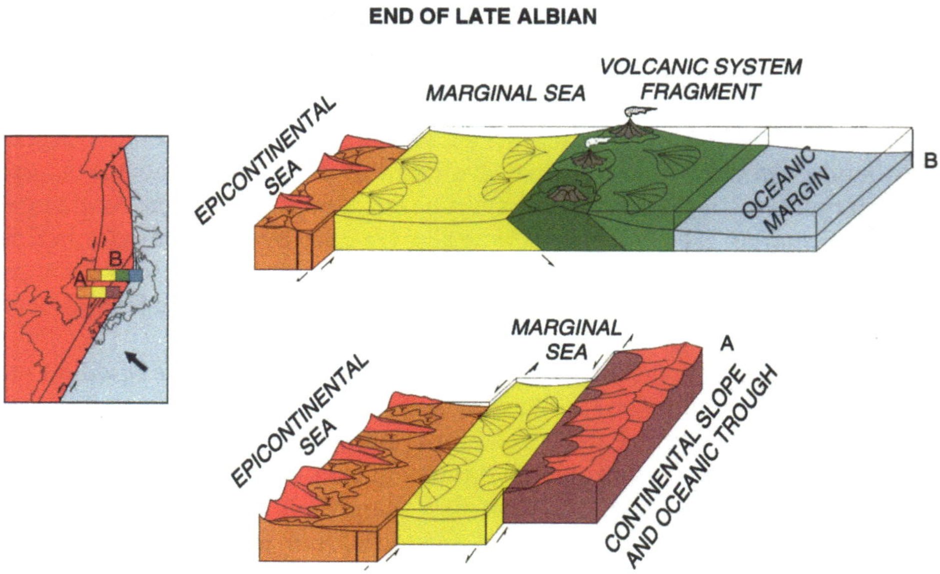

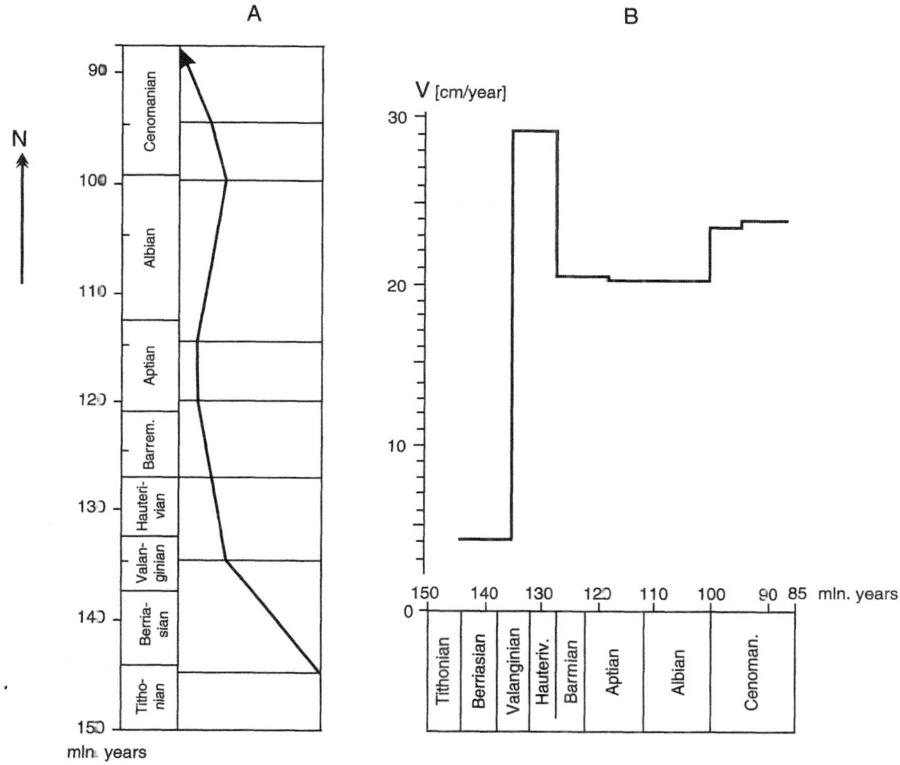

Fig. 12.10: Movement direction (A) and rates (B) of the adjoining Eurasia parts of the Izanagi oceanic plate during Late Jurassic and Cretaceous (after Engebretson et al., 1985).

blocks, and slices of chert, basalt, limestone, and sandstone. At some structural levels, the sandstone-silty claystone alternation has a clear rhythmical aspect. The tectonic-stratigraphical sequence of the terrane assemblages is about 13,000 km thick.

The *Zhravlyovka terrane* is constituted by arkosic sandstones and silty claystones, deposited continuously during the whole Cretaceous, from Berriasian to Late Albian, with a rate of 250–500 mm/Ma. On some levels of the sequence, there occur horizons of a well-expressed flysch. The sedimentary prism disintegrates into a number of large rhythms with a thickness of 1.5–2.5 km, and the total thickness is about 14 km. At the base of the terrane the cherts occur, containing Late Jurassic radiolarians and representing probably, a fragment of the upper part of the oceanic crust (Golozubov et al., 1992; Golozubov and Khanchuk, 1995).

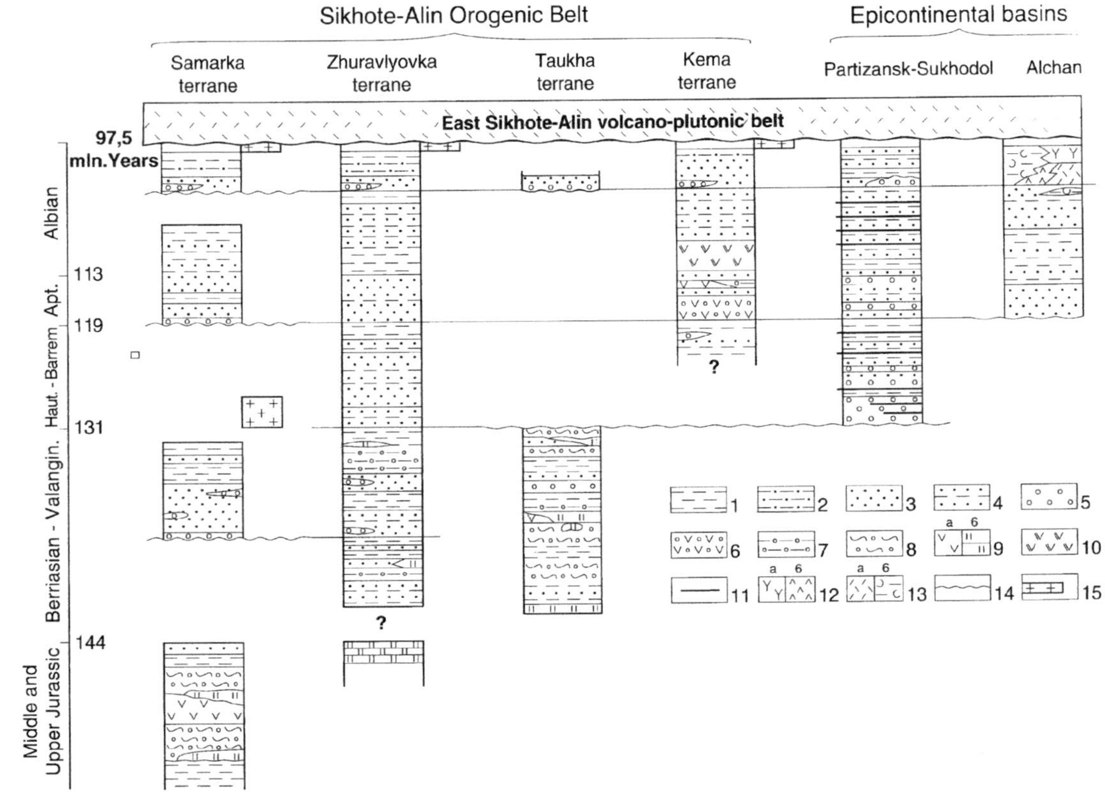

Fig. 12.11: Jurassic and Cretaceous lithological and stratigraphical columns of the Sikhote-Alin orogenic belt and Khanka massif eastern margin. Designations: 1 – siltstones and silty claystones; 2 – sandy siltstones; 3 – sandstones; 4 – sandstone-siltstone alternation; 5 – gravelstones and conglomerates; 6 – volcaniclastic gravelstones; 7 – siltstones with sandstone lenses; 8 – mixed clastics with siltstone matrix, blocks, and flakes of chert, basalt, and sandstone; 9 – Early and Late Paleozoic basalts (a) and chert (b); 10 – Early Cretaceous basalts; 11 – hard coal horizons; 12 – volcanic rocks: intermediate (a) and acid (b); 13 – volcanic rocks: acid (a), tuffs and tuffites (b); 14 – angular unconformity; 15 – age levels of the granitoid penetration.

The *Kema terrane* is composed mainly of Barremian-Albian flysch with different ratios of sandstones (sometimes associated with gravelstones and conglomerates), siltstones, and claystones. In the pebble composition basalt prevails, and the heavy mineral assemblage is represented by minerals characteristic for volcanic island systems. In the middle part of the 6 km sequence, 800 m of basalts and their tuffs have been determined. According to geochemical peculiarities, these basalts correspond to back-arc ones, which were formed at the closing stage of the island arc system (Malinovsky et al., 2002; Simanenko et al., 2004).

In the *Samarka terrane*, a fragments of the Jurassic accretionary prism, the Early Cretaceous marine nearshore terrigenous deposits are local, syn- and post-accretionary, and form three members, deposited during the Valanginian, Aptian-Early Albian, and Middle and Late Albian, respectively. These members form macrorhythms at which base the sandstones (often with basal conglomerates), and in the upper parts of the sequence, silty claystones predominate. At the base of the macrorhythms, there occur signs of erosion and differently expressed angular unconformities.

The Early Cretaceous epicontinental basins have been considered by the examples of the Partizansk-Sukhodol and Alchan basins of the Primorye. The deposits of the *Partizansk-Sukhodol Basin* are situated in the Hauterivian-Albian part of the sequence and they consist of non-marine, essentially nearshore, gray-coloured terrigenous sediments, and coal. The total sequence thickness reaches 2 km, and it is rhythmically constructed. The big rhythms (400–600 m), at the base of which conglomerates and sandstones occur, and at their top silty claystones, coaly claystones, and coal, disintegrate to the rhythm series of the smallest (up to 100 m) thickness. *Alchan Basin* deposits consist of Aptian-Albian sediments and volcanic rocks, laying on the denudation surface of the Khanka massif pre-Mesozoic and Early Mesozoic rocks. The Middle-Late Albian part of the sequence consists of lavas, tuffs and tuffites of intermediate, moderately acid, and acid composition.

Judging from the climatic zonation of the Early Cretaceous flora, the fragments of the active continental margin (accretionary prisms and back-arc basins) were formed considerably farther south than their recent position. The difference constitutes 5–25° latitude (Kimura, 1987; Golozubov et al., 1999; Golozubov, 2004). The interpretation of the margin reconstruction for the Early Cretaceous is shown in Figs. 12.12 and 12.13. The movements along the strike-slip Tan-Lu system began probably, in the Valanginian, and they were most intensive during the Hauterivian-Albian (Chen, 1993; Golozubov and Khanchuk, 1995). Below, the main peculiarities of the Early Cretaceous sedimentation and volcanic processes have been considered for the different geological periods.

In the *Valanginian* in the Samarka terrane the local basins, filled in by terrigenous nearshore-marine sediments with a thick horizon of basal conglomerates,

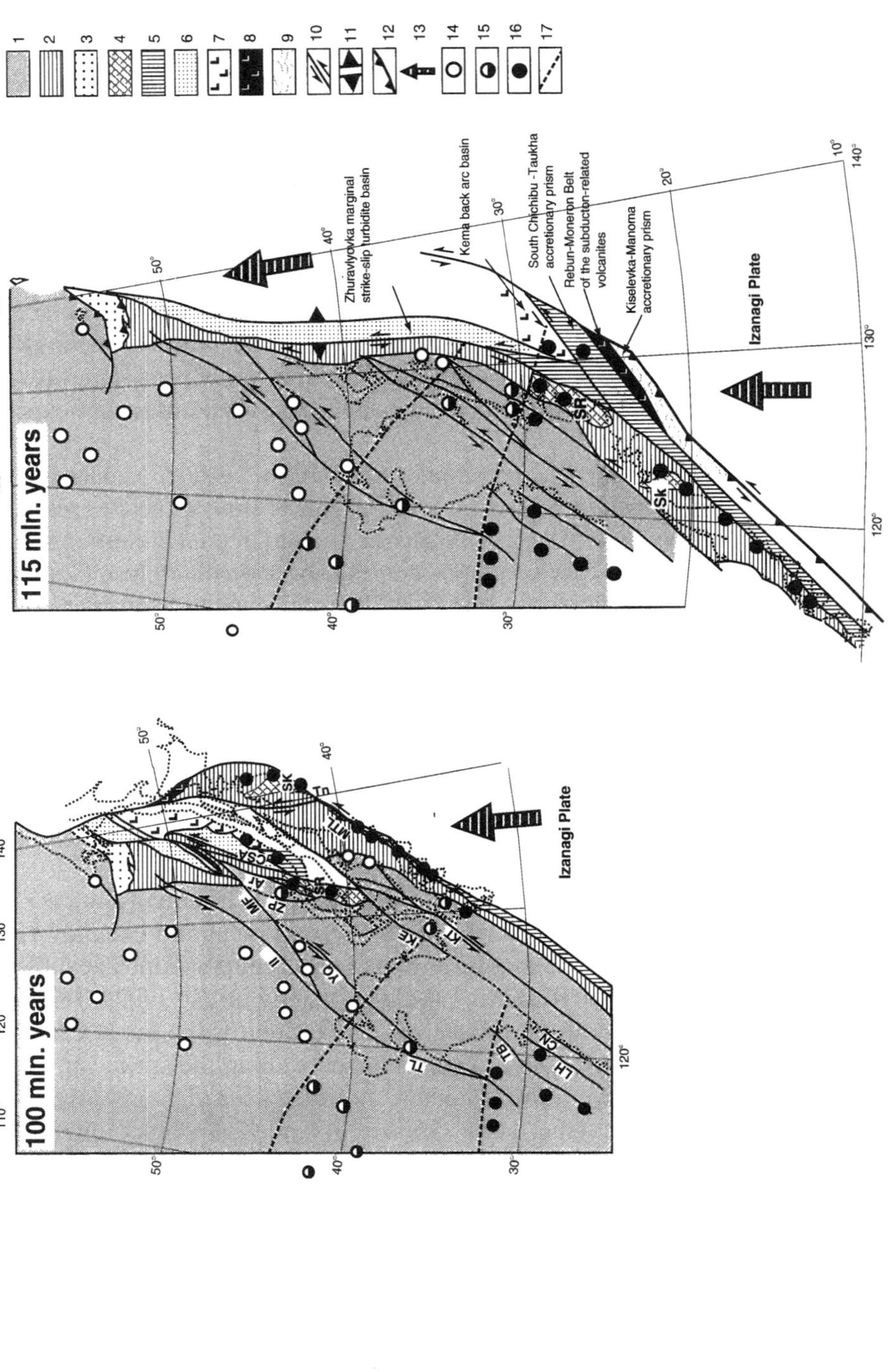

100 mln. years
115 mln. years
Izanagi Plate
Izanagi Plate
Zhuravlyovka marginal strike-slip turbidite basin
Kema back arc basin
South Chichibu -Taukha accretionary prism
Rebun-Moneron Belt of the subducton-related volcanites
Kiselevka-Manoma accretionary prism

overlap with an erosional and angular unconformity the dislocated assemblages of the Jurassic accretionary prism. These basins fix the end of the early stage of fold dislocation, happened in connection with the strike-slip movements (Utkin, 1980, 1989). In the Zhuravlyovka marginal syn-strike-slip basin, the sedimentary cycle accumulated during the Valanginian, at the base of which the sandstones with horizons of conglomerates, and at the top siltstones and silty claystones dominate.

During the *Hauterivian* in the Samarka terrane, early post-accretionary granitoids (Khungary suite) penetrated (Geological Map, 1996). In the continental part of the margin a series of basins of syn-strike-slip extension or contraction developed (Lee and Paik, 1990; Golozubov and Li Dong, 1997; Lee, 1999). In the Zhuravlyovka terrane, this deformation stage has been fixed by an abrupt changing of the sedimentary regime: in the sequence, the role of sandstones abruptly increases and horizons of flysch appear. Especially in the Hauterivian the formation of the Early Cretaceous accretionary prism (Taukha terrane) finished. This period dates one of the early stages of the large-scale movements along the Tan-Lu strike-slip system (Chen, 1993).

In the *Aptian* and *Early Albian* in the Samarka terrane locally new-formed basins, filled by marine nearshore terrigenous deposits developed. In some areas, the Aptian-Albian deposits rest with an erosional and angular unconformity on the folded Valanginian formations. In the Zhuravlyovka marginal syn-strike-slip basin during the Aptian-Albian the next sedimentary cycle accumulated. At its base the sandstones, and at its top the siltstones dominate. In the Kema terrane, a back-arc basalt volcanic activity occurred at that time. In the continental parts of the margins, new-formed syn-strike-slip tensional basins were formed (Alchan

---

Fig. 12.12: Geodynamic reconstruction of the Asian Eastern margin. Designations: 1 – Pre-Jurassic continent; 2, 3 – Jurassic terranes: fragments of accretionary prisms (2) and near-continental strike-slip turbidite basins (3); 4 – fragments of Pre-Mesozoic continent in accretionary prisms: OS – Okrainka-Sergeevka, AK – Abakuma and South Kitakami; 5–9 – Early Cretaceous terrane-fragments of Neocomian accretionary prism (5), strike-slip turbidite basin (6), back-arc basin (7), fore-arc basin (8), Aptian-Albian accretionary prism (9); 10 – left-lateral strike-slip faults of Tan-Lu system, including: II – Ilan-Itun, MF – Mishan-Fushung, AR – Arsenyev, CSA – Central Sikhote-Alin, WP – Western Primorye, TN – Tanakura, YK – Yalu Yang-Quindao, MTL – Median tectonic line, TL – Tan-Lu, KE – Kondju-Yondong, KT – Korea-Taiwan, CHN – Changle-Nanao; 11 – syn-strike-slip divergent zones; 12 – subduction zones; 13 – movement direction of Izanagi plate; 14–16 – flora assemblages: Tetori (14), mixed (15), Rioseki (16); 17 – boundaries of climate zones.

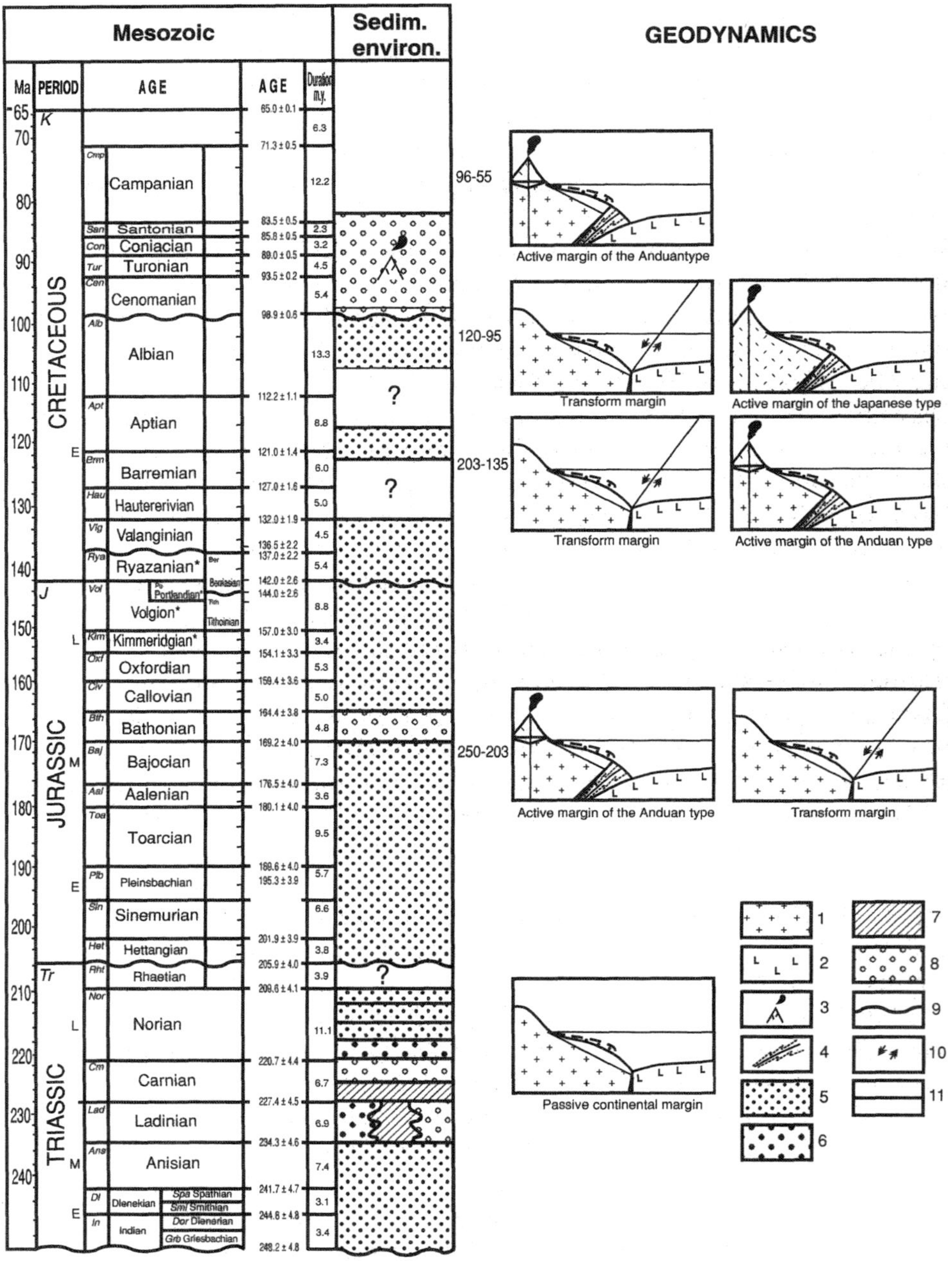

Fig. 12.13: Sedimentation and geodynamic evolution during the Mesozoic (geodynamic settings after Khanchuk and Kemkin, 2003). Designations: 1 – continental plate; 2 – oceanic plate; 3 – volcanic arc; 4 – accretionary prism; 5–8 – sediments: littoral (5), coastal (6), lagoonal (7), non-marine (8); 9 – unconformity; 10 – strike-slip; 11 – other fractures.

Basin), and in the earlier formed basin this stage is determined by the accumulation of the next sedimentary cycle, at the base of which dominate sandstones and conglomerates and at the top siltstones, coaly claystones, and coal.

During the *Middle and Late Albian* in the Samarka and Taukha terranes, the deposits of local basins filled with nearshore marine terrigenous sediments (mainly sandstones and at the base, basal conglomerates) occur with an erosional and angular unconformity on top of the ancient assemblages. In the Zhravlyovska Basin at this time, the next sedimentary cycle deposited, at its base of sandstones with gravelstones and conglomerates, and at the top of siltstones and claystones.

The *Late Albian* is the time of completion of the fold and strike-slip deformation. The intrusion of considerable volumes of granitoid magmas completed the process of the Sikhote-Alin orogenic belt formation.

In the Late Cretaceous, probably, in connection with the changing of movement direction of the Izanagi plate from the south to the north-west (Fig. 12.1), subduction renewed practically on the entire extent of the East Asian margin, and a belt of post-accretionary volcanics developed.

By this means, in the framework of a transform sliding setting, predominating in the margin segment between 30°N and 550°N latitude, in the time interval of 135–100 Ma, the episodes ("impulses") of the strike-slip movements are reflected by the sedimentary cycle accumulation (Zhuravlyovka terrane), or by the formation of strike-slip stretched sedimentary basins, or by volcanic activity manifestations. These impulses (during the Valanginian, Hauterivian, Aptian, and Middle Albian) as a whole, correlate with local changes of the rate and direction of the adjoining Eurasia and oceanic plate movements.

During the Early Cretaceous (Markevich et al., 2000) two types of interaction between the Eurasian continental and Paleopacific oceanic plates have been identified in Sikhote-Alin. The first type of interaction was oceanic plate subduction under the continental plate or volcanic island arc. The second was plate sliding relative to each other, in the case of oblique or parallel sliding relative to the continental border.

In the Middle-Late Jurassic, directly preceding the Early Cretaceous events, along the considerable segment of the East Asia edge, subduction of Andean type directly under the continent took place. The oceanic plate then moved to the north-west. As some part of this plate approached to the deep-water trough, the oceanic sediments were covered by clayey and clastic matter derived from the continent. On the inner slope of the trough these sediments were stripped from the subducting part of the oceanic plate, as well as the submerged rises and mountain fragments. Continental and clastic material diluted the stripped and deformed fragments of the oceanic plate's upper part, and as a result chaotic assemblages were formed, working up "the face" of the so-called accretionary prisms.

Early Cretaceous sedimentation was also controlled by the oceanic plate sliding to the north-east relative to the not mobile Eurasia, produced by oblique subduction. As a result a giant system of marginal continental left-side strike-slips Tan Lu, oriented to the north-northeast, developed. The changing of the geodynamic regimes was connected with the direction change of the oceanic plate movements from north-west to south.

In the *Berriasian-Valanginian* (Fig. 12.7) on the Sikhote-Alin extreme west on the plaeocontinental border, local non-marine deposits have been identified, changing to the east into marine shelf deposits. During that time a rather gentle continental slope existed, which position was controlled by the in the Early Cretaceous active, large strike-slip Central Sikhote-Alin fracture. Eastward 4 km thick clastic deposits accumulated at the foot of the slope. At the same time south-westward of the segment of continental border, an accretionary prism formed due to subduction. A fragment of this segment, moved to the north-east along the Tan-Lu strike-slip system, became fixed in the south-eastern Sikhote-Alin.

In the *Hauterivian* and *Barremian* (Fig. 12.8), the paleogeography strongly changed. In the continental part of the Sikhote-Alin non-marine and marine environments existed as before. Along some fractures of the Tan-Lu system, syn-strike-slip, partly coal-bearing basins formed. At that time, subduction occurred only in the Udyl Lake area of the northern Sikhote-Alin. In the Taukha block marine sedimentation completely ended, while it became represented by a big island or peninsula, partially separating the western part of the ocean as a marginal sea in which thick sediments accumulated, derived in the west from the continent and in the east from the Taukha land. The sedimentation rate was not less than 500 m/Ma as it is presently near the mouth of the biggest rivers, Amur, Indus, Ganges, and Amazon. The high sedimentation rate compensated the sea-floor subsidence, and in such a manner, preserved here constantly a rather shallow-water environment, closely resembling different shelf conditions. This supposition is confirmed by the remains of fauna, inhabiting a not deeper than littoral and sublittoral environment, including shell banks. The remaining part of the basin in the north, opened into the ocean as before without difficulty. At the same time in the open ocean at some distance apart from the continent, the formation of the epioceanic volcanic Udyl island-arc system started. This system, together with the oceanic plate, moved nearer to the continent.

In the *Hauterivian* inside the oceanic plate, northward sliding transform (to the east of the Zhuravlyovka-Amur turbidite basin), lay the Kema (Moneron-Samarga) volcanic island arc and associated Hauterivian-Albian accretionary prism (Kiselyovka-Manoma terrane). The accreted fragments of that prism are represented by cherty-terrigenous successions of the paleo-oecanic plate sedimentary cover, containing in the cherty part of the sequence alkaline volcanics and associated limestones.

In the *Aptian-Albian* in the south of the Sikhote-Alin, the environments remained without essential changes. In the southern part of the marginal sea, bordered in the west by the continent and in the east by the Taukha block, at the same time the deposition of clastic sediments, lacking essentially an admixture of volcanic material, continued. On the continental border the development of the ancient basins continued, and new syn-strike-slip basins were formed (coal-bearing Partizansk, Razdolnoe in the southern Primorye, and Alchan in the north).

The Udyl volcanic island arc system moved close to the continent and at the end of the Early Cretaceous joined it. At the same time in connection with movements along the Tan-Lu strike-slip system, one more (Moneron-Samarga) island arc system developed that advanced towards the ocean volcanic island chain and back-arc marginal sea. The frontal part of this arc is reconstructed on the Sakhalin, Moneron and Hokkaido islands, and the fragments of the back part, situated in the marginal sea near the volcanic islands, have been fixed in the east of the Sikhote-Alin, in the Kema, Samarga, and Kabanya river basins.

At the end of the Early and the beginning of the Late Cretaceous, in the Albian and Early Cenomanian in connection with the subduction and continuing strike-slip movements, the Cretaceous deposits accumulated in different parts of the continental border, were intensively deformed and rumpled into narrow folds, fragments of which can be observed among the numerous newly formed left-side strike-slips of north-eastern trend. The thickness of the sedimentary cover, intruded at that time by numerous granites, considerably increased, and at the sea border a new continental crust, the hard nucleus of the Sikhote-Alin mountain system, was formed, on which in the Late Cretaceous the East Sikhote-Alin volcanic belt was formed, while the movements along the submeridional strike-slips of the Tan-Lu system continued. In the course of the reconstruction of the position of the blocks, moved from south to north by Early Cretaceous (mainly Albian) movements along the Tan-Lu system fractures, it was taken into account that these blocks constantly joined the eastern Asian border. This suggestion is confirmed by the arkosic (granitic-metamorphic) composition of the clastic fragments of the Early Cretaceous and more ancient, Triassic and Jurassic, sandstones of this region. The reconstruction in Fig. 12.12 shows that blocks of the Jurassic and Early Cretaceous rocks of Outer Japan and south-eastern Sikhote-Alin were moved no less than 15–20° latitude from the place, where they were formed (up to 15–30°N). Blocks of Inner Japan and some of the Southern Primorye were also moved in the same direction, but not more than by 5°N latitude.

The distribution of the different types of the Early Cretaceous flora allows exactly to reconstruct the configuration of the East Asia edge during the Early Cretaceous (mainly in Albian time). The paleomagnetic data and the distribution of the Early Cretaceous fauna in this regard are vaguer, but as a whole they do not contradict the here presented model.

In the *Cenomanian,* after the structural reformation of the continental edge of South Asia, stipulated by intensification of the sinistral strike-slip dislocations along the Tan-Lu fracture system, along the eastern edge of the South-Eurasia continent, a new (Late Cretaceous) subduction zone developed. Beginning from that time, an active continental edge of Andean type was renewed. During the Late Cretaceous and Paleocene a lateral row: epicontinental volcanic arc (Eastern Sikhote-Alin belt) – fore-arc basin (Western Sakhalin and Sorachi-Edzo terranes) – accretionary prism (Nabil and Eastern Hidaka terranes), and slightly southward – the Shimanto prism, was formed.

12.6 MESOZOIC SEDIMENTATION AND GEODYNAMIC SETTINGS IN THE TECTONICAL PASSIVE AND ACTIVE AREAS

There were three important geological events in the eastern active area.

The *first event* was the formation of the first accretionary prisms, representing olistostromes, consisting of olistolites of the pre-Jurassic rocks, placed in a Jurassic essentially clayey matrix. These rocks are of oceanic nature and consist of cherts, limestones, basalts, and clayey sediments. The accretionary prisms form the Samarka and Taukha terranes of the fold Sikhote-Alin.

The *second event* was the similar accretion of the Cretaceous prisms, containing pre-Cretaceous oceanic fragments, placed into a Cretaceous matrix. During this event one of the main transgressions occurred on the eastern margin of the Chinese Craton.

The *third event* was the formation of the orogenic continental margin East-Sikhote-Alin volcanic belt, bordering the eastern extremity of Asia and associated with strong magmatic and tectonic activity. Strike-slip movements also occurred along the eastern transform-type margin of the continent. At that geological time the marine epiplatformal sedimentation finished and the continental non-marine sedimentation began. This was accompanied by Late Cretaceous coal-peat accumulation expressed now in the coal-bearing sediments of the South Primorye. Volcanic products of that time were also intruded in the area, forming dikes and sills.

12.7 CONCLUSIONS

During the Mesozoic several sedimentary and geodynamic cycles can be distinguished (Fig. 12.13).

In the *Triassic* the epiplatformal sedimentation in the East Asian continental margin, on the subsided part of the Chinese Craton, continued in the marine shelf, sublittoral and partly non-marine environment.

One large transgressive-regressive cycle has been distinguished when this took place (Fig. 12.13). The transgressive part of the cycle begins with the Induan coarse clastic formations and continued approximately, up to the Middle Olenekian. It is represented by sandstones with coquina lenses and in the lower half of the strata with silty claystones with carbonaceous concretions.

In the Middle Olenekian the regressive part of the cycle began, expressed by the upper half of the silty claystone strata, and upward by the formation of fine-grained sandstones with concretions. The cycle finished with the vertical and horizontal alternation of the Ladinian, Carnian, and Norian marine nearshore and non-marine deposits.

The second large cycle, like the previous one, established during the *Jurassic* and *Early Cretaceous* (Fig. 12.13). It begins in the Lower Jurassic Hettangian with the big regional transgression, expressed by marine littoral sequences, and by the Bathonian non-marine terrigenous and volcanic formations of the Monakino series. The cycle finishes in the Early Cretaceous, represented mainly by the non-marine formations and many fewer nearshore sediments with coals. At the beginning of the Early Cretaceous, mainly non-marine deposits accumulated, and only during short time intervals, nearshore sediments.

In the newly formed intercontinental basins, which appeared along the submeridional Tan-Lu strike-slip system, non-marine, including volcano-sedimentary and volcanic formations accumulated. The epiplatformal cycle corresponds and developed, due to submeridional sinistral strike-slips on the boundary of the continental and oceanic lithospheric plates.

ACKNOWLEDGMENTS

This research was implemented under the financial support of the Russian RFBR grants: projects numbers 02-05-65222, 02-05-65326, 02-05-64038, and 04-05-64061.

13. SEDIMENTARY BASINS OF POLYCYCLIC DEVELOPMENT IN THE SOUTH-EASTERN PART OF THE SIBERIAN PLATFORM: UCHUR-MAISKAYA BASIN AND VILYUISKAYA SYNECLISE

M.V. GOROSHKO

13.1 INTRODUCTION

The Uchur-Maiskaya Basin (Fig. 13.1) is a large long-developed depression structure covering an area of about 100,000 km^2 in the south-eastern part of the Siberian Platform. In the west and the south, the basin is bounded by blocks of the Archean and Early Proterozoic crystalline basement, and in the east, by the Verkhoyansk-Kolyma orogen, which is tectonically an imbricate-thrust system in contact with the Uchur-Maiskaya Basin on the Nelkan thrust and strongly overthrusted on it. In the north and north-east, the basin's boundary runs on the axial part of the Dygdinskiy swell; the thickness of the Riphean deposits behind it is greatly reduced, and they are buried beneath a thick cover of Vendian-Cambrian assemblages.

The basin has a complex fragmental structure, which shows bulges of the pre-Riphean basement (Iduym-Khaikanskiy and Omninskiy) along with different-order troughs, volcano-plutonic zones and central-type magmatic block structures formed during Late Riphean and Mesozoic tectono-magmatic activation of the region (Ket-Kapskaya, Arbarastakhskaya, Tomptokanskaya, and other structures). A big team of geologists of the "Aerogeologiya", "Dalgeologiya", "Taezhgeologiya" enterprises, branch and academic-scientific research institutes were engaged in the study of the stratigraphy, tectonics and magmatism of the Uchur-Maiskaya Basin. The results of many-year investigations of the basin are reflected in a number of collected works and scientific papers of many authors.

The Vilyuiskaya Syneclise, covering an area of over 230,000 km^2, is located in the basins of the lower drainage areas of the Aldan and Vilyuy rivers, on the left bank of the Lena river (Skha Republic, Yakutia). In the basement of the Vilyuiskaya Syneclise occur rocks of the Archean basement; the Nyurbinskiy orogenic belt of Riphean age is also presumed to exist there.

Cyclic Development of Sedimentary Basins
J.M. Mabesocne and V.H. Neumann, editors.

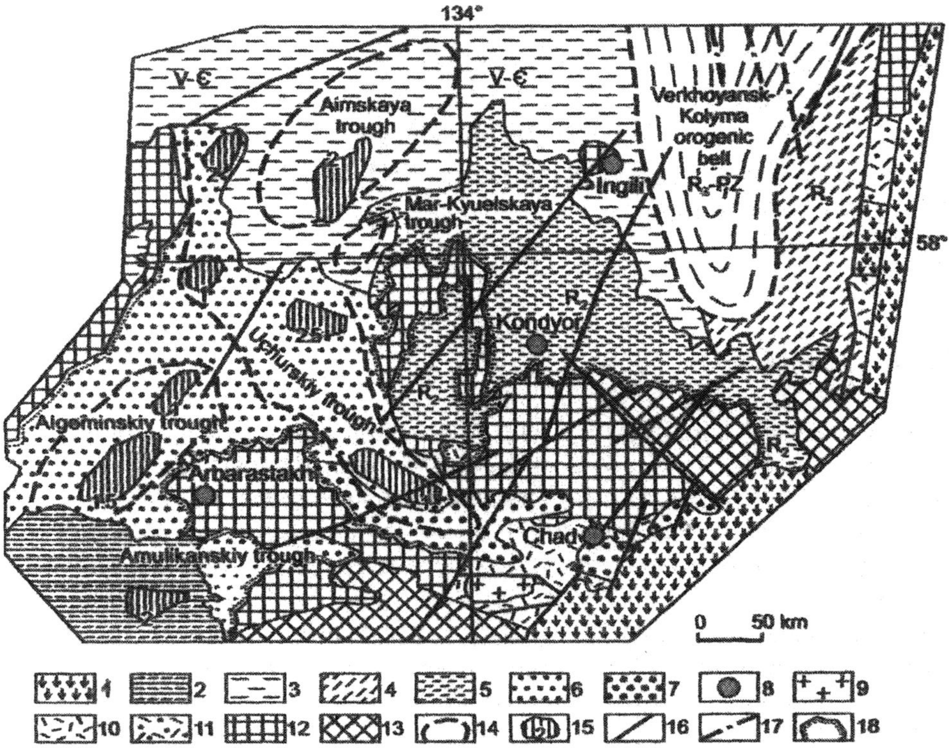

Fig. 13.1: Tectonic structure of the Uchur-Maiskaya Basin. Designations: 1 – Cretaceous volcanics of the Preddzugdzhurskiy trough; 2 – Jurassic coal-bearing deposits of the Tokinskaya Basin; 3 – terrigenous-carbonate deposits of the Vendian-Cambrian plate complex; 4 – terrigenous deposits of the Upper Riphean (Pakhandinskaya and Uiskaya series); 5 – sandstone-bituminous carbonate deposits of the Middle Riphean (Kerpylskaya and Aimchanskaya series); 6 – carbonate terrigenous deposits of the Lower Riphean (Uchurskaya series); 7 – terrigenous deposits of the Lower Riphean (Uyanskaya series); 8 – Late Riphean central-type intrusions; 9 – Early Proterozoic granitoids; 10, 11 – Lower Proterozoic sedimentary-volcanogenic assemblages: Ulkanskaya series (10) and Nelbachanskaya series (11); 12, 13 – Lower Precambrian metamorphic rocks: Lower Proterozoic (12), Archean (13); 14 – structural deposits: basins and troughs; 15 – local submergence of the platform cover (km), according to calculated depths evidence of the upper edges of magnetized bodies; 16 – faults; 17 – thrust zones in platform cover; 18 – zones of structural-stratigraphic unconformities.

13.2 GEOLOGICAL STRUCTURE OF THE SEDIMENTARY COVER OF THE UCHUR-MAISKAYA BASIN

13.2.1 RIPHEAN STRUCTURAL STAGE

The Riphean deposits of the basin have been described by Semikhatov and Serebryakov (1983) as a Siberian hypostratotype of the Riphean overlain by the faunistically characterized Paleozoic (Cambrian). At present, Karsakov et al. (2002) have substantially specified the stratigraphy of the Lower Riphean, as well as studied the relationships between the sedimentary-volcanogenic assemblages of the Upper Karelian and the Riphean, boundary of which is determined at 1670–1600 Ma.

The Uchur-Maiskaya section of the Riphean in the south-eastern part of the Siberian Platform is one of the most complete and amply paleontologically described reference sections of the Upper Proterozoic in northern Eurasia (Semikhatov et al., 2003). It includes Lower, Middle, and Upper Riphean deposits. Its distinctive feature is stratigraphic relations between the Riphean deposits and the underlying youngest pre-Riphean Upper Karelian assemblages (Lower Proterozoic) and the overlying Vendian-Cambrian deposits.

Considering new available data, the Riphean deposits of the south-eastern part of the Siberian Platform have been divided into six series: Uyanskaya, Uchurskaya, Aimchanskaya, Kerpylskaya, Lakhandinskaya, and Uiskaya (Karsakov et al., 2002).

By the Omninskiy bulge of the Archean basement, the Uchur-Maiskaya Basin is divided into two zones: Uchurskaya and Maiskaya (Goroshko et al., 1999). The Uchurskaya zone displays a wide distribution of Lower Riphean deposits and, in the Maiskaya zone, most fully represented are sections of Lower and Middle Riphean platform deposits.

The here accepted stratigraphic scale of the Lower Riphean deposits includes the Uyanskaya and Uchurskaya sedimentary series (Fig. 13.2), separated by regional hiatuses and unconformities, and corresponding to the most important stages of sedimentation. A characteristic feature of these sedimentary units is their speckled character attributed to a high hematite, kaolinite, hydromica, and chlorite content in the cement. The array of structural features of terrigenous rocks is indicative of their formation in shallow-water, coastal-marine zones under conditions of semiarid and sometimes arid climate.

The Uyanskaya series is divided into the Birindinskaya, Konkulinskaya, and Adargaiskaya complexes. The series overlies the Upper Karelian formations of the Ulkanskiy volcanogenic trough, and over large territories, it lies far over the Upper Proterozoic and Archean crystalline basement. The basal layers of the

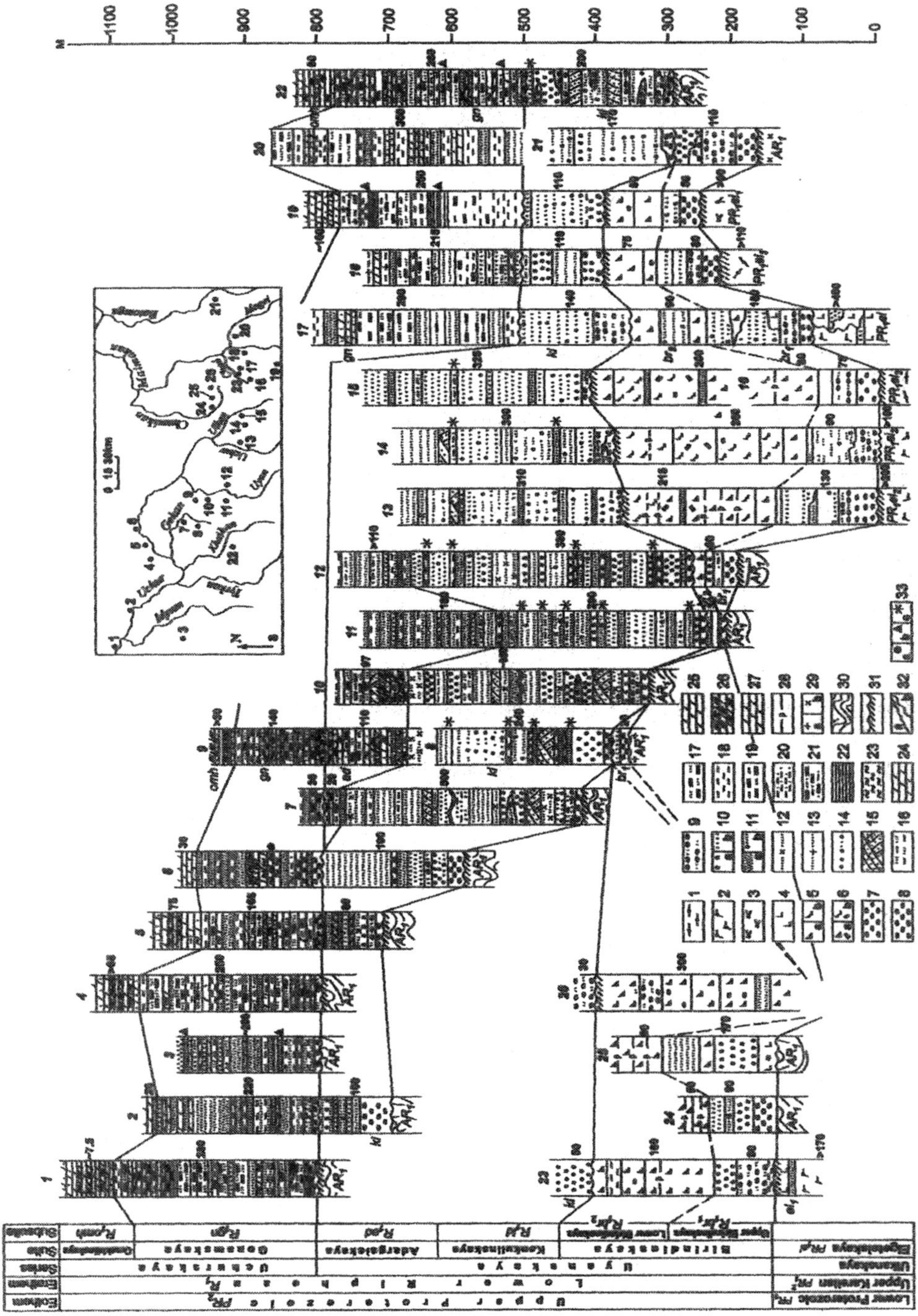

Uyanskaya series, represented by conglomerates of the Birindinskaya complex, overlie with great washout and unconformity the weathering crust of acid effusive rocks from different parts of the section of the Upper Karelian Elgeteiskaya suite in the Ulkanskiy trough (Fig. 13.1). It follows from the analysis of petrochemical nodules that this weathering crust formed under conditions of a hot and humid climate (Karsakov and Guryanov, 1990).

The lower part of the section of the Birindinskaya complex is dominated by medium- and coarse-grained feldspar-quartz and arkosic sandstones. Occasionally, there appear conglomerate intercalations up to 1–3 cm thick, and low-thickness (up to 35 m) basalt flows extending for 2–3 km. The sandstones display wave and ripple marks, and also desiccation fissures filled with basalt, being indicative of hiatuses in the sedimentation. The upper part of complex section is represented by subalkali olivine basalts, leucobasalts, sandstones, and tuffaceous siltstones. In the section the basalt flows (10–100 m thick) generally alternate with interbedded sandstones. The total thickness of the Birindinskaya complex is about 380 m. According to Karsakov and Guryanov (1990), the top of the basalt horizons shows rather thick cinder-like zones, which give evidence of surface conditions of the lava outflows. The accumulation of the Birindinkskaya sediments occurred in an environment of a shallow-water intracontinental basin

---

Fig. 13.2: Comparison of the sections of the Lower Riphean deposits in the Uchurskaya Basin. Designations: 1 – trachyrhyolites; 2 – trachyrhyodacies; 3 – trachydacites; 4 – basalts; 5, 6 – subalkali olivine basalts (5^a) and leucobasalts (5^b): magnophyric (6^a) and almond-shaped (6^b); 7 – quartz conglomerates; 8 – polymicitic conglomerates; 9 – gritstones; 10, 11 – oligomictic sandstones: coarse-grained (10^a), medium-grained (10^b), fine-grained (11^a), inequigranular (11^b); 12–17 – sandstones: quartz (12), glauconite-bearing (13), with "swimming" pebbles of quartz (14), cross-laminated (15), with siltstone and mudstone micro-interbeds and fragments (16), with dolomite cement (17); 18 – siltstones; 19 – siltstones with dolomite cement; 20 – silty sandstones; 21 – silty sandstones with dolomite cement; 22 – mudstones; 23 – opokas; 24–27 – dolomites: pelitomorphic (24), sandy (25), oolitic (26), with stromatolites (27), 28 sandstone interbeds <1 m thick; 29 – Early Archean granitoids: granites (a), quartz diorites (b); 30 – Lower Archean gneisses and crystalline schists; 31 – weathering crusts; 32 – unconformities: stratigraphic (a), angular (b), x – combination of adjacent signs indicates interlamination of rocks of different composition; 33 – pseudomorphs of gypsum (a), rock salt (b), glauconite (c). Figures to the right of the column – complex and sub-complex thickness (m); sign indications – presence in the beds of lenses of gypsum, rock salt, and glauconite pseudomorphs. Location of the sections: see inset.

surrounded by beveled land. This is suggested by the volcanogenic-fragmental composition of the complex: poor sorting of clastic material, predominance of arkoses, desiccation fissures on the bedding planes and raindrop imprints, wide distribution of coarse oblique cross bedding, wave, and ripple marks, and a quick change of the facies.

The Konkulinskaya complex is made up of dark-red, lilac, coarse-bedded feldspar-quartz and arkosic sandstones containing "swimming" pebbles of quartz, and rare gritstone, siltstone and quartz sandstone intercalations. The complex overlies Archean rocks with angular unconformity, and deposits of the Birindinskaya complex transgressively, being separated from both by weathering crusts (Mironyuk, 1986). This weathering crust at the base of the section differs from that at the base of the Birindinskaya complex section by a darker brick-red color. The thickness of the Konkulinskaya complex ranges between 80 and 100, and 380 and 500 m, reaching up to 950 m on the left bank of the Uyan river, after drilling and seismic survey data (Fig. 13.2). North-west of the Uyan river, in the direction of the Givun and Gekan rivers, the thickness of the complex decreases rather strongly, and in the mouth of the Tyrkan river the complex wedges out. The rocks of the Konkulinskaya complex have a certain arrangement of structures, stable over its whole development area. Generally, they display coarse bedding. Platy varieties are characterized by a subhorizontal, gently sinuous bedding; cross bedding is rare. The bedding planes commonly show desiccation fissures filled with different color and granularity sandstones, symmetrical ripple or wave marks with a wavelength of 0.5–2.5 cm and a wavelength of 6–15 cm. The low titanium content is indicative for the formation of the Konkulinskaya sandstones at the expense of weathering crust material. The pink, red, and violet colors of the sandstones are attributed to the dominance of oxidic iron in their cement, attesting a hot climate and an oxidation environment, in a shallow-water weak-salty coastal basin with a considerable supply of clastic material by rivers.

The paleotectonic and paleogeographic settings of the formation of the Birindinskaya and Konkulinskaya complexes are identical.

The Adargaiskaya complex has been distinguished here for the first time. It comprises a unit of speckled carbonate-terrigenous rocks, that in the basins of Adargai brook and Munaly river conformably overlies the red-colored Konkulinskaya sandstones, and that is overlain with washout (Munaly river) or with an angular unconformity (Gekan river) by the red-colored sandstones and conglomerates of the Gonamskaya complex's Uchurskaya series. The thickness of the Adargaiskaya complex is 110–180 m. The lower strata of the Uyanskaya series contain dikes and sills of alkali basaltoids and trachybasalts of the Garyndinskiy complex. The Riphean age of the dikes is demonstrated by superposition on them, of hydrothermal and metasomatic alterations with uranium mineralization, whose U–Pb and Pb–Pb age is about 1325–1300 Ma. Besides the outcrop of the assemblages of the Uyanskaya series at the surface in the south-eastern part of the region, the results of the quantitative interpretation of gravimetric and magnetic survey data suggest protoplatform basins and troughs filled with rocks of the Uyanskaya series overlain by younger deposits in the north-west of the territory, in the basins of the Yurakhite, Aim, Chagdala-Metropolskiy, and Akgoma rivers (Karsakov et al., 1977; Malyshev, 1977; Kosygin et al., 1984). The size of these structures range from 10 km × 30 km to 15 km × 100 km. Their position is in agreement with the trend of the Uchurskaya trough. Taking into account the fact that the thickness of the overlying rocks of the Uchurskaya series, and occasionally of the Vendian-Cambrian does not exceed 100–1500 m, the thickness of the rocks of the Uyanskaya series in the concealed protoplatform troughs and basin may reach up to 1.5–2.5 km (Fig. 13.1).

The Uchurskaya stage of development of the region (figure on front cover) was marked by the development of a vast flat-bottomed basin in the south-western part filled in with shallow-water and extensive littoral sediments of the Gonamskaya, Omakhtinskaya, and Enninskaya complexes.

The rocks of the Gonamskaya complex overlie unconformably deposits of the Adargaiskaya or Konkulinskaya complexes; in places where the latter are absent, they rest directly upon the Archean crystalline basement. A speciality of the rocks of this complex is their intense red coloring, a nearly equal ratio of sandstones (55%), silty sandstones and siltstones (40%); carbonate rocks are subordinate (5%). Besides, in the lower strata of the section of the Gonamskaya complex, in the basin of Berezovyy brook, ditches cutstripped sheet bodies of keratophyres 0.5–6.0 m thick. In the middle reaches of the Mal. Tyrkan river, on the Mal. Tyrkan-Chistaya interfluve, the upper part of the section of the complex (after evidence of the author) contains tuffite and red-colored psammitic tuffite sheets of trachy-rhyodacitic composition. The thickness of the Gonamskaya complex is about 600 m. Its age based on glauconite is 1520–1450 Ma.

The Omakhtinskaya complex overlies conformably the rocks of the foregoing complex. Its lower contact within the Uchurskaya zone is determined by a horizon of stromatolithic and oolitic dolomites and calcareous sandstones. The complex is made up of light-gray, cream quartz sandstones, and less often of quartz-feldspar and calcareous sandstones which are intercalated with stromatolithic, oolitic, and massive dolomites and marls. The thickness of the complex is 250–300 m.

The Enninskaya complex occurs with a small angular unconformity on the rocks of the Omakhtinskaya complex. Its base exhibits lilac-red and reddish-gray feldspar-quartz sandstones, and less often arkosic sandstones with frequent intercalations of gray, pink-gray quartz sandstones. Occasionally, dolomite interbeds of low thickness occur. The thickness of this sequence is 40–50 m. The sequence is overlain by light-gray and pink-gray quartz and quartz-like sandstones with rare dolomitic intercalations of low thickness (0.1–0.3 m). Its total thickness is 80–90 m. The section of the Enninskaya complex is topped by a sequence of light-gray, pink-gray quartz sandstones with frequent interbeds of low thickness containing greenish-gray and dark gray siltstones and calcareous mudstones. The thickness of this sequence makes 30–40 m. The maximum thickness of the Enninskaya rocks in the Uchurskiy trough does not exceed 180 m.

The rocks of the Uchurskaya series are transgressively cut from east to west by Vendian deposits. These latter occur directly on the Archean basement on the left bank of the Uchur river and on the Gonamskiy watershed. The rocks of the Uchurskaya series in the eastern part of the Aldan shield are characterized by a variability of thickness and facies composition of the deposits. The maximum thickness (up to 600 m) of the series is observed in the central zone of the Algominskiy trough and in the Uchurskiy trough. On the western slope of the Central Aldan uplift and on the slopes of the Idyum-Khaikanskiy bulge of the basement, the thickness of the complex is reduced to 300 m, whereas the carbonate rocks of the Omakhtinskaya complex are almost completely replaced by sandstones. The rocks of the Uchurskaya series formed under other conditions than those of the Uyanskaya series, which is indicated by pseudomorphs of rock salt, glauconite grains, thinner-bedded intercalations of terrigenous and carbonate rocks, the finer-grained character of fragmental material, the more significant role of siltstones and mudstones in its section as compared with the Adargaiskaya complex, and the more significant role of dolomite in the cement. All this is indicative of a sediment accumulation under marine conditions accompanied by the general subsidence of the basin. Considerable amounts of glauconite demonstrate its accumulation in a colder than in Uyanskaya time, basin, 30–1000 m deep, under conditions of a weakly reducing environment.

In the southern part of the Uchurskaya zone, the base of the Riphean deposits dips north-west at an angle of 3–5°. A similar dip, however, in a south-western direction, is noted for the north-eastern limb of the trough adjoining the Omninskiy bulge of the Archean crystalline basement. Dolerite and gabbro dolerite dikes of the Sivaglinskiy suite are common among the rocks of the Uchurskaya series in the basins of the Gonam and Algama rivers. The dikes are generally NW-striking, and the occurrence is almost horizontal. The K–Ar absolute age of the dike bodies is 1350 ± 40 Ma.

In the Maiskaya zone, Lower, Middle and Upper Riphean deposits are distinguished. With respect to the compositional features of the section and stromatolithic characteristics, the Riphean deposits of the trough are divided into five series: Bilyakchanskaya, Aimchanskaya, Kerpylskaya, Lakhandinskaya, and Uiskaya.

The Bilyakchanskaya series was introduced in the stratigraphic scale of the Riphean in 2000, by decision of the Far East Interdepartmental Commission, and was correlated with the Uchurskaya series of the Lower Riphean. It is distinguished in the extreme south-east of the territory, in the Early Proterozoic Bilyakchanskaya volcanogenic trough, and is represented by sandstones, siltstones, dolomites, mudstones, limestones, gritstones, conglomerates, cherty rocks, and quartz-chlorite-sericite schists. The thickness of the series is 500–2000 m.

The Middle and Upper Riphean deposits are represented by marine, lagoonal-marine, and less often by continental rocks: dolomites, limestones, marls, sandstones, siltstones, and schists.

The Aimchanskaya and Kerpylskaya series of sedimentary rocks are referred to the Middle Riphean (Fig. 13.3). Prior to the accumulation of the sediments of the Aimchanskaya series, the territory of the Uchur-Maiskaya Basin experienced a general uplift accompanied by active washout of the subjacent rock units. The Middle Riphean deposits of the Aimchanskaya series occur on the northern and north-eastern slopes of the Batomgskiy uplift, and on the eastern slope of the Omninskiy uplift, being a large transgressive sedimentary rhythm consisting of two complexes: the terrigenous Talynskaya and the overlying terrigenous-carbonate Svetlinskaya complexes. The basal horizons of the Talynskaya complex include light-gray, medium- and coarse-grained quartz sandstones, occasionally poorly glauconitic sandstones holding lenses of quartz gritstones, sheets of ferruginous sandstones, and in the upper part interbedded siltstones (Semikhatov and Serebryakov, 1983). These are covered by dark siltstones, sandstones of quartz and quartz-feldspar compositions, and rare mudstones. All the above-referred rocks display a wide range of shallow water structures, among which desiccation fissures are most common in the sheets of sandstone-siltstone alternation. The thickness of the Talynskaya complex increases from west to east from 80 m in the basin of the Chumikan river to 980 m in the basin of the Sev. Ui river (Semikhatov and Serebryakov, 1983). The K–Ar glauconite-based age of the complex is 1230–1210 Ma.

The lower rock mass of the Svetlinskaya complex is made up of massive dolomites. Its middle unit is represented by dark, sinuous-bedded, fine platy siltstones comprising sheets of gray, fine-grained quartz-feldspar sandstone. Abundant desiccation cracks and wave marks are discernible on the surface of the sheets. The section is topped by a dolomite unit, which does not differ substantially from that observed in the lower part of the section. Maximum sedimentation for the Aimchanskaya development stage of the Uchur-Maiskaya Basin occurred in its south-eastern part, in the basin of the Chelasin river, where the thickness of the Svetlinskaya complex reaches 1000 m, and in the middle reaches the Maya river. To the north-west, the thickness of the complex

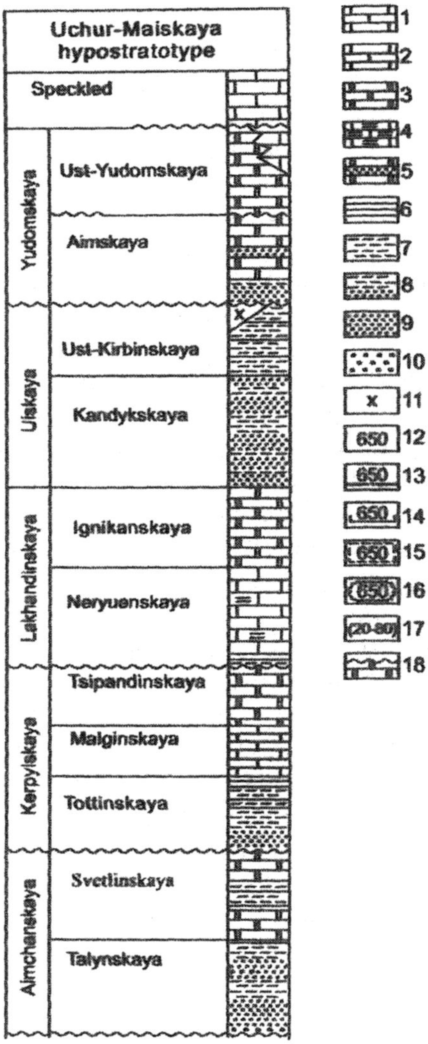

Fig. 13.3: Composition of Middle and Late Riphean, and Vendian-Cambrian sediments from the Uchur-Maiskaya Basin. Designations: 1–11 – dominating rock types: limestones (1), algal-bedded speckled limestones (2), dolomites (3), mudstone alternation with dolomites (4) and limestone (5), mudstones, clay slates (6), siltstones (7), gradation of sandstone into siltstone (8) sandstones (9), gritstones, conglomerates, intrusive assemblages (10), ultrabasic alkali rocks (11); 12–16 – isotope age: K–Ar method [based on glauconite (12), based on bulk samples of clay rocks (13)], based on bulk samples of magmatic rocks (14), Pb-isochrone age based on dolomites, plot with concordia (15), Rb-Sr isochrone age based on clay rocks (16); 17 – thickness, m. (18) Unconformity.

decreases abruptly to 40–60 m. The western sections differ from the eastern ones in the high fragmental rock content. An increase of the amount of sandstones and the appearance of conglomerates in the Talynskaya complex in the Batonga and Chelasin interfluve were related to a short-range activation of the uplifts south of the study region. The K–Ar glauconite-based age of the Svetlinskaya rocks is 1210 Ma.

The end of the Aimchanskaya time was marked by a short-range drying of the territory and crust formation, followed by the extensive Kerpylskaya transgression, which substantially overlaps the Aimchanskaya sedimentation basin (Semikhatov and Serebryakov, 1983). The Kerpylskaya series is represented by the Kondyorskaya, Ominiskaya, Malginskaya, and Tsipnadinskaya complexes.

A 7–15 m thick weathering crust has been mapped at the base of the Kondyorskaya complex, in the eastern framing of the Ominiskaya uplift of the basement. This weathering crust is made up of breccias, grus, clay, and siliceous-carbonate rocks. The deposits of the complex are stripped in the extreme south-west of the Maiskaya zone, on the Ignikan and Elgekan interfluve. The lower part of the complex is composed of different-grained white quartz sandstones with interbedded siltstones and conglomerate lenses of reduced thickness, non-persistent in strike. The upper part of the Kondyorskaya complex is mainly represented by dolomites and to a lesser degree, by siltstones and sandstones. The thickness of the complex is 950 m, and the K–Ar glauconite-based age is 1170–1070 Ma.

The Omninskaya complex made up of mudstones and siltstones, with lenses of calcareous conglomerates, cherty rocks, and quartz sandstones crops out in the Omnya-Uorgalan-Ichas interfluve, on the left watershed of the Uorgalan river and along the outer slope of the Kondyor Range. The thickness of the complex is 400–740 m. The K–Ar glauconite-based age of the rocks of this complex is 1020–970 Ma.

The limestones of the Malginskaya complex rest conformably on top of the Omninskaya siltstones. Its lower strata are represented by peculiar ferruginous-siliceous-carbonate, siliceous-ferruginous and ferruginous-siliceous rocks and conglomeratic limestones marking a change in sedimentation conditions. These rocks are overlain by speckled, followed by gray clayey, occasionally dolomitized limestones capped by a horizon of clay-calcareous dolomites, bituminous dolomites, and limestones with combustible shale interbeds up to 15–20 m thick. The absolute age of the Malginskaya limestones determined by the K–Ar technique based on glauconite is 1000 Ma, and by the Pb–Pb isochrone, 960 Ma.

The Malginskaya complex is conformably overlain by the assemblages of the Tsipandinskaya complex. They form the Oldoguya-Kerpyl watershed. The complex outcrops were detected also in the basin of the Bolshaya Kira and Ignikan rivers, where the assemblages are traced as a narrow band in north-western direction towards the nearmouth part of the Sev. Ui river. Only the lower part of the Tsipandinskaya complex which forms an 80–100 m topographic scarp, is ubiquitously exposed. The complex is composed of gray, pink, pink-gray, fine- and micro-grained, thick-platy massive dolomites, and in the upper part by bituminous dolomites. Some dolomite horizons are silicified, occasionally calcified. The thickness of the complex is 250–400 m. The Pb–Pb isochrone shows an absolute age for the Tsipandinskaya rocks of 980 Ma; the K–Ar technique yields 950 Ma based on glauconite.

The transgressive character of the Kerpylskaya complex, which is reflected in the geochemical parameters of its rocks, has predetermined the influence of the wash-down source in the Upper Kerpylskaya sediment to be evident (incidentally, only to a small degree) only in the extreme south-western section of the Malginskaya complex. They are marked by an increased content of clayey matter and hydrous ferric oxide. The Tsipandinskaya dolomites again are actually devoid of terrigenous admixture.

A substantial hiatus in sedimentation has been established between the subjacent Middle Riphean Kerpylskaya series and the overlying Upper Riphean Lakhandinskaya series; the hiatus is registered by the development of a weathering crust represented by grus, dolomite breccias with a loamy binding mass and clay-ferruginous bauxite-bearing rocks up to 0.4 m thick (Mironyuk, 1986).

The Lakhandinskaya series is divided into the Kumakhinskaya, Milkonskaya, Nelkanskaya, and Ignikanskaya complexes. The total thickness of rock series is 280–775 m.

The Kumakhinskaya complex consists of mudstones with subordinate interbedded stromatolite and oncolite dolomites and limestones. Its thickness is 70–110 m, and the K–Ar age based on glauconite 930 Ma.

The Milkonskaya complex is made up of limestones, often stromatolithic, and less frequently by oncolite or weakly bituminous limestones. The thickness of the complex increases from north-west (130 m) to south-east (up to 300 m). The K–Ar glauconite-based age is 950–780 Ma.

The composition of the Nelkanskaya complex is dominated by mudstones, often ferruginized, with lenses and concretions of brown iron ore. Siltstones and sandstones are present in a lesser quantity. A persistent guide horizon of stromatolithic limestones is present in the middle part of the complex. The aggregate thickness of the Nelkanskaya sequence increases eastward from 120 to 160 m. Its K–Ar glauconite-based age is 970–920 Ma.

The Ignikanskaya complex consists mainly of dolomitic limestones, calcareous dolomites, dolomites, and limestones with oncolites and katagraphs; less frequent are stromatolithic and bituminous differences. In the lower part of the complex an interbed of cherry-colored mudstones is observed. The total thickness of the complex over a major part of the territory is 180 m; in the basin of the Chuminda river it increases to 270 m. The glauconite-based age of the Ignikanskaya rocks determined by the K–Ar method is 870 Ma, by the U–Pb isochrone, 820 ± 10 Ma, and by the Pb–Pb isochrone, 800 ± 90 Ma.

The vertical change in the chemical composition of the Lakhandinskaya carbonate deposits from dolomites to mixed dolomite-calcareous rocks and pure limestones and again back to essentially dolomite, mainly fragmental units in the complex, and also some geochemical parameters of the sediments characterize the Lakhandinskaya series as a transgressive–regressive complex.

The rocks of the Lakhandinskaya series are overlain without any visible hiatus by rocks of the Uiskaya series, represented by the Kandykskaya and Ust-Kirbinskaya complexes (Fig. 13.3). The thickness of the deposits of this series increases from south to north.

The Kandykskaya complex consists of terrigenous rocks – siltstones, sandstones, and mudstones, overlying without visible unconformity the bituminous dolomites of the Ignikanskaya complex in the Lakhandinskaya series. The Lower Kandykskaya deposits on the eastern slope of the Uchur-Maiskaya Basin contain a considerable amount of sheet bodies of diabases, among which occasionally occur effusive facies. The thickness of the complex is about 1100 m.

The sandstones of the Kandykskaya complex are conformably overlain by the Ust-Kirbinskaya complex, represented by siltstones, mudstones, and sandstones. The rocks crop out in the basins of the Algaya, Nayum, Chandykan, Ikachan, and Uikan rivers. The maximum thickness of the complex reaches 1000 m.

13.2.2 VENDIAN-CAMBRIAN STRUCTURAL STAGE

After the considerable hiatus caused by the uplifting and drying of the territory of the Uchur-Maiskaya Basin, terrigenous-carbonate sedimentation continued in it in the Vendian-Cambrian. In the central part of the basin, the Vendian-Cambrian submergence is of inherited character being developed on the flanks of the Riphean structures. By and large, the Vendian-Cambrian deposits are more developed on the Aldan-Lenskaya plate than the Riphean deposits (Mironyuk, 1986).

The Vendian deposits are represented by the Yudomskaya series, 95–300 m thick, divided into the Aimskaya and the Ust-Yudomskaya complexes. The deposition of the Yudomskaya series was preceded by the largest pre-Yudomian hiatus in sedimentation accompanied with the formation of ultrabasic-alkali intrusions against a background of general inversion. A 0.1–1.2 m thick weathering crust at the base of the series is represented by the lithified fragmental gibbsite-bearing assemblages: grus-clay and carbonate-clay rocks, grus, mudstones, and siltstones.

The Aimskaya complex of the Yudomskaya series is made up of mudstones, dolomites, dolomitic limestones, and siltstones. The presence of sandstones (mainly at the base), including quartz sandstones, and of conglomerates is characteristic. The composition of the complex is strongly changeable; fragmental rocks prevail, especially in its lower part.

The upper Ust-Yudomskaya complex is represented by fine-grained dolomites and dolomitized limestones. It lays on the lower complex with traces of insignificant washout registered by a horizon, non-persistent in thickness, of medium-and coarse-grained, obliquely laminated sandstones, including dolomitic sandstones. The presence of stromatolithic and oncolitic dolomites is characteristic.

According to evidence from Semikhatov et al. (2003), the age of the Upper Yudomian deposits in the $^{207}Pb/^{204}Pb$–$^{206}Pb/^{204}Pb$ coordinates on nine points is 553 ± 23 Ma. On the one hand, it is in good agreement with the U–Pb age of zircons from the middle part of the Nemakit-Doldykskiy stage of the Olenk uplift (North Siberia), and on the other hand, with the U–Pb age of the Vendian upper horizons in Belomorye in the north of the East European platform (less than 555.3 Ma).

Between the deposits of the Yudomskaya series and Cambrian rocks, there was a hiatus in sedimentation evidenced by the formation of weathering crusts, represented by lithified grus-clay and clay rocks, breccias, sandstones, and siltstones, with a thickness of 1.5–4 m. The Cambrian deposits are made up of colored limestones and dolomites (speckled sequence), clayey and bituminous limestones intercalated with siliceous limestones, and marls (Inikanskaya complex), with a total thickness of 72–213 m.

The rocks of the Vendian-Cambrian structural stage fill wide, not very deep troughs and narrow fault-line synclinal folds. The largest structures are the Aimskaya, Bolshemarkyuelskaya, and Malomarkyuelskaya troughs with a subsidence depth of the pre-Vendian basement of no more than 300 m.

13.2.3 JURASSIC

The distribution of the Jurassic deposits of the Bokurskaya series is limited on the Kep-Kap Range (Mironyuk, 1986). These deposits make up the remainders of several small (tens of km^2) calderas within the bounds of volcano-plutonic polygnenetic structures, most exposed on the Bokur, Dugaiyan, and Yarmarka rivers. Data from Mironyuk (1986) demonstrate that one can conventionally distinguish three parts in the section, although the relations between them have not been well established.

The lower part of the Jurassic deposits (<250 m) is made up of oligomictic sandstones, tuffaceous sandstones, sometimes with lenses of small-pebble conglomerate-like sandstones and tuff conglomerates, containing layers, less often horizons (up to 120 m), of tuffs, agglomerates, andesites, and trachyandesites. The content of volcanogenic material in the section is changeable.

The middle part (<450 m) consists of horizons (100–200 m) of tuffs, clastic lavas, agglomerates (including lapilli agglomerates), and trachyandesites, among which trachyte, sandstone, and tuff sandstone beds are noted with a thickness of several meters.

The upper part (<600 m) is characterized by a complete absence of sedimentary deposits and a nearly equal content of effusive and pyroclastic layers with a petrographic composition of variegated trachyandesites, trachybasalts, tephrites, trachytes, and phonolites. Some sections exhibit unclear rhythmicity, however, especially in agglomerate-tuff facies, rock stratification by composition of magmatic material fails. This may be due to the multifocus character of the volcanic manifestation.

The flora from the sandstones gives evidence of sedimentation during the Jurassic. The radiometrical age of the tephrites is 159 $\pm$ 4 Ma, of the trachyte, 135 $\pm$ 4 Ma (Mironyuk, 1986).

13.2.4 DEVELOPMENT STAGES

Five essential development stages are distinguished in the Uchur-Maiskaya Basin, each being preceded by restructuring of the territory. These are the pre-Uyanskaya, pre-Uchurskaya, pre-Aimchanskaya, pre-Kerpylskaya, and pre-Yudomskaya boundaries. The first stage is geologically the most significant (age 1670 $\pm$ 24 Ma). From all viewpoints, it corresponds to the lower boundary of the Riphean, as has been accepted by most geologists.

The pre-Aimchanskaya boundary with a K–Ar age of 1350 $\pm$ 50 Ma was characterized by the following features: (1) dying-off of the downwarping zone in the basin of the Uchur and Bol. Aim rivers, and the formation of the pre-Uchurskaya structure on its place; (2) subsequent shift of the sedimentation area onto the eastern part of the basin (beginning of the pre-Maislaya structure).

The pre-Kerpylskaya restructuring and the corresponding geologial boundary (K–Ar age about 1200 $\pm$ 20 Ma) were linked to involvement in the downwarping of the western and central part of the Maiskaya zone, which had been included into the area of stable uplifts at the Aimchanskaya stage.

The pre-Yudomskaya boundary with an age of ca. 650–640 Ma, separated the formation stage of the proper Uchur-Maiskaya Basin from the stage of development of the most extensive marine transgression; within the framework of this transgression, the Aldan shield (that earlier had got a stable tendency toward uplifting) and its northern continuation subsided below sea level.

The Jurassic stage corresponds to activation of Mesozoic tectonic movements, which in fact, embraced the whole region.

13.3 VILYUISKAYA SYNECLISE

The sedimentary deposits filling the Vilyuiskaya Syneclise are subdivided into four stages: Vendian-Lower Paleozoic, Middle Paleozoic, Upper Paleozoic, and Mesozoic (Gusev et al., 1985; Parfenov and Kuzmin, 2001).

13.3.1 VENDIAN-LOWER PALEOZOIC STRUCTURAL STAGE

Drilling up to the base of the Vilyuiskaya Syneclise has determined Vendian, Cambrian, Ordovician, and Silurian deposits with a total thickness of up to 5000 m (Parfenov and Kuzmin, 2001) (Fig. 13.4).

The Vendian was the onset of a large transgression on the Siberian continent, the maximum of which was marked by the Early-Middle Cambrian. In the Late Cambrian, Ordovician, and Silurian, gradual regression of the sea took place, ending in the Devonian with a general uplifting of the eastern part of the platform and washout. The structural plan of the Vendian-Lower Paleozoic structural stage is defined by wide plane sedimentary basins formed as a result of thermal subsidence of the eastern part of the Aldan shield.

The Vendian deposits transgressively overlie different parts of the Riphean section, and in many regions they rest directly on the Precambrian crystalline basement (Krasny, 1981). The deposits show a similar texture over the whole study area, and are represented by shallow-water marine quartz sandstones and overlying dolomites and limestones, which make up a major part of the section. The thickness of the Vendian deposits is 200–500 m. The age of the glauconites from these deposits ranges between 650 and 530 Ma.

The Cambrian deposits with washout and conglomerates at the base, overlie the Vendian deposits, and are represented by all types of sediment. In the Early Cambrian, the entire territory of the eastern part of the Siberian platform was located in the zone of hot arid climate (Bulgakova, 1996). Then, on the site of the Vilyuiskaya Syneclise, was located a "black shale" bay-like basin closed in the north-west and open in the south-east. Its maximum depths are determined as exceeding the level of carbone accumulation (Bulgakova, 1996). The sedimentation was of exclusively autochthonous character. The source material was the product of (1) biochemogenic and biogenic processes (algal, microphytolithic, oncolitic limestones and dolomites); (2) washout of volcanic and rift complexes, bottom sediments of carbonate littoral zones (clay substance, different-size

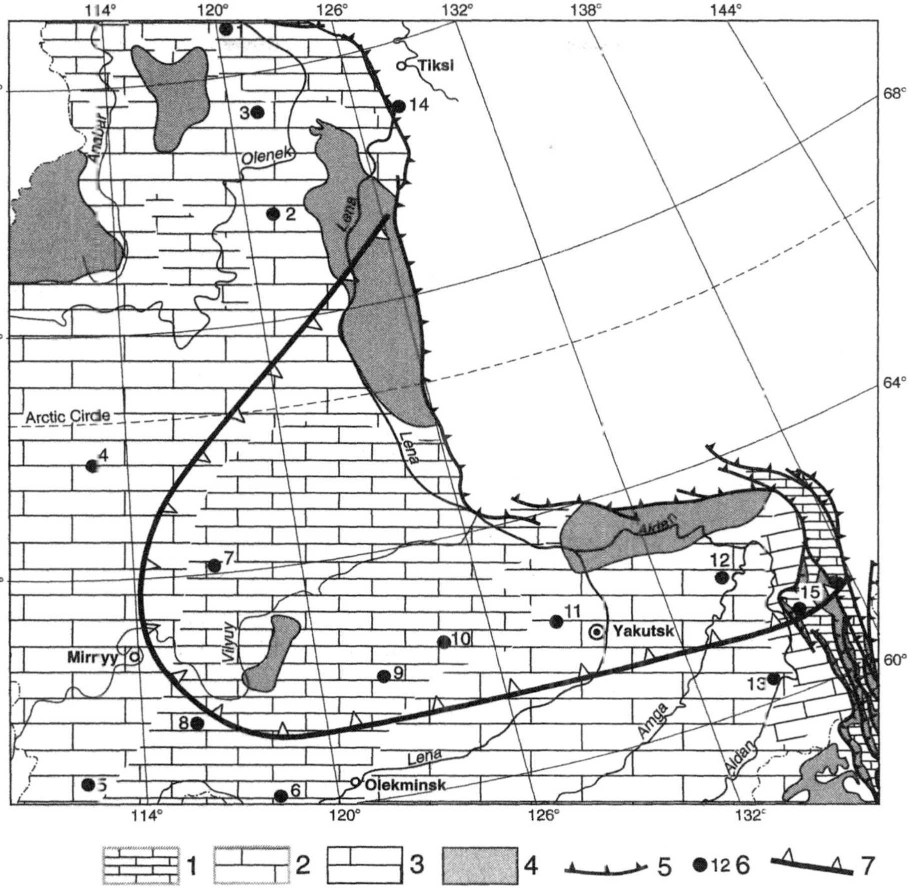

Fig, 13.4: The Vendian-Lower Paleozoic structural stage in the eastern part of the Siberian platform (after Parfenov and Kuzmin, 2001). Designations: 1 – Vendian-Lower Paleozoic deposits within the limits of the Verkhoyansk fold-and-thrust belt; 2 – sedimentary basins with Vendian-Lower Paleozoic deposits, 1.5–5.0 km thick [Nizhneolenekskiy, Sukhanskiy, Central Yakutskiy, Aldan-Maiskiy]; 3 – relatively uplifted areas with Vendian-Lower Paleozoic deposits with a thickness of mainly <1.5 km; 4 – areas of contemporary absence of Vendian-Lower Paleozoic deposits, where these deposits did not accumulate or were subsequently eroded; 5 – front of Phanerozoic orogenic belt; 6 – location of reference stratigraphic sections; 7 – boundaries of the Vilyuiskaya Syneclise.

clastics); (3) hydrothermal processes (silica, phosphorous, vanadium, copper and other elements); and (4) mass development of phytoplankton (sapropelic organic substance). The Early Cambrian sedimentation resulted in the formation of a unit of combustible shales, bituminous limestones, and mudstones with an increased phosphorous and vanadium content. These deposits are recognized as oil-source deposits, with a thickness of 400 to 1200–1400 m (Frolov et al., 1987; Kashirtsev, 1994).

The Middle Cambrian deposits are represented by greenish-gray and yellowish-gray limestones, speckled marls, dolomites, and clayey limestones, 380 to 640 m thick.

There was a hiatus in sedimentation between the Middle and Upper Cambrian, when the Lower and Middle Cambrian sediments were subjected to washout.

The Cambrian section ends with a unit of speckled, mainly red-brown marls, and clayey dolomites, with interbeds of clay, mudstone and gypsum. Siltstones, sandstones and limestones are inconsiderable. Characteristic for the rocks are structures of shallow water, ripple marks, wave marks, desiccation cracks, mud streams, etc. on the bedding surface. Toward the center of the basin, the quantity of gypsum and anhydrite increases. The thickness of the deposits is 170–250 m; on the sides of the syneclise, up to 300–310 m in its central part.

In the Early Ordovician, the area of the Vilyuiskaya Syneclise was covered with a shallow epicontinental sea. The Lower Ordovician shallow-water deposits lie conformably on the Upper Cambrian deposits. The basal horizons of these deposits are represented by gray massive sandstones. These are overlain by thin-bedded and oolitic dolomites, occasionally bituminous. The thickness of the Lower Ordovician deposits is 80–120 m.

The Middle Ordovician deposits occur conformably on top of the Lower Ordovician sediments. The principal rocks are sandy-siltstone dolomites, and clayey dolomites with sandstone and siltstone interbeds (80–120 m).

The Upper Ordovician sediments overlie with washout the Middle-Lower Ordovician beds. They are known at some places on the western side of the Vilyuiskaya Syneclise. The section begins with red-colored shelly conglomerates and clays with glauconite and iron oxides. The above-occurring section is represented by speckled mudstones and sandstones with thin limestone interbeds and abundant fauna. Their thickness is 35–80 m.

The Lower Silurian deposits in the Vilyuiskaya Syneclise are represented by the Llandoverian and Wenlockian stages. They occur on top of the Upper Ordovician assemblages with a hiatus in sedimentation (Krasny, 1981). The Llandoverian stage is made up of gray, cloddy-clay limestones with interbedded siltstones and marls. Intraformational flat-pebble limestone conglomerates are characteristics. The deposits contain numerous fragments of corals and brachiopods, yielding a Middle-Late Llandoverian age. The thickness of the stage is 120–300 m. The 100–300 m thick Wenlockian deposits are represented by light gray limestones and dolomites, with rare interbeds of greenish siltstones and intraformational limestone conglomerates.

In the Ludlovian, a plane low continental land formed on the place of the Vilyuiskaya Syneclise as a result of a sharp fall of the sea level; this continental realm remained for a long time (till the Middle Devonian) an area of denudation.

13.3.2 MIDDLE PALEOZOIC STRUCTURAL STAGE

The structural plan of the Middle Paleozoic stage comprising assemblages of the Middle and Upper Devonian, and of the Lower Carboniferous, is in many respects similar to the Vendian-Cambrian stage (Fig. 13.5). It manifests itself widely in the rifting processes at the eastern margin of the syneclise (Gaiduk, 1988). The Middle-Upper Devonian and Lower Carboniferous sedimentary and volcanogenic-sedimentary assemblages fill linear grabens-aulacogens, which begin at the front of the Verkhoyansk fold-and-thrust belt and wane in the interior of the platform. Fault-bounded horsts, and extended (for hundreds of kilometers) belts of basalt dikes and sills are associated with the aulacogens. Kimberlite fields, diatremes, and massifs of alkali-ultrabasic rocks with carbonatites are characteristics. Beyond the boundaries of the aulacogens, Middle Paleozoic deposits are absent over a major part of the eastern Siberian platform. This is attributed, first, to the fact that sedimentation took place largely within the actively downwarping aulacogens, and second, to the deep washout and erosion of the adjacent platform parts.

The Middle Paleozoic aulacogens, located at the base of the Vilyuiskaya Syneclise, are most investigated by drilling and seismic prospecting. The Kempendyaiskaya, Lindenskaya, Syangdinskaya, Sarsankaya, Tangarynskaya, and Ytyakttinskaya rift basins and a number of longitudinal and transversal uplifts separating them are distinguished here. The greatest thickness of the Middle Paleozoic deposits is in the Kempendyaiskaya Basin is <6000 m, and in the other depressions their thickness is estimated as 2000–3000 m.

The active downwarping of the rift basins was preceded by the formation of an arched uplift and mass basalt outflows at the beginning of the Late Devonian, involving a vast territory (about 400 km across). Basalt covers with a thickness of several tens of meters occur intercalated with interbedded sandstone, siltstone, and limestone, of reduced thickness. The formation of dike belts is referred to this time. The Upper Devonian and Lower Carboniferous deposits filling the rift basins are represented by speckled and gray-colored sandstones, siltstones, limestones, marls, gypsum, and anhydrite. Acid tuff interbeds are also characteristic. A thick unit (<1000 m) of rock salt is established at the base of the Upper Devonian in the Kempendyaiskaya Basin.

The tectonic structures of the rift zone are determined by one-sided grabens and horsts, which are separated by normal faults with displacement amplitude

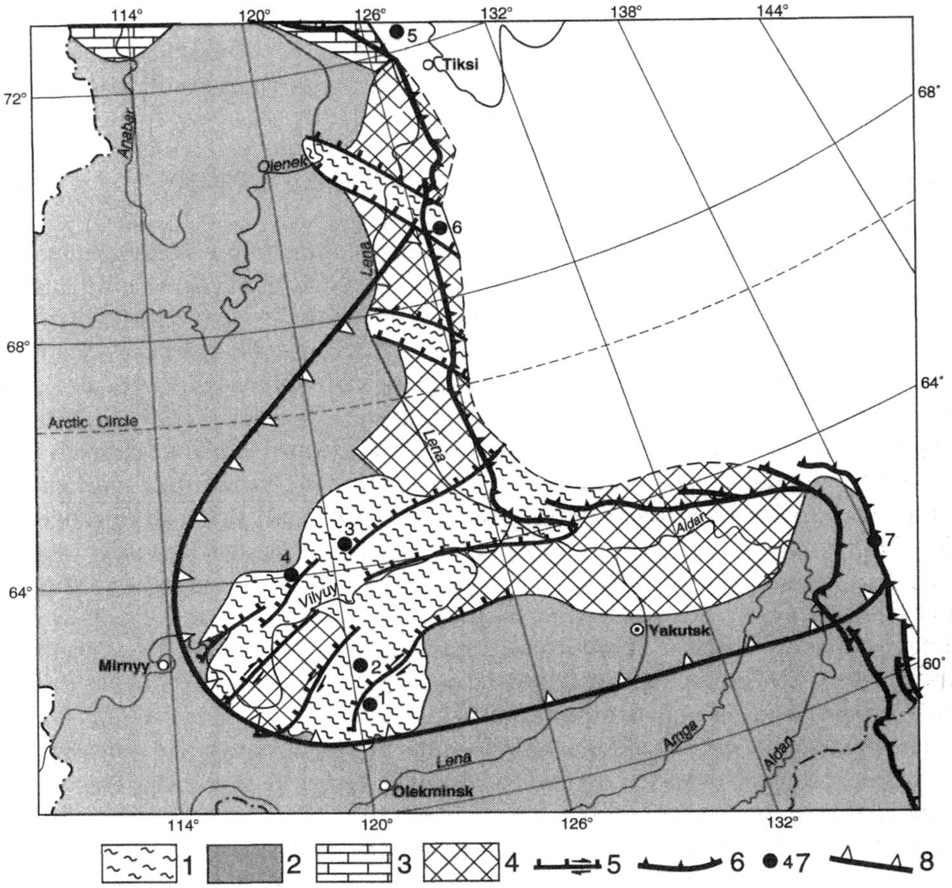

Fig. 13.5: The Middle Paleozoic structural stage in the eastern part of the Siberian platform. Designations: 1 – rifts filled with red-colored terrigenous evaporite deposits, with a thickness of up to 6 km [Ygattinskiy, Kempendyaiskiy, Kyutyungdinskiy]; 2 – rift assemblages within the Verkhoyansk fold-and-thrust belt; 3 – sedimentary basins with marine deposits with a thickness of up to 1.5 km; 4 – areas of greatest uplifts associated with rifts [Govornovskoye, Dzhardzhanskoye, Yakutskoye, Suntarskoye]; 5 – normal faults and strike-slip faults; 6 – fronts of Phanerozoic orogenic belts; 7 – location and numbers of reference stratigraphic sections; 8 – boundaries of the Vilyuiskaya Syneclise.

of several tens of meters in 3000 m (Gaiduk, 1988). Fault-line structures developed actively in the Late Devonian-Early Carboniferous, which are overlain in a mantle-like way by deposits of the Middle-Upper Carboniferous and Permian.

The dikes, sills, chonoliths and volcanic pipes of Middle Paleozoic basites occur among the carbonate deposits of the Cambrian, Ordovician, and less often, of the Silurian. The thickness of the dikes is 6–8 m, their extension is several tens of kilometers. The established maximum thickness of the sills is 120–140 m. According to drilling evidence, the total thickness of lava flows is hundreds of meters.

13.3.3 UPPER PALEOZOIC STRUCTURAL STAGE

The Upper Paleozoic structural stage (Fig. 13.6) within the Siberian platform includes Middle-Upper Carboniferous and Permian deposits, being represented by a thick complex of coastal-continental, deltaic, and less often, marine sediments, which predetermined their considerable lithologic-facial variability both in terms of area and section. In the central part of the Vilyuiskaya Syneclise, the Permian and the uppermost strata of the Carboniferous have a reliable thickness of over 3600 m (Parfenov and Kuzmin, 2001). The thickness of the Upper Paleozoic deposits toward the near-wall zones of the syneclise becomes substantially reduced (down to 1000–550 m), while the section becomes sandier.

The Upper Tournaisian-Middle Visean rocks occur with angular unconformity and with conglomerates at the base, on the riftogenic assemblages of the Middle-Upper Devonian and Lower Tournaisian, and occasionally on the Silurian deposits (Parfenov and Kuzmin, 2001). The Upper Carboniferous sediments consist mainly of sandy rocks and low-thickness bands of siltstone and mudstone intercalations, up to 190 m thick. The Lower Permian deposits rest unconformably on various horizons of Upper Devonian-Lower Carboniferous sequences. They are made up of carbonate milonitized rocks with kaolin, siltstones, and friable sandstones; the thickness of these deposits is up to 150 m (Krasny, 1981). The Upper Permian sediments in the syneclise have been disclosed by holes during oil drilling. They occur in the depth interval of 2600–3800 m, and are represented by sandstones, siltstones, and mudstones.

Fragmental material in the sedimentary basin was supplied by large rivers of the modern Lena river type. The largest of them is presumed to have run through the region of the Vilyuiskaya Syneclise, and its sources, judging by the character of the detrital material of terrigenous rocks, were located in the Baikal mountain area. Two other river systems have been traced at the northern and southern margins of the Siberian platform, respectively. The main source rocks for the formation of the Upper Paleozoic structural stage were products of acid intrusive and effusive rocks, and ancient crystalline schist destruction.

Based on reference drilling (Hole 27), numerous beds and interbeds of bituminous coal with a thickness of several centimeters to 2–3 m have been detected

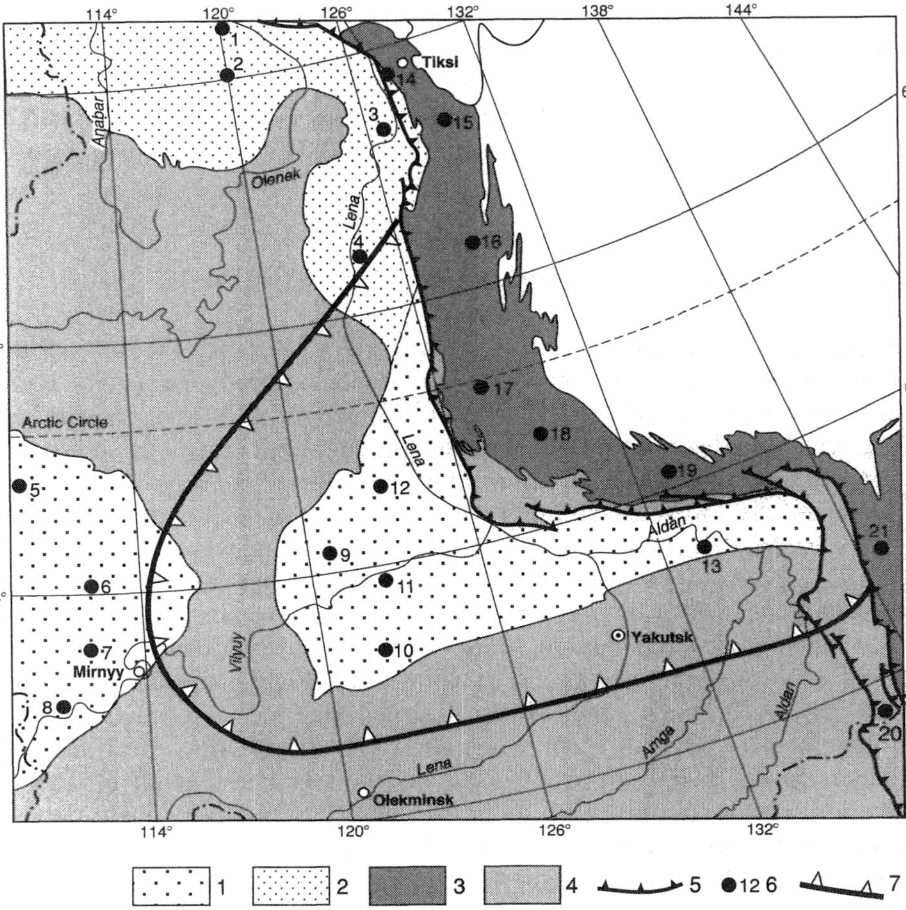

Fig. 13.6: The Upper Paleozoic structural stage in the eastern part of the Siberian platform. Designations: 1 – mainly continental deposits, thickness up to 3 km [Vilyuiskiy Basin (E), Tungusskiy Basin (W)]; 2 – mainly coastal-marine deposits, thickness up to 3 km [Leno-Anabarskiy Basin (N)]; 3 – marine deposits, thickness over 3 km [Verkhoyansk fold-and-thrust belt]; 4 – areas of contemporary absence of Upper Paleozoic deposits; 5 – fronts of Phanerozoic orogenic belts; 6 – location and numbers of reference stratigraphic sections; 7 – boundaries of the Vilyuiskaya Syneclise.

in the terrigenous units of the Permian and Upper Carboniferous deposits. Generally, these units often contain enclosures of disseminated carbonized plant material (Frolov et al., 1987). By petrographic composition, all studied coals belong to the vitrinite type, in most cases with a small (up to 3–4%)

leptinite admixture and generally with a more substantial fusinite content, whose total amount increases downsection, sometimes reaching 25–30% of the sample's organic substance (Frolov et al., 1987). These authors made also an attempt of accessing the paleotemperature regime of the Permian-Carboniferous unit in the Vilyuiskaya Syneclise. Assuming the temperature of transition of brown coals to long-flame coals being 90 °C, and the temperature of transition of lean coals to semi-anthracites being 220 °C, the paleotemperature gradient was 3.2 °C for 100 m, which is somewhat higher than the modern temperature gradient of 2.9 °C for 100 m. At the same time, comparing the modern temperature (140 °C) at a depth of 500 m and the transformation temperature of coal substance at this depth (200 °C), it can be suggested that the Upper Paleozoic rock mass in the central part of the syneclise cooled by 80 °C after maximum warming-up.

In connection with a high degree of coal substance catagenesis, one should be more careful when approaching the problem of the prospects of industrial gas potential of the deep horizons of the sedimentary cover. This conclusion is also supported by the extremely low competency-filtration properties of the Carboniferous and Permian deposits. The degradation of the reservoir properties of terrigenous rocks is also attributed to their high catagenetic transformation.

13.3.4 MESOZOIC STRUCTURAL STAGE

The deposits of the Mesozoic structural stage in the Vilyuiskaya Syneclise (Fig. 13.7) are most widespread. At the moment of formation of Mesozoic deposits, the basement of the structure was characterized by a complicated relief, and the presence of local depressions and uplifts. The largest uplift, Yakutskoe, is made up of Cambrian deposits, which are overlain by Mesozoic sediments. The depth of occurrence of the basement is 14 km.

Triassic deposits, with conglomerates of low thickness at the base, occur unconformably on top of the Permian rocks. In the Early Triassic, the Vilyuiskaya sedimentary basin was a shallow-water bay-like sea. The marine basin was contiguous with land, which remained rather uplifted in the south and southwest, being an area of removal. Mainly silty-clayey, often tufogenic deposits formed over a large part of the basin, which facially corresponds to sediments of shallow-water and medium zones of the shelf. Following the accumulation of sediments, volcanism manifested itself in the area of the syneclise. It was accompanied with basic lava outflows and ash material ejection. The outflow of effusives occurred in a submarine environment. The thickness of the deposits increases naturally from the sides toward the central part of the depression. The sedimentary rocks are characterized by speckled coloring: green, greenish-gray,

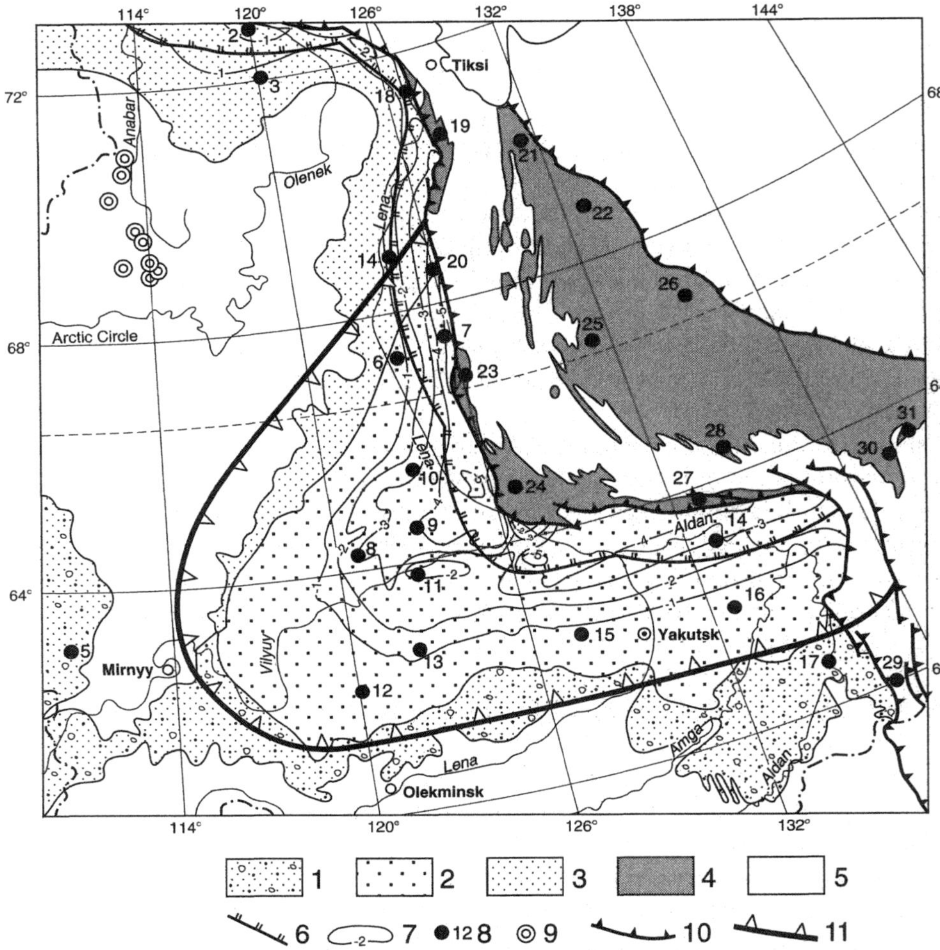

Fig. 13.7: The Mesozoic structural stage in the eastern part of the Siberian platform (after Parfenov and Kuzmin, 2001). Designations: 1 – continental deposits of the Tungusskiy Basin; 2 – mainly continental deposits, thickness up to 6 km, with coastal-marine deposits in the lower part of the section (Vilyuiskiy Basin); 3 – sedimentary basins: mainly coastal-marine deposits, thickness up to 1.5 km [Leno-Anabarskiy Basin]; 4 – marine deposits with a thickness of over 3 km [Verkhoyansk fold-and-thrust belt]; 5 – areas of contemporary absence of Mesozoic deposits; 6 – boundary of the Priverkhoyansk foredeep; 7 – contour lines of the bottom of Jurassic deposits; 8 – reference stratigraphic sections and their numbers; 9 – ultrabasic-alkali rocks and carbonatites; 10 – fronts of the Phanerozoic orogenic belts; 11 – boundaries of Vilyuiskaya Syneclise.

red-brown, gray, and dark-gray, and the mudstones by a nearly black color. This black color of the rocks is attributed to an arid climate in Early Triassic time. The thickness of these deposits is 20–130 m. Second to volcanogenic deposits in accumulation were mainly sandy sediments of the shallow marine shelf. The most intense supply of terrigenous material was from the south. This was responsible for the formation of numerous thick sheets and bands of intraformational conglomeratic breccias in the southern part of the Vilyuiskaya marine basin. During the period of slackening of the erosion processes in the denudation areas, silty-clay bands with a thickness of <40 m accumulated in the sedimentary basin. The thickness of the deposits was 240–250 m.

The Lower Jurassic deposits of the Hettangian and Sinemurian stages rest with washout on the Paleozoic and Triassic deposits, and are overlain with washout by deposits of the Upper Pliensbachian. The sediments consist of sandstones and sands with subordinate interbedded conglomerates, coarse gravel, siltstones, and clays. Carbonized plants and thin interbeds and lenses of humus coals are characteristics. The lower part of the deposits is ubiquitously made up of coarser rocks than the upper part. Finds of fresh-water bivalves and plant imprints are known in the lower part (Krasny, 1981). The thickness of the deposits is 10–320 m. The sediments of the Pliensbachian stage are represented by sandstones and sands alternating with siltstone-clay bands, containing ammonite, bivalve, and gastropod remains, microfauna, spores, and pollen. The unit is 20–97 m thick. The Toarcian deposits consist of clays and mudstones containing interbeds of siltstones, less often of sands, sandstones, and organogenous fragmental limestones, attaining a thickness of 30–50 m.

The Middle Jurassic Aalenian in the south-west of the syneclise includes sandstones with siltstone intercalations, containing bivalve remains. In the north-western region of the syneclise, the Aalenian includes sands and sandstones, often ferruginzed, with siltstone and clay interbeds, also with bivalve remains. The thickness of these beds is 10–50 m. The Bajocian-Bathonian sediments are represented by marine assemblages, including sands and sandstones containing intercalations of siltstone and clay layers and bands, low-thickness coal interbeds, and fragments of charred and stone wood. Their thickness is 120–130 m. Sands and sandstones with Callovian brachiopods occur on the washed-out surface of the deposits. The maximum thickness of the Callovian is 17–20 m.

The Callovian and Bathonian deposits are transgressively overlain by Upper Jurassic Volgian (Tithonian) layers represented by a coal-bearing unit. The base contains sandstones with siltstone lenses and coal interbeds of low thickness (40–140 m). The middle part is represented by alternation of mudstones, siltstones, sandstones, coaly shales, and coal interbeds, also of low thickness (50–150 m). The upper part (Krasny, 1981) comprises sandstones alternating with bands of siltstone, mudstone, and coal seams (250–380 m).

The sediments of the Lower Cretaceous Berriasan-Barremian stage rest conformably on top of the Upper Jurassic coal-bearing deposits. They are made up of greenish-gray sandstones with interbeds and bands of siltstone, mudstone, and coals alternation (Krasny, 1981), with a thickness of 800–1300 m. The rocks of the Aptian stage conformably overlie the Berriasian-Barremian deposits. They are made up of mainly cross-bedded sandstones, and contain interbed sand bands of alternating siltstone and mudstone. Coals occur as rare lenses and interbeds of reduced thickness. In the sections noted for increased siltstone-pelitic rock content, coals are more frequent, sometimes reaching an effective thickness. Rounded concretions of sideritized sandstones are characteristics. The thickness of the stage is 550–1150 m. The deposits of the Albian stage (1250–1600 m) conformably overlie the Aptian layers. They consist of kaolinitized sands and sandstones, alternating with siltstones, mudstones, and coal seams, whose thickness occasionally reaches 15 m. Upward in the section, occur siderite concretions, fragments of charred and stone wood, and rare conglomerate interbeds and lenses with various pebbles, including of local rocks. Thick lenses of siltstone and clay containing fern, conifer, and angiosperm imprints remain confined to the uppermost part.

Upper Cretaceous Cenomanian-Maastrichtian rocks in the central and western parts of the Vilyuiskaya Syneclise are represented by a unit of weakly kaolinitized sand sheets with interbeds and lenses of sandy clays and siltstones in the central and upper parts of the section (600 m). They overlie with washout the subjacent deposits and contain interbeds and lenses of conglomerates, coarse gravel and gritstones, pebble and boulder inclusions, and charred fragments of wood. The deposits of the already Early Tertiary Danian stage top the section of Cretaceous deposits in the Vilyuiskaya Syneclise. They are made up of white, strongly kaolinitized sands and are, probably, the product of dense redeposition of weathering crusts. The thickness of these deposits does not exceed 100 m.

Triassic and Lower and Middle Jurassic deposits ($<$700 m) are represented mainly by marine shallow-water sandy-clayey assemblages up to 200 m thick, which are replaced by continental sands and conglomerates at the north-western and south-eastern sides of the Vilyuiskaya Syneclise. The Upper Jurassic and Lower Cretaceous sediments are made up of mainly continental sandstones, ashes, siltstones, clays, and coals. Their maximum thickness ($<$4500 m) occurs near the Verkhoyansk fold-and-thrust belt, and becomes strongly reduced (up to several hundreds of meters) toward the platform (Parfenov, 1984; Parfenov and Kuzmin, 2001). The lower part of the section contains shallow-water coastal-marine deposits. The change of marine deposits into continental deposits in the main part of the basin took place in the Late Jurassic, and in the north in the earliest Early Cretaceous. In the central part of the trough dominate facies of lacustrine-alluvial plains; the former are changed of facies from a marine alluvial

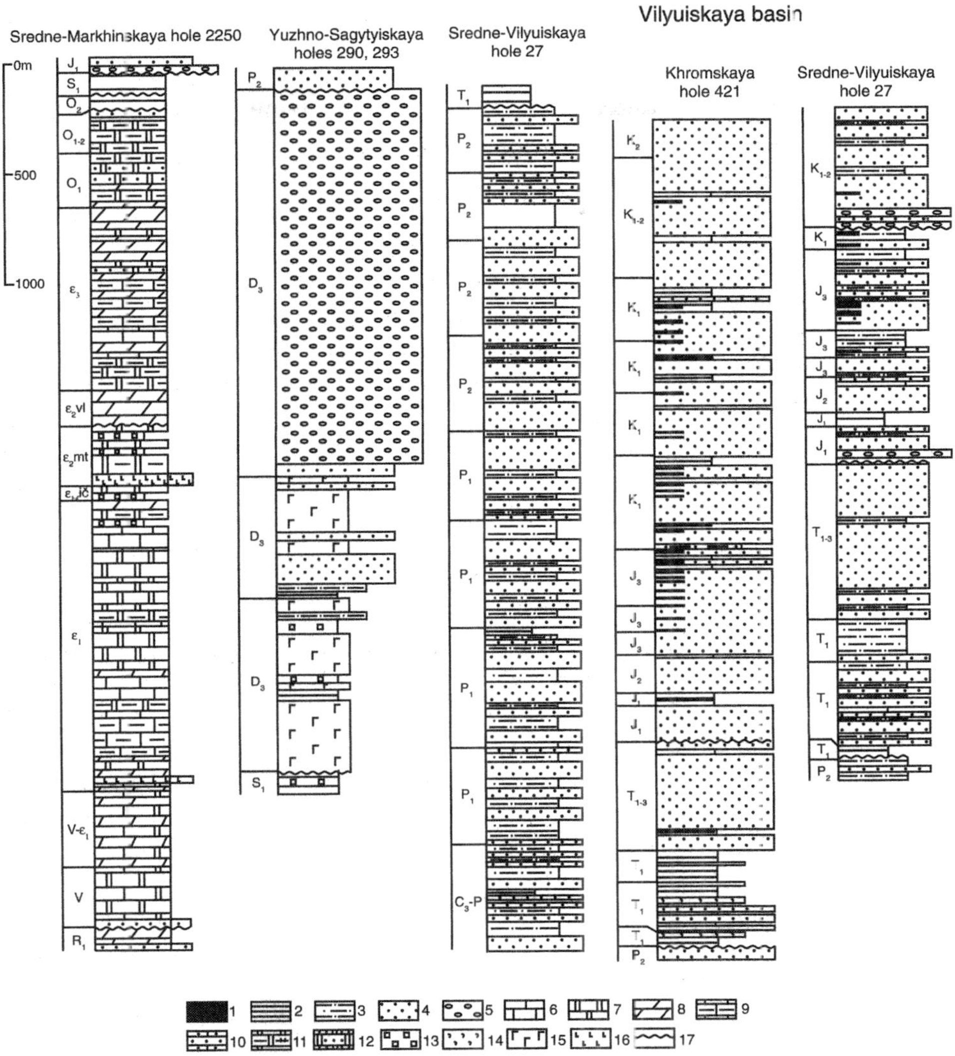

Fig. 13.8: Examples of sections structure of different-age sediments from the Vilyuiskaya Syneclise, after borehole data. Designations: 1 – coal; 2 – mudstones; 3 – siltstones; 4 – sandstones; 5 – conglomerates; 6 – limestones; 7 – dolomites; 8 – marls; 9 – bituminous limestones; 10 – clayey limestones; 11 – sandy limestones; 12 – clayey dolomites; 13 – sandy dolomites; 14 – rock salt; 15 – rhyolite and dacite tuffs; 16 – basalts; 17 – unconformable mode of rock occurrence.

plain to coastal shallow water. The Upper Cretaceous deposits (up to 1000 m thick) consist of coarse-grained sands and sandstones, and kaolinic clays; interbedded brown coals, lignites, and conglomerates are characteristics.

In the Late Jurassic and Early Cretaceous, the trough was filled almost exclusively at the expense of wash-down from the Siberian platform. The principal source of removal was in the south of the platform, in the region of the Aldan-Stanovoy shield. There were no mountains on the place of the Verkhoyansk fold-and-thrust belt. These were formed only in the Late Cretaceous; however, they were not high, since the Upper Cretaceous sediments are characterized by an oligomictic composition of the fragmental material, which was, apparently, supplied from the east (Bulgakova, 1996).

The summarized section of the deposits of the Vilyuiskaya Syneclise from drilling evidence is shown in Fig. 13.8.

14. OVERVIEW

J.M. MABESOONE, A.M. ABED, P.K. DASGUPTA, F.R. ETTENSOHN, M.V. GOROSHKO, P.V. MARKEVICH, D. PAVELIĆ, AND O.I. SUPRUNENKO

14.1 INTRODUCTION

In the foregoing chapters, the authors have endeavored to explain that most geological phenomena of the Earth occur in a repeated cyclical succession. In spite of the fact that the earth's history shows a one-way evolution, passing through various development stages, called permobile, stabilization, transition, and "modern", the geological events within these stages follow each other rhythmically or cyclically since the Archean. The causes of this cyclicity have been presented by Golubev in Chapter 2. The effects of the cyclical events in the earth's geological evolution and on the sedimentary basin formation have been the subjects of the next chapter 15. The results of this have been applied for a number of sedimentary basins in the world, presented by authors specialized in the study of the respective basins, as well as a few examples taken from literature. Because such cyclic development of sedimentary basins can be traced with examples since Mesoproterozoic times, a full integration over many areas has still to be made.

14.2 ANDES FUEGINOS

Recently, Olivero and Martinioni (1998) published a revision of the geological history of the Andes Fueginos in southern South America.

The Alpine geology and tectonics of the Andes Fueginos exerted? a strong control on the stratigraphic evolution of the region. Seven major stratigraphic units reflecting the main features of the evolving tectonic regimes can be differentiated in the area. These are from base to top: (1) Paleozoic-Jurassic basement rocks, (2) Late Jurassic Lemaire Formation, (3) Early Cretaceous Yahgan-Beauvoir formations, (4) Late Cretaceous Cerro Matrera Formation, (5) latest Cretaceous-Paleocene Rio Claro Formation, (6) Eocene La Despedida Group, and (7) late Eocene-Oligocene Cabo Peña-Rio Leona formations.

At the southern main axis of the Andes, the Paleozoic-Jurassic basement (1), consisting of garnet, quartz-sericite, and chlorite schists, greenstones and amphi-

Cyclic Development of Sedimentary Basins
J.M. Mabesoone and V.H. Neumann, editors.

bolites, is only exposed in a small area of the south-eastern Argentinean Andean region. The basement rocks are unconformably overlain by a complex submarine unit of rhyolitic lavas and domes, acidic volcaniclastic breccias, tuffs, conglomerates, turbidites, and slates, and basaltic rocks of the Late Jurassic Lemaire Formation (2). This latter unit is unconformably covered in the south-eastern Andean area by Early Cretaceous (3), deep-marine black slates, andesitic volcaniclastic turbidites, and tuffs of the Yaghan Formation, which grades laterally to the north into the Beauvoir Formation. These three latter units form a distinct tectonic-stratigraphic interval of the Central Cordillera of the Andes Fueginos. The basement rocks are loosely referred to as an accretionary prism of deep-marine sedimentary and basaltic rocks on the panthalassic margin of Gondwana. The Lemaire volcanic rocks are interpreted as a syn-rift unit formed during the main extensional tectonic regime related to the initial opening of the South Atlantic Ocean. The Yahgan-Beauvoir formations are interpreted as post-rift units, representing the first an andesitic, volcaniclastic apron, filling a deep-marine marginal basin, while the second to the north interfingers with the first, with slope and platform mudstones. During the Late Cretaceous, the compressional Patagonidic Orogeny resulted in the tectonic inversion and closure of this Rocas Verdes Marginal Basin, peak metamorphism, folding, and initial uplifting of the Andes Fueginos.

The Rio Claro Formation (5) represents the first molassic deposits of the foreland stage of evolution of the cordillera. By the latest Cretaceous-earliest Paleogene, the already uplifted central Andean rocks became exposed to subaerial erosion, and the lowest Danian part of the sediments of this formation presents the first clear evidence of an Andean clastic provenance. The Late Cretaceous Cerro Matrero Formation (4) and equivalent units represent the sediment accumulation during the first pulse and rapid uplifting and cooling during the tectonic inversion, with slope and platform black shales, marls, and some sandstones and limestones. During the earliest Paleogene, northward thrust propagation involved deformation of the Rio Claro Formation and older units as well as migration of the depocenters to the north.

The Eocene foreland molasse of the La Despedida Group (6) rest unconformably on the Rio Claro Formation but during the northward migration of the deformation it was itself involved in the thrust and fold belt. Together with the Cabo Peña, Rio Leona and equivalent formations (7), they define three main unconformably bounded successions composed of deep- to shallow-marine conglomerates, sandstones, shales, and limestones, deposited in the Austral and Malvinas basins. During the Eocene an important compressional event resulted in uplifting of deeper rocks and a northward, basement-involved, thrust transport of the central Andean schists or Late Jurassic Lemaire rocks over Cretaceous or Paleogene units, respectively. The compressional orogenetic phase finished by the end of the Paleogene, and the mostly Oligocene Cabo Peña-Rio

Leona formations are represented by subhorizontal beds. A clear evidence of initiation of a strike-slip regime is found in the latest Paleogene, suggesting a left-lateral offset of the order of 25–30 km of Cretaceous-Paleogene rocks along the Magallanes-Fagnano fault system.

The Quaternary geology of Tierra del Fuego reveals the development of geological processes, as a consequence of the ice ages, ash fall, lacustrine sedimentation, peat accumulation, and relative sea-level fluctuations.

14.3 ANDES FORELAND: PATAGONIA

The formation and evolution of the southern South American sedimentary basins started in the Late Triassic (Spalletti and Franzese, 1996).

In the Late Triassic, Patagonia was an almost positive land. Narrow and isolated continental rifts filled with volcaniclastic sediments developed in NW Patagonia and in the Deseado Massif (south-central Patagonia). Calc-alkaline intrusions (Central Patagonian Batholith) and the Comallo volcanics are emplaced at the NW end of the Gastre Fault System.

During the Early Jurassic, shallow- to deep-marine deposits related to a Paleopacific transgression are recognized in the Neuquen and the Pampa de Agnia basins. Near the boundary between the early and middle Jurassic, a dominantly acidic volcanism (Marifil Complex) covered large areas of northern Patagonia. The older evidence of an Andean magmatic arc occurs at the southern margin of the Pampa de Agnia depocenter.

For the Bathonian-Callovian transition (165 Ma), most of the Patagonian region to the south of the Gastre Fault System was characterized by the Chon Aike and Tobifera bimodal volcanics. In central and southern Patagonia, several NW-SE and NNW-SSE trending grabens formed as a result of widespread extensional tectonism. Transcurrent displacement along the Gastre Fault System controlled the Caoadun Asfalto depocenter (north-central Patagonia), characterized by fluvial and lacustrine facies associations.

In the Late Jurassic (150 Ma), significant paleogeographic changes took place. The Andean magmatic arc reached 50°S and the silicic Tobifera volcanism became restricted to SW Patagonia, whereas submarine rhyolite flows intercalate wit deep-marine siliciclastics. The Rio Mayo-San Jorge and the Magallanes basins are entirely developed. The early rift continental deposits of the San Jorge Basin laterally grade into continued shallow-marine sediments of the intra-backarc Rio Mayo Basin. Shallow-marine facies in most of the Magallanes Basin indicate the onset of widespread extension, and to the west, deep-marine deposits suggest an effective connection between the Magallanes Basin and the Pacific Ocean.

No major paleogeographic changes are recorded for the Valanginian-Hauterivian (135 Ma). However, the Andean magmatic arc extended along the whole western Patagonian margin, though the Pacific connection of the Magallanes Basin still persisted.

The Aptian (120 Ma) was a time of transition. Continental red beds are widespread in the Neuquen and San Jorge basins. The topographic barrier of the magmatic arc produced the closure of the Rio Mayo Basin. To the south, several paths through the volcanic chain connected the Magallanes Basin with the Pacific Ocean.

The Albian panorama (105 Ma) is characterized by a positive volcanic chain (Andean Magmatic Arc), which separated Patagonia from the Pacific Ocean. Fluvial red beds are widespread in the Neuquen Basin, and in the San Jorge Basin, extended fluvial systems terminate in deltaic-lacustrine depocenters. The Magallanes Basin passes through an early foreland stage, closing the marginal basin, and from the north-west corner a very active deltaic system prograded towards the south where prodelta and deep-marine shales are associated to axial (N–S oriented) turbidite systems.

During the Cenomanian-Turonian (90 Ma), the Neuquen and San Jorge basins are integrated in a single continental depocenter demonstrated by continental red beds. In north-eastern Patagonia, the newly opened Colorado rift is also filled in with continental sediments. Along the western margin of southern South America, the topographic barrier of the Andean magmatic arc separated Patagonia from the west and a progressive migration of the depocenter to the east.

The Campanian-Maastrichtian (75 Ma) shows a widespread transgressive episode, embracing the Colorado Basin and the North Patagonian shallow platform. The San Jorge Basin became again a large and isolated depocenter. A general regression is recorded in the Magallanes Basin, caused by both a renewed uplift of the Andean margin and a marked NNW-SSE fluvial-deltaic progradation.

The Late Cretaceous, a time of relative tectonic quiescence in southern South America, records the maximum flooding of the area by Atlantic waters. Plate tectonic reorganization in the Paleogene introduced new patterns of Andean deformation and magmatism, and coincided with a general marine regression. Magmatic arc buildup and renewed marine flooding followed in the early Neogene, with the late Neogene being dominated by pronounced shortening, Cordilleran uplift, and marine withdrawal.

14.4 GREATER AMAZONAS BASIN

The E–W trending Greater Amazonas Basin has been implanted upon the Amazon Craton, dividing it into a northern Guyana shield and a southern

Central Brazilian shield. Due to the fact that the main traces of the formerly contiguous basement remained evident, the basin became subdivided into four sub-basins, separated by structural highs, each sub-basin with its proper identity. The westernmost sub-basin, called Acre Basin, is separated in the east from the Solimões sub-basin by the Iquitos arch. The Purus arch separates the Solimões sub-basin from the so-called Amazonas sub-basin, and this latter is separated from the Amazon Mouth sub-basin by the Gurupá arch.

The Acre Basin belongs still to the Andean foreland basins and has been strongly affected by the Andean Orogeny. Its sedimentary fill is represented by: (1) a Late Carboniferous-Permian section consisting of a siliciclastic-carbonate sequence, with basal conglomerates, sandstones, mudstones, limestones, and evaporites, representing a submergent episode with fluvial-alluvial and neritic environments; (2) a restricted Jurassic section, with sandstones, siltstones and evaporites, also from a submergent episode, and deposited in fluvial, lacustrine, and coastal sabkha realms; (3) an upper Late Cretaceous-Cenozoic section, deposited under oscillatory-emergent circumstances, under the influence of the Andean orogeny, with sandstones, microclastics, and some calcarenites, accumulated in fluvial-deltaic and fluvial-lacustrine depositional environments.

The Solimões sub-basin shows a Precambrian basement section composed of crystalline rocks, covered by a Vendian continental rift basin fill of sandstones and shales. The upward following section of Late Silurian-Early Devonian age, from an oscillatory-emergent episode, is represented by shales with sandstone intercalations, accumulated in a neritic to transitional realm. The Middle-Late Devonian to Early Carboniferous sequence, with submergent character, shows fine- to medium-grained, siliciclastic rocks and a few carbonate sediments from a neritic-marine to transitional environment as well as glacial-marine diamictites. An Upper Carboniferous-Permian section, representing an oscillatory-emergent period, shows from the base upwards sandstones, a thick carbonate-evaporitic sequence with a cyclic association of alternating shales and salts, and finishing with siltstones and shales. These sediments accumulated in fluvial-eolian, restricted marine to neritic, and fluvial-lacustrine environments, respectively. The uppermost section of Late Cretaceous-Cenozoic age is composed of sandstones at the base and somewhat sandy claystones at the top, deposited in oscillating fluvial to fluvial-lacustrine realms.

In the proper Amazonas sub-basin, five complexes of different tectonic-sedimentary character have been recognized. The lowermost Vendian section, accumulated under oscillatory-emergent circumstances, covers the crystalline basement rocks with fluvial-alluvial rift basin conglomerates at the base, and tidal-plain carbonate-fine clastics at the top, already under subsiding conditions. The upward following Silurian-Early Devonian section is composed of alternating sandstone and mudstone sequences, from neritic, possibly glacial-marine, and again neritic environments, representative for an oscillatory-emergent episode. The third complex is of Middle Devonian-Early Carboniferous age, of submergent character and composed of mudstones, siltstones, shales and a few diamictites, accumulated in chiefly marine and littoral environments, of neritic, fluvial-deltaic and anoxic marine character with a certain glacial influence. The following upward Late Carboniferous-Permian section shows sandy-silty siliciclastic, carbonate, and evaporitic deposits, from a number of alternating environments (fluvial-eolian, neritic, restricted marine, hypersaline lacustrine and fluvial-lacustrine), characterizing the oscillatory-emergent character of the depositional episode. The fifth and upper complex is almost equal to those of the Acre and Solimões basins, representing the oscillatory-emergent Late Cretaceous-Cenozoic tectonic-sedimentary episode, with Late Cretaceous-Tertiary fluvial sandstones at the base, and a few fluvial-lacustrine clays at the top.

The Amazon Mouth sub-basin is the youngest of the Greater Amazonas sub-basins, essentially formed when the Amazon river directed its drainage towards the Atlantic Ocean. A thin volcano-sedimentary sequence, composed of sandstones and basic igneous rocks of latest Permian-Early Triassic age (186–222 Ma) has still been determined underneath the thick Late Cretaceous-Cenozoic accumulations. Various lithostratigraphic units have been distinguished since Aptian until Holocene. The Aptian-Albian section represents the opening Equatorial Atlantic rift, and its infilling with alternating sandstones and shales. The passive margin section, since Cenomanian shows a neat distinction between the coastal, central, and deep-marine areas. In general, the near-coastal sections show sandstones and sands, the central section different-type limestones and marls, and the deep-water section shales, claystones, and clays.

All sedimentary complexes occurring in the Greater Amazonas Basin are separated between them by erosional unconformities and hiatuses. The proper complexes correspond neatly to the sediment deposition during the different tectonic-sedimentary episodes, varying between submergent and oscillatory-emergent in a cyclic succession.

14.5 NE BRAZILIAN BORBOREMA AND SÃO FRANCISCO PROVINCES

The formation and following evolution of the sedimentary basins, formed within the NE Brazilian Borborema and São Francisco tectonic provinces, show that their origin occurs chiefly during submergent tectonic-sedimentary episodes when subsidence prevails. Two types of basins series can be recognized in the area: (1) marginal basins where the provinces became amalgamated to neighboring cratons, and (2) intracratonic and intracontinental basins developed in tectonically weak zones within the proper provinces.

The marginal basins formed at the end of the Paleoproterozoic during collage processes in convergent settings, and became later often reactivated during periods of crustal extension in these marginal belts, having successor basins implanted upon them. In the Cretaceous, marginal basins developed again, now alongside the opening Atlantic Ocean rift.

The intracratonic and intracontinental basins formed through intraplates stresses, causing pull-apart movements in weak crustal zones, as transtensional basins in hybrid settings. These latter basins formed cyclically since Mesoproterozoic, either as rift basins, or as centroclines, these chiefly in Mesoproterozoic times, or as syneclises, these essentially in the Neoproterozoic and Phanerozoic, depending on the intensity of the tectonic activity.

The character of the lithic infillings of these sedimentary basins reflects their supposed origin. Record of oceanic crust has only been reported from the Paleoproterozoic marginal basins whence their formation. From later periods, the sedimentary sequences, chiefly of clastic composition, point only to basins implanted upon continental crust. The sedimentary sequences deposited in the different basins appear to be surprisingly uniform, with nearly always the same type of sediment succession. During submergent episodes, the accumulated sediments are chiefly rather fine clastics and non-clastics deposited in marine and littoral environments (thalassocratic phase). The sediment sequences accumulated during oscillatory-emergent episodes are essentially continental terrigenous, various-sized siliciclastics (geocratic phase).

Three sedimentary basins of the study area have been selected for a more detailed consideration of their cyclic development.

The rather well-known Sergipe belt marginal basin between the Borborema Province and the São Francisco Craton started its formation in the beginning of the Mesoproterozoic, and finished its proper history in the beginning of the Phanerozoic. During this period the basin became cyclically reactivated. With exception of the last deposited sequence, the other sediment successions have been metamorphosed, the oldest ones stronger than the youngest; however, the sediment properties remained recognizable. The deposited lithic sequences are represented by five lithostratigraphic groups and formations during the Early Stenian, Early Statherian, Tonian, Late Cryogenian (with distinction between distal and marginal sections), and Cambrian-Early Ordovician (molassic deposits), respectively. The accumulated sediments are different-grade metamorphosed siliciclastics of various sizes and limestones, however almost not studied in more detail about their original character, with exception of the two youngest sections.

The Parnaíba intracontinental syneclise is found within the Borborema Province. The basins shows a spectrum of sedimentary environments, from marine to continental, in various stages which correspond to the cyclic tectonic-sedimentary episodes, of which six are represented in the basin. The lowest sequence, of Late Neoproterozoic age, consists of siliciclastic sediments and ignimbrites, with a molassic character. The Cambrian-Early Ordovician sequence, of submergent character, was found only in one well, also with probably marine siliciclastic deposits. The chiefly Silurian Serra Grande Group sediments deposited during an oscillatory-emergent period, are represented by diamictites, sandstones and shales, denoting glacial, fluvio-glacial, periglacial, neritic marine and fluvial environments of a complete transgressive–regressive cycle. The Middle Devonian-Early Carboniferous Canindé Group consists of alternating sandstones and microclastic sedimentary units, of a new transgressive–regressive cycle representing the major marine ingression in the basin; the depositional environments were rather cold climate, shelf, tidal, littoral and fluvio-deltaic realms with frequent storm-deposited tempestites. The next sequence of Late Carboniferous-Early Triassic age, is composed of the clastic-evaporitic complex of the Balsas Group, deposited during an oscillatory-emergent episode, in alternating desertic, lacustrine, littoral, sabkha, and neritic environments. The sixth Jurassic-Early Cretaceous sequence, although of submergent character, is represented by a rather thin section of fluvio-lacustrine to desertic sandstones, siltstones and shales at the base, intercalated between two basalt extrusions (Early Jurassic and Early Cretaceous, respectively), and a basin subsidence cycle as a consequence of the rupture of the Equatorial Atlantic margin, characterized by fluvio-lacustrine sandstones and claystones and shales, limestones and anhydrite deposited during a short-lived marine ingression. The uppermost and last oscillatory-emergent episode is recorded by Cenozoic relief-correlated fluvial sandy sediments.

The so-called Afrobrazilian depression is suggested to have existed between Middle Paleozoic and Early Mesozoic times in the present-day Atlantic coastal area. It was almost probably an intracontinental sag basin upon a rifted basement within the Gondwana supercontinent. Its preserved deposits are of marine and continental origin, accumulated during the Middle Devonian-Early Carboniferous submergent and Late Carboniferous-Triassic oscillatory-emergent episodes. The sedimentary record of the submergent episode consists of basal conglomerates, sandstones, shales, and a few limestones at the top. The record of the next geocratic episode is also composed of various-type siliciclastic sediments and some silica. The depositional environments were alternating marine and continental. The depression as such ceased to exist when during the Late Jurassic-Early Cretaceous submergent episode the Atlantic rift opened and the ocean formed. As a consequence, a series of Atlantic margin successor basins developed and their deposits accumulated on top of the Afrobrazilian sequences. The Jurassic-Late Cretaceous (Santonian) period shows the most complete sedimentary sequences deposited in four phases: pre-rift (Callovian-Tithonian), syn-rfit (Berriasian-Aptian), proto-oceanic gulf (Aptian-Albian) and open-ocean drift (Albian-Coniacian/Santonian), during which a broad spectrum of different-type sediments accumulated recorded by clastic and non-clastic deposits. From the Campanian onwards the episode turned oscillatory-emergent again; diverse sequences of relief-correlated sediments accumulated, in the following system tract from land seaward: fluvial – fan-delta – carbonate shelf – clastic slope, with their respective sediment types: sandstones, claystones, and limestones.

The cyclic behavior of earth's geologic phenomena is thus also reflected in the sedimentary basin formation and further evolutionary history as well as in the sediment type accumulation. This was demonstrated for the NE Brazilian Borborema and São Francisco tectonic provinces since the beginning of the Mesoproterozoic until the present-day Holocene.

14.6 APPALACHIAN FORELAND BASIN

Orogenies appear to occur in pulses or phases, called *tectophases*, which are focused at certain times and places along an orogenic belt for duration in the order of 10^6–10^7 years. On the Appalachian margin, tectophase occurrence was apparently mediated by successive convergence with continental promontories, but more importantly, each tectophase represents generation and migration of a deformational load. The flexural consequences of the loading and subsequent relaxation produced generalized, third-order, transgressive–regressive, sedimentary cycles in the resulting foreland basin that can be explained through a seven-part model. Inasmuch as the Appalachian foreland basin was the product of a series of orogenies and included tectophases, the sedimentary record in the basin is a series of stacked, unconformity-bound sequences, reflecting tectophase cycles of unequal periodicities. These cycles are herein referred to as *foreland tectophase cycles* or *sequences*.

Thirteen such cycles in the composite, Appalachian foreland basin record the Middle Ordovician to Permian (?), convergence history of the Laurentian/Laurussian margin during four orogenies in two, second-order orogenetic cycles. Early convergence on the margin seems to largely reflect subduction-type orogenies, whereas the later stages were collisional, and the sedimentary manifestation of cycles from both types is distinctive. The eleven, early, Appalachian tectophase cycles are dominated by similar sequences of largely marine lithologies, the most prominent of which are dark, marine shales, indicating major episodes of sediment starvation in rapidly subsiding, deep basins. In contrast, the final two Alleghanian tectophase cycles are dominated by marginal-marine to terrestrial clastic sediments, deposited in what were probably broad, shallow, foreland basins that were overfilled or close to overfilling. Other differences reflect the symmetry of orogenies relative to climatic and tectonic processes and the superimposition of glacio-eustatic fluctuations. Although regional attributes like continental promontories may control the development of tectophases and resulting tectophase sequences in the Appalachian Basin, the fact that orogenic, epeirogenic, and eustatic processes are related at interregional and global scales could mean that such cycles and their bounding unconformities may also have correlation value at these scales.

14.7 WEST IRELAND

Western Ireland belongs geologically to the Caledonian orogenic belt that also occurs in Scotland and Scandinavia, and, therefore, has not been included in the paragraph about the NW European sedimentary basins. Here is sketched the tectono-sedimentary history of the western Irish Dingle, Munster and Clare basins.

The Dingle Basin is found in County Kerry and preserves an Ordovician to Carboniferous sediment succession including the most complete Old Red Sandstone magnafacies in SW Ireland. The oldest part of this sequence was deposited under a transgressive tectonic regime between Avalonia and Laurentia. The younger sequences were deposited in basins of the northern margin at the Munster Basin of South Kerry and Cork. The Old Red Sandstone megafacies of the Dingle Peninsula is structurally constrained between two ENE-trending lineaments south of the Iapetus Suture Zone. The area is bounded to the north by the North Kerry Lineament and to the south by the Dingle Bay Lineament. The basin was subjected to compressive and transpressive tectonic regimes during the Caledonian and Variscan deformations; these two phases were separated by a period of Middle-Late Devonian extension.

Stratigraphically, the following major units can be recognized in the Dingle Basin:

Early Ordovician, Annascaul Formation – deep-water clastics and volcanics;

Early-Late Silurian, Dunquin Group – shallow-marine, shelf and ephemeral-fluvial clastics, and volcanics;

Late Silurian-Early Devonian, Dingle Group – clastics accumulated in barrier island, ephemeral lacustrine, ephemeral fluvial terminal fan, perennial braided river to alluvial fan environments;

Early Devonian, Smerwick Group – ephemeral fluvial sinuous stream, epehemeral fluvial terminal fan, and erg to erg-margin clastics;

Middle Devonian, Caherbla Group – clastic sediments deposited in perennial and ephemeral braided rivers, alluvial fans, and erg to erg-margins;
Middle Devonian, Pointagare Group – deposited in the same environments as the Caherbla Group clastics;
Late Devonian, Carrigduff Group – perennial braided river and alluvial fan clastics;
Late Devonian, Ballyroe Group – clastic deposits from perennial and braided river, alluvial fan, erg-margin, and tidal incursion realms;
Late Devonian, Slieve-Mish Group – perennial braided river, sinuous stream, and alluvial fan clastics;
Early Carboniferous, Tralee Group – clastics.

In the South Munster Basin of South Kerry and Cork, the sedimentary record is represented by a transgressive rock sequence of Late Devonian-Early Carboniferous age. At the base, a predominantly sheltered, shallow-marine sequence developed across the basin, in the east sand-dominated, in the west mud-dominated. Upward this sequence is covered by a transgressive mudstone unit that passes into sand-dominated sheltered facies in the west, and a muddy open-shelf sequence in the east.

The Clare Basin of County Clare shows spectacular exposures of turbidite, slope, and deltaic depositional systems, which accumulated in an Upper Carboniferous (Namurian) basin. The Clare Basin was an intracratonic basin formed on continental crust that underwent active extension during the late Devonian and early Carboniferous. Sediment supply to the basin via fluvial systems was substantial and comprised a mixed load of clay, silt, and sand. The lithostratigraphy of the Namurian succession recognizes two groups of strata as follows:

Shannon Group – Clare Shale Formation, with deep basin, condensed black shales; Ross Sandstone Formation, a sandstone-dominated, axial, deep basin turbidite system; Gull Island Formation, represented by a fine-grained, unstable progradational slope system.
Central Clare Group – comprises five major fine-dominated unstable river delta system cyclothems.

14.8 NORTH-WEST AND CENTRAL EUROPE

North-western and Central Europe underwent a long geological evolution during which its megatectonic setting changed repeatedly. In the course of its geological history, a number of genetically different sedimentary basins formed; some of these were stacked on top of each other while others were partly destroyed by subsequent events.

The complexity of the structure of Central and Western Europe is for a large part due to the rapid succession of four important orogenic events, each belonging to a major orogenic period: the Cadomian (Baikalian), Caledonian, Variscan, and Alpine Orogenies. The four successive mountain chains have partly occupied the same space, resulting in an intricate pattern of rocks formed or deformed during each orogeny. Due to this multiple orogenesis, the presence of older Precambrian rocks is difficult to demonstrate.

Low-grade Proterozoic sediments and volcanics are known from the Armorican and Bohemian Massifs. Crystalline basement rocks predating these supracrustals have been reported only from these massifs. Other signs of Precambrian events are zircon ages of more than 2000 Ma.

The Cadomian orogeny has been demonstrated in England, the Armorican Massif, and the Bohemian Massif, either by the unconformity of Cambrian on folded basement, or by geochronological methods. The folding is accompanied by regional metamorphism, of varying grade up to granulite facies. Post-orogenic Cadomian granitic magmatism occurs in the Armorican and Bohemian Massifs.

The Caledonian orogeny occurs besides the main belt in Scandinavia, Scotland, Ireland, and Wales, in the Ardennes and a zone from northern Germany towards Poland and Rumania. South of this zone, no Caledonian folding based on geological evidence as unconformities can be ascertained, although numerous radiometric data on metamorphic and igneous rocks indicate a thermal event during the period corresponding to the Caledonian orogeny.

The Variscan orogeny is much better known although many problems still remain unsolved. Folding, metamorphism, and granitic activity started in the Devonian, but most widespread and extensive folding, low P/T type metamorphism and granitic intrusion is of Late Carboniferous age.

During Early Devonian times, much of north-western Europe was occupied by Caledonian fold belts masking the suture between the North America-Greenland and the Fennoscandian-Russian plates. Remnants of a Lower Paleozoic continental shelf sequence are preserved in the Baltic. The London-Brabant Massif formed a Caledonian intramontane stable platform. The Mid-European and North German-Polish Caledonides grade southwards and eastwards into the Variscan geosynclinal system, the main elements of which are the Central Armorican-Saxothuringian and the Averno-Lugian-Moldanubic geanticline.

Devonian transcurrent movements of major proportions between the North America-Greenland and Fennoscandian-Russian plates caused the subsidence of the fault-controlled Old Red basins in the Arctic-North Atlantic domain. In the area of the Mid-European Caledonides, the Cornwall-Rhenish Basin subsided during the Devonian and Early Carboniferous; this was accompanied by extension tectonics and a distinctly bimodal rift volcanism. Rift tectonics also characterized the Carboniferous development of the Norwegian-Greenland sea and the northern parts of the British islands.

During the Early Carboniferous, underthrusting along the margins of the Averno-Lugian geanticline marked the onset of the Variscan orogeny. By late Visean times underthrusting also set in along the margin of the Normanian-Mid German High; by this time the Cornwall-Rhenish rift had become inactive and the Variscan foredeep basin developed in its place.

Whereas in Europe the Variscan orogeny came to a close during the Late Westphalian, crustal shortening persisted in the Appalachians and Urals until Mid-Permian times. A transform fault system linked these fold belts and caused in Europe the emplacement of a complex shear fault system and the subsidence of related grabens and troughs. This was accompanied by widespread volcanism. Extinction of this volcanism during the Late Permian followed the consolidation of the Appalachians and the Urals and the final suturing of the Pangea mega-continent.

In North-Western Europe the Permian post-orogenic uplift and partial collapse of the Variscan fold belt went parallel with the subsidence of two new intracratonic basins. Rifting in the Arctic-North Atlantic led to the opening of a first seaway between the Arctic and the NW European basins during late Permian times.

The Triassic development of North-Western Europe, the Arctic-North Atlantic, and the Tethys was dominated by regional crustal extension. A new set of grabens and troughs started to subside, whereby fractures emplaced during the early Permian were partly reactivated.

The Jurassic split-up of the Pangean continent led to the opening of the Tethys as well as the progressive rifting in the Arctic-North Atlantic; this was accompanied by the development of major rift systems in North-Western Europe. Of these, only the North Sea graben experienced a short-lived volcanic phase.

With the Lower Tertiary onset of seafloor spreading in the northern North Atlantic and in the Norwegian-Greenland Sea the Mesozoic rifts of North-Western Europe became inactive and started to subside regionally.

Alpine suturing of Eurasia and Africa during the Late Cretaceous and Early Tertiary was accompanied by compression and inversion of the NW European Mesozoic troughs located at a distance of up to 800 km to the north of the Alpine front. However, the Atlantic seaboard, as well as the North Sea Basin, continued to subside during the Tertiary.

The sedimentary infillings of the basins developed in West and Central Europe since Middle Devonian, are in accordance with the general thalassocratic and geocratic phases that correspond to the submergent and oscillatory-emergent tectonic-sedimentary episodes, respectively. During the Early Variscan submergent phase, most of West and Central Europe constituted a marine environment with limestone sedimentation; continental sandstone deposits prevail in the present North Sea area and the British islands. Towards the end of the episode, the realm turned paralic, with retreat of the sea, and finally continental, with the abundant deposition of coal cyclothems. The Late Variscan oscillatory-emergent phase is represented mainly by continental sandstones and shales. In South-Central Europe, through influence of the developing Tethys Ocean, an important transgression resulted in the accumulation of marine limestones. Towards the end of the episode, this transgression advanced in a westward direction, accompanied by the carbonate deposition. The Early Alpine submergent phase shows an advance of the marine environment over the area, culminating in the well-known Late Cretaceous marine limestones. From then on, during the Cenozoic oscillatory-emergent period, the area became ever more continental, with the exception of the North Sea rift basin in which an alternation of continental and marine, chiefly fine clastic sediments became deposited.

14.9 DUERO BASIN (SPAIN)

The Duero Basin occupies a large area in the north-west of the Iberian Peninsula. It has an approximately quadrangular shape, and three of its four corners are the sites of distinctive sub-basins, which extend outwards from the main basin. The different margins of the sub-basins and the main basin have distinctive histories of tectonic and sedimentary evolution.

The tectonic activity in the basin during most of the Tertiary was determined by a transpressive regime that reactivated Hercynian faults. The record of the Alpine Orogeny is complex because the sedimentary record indicates a compressive regime in the source areas coeval with the extensional to transpressive regime indicated by normal or strike-slip faults. This duality is due to the geotectonic position of the area between two compressive areas, the Cantabrian Range and the Central System, and the extensional Atlantic margin.

The sedimentary infilling of the Duero Basin has been divided into three tectonosedimentary complexes, composed of several tectosedimentary units. The Upper Cretaceous-Paleocene pre-orogenic complex consists of siliciclastic, carbonate, and evaporitic deposits arranged in a fining-upward sequence. Silicification and lateritic weathering profiles of a seasonal tropical climate prevail. The depositional environments ranged from terrestrial in the west to marine in the east.

The next syn-orogenic complex is of Eocene-Miocene age. It is composed chiefly of siliciclastic sediments forming generally a coarsening-upward sequence that records the uplift of the basin borders. The deposited sediments are mineralogically immature, being mainly arkoses, lithic arkoses, and polymictic conglomerates. They accumulated in alluvial fan and fluvial realms, under arid subtropical climatic circumstances. In the south-east and north-east corners of the basin marine carbonates are found.

The Miocene-Holocene post-orogenic complex is the youngest depositional section of the Duero Basin. It consists of siliciclastic, carbonate, and evaporitic deposits, which form a fining-upward sequence, recording a new extensional stage. The complex progressively overlaps previous units and the basin borders. The sediments deposited in alluvial fan, fluvial and lacustrine environments which filled a basin with a shape roughly similar to the present Duero Basin. The profiles reveal a shift from arid (near the base) to humid (towards the top) mediterranean climate.

Gravel sheets previous to the fluvial incision (*rañas*) cover large areas of the basin and its borders. They formed diachronously since the end of the Oligocene.

14.10 NORTH CROATIAN BASIN

The sedimentary succession of the Neogene North Croatian Basin is a result of a very complex basin evolution. Since the basin was generated by continental rifting processes, tectonic and thermal subsidence strongly controlled the sedimentation. A position of the basin within the Central Paratethys provided specific influence of global sea-level changes because a connection of the Central Paratethys with the Mediterranean and Indo-Pacific Ocean was several times established and ceased ending with total basin isolation. The consequence was an evolution of fresh-water, marine and brackish, and fresh-water environments at its end.

The basin evolution represents a first-order cycle, which comprises three second-order cycles separated by major regional unconformities. Six third-order cycles can also be distinguished. They show a transgressive–regressive tendency as result of a very complex interaction of tectonic uplifting and subsidence, global and regional sea-level changes, thermal subsidence, change of climate, and delta progradation.

14.11 RUSSIAN ARCTIC SHELF BASINS

Along the Eurasian Arctic Ocean passive continental margin, a number of sedimentary basins developed in Russian territory: Barents-North Kara, South Kara, Laptev Sea, and basins of East Siberian and Chukchi Seas. Their characteristics are in many respects determined by the rather young age of the Arctic Ocean. Besides this young age, the moderate areal extension of the deep ocean floor correlates with the great shelf width and the very thick sedimentary sequences in the shelf basins.

Due to poor and irregular geological and geophysical data, the cyclic development of the basins and their sedimentary sequences is still being studied. Within the Russian Arctic shelf sedimentary basins the processes of platform development, rifting, compression-extension and uplift-subsidence migrated from west to east in a complicated way. The most important event in the history of the Phanerozoic sedimentation was the transition from carbonate-terrigenous deposition through most of the Paleozoic to terrigenous deposition from Late Permian to Mesozoic and Cenozoic. The age of the sedimentary sequences ranges from west eastward from Precambrian to Cenozoic in the Barents Sea to Late Cretaceous-Cenozoic in the basins of the East Siberian and Chukchi Seas.

In the Barents Sea-North Kara Basin the sedimentary sequence can be divided into three lithostratigraphic complexes: Vendian-Lower Permian terrigenous-carbonate, Upper Permian-Triassic terrigenous and Jurassic-Cretaceous terrigenous. The Vendian and Cambrian deposits are mainly terrigenous and dolomite; the Ordovician-Early Devonian ones are carbonaceous and terrigenous-carbonate with vertical and lateral facies variability. The Upper Devonian-Early Permian deposits are essentially carbonates with some evaporites. The Upper Permian-Triassic complex is a rift infilling with continental to shallow-marine sediments. The Jurassic-Cretaceous deposits are predominantly marine with small environmental oscillations.

The South Kara Basin is filled with a thick sequence of strata from Permian to Quaternary in age, with the following parts: rift-related terrigenous strata (Permian-Lower Triassic), general basin subsidence related, rhythmically alternating sandstones, siltstones and claystones (Triassic), and downwarping of the basin (Jurassic-Quaternary) with chiefly marine clastic deposits of marine and littoral origin.

The Laptev Sea Basin is a large centrocline open to the Eurasia Basin. In the sedimentary fill, two stages can be distinguished. The lower stage lasted from Late Proterozoic to Late Mesozoic, formed in different geodynamic and sedimentary settings, having suffered repeated reworking. Vendian and Cambrian strata are clastics and carbonates; Ordovician-Middle Devonian sediments are mainly carbonates. Upward the carbonates become substituted by marine and paralic terrigenous deposits, which become dominating in the Mesozoic. The upper stage comprises sediments from Aptian to Holocene. The sediments are mainly non-marine clastics with coal and widespread volcanics. Their deposition occurred in oscillatory environments, often separated by erosional unconformities.

The sedimentary basins of the East Siberian and Chukchi Seas are shelf basins, with their sedimentary complexes subdivided into a Paleozoic-Early Mesozoic and an Early Cretaceous-Cenozoic stage. The deposits of the lower stage are terrigenous clastics and carbonates, with magmatic intercalations. The Late Cretaceous-Cenozoic sediments accumulated in alternating marine and fresh-water environments, being represented by clastic deposits and lignite.

The different basins are separated by structural highs and fold systems. The South Kara Basin appears to be the northern and most subsided structural depression.

14.12 SOUTH-EAST SIBERIAN PLATFORM

In the south-eastern part of the Siberian platform, the Uchur-Maiskaya Basin is a large, extended depression, with a complex fragmental structure with bulges of the Pre-Riphean basement, different-order troughs and volcano-magmatic structures. The sedimentary cover may be divided into two structural stages: Riphean and Vendian-Cambrian.

The Uchur-Maiskaya section of the Riphean is one of the most completely described reference sections of the Neoproterozoic in Northern Eurasia, including Lower, Middle, and Upper Riphean deposits. These have been divided into six series:

The Uyanskaya series, consisting of three complexes – (a) at the base there occur conglomerates passing upward into sandstones deposited in a shallow-water intracontinental basin; (b) the following complex is also composed of sandstones, with reddish colors, deposited in a shallow-water, brackish coastal basin; (c) the uppermost complex shows carbonate-terrigenous rocks.

The Uchurskaya series, with sedimentation in a vast flat basin filled with shallow-water marine and littoral deposits united into three complexes, accompanied by a general subsidence of the basin – (a) the lowermost complex shows a nearly equal ratio of sandstones and siltstones; (b) the middle complex is made up of different-type sandstones, intercalated with dolomites and marls; (c) the upper sequence consists also of different-type sandstones.

The Aimchanskaya series of Middle Riphean age, is a large sedimentary transgressive rhythm consisting of two complexes – (a) the lower sequence is of terrigenous origin and composed of sandstones and siltstones; (b) the overlying complex is of marine origin, recorded by a lower and upper dolomite section and a middle sandstone section.

The Kerpylskaya series, with a transgressive marine character, has been subdivided into four complexes, with a weathering crust at the base – (a) the lowermost complex is made up by sandstones with siltstone interbeds and chiefly dolomites at the top; (b) the following section is composed of mudstones and siltstones; (c) limestones characterize the next complex, which are often dolomitized; (d) the upper sequence consists chiefly of dolomitic limestones.

In the Lakhandinskaya series, also of marine origin, four complexes are distinguished, all composed of mudstones and different-type dolomites and limestones.

The Uiskaya series, consisting of two complexes, is composed – (a) at the base of clastic terrigenous rocks, and (b) at the top also of the same rock types.

The Vendian-Cambrian structural stage overlies unconformably the Riphean sequences, after a considerable hiatus of uplifting. During this stage, basin subsidence continued. The Vendian is represented by the marine Yudomskaya series, with two complexes – (a) the lower sequence consists of fine-grained clastics, dolomites and limestones; (b) the upper section is also recorded by carbonate rocks. A hiatus, characterized by weathering crusts, separates the Vendian from the Cambrian series. The latter deposits are made up of marine limestones and dolomites.

After a long time of non-sedimentation, deposition was retaken in the Jurassic, corresponding to the reactivation of Mesozoic tectonic movements. The sedimentary record is represented by oligomictic sandstones, with a considerable volcanic component. Their thickness is reduced in relation to the rest of the Jurassic which is essentially represented by effusive and pyroclastic rocks.

The Vilyuiskaya syneclise shows on top of the Precambrian basement rocks, four stages: Vendian-Lower Devonian, Middle Devonian-Lower Carboniferous, Upper Carboniferous-Permian, and Mesozoic.

The Vendian-Lower Devonian episode shows in the beginning a large transgression until the Middle Cambrian, followed by a gradual regression. The Vendian deposits overlie transgressively the basement, and are composed of shallow-water marine sandstones and limestones. Carbonate sediments and mudstones characterize the Cambrian. The epicontinental sea of the Ordovician resulted in the deposition of sandstones, siltstones, and dolomites. The Silurian sequence consists also of limestones and dolomites as well as siltstones and marls. The Lower Devonian appears to have been a denudational period.

The Middle Paleozoic stage resembles in many respects the Vendian-Cambrian stage. Wide rifting processes were frequent and determined the graben fills of volcanogenic-sedimentary assemblages. The sediments are chiefly sandstones, siltstones, limestones, marls, and evaporites, mainly of marine origin.

The Upper Carboniferous-Permian stage is represented by a thick complex of coastal-continental, deltaic, coastal-marine sandstones, and microclastic rocks. Interbedded there occur bituminous coal seams.

The Mesozoic deposits in the Vilyuiskaya syneclise are most widespread. The Triassic environment was essentially a shallow-water bay-like sea, in which dark-colored mudstones accumulated, accompanied by basic lava outflows. Jurassic deposits are conglomerates, sandstones, and microclastics of marine origin, turning upward into a more continental environment. The Cretaceous sedimentary sequences are also essentially clastic, often with rather thick coal seams, of continental origin. The marine regression took place in the Late Jurassic.

14.13 EASTERN MARGIN OF CHINESE CRATON

Two big sedimentary and geodynamic cycles can be distinguished in the Mesozoic of the eastern margin of the Chinese Craton.

During the Late Paleozoic-Early Mesozoic oscillatory-emergent episode, the first major transgressive–regressive cycle occurred in the Triassic in the Sikhote-Alin mobile belt, in spite of this period being essentially geocratic. The rather rapid transgression started in the Induan, with the deposition of mainly clastic sediments, coarse-grained at the base and fining-upward up to silty claystones. The regressions started in the Middle Olenekian, when the depositional sequence coarsened upward into fine-grained sandstones. This regression was rather slow, with various oscillations with changes in the extent of the sea, its level and depth, and the coastline configuration, and lasted until the Norian. The rest of the episode is characterized by an alternation of marine, nearshore, and non-marine deposits recording the oscillations; this phase lasted until about the Early Jurassic.

In the Hettangian the second major regional transgression started, lasting until about the Albian. The resulting deposits are again mainly clastic, predominating the sandstones and claystones. The marine depositional environments were not very deep, and in the area, many littoral sequences are found in which coals are sometimes frequent.

Due to the active plate tectonics of the area, with subduction and strike-slip movements between the Asian continental plate and the Pacific oceanic plate, thick sedimentary prisms could accumulate. In most terranes volcanic-sedimentary and volcanic formations occur. The strong tectonic activity folded many sequences, and separated them into diverse terranes, which became many times dislocated.

On the eastern margin of the Chinese Craton, a great number of Mesozoic-Cenozoic folded terranes have thus been recognized. They are mainly composed of Early Jurassic-Paleocene accretionary prisms, turbidite basin of a transform margin, volcanic island arc, and volcanic belt. During the Mesozoic, the sedimentation took place in an epiplatform environment on the subsiding margin.

The different shelf sediments accumulated in shallow water, not deeper than sublittoral, and also in fresh-water realms. Interruptions in the sedimentation and alternations of marine, littoral and continental environments mark the geodynamic events during the Mesozoic, tectonically active area.

In the Sea of Japan region, several stages of geodynamic evolution have been distinguished: (1) Pre-Jurassic passive margin stage, (2) Jurassic margin of Andean type with transform margin, (3) Tithonian-Early Cretaceous transform margin combined with the Andean-type active margin, (4) Hauterivian-Albian transform margin in combination with a Japanese-type active margin, and (5) a Cenomanian-Paleocene active margin of Andean type.

14.14 PROTEROZOIC CUDDAPAH BASIN (INDIA)

Records of sedimentary cycles mostly owing to overprinting by geological events are limited in the Precambrian. Preservation of cyclicity in stratal assemblages of Paleoproterozoic to Neoproterozoic of the Cuddapah Supergroup, Peninsular India, provides a case history of Proterozoic cyclic deposition. Despite identification of some allogenetically controlled periodic sequences, it is however, difficult to interpret such periodicities in cyclic stratal assemblages.

Cyclic hierarchy based on interplay among various controlling factors are documented from the Cuddapah stratigraphic record. Cyclic bedding, dicyclic bedding, rhythmic bedding, platformal cycles, and orbitally forced cycles are some of the salient cyclicities.

The initiation of Cuddapah sedimentation is marked by deep-seated fault-controlled cyclicities of alluvial fan complexes in the Gulcheru Formation in response to extensional basin tectonics. Switching between arid to humid climatic influence is allocyclic and orbitally forced with attendant cyclic arrangements of eolian and braided fluvial deposits. The transition facies records the span of time needed for changing depositional styles from the extrabasinal Gulcheru to the intrabasinal Vempallae sedimentation. Cyclic deposits of lagoonal, shallow-marine, and associated terrestrial volcaniclastics characterizing the transition facies attest to the oscillatory emergence induced by the pulsatory nature of the basin.

Upward shallowing platform carbonate cycles under variable paleoenvironmental dynamics suggest that cyclic sedimentation in the Vempallae Formation occurred in response to low-amplitude (Milankovich-band) sea-level oscillations.

A trappean-intertrappean sequence with signatures of bimodal volcanism and unmodified mantle xenoliths, rarely reported from the Paleoproterozoic, designates the basal part of the Pulivendla Formation and corroborates reactivation of deep-seated faults. Boulder beds and repetition of couplet facies testify episodic interference by extremely high-energy depositional conditions in the upper part of the formation. This was brought by frequent redistribution of topography of the provenance area and the basin itself in response to regional extension.

Repetitive generation of silicic magma at shallower depths and its intermittent emplacement deposited cyclic ignimbrites and fluvial and related facies in the Tadpatri Formation. This is overlain by cyclic fluvial sandstone-mudstone facies of the Gandikota Formation.

The Nallamalai Formation demonstrates a cyclic association of turbidites with clastic facies, debrites, felsic to silicic volcaniclastics, seismites, seismoturbidites, and tsunamiites owing to variable genetic controls under the influence of extensional tectonics.

The Egalapenta Member of the Srisailam Formation depicts three discrete cyclic deposition types of fluvial to shallow-marine in the lower part, coastal sabkha-eolian sandsheet in the middle and climbing dunes (draa) to interdune forming a part of an erg in the upper part. Such cycles, marking the closing phase of the Cuddapah sedimentation owes its origin to eustatic fluctuations of sea-level and/or orbitally controlled climatic variations.

14.15 ARABIAN-NUBIAN SHIELD

By about 1200 Ma, the Pan-African thermal event started to affect the Mozambique Craton and led to its separation into an ocean in the area occupied now by the Arabian-Nubian Shield. Island arc formation, subduction, collision, and suturing continued until the cratonization of this Arabian-Nubian Shield at about 640 Ma. Intrusive, volcanic, and sedimentary rocks were forming throughout this period in accordance with the geological setting. By about 640 Ma, the newly born craton was affected by nine long-term cycles of emergent, oscillatory and submergent character. Each cycle is defined by two interregional unconformities at its base and top. The main reason for this cyclicity is tectonic on the craton and/or global scale with minor modifications due to eustasy. The duration of the cycles is variable, but the submergent cycles are usually longer than the emergent ones. These cycles are mostly first-order cycles and partly second-order cycles.

14.16 KAROO BASIN (SOUTH AFRICA)

The main Karoo Basin of South Africa is a Late Carboniferous-Middle Jurassic retroarc foreland fill, developed in front of the Cape Fold Belt, in relation to subduction of the Paleo-Pacific plate underneath the Gondwana plate. The Karoo sedimentary fill corresponds to a first-order sequence, with the basal and top contacts marking profound changes in the tectonic setting, i.e. from extensional to foreland and from foreland to extensional, respectively, thus forming a cycle.

Sedimentation within the Karoo Basin was closely controlled by orogenic cycles of loading and unloading in the Cape Fold Belt. During orogenic loading, episodes of subsidence and increase in accommodation adjacent to the orogen correspond to episodes of uplift and decrease in accommodation away from the thrust fold belt. During orogenic unloading the reverse occurred. As a consequence, the depocenter of the Karoo Basin alternated between the proximal region, during orogenic loading, and the distal region, during orogenic unloading. Orogenic loading dominated during the Late Carboniferous-Middle Triassic interval, leading to the accumulation of thick foredeep sequences with much thinner forebulge correlatives. The Late Triassic-Middle Jurassic interval was dominated by orogenic unloading, with deposition taking place in the distal region of the foreland system and coeval bypass and reworking of the older foredeep sequences.

The lithostratigraphic sequences of the Karoo Supergroup represent the following successions: (1) Late Carboniferous-Early Permian Dwyka Group, of continental glaciation in the NE, and floating icebergs and deposition in a marine environment in the S, with diamictites, fluvioglacial sandstones and rhythmites; (2) upper Early to lower Late Permian Ecca Group, represented by epicontinental sea, turbidite fan complexes, and paraglacial deltas and braidplains, composed of a proximal mudstone section and a distal sequence of mudstones, siltstones, and sandstones, with occasional coal seams; (3) upper Late Permian to early Middle Triassic Beaufort Group, from aggradational floodplain, delta distributary channel, meandering channel and lake realms, with mudstones and siltstones interbedded with subordinate sandstones in the lower section, and aggradational meandering floodplain and ephemeral braided stream environments, of lower sandstones and upper siltstones and mudstones with interbedded sandstones, in the upper section; (4) after a time gap, Late Triassic to Early Jurassic units: (a) Molteno Formation, of perennial braided stream sandstones, with subordinate siltstones, mudstones and coal; (b) Elliott Formation, deposited by ephemeral streams and floodplain playas of a dry climate, represented by red bed sandstones interbedded with red to gray mudstones; (c) Clarens Formation, with sheet-like sandstones deposited by sand dunes, in wadis and playas, and by mass flows, under arid climatic conditions. Moreover, on top, there occur still the basalts and dolerites of the Middle Jurassic (about 183 $\pm$ 2 Ma) Drakensberg Group volcanics.

Table 14.1: Integration of sedimentary basin cyclic sequences and tectonic-sedimentary episodes.

Ga ± 10 Ma		ages	tectonic-sedimentary episode	Andes Fueginos	Patagonia	Amazonas	NE Brazil	Appalachian	W Ireland	NW Europe	Duero (Spain)	N Croatia	Arctic	SE Siberia	Eastern margin Chinese craton	India	Arabia-Nubia	Karoo (South Africa)
0	Phanerozoic	Cenozoic-Late Cretaceous	oscillatory – emergent	x	x	x	x			x	x	x	x				x	
0.08		Late Cretaceous-Middle Jurassic	submergent	x	x	x	x			x			x	x	x		x	
0.18		Middle Jurassic-Early Permian	oscillatory – emergent	x	x	x	x		x	x			x	x	x		x	x
0.28		Early Permian-Middle Devonian	submergent			x	x	x	x	x			x	x			x	x
0.38		Middle Devonian-Middle Ordovician	oscillatory – emergent			x	x		x				x	x			x	
0.48		Early Ordovician-Cambrian	submergent				x		x				x	x			x	
0.58	Neoproterozoic	Vendian	oscillatory – emergent			x	x						x	x			x	
0.68		Late Cryogenian	submergent				x							x				
0.78		early Cryogenian	oscillatory – emergent				x							x				
0.88		Tonian	submergent				x							x				
0.98	Mesoproterozoic	Late Stenian	oscillatory – emergent				x									x		
1.08		Early Stenian	submergent				x									x		
1.18		Late Ectasian	oscillatory – emergent				x									x		
1.28		Early Ectasian	submergent				x									x		
1.38		Late Calymmian	oscillatory – emergent				x									x		
1.48		Early Calymmian	submergent				x									x		
1.58		Late Statherian	oscillatory – emergent				x									x		
1.68		Early Statherian	submergent				x									x		
1.78																		
1.88		Orosirian	?													x		
1.98																x		
2.08																		

The out of phase history of base-level changes generated contrasting stratigraphies between the proximal and distal regions of the foreland system separated by a stratigraphic hinge line. The patterns of hinge line migrations show the flexural peripheral bulge advancing towards the craton during the Late Carboniferous-Permian interval in response to the orogenic front. The orogenward migration of the foreland system recorded during the Triassic-Middle Jurassic may be attributed to piggyback thrusting accompanied by a retrogradation of the center of weight within the orogenic belt during orogenic loading (Early-Middle Triassic) or to the retrogradation of the orogenic load through the erosion of the orogenic front during times of orogenic unloading (Late Triassic-Middle Jurassic).

The Permo-Carboniferous to Jurassic aged rocks of the main Karoo Basin are world renowned for the wealth of synapsid reptiles and early dinosaur fossils, which have allowed a 10-fold biostratigraphic subdivision of the Karoo Supergroup to be erected. The role of fossils in interpreting the development of the Karoo Basin is not, however, restricted to biostratigraphic studies. Integrated sedimentological and paleontological studies have helped in more precisely defining a number of problematical formational contacts within the Karoo Supergroup, as well as enhancing paleoenvironmental reconstructions, and basin development models.

14.17 CONCLUDING REMARKS

The above given examples show that in many sedimentary basins cyclic sequences have been distinguished and interpreted. However, it is also shown that full integration of such cyclic development into the scheme of alternating tectonic-sedimentary episodes of submergent or oscillatory-emergent character has not yet been made in most of the studied basins. Therefore, in Table 14.1 such integration has been presented. It appears that the majority of the basins show only Phanerozoic sequences, and only a few deal with the Neoproterozoic. Moreover, a study of Mesoproterozoic and late Paleoproterozoic basins does almost not exist, maybe due to the fact that sedimentological studies of these basins have hardly been made. It is hoped that this could be remedied in near future.

15. CONCLUSIONS

J.M. MABESOONE AND V.H. NEUMANN

15.1 INTRODUCTION

It was the aim of this book to emphasize the long-term geological cycles throughout the Earth's history, mainly directed to the development of sedimentary basins and their lithic infilling.

Sedimentary basins originate commonly during certain periods of the geologic history. These periods occur in a cyclic succession of tectonic-sedimentary episodes with alternating submergent and oscillatory-emergent character. The basins form chiefly during submergent episodes when subsidence prevails. This subsidence triggers the formation of a depression that becomes a depositional sedimentary basin when enough differences in the surrounding relief are reached. The basins are generally centroclines or syneclises upon a rifted basement. When tectonic activity during submergence is quite strong, rift basin formation substitutes commonly centrocline development. The character of the lithic infillings of these sedimentary basins reflects their supposed origin. Throughout a submergent episode, mainly coastal and marine fine-clastic and carbonate deposits accumulate during subsidence (*thalassocratic phase*); under favorable climatic circumstances also evaporites may precipitate. The oscillatory-emergent episode is characterized by coarse- to medium-grained, often conglomeratic, continental clastic deposits, accumulated during uplift and significant relief development (*geocratic phase*); when a sufficiently high relief is present, glaciations may occur, leaving their respective deposits. The individual depositional sequences accumulated during the respective episodes, are always separated by erosional unconformities. In addition, after Yeh Lien-tsun (1977), there exists also a cyclicity in the development of sedimentary ores, with the formation of oolitic iron, manganese, and phosphorite, preferably during transgressional submergent episodes, and of bauxite, coal, copper in sandstones and evaporites, during regressional oscillatory-emergent episodes.

However, not many regional studies have been made on this subject. Most of the here presented examples deal with shorter-term cycles within only a few long-term ones. Therefore, the results of these regional studies have been taken together into long-term cycles, which seem to affect all of world's sedimentary basins (see Table 14.1).

Cyclic Development of Sedimentary Basins
J.M. Mabesoone and V.H. Neumann, editors.

15.2 CAUSES OF CYCLICITY

Cyclicity in the formation of sedimentary basins cannot be understood without the driving forces and mechanisms of geotectonics. Long-term stratigraphical cycles in sedimentary basins are related to the assembly and dispersal of supercontinents over an episodicity of 200–500 Ma, as a result of tectonism, also including the effects of long-term changes in plate-kinematic patterns. The sequences of comparable time duration may be correlated with each other within an area as large as two or more adjacent tectonic plates, dominated by the same major tectonic events, as e.g. large-scale collision or rifting.

The cyclicity of global sedimentation is correlated with the motion of the solar system around the Galactic core and associated variations in the earth's kinematics and physical fields. The sequential approach of initial phases of the sidereal and anomalisitic periods of the solar system constitutes geodynamic megacycles of 1550–1700 Ma duration. Undulatory change of amplitude of the geodynamic activity of the earth's cycles marks the boundaries of Priscoan, Archean, Proterozoic, and incomplete Phanerozoic (see Fig. 2.3). Extreme points of cycles and megacycles are periods of successive evolution of the geosphere and contemporaneous complication of the biosphere.

Cyclic geodynamic activation is determined by rotation rate variations of the earth, and resonance gravitational effects by the bodies of the solar system. The global sedimentation structure is controlled by the earth's rotational network of the planetary jointing and crustal heterogeneities (Fig. 2.1). Tectonic-magmatic activation causes crustal vibrations, while a substantial sediment supply changes the level of the World Ocean and influences global transgressions and regressions. Simultaneous variations of the geophysical fields are expressed in gravity variations at the earth's surface and changes in sedimentation rate. Cyclic deceleration of the earth rotation and intensification of tectonic-magmatic processes result in sedimentation rate acceleration during the Phanerozoic.

15.3 CYCLICITY OF EARTH'S GEOLOGICAL PHENOMENA

As Golubev (Chapter 2) stated, cyclicity of earth's geological phenomena has a galactic cause. This means that most processes affecting the planet are recurrent in time and space. However, although modern processes are our keys to understanding the earth's geological evolution, the magnitude and rates of these processes may have been different in the past. Through geological time, the earth underwent a number of changes in the fundamental controls of its evolution. Nevertheless, most phenomena occur in a repeated succession that appears to be cyclic within a rather restricted time span. The underlying control of all processes is plate tectonics. Periodical uplifts and downwarps of the earth's crust originate the tectonic cycles, often with a duration of more than one geological period. As a consequence, sedimentary basin formation and further evolution is also controlled by tectonics at every stage of the cycle.

The paradigm of submergent and oscillatory-emergent tectonic-sedimentary episodes gives the answer for most of the problems involving the formation of sedimentary basins, their lithic infilling and their post-depositional history.

15.4 FINAL REMARKS

Important is the awareness among sedimentologists and stratigraphers that there exists order and system in stratigraphic phenomena. This order and system is manifested by the repeated and recurrent associations of materials and geometric forms that can be discerned among the rocks of the earth's crust.

It is still arguable that the attempt to discern a pattern of repetitive sedimentation in successions has led to mere misinterpretation and difficulties than it has provided enlightenment. Nevertheless, such a search is probably inevitable at some stage in sedimentary studies and efforts will no doubt continue.

Besides the still very real difficulties involved in absolute dating and correlation, there remains the fundamental objection that regular periodicity of the controlling mechanism is assumed. Even when this is the case, there is no guarantee that such regularity will be expressed in the sedimentation.

Whatever one's ultimate conclusions about the regularity or even the reality of sedimentary cycles, their study, perforce, directs attention to the relatively neglected subject of stratigraphic relationships, that is, the distribution of facies through space and time. This, in effect the study of changing environment and their causes, is the very sense of geology.

We may finish with the often-cited affirmation of Zhemchuzhnikov (1958): "Those who accept rhythms in nature will find it even where it is rather indistinct, and they will arrive at proper conclusions. Those who do not want to, will not find it even where it is obvious."

But we must never forget the admonition of Zeller (1964): "Science, to an extent matched by no other human endeavor, places a premium upon the ability of the individual to make order out of what appears disordered. Therefore, the scientist more than anyone else, needs to maintain his objectivity about his work, and perhaps even more vigorously about himself."

REFERENCES

Abed, A.M., 1982, Depositional environments of the Early Cretaceous Kumub (Hathira) sandstones, North Jordan, Sedim. Geol., 312, 267–279.

Abed, A.M., 1994, Shallow marine chert-phosphorite-palygorskite association, Late Cretaceous Amman Formation, Jordan, in: A. Iijima, A.M. Abed and R. Garrison (eds.), Siliceous, Phosphatic and Glauconitic Sediments of the Tertiary and Mesozoic, VSP, Zeist, The Netherlands, pp. 205–224.

Abed, A.M., 2000, Geology of Jordan, Jordanian Geol. Assoc., Amman (Jordan), 570 pp. (in Arabic).

Abed, A.M., I. Makhlouf, B. Amireh and B. Khalil, 1993, Upper Ordovician glacial deposits in southern, Jordan, Episodes, 16, 316–328.

Aguiar, G.A., 1971, Revisão geológica da bacia paleozóica do Maranhão, 25° Congr. Brasileiro Geol., São Paulo (SP), Anais, 3, 113–122.

Aguirar, G.A., R.P. Bahia and W.M. Rezende, 1966, Prospecções geológicas e geofísicas efetuadas pela Petrobrás na Foz do Amazonas, in: Proc. 7th Guiana Geol. Conf., Paramaribo (Surinam), Verhandelingen Kon. Nederl. Geol. Mijnbouwkundig Genootschap, 27, 13–21.

Al-Jallal, I.A., 1995, The Kuff Formation: its regional reservoir potential in Saudi Arabia and other Gulf countries; depositional and stratigraphic approach, in: Al Husseini (ed.), Middle East Petroleum Geosciences Geo 94, Bahrein, Gulf Petrolinks, 2, 103–119.

Al-Laboun, A.A., 1986, Stratigraphy and hydrocarbon potential of the Paleozoic succession in both Tabuk and Widyan basins, Arabia, in: M.T. Halbouty (ed.), Future Petroleum Provinces of the World, Am. Assoc. Petroleum Geol., Memoir 40, pp. 399–425.

Allen, J.R.L., 1982, Sedimentary structures: their character and physical basis, Developments in Sedimentology (Elsevier, Amsterdam), 30A and 30B, pp. 593 and 663.

Allen, J.R.L., 1983, Studies in fluviatile sedimentation: bars, bar complexes and sandstone sheets (low-sinuosity braided streams) in the Brownstone (L. Devonian), Welsh Borders, Sedim. Geol., 33, 237–293.

Almeida, F.F.M., 1986, Distribuição regional e relações tectônicas do magmatismo pós-paleozóico no Brasil, Rev. Brasileira Geociências, 16, 325–349.

Almeida, F F.M. and Y. Hasui, 1984, O Pré-Cambriano do Brasil, Edit, Edgard Blücher Ltda, São Paulo, 378 pp.

Almeida, F.F.M. and N. Litwinski, 1984, Província Mantiqueira: setor setentrional, in: F.F.M. Almeida and Y. Hasui (coord.), O Pré-Cambriano do Brasil, Edit, Edgard Blücher Ltda, São Paulo, pp. 282–307.

Al-Shanti, A.M., 1993, Geology of the Arabian Shield, King Abdulaziz University, Jeddah (Saudi Arabia), 196 pp. (in Arabic).
Alsharhan, A.S. and C.G.S.C. Kendall, 1994, Depositional setting of the Upper Jurassic of the Arabian Gulf Hith anhydrite: Na analog to Holocene evaporites of the United Arab Emirates and Lake MacLeod of Western Australia, Am. Assoc. Petroleum Geol. Bull., 78, 1075–1096.
Alsharhan, A.S. and A.E.M. Nairn, 1995, Tertiary of the Arabian Gulf, sedimentology and hydrocarbon potential, Palaeogeogr., Palaeoclim., Palaeoecol., 114, 369–388.
Alsharhan, A.S. and A.E.M. Nairn, 1997, Sedimentary Basins and Petroleum Geology of the Middle East, Elsevier, Amsterdam.
Al Silwadi, M.S., A. Kirkham, M.D. Simmons and B.N. Twombley, 1996, New insight into regional correlation and sedimentology, Arab Formation (Upper Jurassic), Offshore Abu Dhabi, Geo Arabia, 1, 6–27.
Amireh, B. and A.M. Abed, 2000, Surface and subsurface occurrence of the new Infracambrian Umm Ghadah formation in Jordan, tectonic implications, Dirasat, Pure Science, 27, 143–168.
Andrews, I.J., 1991, Paleozoic lithostratigraphy in the subsurface of Jordan, Subsurface Geological Bulletin, Vol. 2, NRA, Amman, Jordan, 75 pp.
Apesteguía, S., 2002, Greater Gondwana and the Kawas Sea coastal tetrapod fauna (Campanian-Maastrichtian), Bol. 6° Simp. Cretáceo Brasil, 2° Simp. Cretácico América del Sur, São Pedro (SP. Brasil), pp. 143–147.
Avanić, R., 1997, Facies analysis of the Middle Miocene deposits of the southeastern part of the Medvednica Mountains, unpublished MSc thesis, University of Zagreb, Zagreb, 54 pp. (in Croatian).
Avetisov, G.P., 1982, Deep structure of Novosibirsk Islands and adjacent offshore by seismic data, Sib. Geologia, 11, 113–122 (in Russian).
Avetisov, G.P., 1991, Hypocenters and focal mechanisms of earthquakes in Lena river delta and adjacent areas, Volcanol. Seismol., 6, 59–69 (in Russian).
Avetisov, G.P. and Yu.B. Guseva, 1991, Deep structure of the Lena river delta region by seismologic data, Sov. Geologia, 4, 73–80 (in Russian).
Bajraktarević, Z. and D. Pavelić, 2003, The Karpathian in Croatia, in: Brzobohatý, M. Cicha, M. Kováč and F. Rögl (eds.), The Karpathian. A Lower Miocene Stage of the Central Paratethys, Masaryk University, Brno, pp. 141–144.
Baklanova, L.S., I.V. Buriyi and I.A. Vasilyeva, et al., 1971, Geology of the South Primorye Mesozoic depressions: structure of the Mesozoic depressions zone, Leningrad, Trudy VNIGRI Vypusk, 302, 38–172 (in Russian).

Balcer, R.A., 1984, Stratigraphy and depositional history of the Pantano Formation (Oligocene-early Miocene), Pima County, Arizona, Tucson (Arizona), unpublished MSc dissertation, Tucson University, 107 pp.
Bandel, K. and H. Khoury, 1981, Lithostratigraphy of the Triassic in Jordan, Facies, 4, 1–26.
Banerjee, I., 1974, Barrier coastline sedimentation model and the Vindhyan example, in: A. Dey (ed.), Contributions to the Earth and Planetary Sciences, Golden Jubilee Volume, Quart. J. Geol. Miner. Metal. Soc. India, pp. 101–127.
Bangert, B., H. Stollhofen and V. Lorenz, 1997, Permo-Carboniferous tuff layers in the Kalahari-Karoo Basin (Southern Namibia) – volcaniclastic deposits as markers for initial South Atlantic rifting? GAEA heidelbergensis, 3; 18th Regional European Meeting Sedim., Heidelberg (Germany), Abstracts, 55–56.
Baptista, M.B., O.P.G. Braun and D.A. Campos (coord.), 1984, Léxico estratigráfico do Brasil, Brasília, Min. Minas Energia, Dept. Nac. Producão Mineral, 560 pp.
Barretto, P.M.C., 1968, O Paleozóico da Baica de Jatobá, Bol. Soc. Brasileira Geol., 17, 29–45.
Bates, R.L. and J.A. Jackson, 1987, Glossary of Geology (3rd ed.), Am. Geol. Inst., Alexandria (Virginia), 788 pp.
Baturin, D., A. Vinogradov and A. Junov, 1991, Tectonics and hydrocarbon potential of the Barents Megatrough, Am. Assoc. Petroleum Geol. Bull., 75, 1404.
Beaumont, C., 1981, Foreland basins, Geophys. J. R. Astron. Soc., 65, 291–329.
Beaumont, C., G.M. Quinlan and J. Hamilton, 1987, The Alleghanian orogeny and its relationship to the evolution of the eastern interior, North America, in: C. Beaumont and A.J. Tankard (eds.), Sedimentary Basins and Basin-Forming Mechanisms, Canadian Soc. Petroleum Geol., Memoir 12, pp. 425–445.
Beaumont, C., G.M. Quinlan and J. Hamilton, 1988, Orogeny and stratigraphy: numerical models of the Paleozoic in the eastern interior of North America, Tectonics, 7, 389–416.
Beerbower, J.R., 1964, Cyclothems and cyclic depositional mechanisms in alluvial plain sedimentation, Kansas State Geol. Survey Bull., 169 (1) 31–42.
Beloussov, V.V., 1962, Basic Problems in Geotectonics, McGraw-Hill Book Co., New York, 809 pp.
Bender, F., 1974, Geology of Jordan, Gebr. Borntraeger, Berlin, 196 pp.
Bentor, Y.K., 1985, Crustal evolution of the Arabo-Nubian Massif with special references to the Sinai Peninsula, Precambr. Res., 28, 1–74.
Bertels, A., 1989, Paleoambientes de la Formación Río Mayer (Hauteriviano-Barremiano), Cuenca de Magallanes, República Argentina, Contrib. Simp.

Cretácico América Latina, Buenos Aires (Argentina), A: Eventos y Registro sedimentário, 45–56.
Berthou, P.Y. and P. Bengtson, 1988, Stratigraphic correlation by microfacies of the Cenomanian-Coniacian of the Sergipe Basin, Brazil, Fossils and Strata, 21, 88 pp.
Bertrand, M., 1886–1887, La chaîne des Alpes et la formation du continent européen, Bull. Soc. géol. France, 3me série, 15, 423–447.
Beurlen, K., 1961, Die paläogeographische Entwicklung des südatlantischen Ozeans, Nova Acta Leopoldina, NF 24, no. 154, 36 pp.
Beurlen, K., 1970, Geologie von Brasilien, Gebr. Borntraeger, Berlin; Beiträge zur regionalen Geologie der Erde, Vol. 9, 444 pp.
Bhattacharji, S., 1986, Astenospheric upwelling, lineament activation, magmatic episodes and ore mineral localization in Proterozoic basin evolution on the Archaean Indian shield, Int. Basement Tect. Assoc., Publ. S, 187–200.
Bhattacharji, S. and R.N. Singh, 1984, Thermo-mechanical structure of the southern part of the Indian Shield and its relevance to Precambrian basin evolution, Tectonophysics, 105, 103–120.
Biddle, K.T., M.A. Uliana, R.M. Mitchum Jr., M.G. Fitzgerald and R.C. Wright, 1986, The stratigraphic and structural evolution of the central and eastern Magallanes Basin, Southern South America, in: P.A. Allen and P. Homewood (eds.), Foreland Basins, Vol. 8, IAS Spec. Publ., pp. 41–61.
Binda, P.L., 1981, The Precambrian boundary in the Arabian Shield: a review, King Abdulaziz Univ., Jedda, Faculty Earth Sci. Bull., 4, 107–120.
Biswas, A., 2000, Facies analysis and depositional milieu of the Proterozoic Egalapenta Member, Srisailam Formation, in parts of the Krishna Gorge West, Cuddapah Basin, India, unpublished PhD thesis, University of Calcutta, Calcutta, 119 pp.
Bjorlykke, K., 1985, Glaciations, preservation of their sedimentary record and sea level changes – a discussion based on the Late Precambrian and Lower Paleozoic sequence in Norway, Palaeogeogr., Palaeoclim., Palaeoecol., 51, 197–207.
Blair, T.C., 1985a, Depositional chronology, sedimentary processes and the resulting vertical stratification sequences in the Roaring river alluvial fan, Rocky Mountain National Park, Colorado, in: R.M. Flores and M.D. Harvey (eds.), Field guide book to modern and ancient fluvial systems in the United States, Fort Collins (Colorado), Proc. 3rd International Fluvial Conference, pp. 96–101.
Blair, T.C., 1985b, Nonmarine sedimentological response to tectonism in rift basins: a comparison of the Jurassic Todos Santos Formation, Chiapas, Mexico, with Quaternary deposits in Death Valley in California, Soc. Econ. Paleont. Min., Annual Mid-Year Meeting, Abstracts, 2, 11–12.

Blair, T.C., 1987, Tectonic and hyrdrologic controls on cyclic alluvial fan, fluvial and lacustrine rift-basin sedimentation, Jurassic-Lowermost Cretaceous Todos Santos Formation, Chiapas, Mexico, J. Sedim. Petrol., 57, 845–862.

Blair, T.C., 1999, Cause of dominance by sheet flood vs. debris flow processes on two adjoining alluvial fans, Death Valley, California, Sedimentology, 46, 1015–1028.

Blair, T.C. and J.G. Mcpherson, 1994, Alluvial fan processes and forms, in: A.D. Abrahams and A.J. Parsons (eds.), Geomorphology of Desert Environments, Chapman and Hall, London, pp. 354–402.

Blatt, H., G. Middleton and R. Murray, 1972, Origin of Sedimentary Rocks, Prentice-Hall Inc., Englewood Cliffs (New Jersey), 634 pp. [Chapter 20: Major external controls of sedimentation, pp. 585–607].

Bluck, B.J., 1981, Upper Old Red Sandstone, Firth of Clyde, in: T. Elliot (ed.), Field guides to modern and ancient fluvial systems in Britain and Spain, Keele (U.K.), Proc. 2nd Conf. Fluvial Sedim., University of Keele, Int. Assoc. Sedim., pp. 5.14–5.20.

Bochkarev, V.S., 1985, Mesozoic stage of the West Siberian hemisyneclise development, in: General Problems in Geology, Tyumen, pp. 21–34 (in Russian).

Bonaparte, J.F., 1996, Cretaceous Tetrapods of Argentina, Münchner Abh. Geowiss. A. Reihe Geol. Paläont., 30, 73–130.

Boserio, I.M., C. Kadellos and H. Priebe, 1995, Cambro-Ordovician tectonostratigraphy and plays in the south Oman basin, in: M.I. Al-Husseini (ed.), Middle East Petroleum Geosciences Geo 94, Gulf Petrolinks, Bahrain, 2, 203–213.

Boucot, A.J., 1962, Appalachian Siluro-Devonian, in: K. Coe (ed.), Some Aspects of the Variscan Fold Belt, Manchester University Press, Manchester; pp. 155–163 [Chapter 10].

Boucot, A.J., M.T. Filed, R. Fletcher, W.H. Forbes, R.S. Naylor and L. Pavlides, 1964, Reconnaissance bedrock geology of the Presque Isle Quadrangle, Maine, Maine Geol. Survey, Quadr. Mapping Ser., 2, 123 pp.

Bouma, A.H., 1996, Review of fine-grained submarine fans and trubidite systems, U.S. Department of Energy, Lecture Notes, 73 pp.

Braga, J.A.E. and J.C. Della Fávera, 1978, Sequências deposicionais nas bacias brasileiras – determinação, integração regional e sua utilização nas pesquisas de hidrocarbonetos, Rio de Janeiro, Petrobrás/DEXPRO, internal report, 70 pp.

Brandão, J.A.S.L. and F.J. Feijó, 1994, Bacia da Foz do Amazonas, Bol. Geociências Petrobrás, 8, 91–99.

Brito Neves, B.B., W. Teixeira, C.C.G. Tassinari and K. Kawashita, 1994, A contribution to the subdivision of the Precambrian in South America, Rev. Brasileira Geociências, 20, 267–276.

Brouwer, A., 1984, Een kleine Aardgeschiedenis, Rijksuniv. Leiden, Afscheidscollege, 28 pp.

Brown, G.F., R.O. Jackson, R.G. Bogue and E.L. Bergy Jr., 1963, Geological map of the northwestern Hijaz quadrangle, Kingdom of Saudi Arabia, U.S. Geol. Survey, Miscellaneous geologic investigation map, 1-240A, scale 1:500,000.

Brown, G.F., D.L. Schmidt and C. Huffman Jr., 1989, Geology of the Arabian Peninsula, shield area of western Saudi Arabia, U.S. Geol. Survey, Prof. Paper, 560-A.

Bruhn, C.H.L. and M.A.S. Moraes, 1988, Turbiditos brasileiros: caracterização geométrica e faciológica, 35° Congr. Brasileiro Geol., Belém (PA), Anais, 2, 824–838.

Bucher, W.H., 1933, The Deformation of the Earth's Crust, Princeton University Press, Princeton (New Jersey), 518 pp.

Bulgakova, M.D., 1996, Paleogeography of Yakutiain the Early-Middle Paleozoic. Yakutsk Center, SO RAN, Yakutsk, 70 pp. (in Russian).

Buriyi, I.V., 1997, Paleogeography of Triassic sedimentation in South Primorye; Late Paleozoic and Early Mesozoic Circum-Pacific events: biostratigraphy, tectonics and ore deposits of Primorye, Mem. Géol. (Lausanne), 30, 35–44.

Burollet, P.F., 1982, Intracratonic and pericratonic basins in Africa, 11th Int. Sedim. Congr., Hamilton (Ont., Canada), Symp. 12, 14 pp. (separate).

Cagnolatti, M., G. Covellone and L. Svave, 1989, Análises secuencial del Cretácico inferior em Tierra del Fuego, Contrib. Simp. Cretácico América Latina, Buenos Aires (Argentina), A: Eventos y Registro sedimentario, 29–44.

Cairncross, B., 2002, The Permian coal deposts of Africa, south of the Sahara – location, tectonic setting and depositional environments, 16th Int. Sedim. Congr., Johannesburg (South Africa), Abstracts, 11–12.

Campbell, C.V., 1967, Lamina, lamina set, bed and bedset, Sedimentology, 8, 7–26.

Cant, D.J. and R.G. Walker, 1978, Fluvial processes and facies sequences in the sandy braided South Saskatchewan river, Canada, Sedimentology, 25, 625–648.

Caputo, M.V. and J.C. Crowell, 1985, Migration of glacial centers across Gondwana during Paleozoic era, Geol. Soc. Am. Bull., 96, 1020–1036.

Caputo, M.V. and E.C. Lima, 1984, Estratigrafia, idade e correlação do Grupo Serra Grande, Bacia do Parnaíba, 33° Congr. Brasileiro Geol., Rio de Janeiro (RJ), Anais, 2, 740–753.

Caputo, M.V., R. Rodrigues and D.N.N. Vasconcelos, 1972, Nomenclatura estratigráfica da Bacia do Amazonas, 26° Congr. Brasileiro Geol., Belém (PA), Anais, 3, 35–46.

Catuneanu, O., P.J. Hancox and B.S. Rubidge, 1998, Reciprocal flexural behaviour and contrasting stratigraphies: a new basin development model for the Karoo retroarc foreland system, South Africa, Basin Res., 10, 417–439.

Cecil, C.B. and F.T. DuLong, 2003, Precipitation models for sediment supply in warm climates, in: C.B. Cecil and N.T. Edgar (eds.), Climate Controls on Stratigraphy, SEPM (Soc. Sedim. Geol.), Spec. Publ., 77, pp. 21–27.

Cecil, C.B., T.S. Ahlbrandt and M.T. Heald, 1991, Paleoclimatic implications for the origin of Paleozoic quartz arenites in the Appalachian Basin, Geol. Soc. Am. Abstracts Progr., 23, A72.

Cecil, C.B., F.T. DuLong, R.R. West, R. Stamm, B. Wardlaw and N.T. Edgar, 2003, Climate controls on the stratigraphy of a Middle Pennsylvanian cyclothem in North America, in: C.B. Cecil and N.T. Edgar (eds.), Climate Controls on Stratigraphy, SEPM (Soc. Sedim. Geol.), Spec. Publ., 77, pp. 151–180.

Cecil, C.B., D.K. Brezinski and F. DuLong, 2004, The Paleozoic record of changes in global climate and sea level: Central Appalachian Basin, in: S. Southworth and W. Burton (eds.), Geology of the National Capital Region – filed trip guidebook, U.S. Geol. Survey Circular, 1264, 77–135.

Chakraborty, C. and A. Bhattacharyya, 1996, Fan-delta sedimentation in a foreland moat: Deoland Formation, Vindhyan Supergroup, Son Valley, in: A. Bhattacharyya (ed.), Recent Advances in Vindhyan geology, Bangalore, Geol. Soc. India, Memoir 36, pp. 27–48.

Chan, M.A., 1989, Erg margin of the Permian White Rim Sandstone, SE Utah, Sedimentology, 36, 235–251.

Chen, Peiji, 1993, Timing of displacement along Tancheng-Lujiang fault zone and the migration of Late Mesozoic volcanism in Eastern China, in: Xu Jiawei (ed.), Tancheng-Lujiang wrench fault system, John Wiley & Sons, New York, pp. 105–113.

Chesnut Jr., D.R., 1989, Stratigraphic framework of the Pennsylvanian-age rocks of the Central Appalachian Basin, eastern U.S.A, in: J. Yugan and L. Chun (eds.), Compte Rendu 11me Congr. Int. Stratigraphie Géol. Carbonifère, Vol. 2, Nanjing University Press, Nanjing, pp. 1–19.

Chesnut Jr., D.R., 1992, Stratigraphic and structural framework of the Carboniferous rocks of the Central Appalachian Basin in Kentucky, Kentucky Geol. Survey, Ser. XI, Bull., 3, 42 pp.

Chesnut Jr., D.R., 1994, Eustatic and tectonic control of deposition of the Lower and Middle Pennsylvanian strata of the Central Appalachian Basin, in:

J.M. Dennison and F.R. Ettensohn (eds.), Tectonic and eustatic controls on sedimentary cycles, SEPM (Soc. Sedim. Geol.), Concepts Sedim. Paleont., 4, 51–64.
Chesnut Jr., D.R., 1996, Geologic framework for the coal bearing rocks of the Central Appalachian Basin, Int. J. Coal Geol., 31, 55–66.
Clifton. H.E. (ed.), 1988, Sedimentologic consequences of convulsive geological events, Geol. Soc. Am., Spec. Paper 229, 157 pp.
Coakley, B. and M. Gurnis, 1995, Far-field tilting of Laurentia during the Ordovician and constraints on the evolution of a slab under an ancient continent, J. Geophys. Res., 100, 6313–6327.
Coe, A.J. (ed.), 2003, The Sedimentary Record of Sea-Level Change, The Open University Press, Cambridge, 288 pp.
Cole, D.I., 1992, Evolution and development of the Karoo Basin, in: M.J. De Wit and I.G.D. Ransome (eds.), Inversion Tectonics of the Cape Fold Belt, Karoo and Cretaceous Basins of South Africa, Balkema, Rotterdam, pp. 87–99.
Coleman, R.G., 1984, Ophiolite and tectonic evolution of the Arabian Peninsula, in: I.G. Gass, S.J. Lippard and A.W. Shelton (eds.), Ophiolites and Oceanic Lithosphere, Geol. Soc. London Spec. Publ. 17, 359–366.
Collinson, J.D. and D.B. Thompson, 1982, Sedimentary Structures, George Allen and Unwin Publ., London, 194 pp.
Collinson, J.D., O. Martinsen, B. Bakken and A. Kloster, 1991, Early fillmof the Western Irish Namurian basin: a complex relationship between turbidites and deltas, Basin Res., 3, 223–242.
Condie, K.C., 1992, Plate Tectonics and Crustal Evolution, Butterwarth Heinmann, 282 pp.
Cordani, U.G. and B.B. Brito Neves, 1982, The geologic evolution of South America during the Archaean and Early Proterozoic, Rev. Brasileira Geociências, 12, 78–88.
Cordani, U.G., B.B. Brito Neves, R.A. Fuck, R. Porto, A. Thomaz Filho and F.M.B. Cunha, 1984, Estudo preliminar de integraçaõ do Pré-Cambriano com os eventos tectônicos das bacias sedimentares brasileiras, Rio de Janeiro, Petrobrás/CENPES, Ciência-Técnica-Petróleo, Vol. 15, 70 pp.
Costa, J.B.S., Y. Hasui and R.V.L. Pinheiro, 1992, Bacias Sedimentares, Edit, Univ., Univ. Fed. Pará, Belém (PA), 116 pp.
Crawford, A.R. and W. Compton, 1973, The age of the Cuddapah and Kurnool Systems, South India, J. Geol. Soc. Australia, 19, 453–464.
Cross, T.A. and R.H. Pilger Jr., 1978, Tectonic controls of Late Cretaceous sedimentation, western interior, USA, Nature, 274, 653–657.
Csató, I., 1993, Neogene sequences in the Pannonian basin, Hungary, Tectonophysics, 226, 377–400.

Cunha, P.R.C., F.G. Gonzaga, L.F.C. Coutinho and F.J. Feijó, 1994, Bacia do Amazonas, Bol. Geociências Petrobrás, 8, 47–55.
Cunningham, D.W., I.W.D. Dalziel, Tung-Yi. Lee and L.A. Lawver, 1995, Southernmost South America-Antarctic peninsula relative plate motions since 84 Ma. Implications for the tectonic evolution of the Scotia Arc region, J. Geophys. Res., 100, 8257–8266.
Daemon, R.F., 1976, Correlação bioestratigráfica entre os sedimentos do Siluriano, Devoniano e Carbonífero Inferior das bacias do Amazonas, Parnaíba e Paraná, 29° Congr Brasileiro Geol., Ouro Preto (MG), Anais, 2, 189–194.
Daemon, R.F. and C.J.A. Contreras, 1971, Zoneamento palinológico da Bacia do Amazonas, 25° Congr. Brasileiro Geol., São Paulo (SP), Anais, 3, 79–88.
Dallmeyer, R.D., 1986, Polyphase terrane accretion in the southern Appalachians, Geol. Soc. Am. Abstracts Progr., 18, 579.
Dana, J.D., 1873, On some results of the earth's contraction from cooling, including a discussion of the origin of mountains and the nature of the earth's interior, Am. J. Sci., 3rd ser., Vol. 5, pp. 423–443; Vol. 6, pp. 6–14, 104–115, 161–172.
Dana, J.D., 1880, Manual of Geology (3rd ed.), Ivison, Blakeman & Co., New York, 911 pp.
Dasgupta, P.K., 1986, Picrite borne mantle xenoliths from Cuddapah Basin, South India, Indian J. Earth Sci., 13, 333–338.
Dasgupta, P.K., 1996, Cyclicity in the Proterozoic Cuddapah Supergroup and its bearing on basin evolution, India, Int. Geol. Congr., Beijing (P.R. China); Session 3–6 (Sedimentation in intracratonic basins) 3-6-5, 01214, 0925, Abstracts, 2, 198.
Dasgupta, P.K., 1998, Seismic signatures from thin bedded olistostromes and associated turbidites: Cumbum Shale, Cuddapah Basin, India, 15th Int. Sedim. Congr., Alicante (Spain), Abstracts, 210–212.
Dasgupta, P.K. and A. Sengupta, 1997, Cyclicity in the transition facies of the Cuddapah Supergroup and role of oscillatory movement in basinal evolution, Govt. India, Department of Science and Technology, unpublished project report, 140 pp.
Davies Jr., A., 1983, Depositional Systems: A Genetic Approach to Sedimentary Geology, Prentice-Hall, Inc., Englewood Cliffs (New Jersey), 669 pp.
Davies, B.F., 1984, Strain analysis of wrench faults and collision tectonics of the Arabian Nubian Shield, J. Geol., 92, 37–54.
De Boer, P.L. and D.G. Smith (eds.), 1994, Orbital Forcing and Cyclic Sequences, Int. Assoc. Sedim., Spec. Publ. 19, 559 pp.
Delaloye, M. and J. Wagner, 1984, Ophiolites and volcanic activity near the western edge of the Arabian Plate, in: J.E. Dixon and A.H.F. Robertson

(eds.), The Geologic Evolution of the Eastern Mediterranean, Geol. Soc. London Spec. Publ. 17, pp. 225–233.
DeLaune-Myers, M., 1984, Evolution of a passive continental margin: Baer-Bassit, NW Syria, in: J.F. Dixon and A.H.F. Robertson (eds.), The Geologic Evolution of the Eastern Mediterranean, Geol. Soc. London Spec. Publ. 17, pp. 151–159.
Delfour, J., 1970, La groupe de J'balah, une nouvelle unité du bouclier Arabe, Bureau Rech. Géol. Min. France, sér. 2, sec. 4, pp. 19–32.
Delfour, J., 1977, Geology of the Nugrah quadrangle. Kingdom of Saudi Arabia DGMR, Geologic Map GM-28, sheet 25E, scale 1:200,000.
Dennison, J.M., 1976, Appalachian Queenston Delta related to eustatic sea-level drop accompanying Late Ordovician glaciation centered in Africa, in: M.G. Bassett (ed.), The Ordovician System, Univ. Wales, Natl. Museum Wales, Cardiff, Proc. Paleont. Assoc. Symp., Birmingham, 1974, pp. 107– 120.
Dennison, J.M. and J.W. Head, 1975, Sea level variations interpreted from Appalachian Basin Silurian and Devonian, Am. J. Sci., 275, 1089–1120.
Denny, C.S., 1965, Alluvial fans in the Death Valley region, California, U.S. Geol. Survey, Prof. Paper 466, 62 pp.
Devuyst, F.X. and A. Lees, 2000, Precursor muds associated with the initial development of Waulsortian buildups in western Ireland, 20th IAS Regional Meeting Sedim., Dublin (Ireland), Abstracts, 90.
Devyatov, V.P., A.M. Kazakov and G.V. Kasatkina, 1994, Problems of the Lower and Middle Jurassic stratigraphy of West Siberia, Geol. Geophys., 12, 3–17 (in Russian).
Dewey, J.F. and K.C.A. Burke, 1974, Hot spots and continental break-up: implications for collisional orogeny, Geology, 2, 57–60.
Dewey, J.F. and W.S.F. Kidd, 1974, Continental collisions in the Appalachian Caledonian belt: variations related to complete and incomplete suturing, Geology, 2, 543–546.
Dhoundial, D.P., 1987, Purana basins of Peninsular India, in: Purana basins of Peninsular India, Bangalore, Geol. Soc. India, Memoir 6, pp. 1–4.
Dias, J.L., 1991, Análise estratigráfica e evolução da fase "rift" nas bacias das margens leste e sudeste do Brasil, MSc dissertation, Univ. Fed. Rio de Janeiro, Rio de Janeiro (RJ), 145 pp. (manuscript).
Dickinson, W.R., 1974, Plate tectonics and sedimentation, in: W.R. Dickinson (ed.), Tectonics and Sedimentation, Soc. Econ. Paleont. Min. Spec. Publ. 22, pp. 1–27.
Dibner, V.D. (ed.), 1998, Geology of Franz Josef Land, Norsk Polarinst., Meddedelser 146, 190 pp.
Dino, R., 1994, The Marajó basin, in: G. Beurlen, D.A. Campos and M.C. Viviers (eds.), Stratigraphic Range of Cretaceous Mega- and Microfossils of

Brazil, Univ. Fed. Rio de Janeiro, Inst. Geociências, Rio de Janeiro (RJ), pp. 324–332 [Chapter 3.8].

Dino, R. and N. Uesugui, 1994, The Amazonas and Solimões basins, in: G. Beurlen, D.A. Campos and M.C. Viviers (eds.), Stratigraphic Range of Cretaceous Mega- and Microfossils of Brazil, Univ. Fed. Rio de Janeiro, Inst. Geociências, Rio de Janeiro (RJ), pp. 392–399 [Chapter 4.4].

Dodin, D.A. and V.S. Surkov (eds.), 2002, The Russian arctic: geological history, mineragenesis, environmental geology, VNIIOkeangeologia, St. Petersburg, 960 pp. (in Russian).

Donaldson, A.C. and C. Eble, 1991, Pennsylvanian coals of central and eastern United States, in: H.J. Gluskoter, D.D. Rice and R.B. Taylor (eds.), Economic Geology, U.S. Boulder, Geol. Soc. Am., The Geology of North America, P-2, pp. 523– 546.

Donaldson, A.C. and R.C. Shumaker, 1981, Late Paleozoic molasse of the central Appalachians, in: A.D. Miall (ed.), Sedimentation and Tectonics in Alluvial Basins, Geol. Assoc. Canada Spec. Paper 23, pp. 99–124.

Dorobek, S.L. and J.F. Read, 1986, Sedimentology and basin evolution of the Siluro-Devonian Helderberg Group, central Appalachians, J. Sedim. Petr., 56, 601–613.

Dorofeev, V.K., M.G. Blagoveschenskyi, A.N. Smirnov and V.I. Ushakov, 1999, Novosibirskie Islands, Geological structure and mineragenesis, VNIIO-keangeologia, St. Petersburg, 130 pp. (in Russian).

Dott Jr., R.H., 1988, An episodic view of shallow-marine clastic sedimentation, in: P.L. De Boer, A. van Gelder and S.D. Nio (eds.), Tidal-Influenced Sedimentary Environments and Facies, Reidel, Dordrecht, pp. 3–12.

Doyle, P., D.G. Poiré, L.A. Spalletti, D. Pirrie, P. Brenchley and S.D. Matheos, 1998, Paleoenvironmental analysis of the Vaca Muerta and Chachao Formations of the Neuquén Basin in West-Central Argentina: implications for palaeo-oxygenation studies, 15th Int. Sedim. Congr., Alicante (Spain), Abstracts, 303–304.

Drachev, S.S., A.B. Elistratov and L.A. Savostin, 2001, Structure and seismic stratigraphy of the East Siberian Sea shelf along seismic line "Indigirka Bay – Jannetta Island", Proc. Acad. Sci., 377, 521–525 (in Russian).

Drachev, S.S., L.A. Savostin, V.G. Groshev and I.E. Bruni, 1998, Structure and geology of the continental shelf of the Laptev Sea, Tectonophysics, 298, 357– 393.

Driese, S.G. and R.H. Dott Jr., 1984, Model for sandstone-carbonate "cyclothems" based on upper member of Morgan Formation (Middle Pennsylvanian) of northern Utah and Colorado, Am. Assoc. Petroleum Geol. Bull., 68, 574–597.

Drukman, Y., 1984, Evidence for Early-Middle Triassic faulting and possible rifting from Helez Deep Borehole in the coastal plain of Israel, in: J.E. Dixon and A.H.F. Robertson (eds.), The Geologic Evolution of the Eastern Mediterranean, Geol. Soc. London Spec. Publ. 17, pp. 203–212.
Drury, S.A. and R.W. Holt, 1983, The tectonic framework of the South Indian Craton: a reconnaissance involving Landsat imagery, in: Sinha-Roy, S. (ed.), Structure and Tectonics of Precambrian Rocks in India, Hindusthan Publ. Corp. Press, New Delhi, pp. 178–185.
Einsele, G., 1992, Sedimentary Basins, Springer-Verlag, Berlin, 627 pp.
Einsele, G., W. Ricken and A. Seilacher (eds.), 1991, Cycles and Events in Stratigraphy, Springer-Verlag, Berlin, 955 pp.
Eiras, J.F., C.R. Becker, E.M. Souza, F.G. Gonzaga, J.G.F. Silva, L.M.F. Daniel, N.S. Matsuda and F.J. Feijó, 1994, Bacia do Solimões, Bol. Geociências Petrobrás, 8, 17–45.
Elliott, T., A.J. Pulham and S.J. Davies, 2000, Sedimentology, sequence stratigraphy and spectral gamma ray expression of turbidite, slope, and deltaic depositional systems in an Upper Carboniferous basin-fill succession, western Ireland, in: J.R. Graham and A. Ryan (eds.), Field Trip Guidebook, Vol. 1, 20th IAS Regional Meeting, Dublin (Ireland), pp. 1–40.
Engebretson, D., A. Cox and R.G. Gordon, 1985, Relative motions between oceanic and continental plates in the northern Pacific basin, Geol. Soc. Am., Spec. Paper 206, 1–59.
Englund, K.J., H.H. Arndt and T.W. Henry, 1979, IX ICC Field Trip No. 1; Proposed Pennsylvanian stratotype, Virginia and West Virginiana, Am. Geol. Inst. Selected Guidebook, Ser. 1, 138 pp.
Ettensohn, F.R., 1985a, The Catskill Delta complex and the Acadian orogeny: a model, in: D.L. Woodrow and W.D. Sevon (eds.), The Catskill Delta, Geol. Soc. Am. Spec. Paper 201, 39–49.
Ettensohn, F.R., 1985b, Controls on the development of Catskill Delta basin-facies, in: D.L. Woodrow and W.D. Sevon (eds.), The Catskill Delta, Geol. Soc. Am. Prof. Paper 201, 65–77.
Ettensohn, F.R., 1987, Rates of relative plate movement during the Acadian orogeny based on the spatial distribution of black shales, J. Geol., 95, 572–582.
Ettensohn, F.R., 1991, Flexural interpretation of relationships between Ordovician tectonism and stratigraphic sequences, central and southern Appalachians, U.S.A, in: C.R. Barnes and S.H. Williams (eds.), Advances in Ordovician Geology, Geol. Survey Canada Paper 90-9, pp. 213–224.
Ettensohn, F.R., 1993, Possible flexural controls on the origins of extensive, ooid-rich, carbonate environments in the Mississippian of the United States,

in: B.D. Keith and C.W. Zuppan (eds.), Mississippian Oolites and Modern Analogues, Am. Assoc. Petroleum Geol. Studies Geol., 35, 13–30.

Ettensohn, F.R., 1994, Tectonic control on formation and cyclicity of major Appalachian unconformities and associated stratigraphic sequences, in: J.M. Dennison and F.R. Ettensohn (eds.), Tectonic and Eustatic Controls on Sedimentary Cycles, SEPM (Soc. Sedim. Geol.), Concepts in Sedimentology and Paleontology, Vol. 4, pp. 217–242.

Ettensohn, F.R., 1998, Compressional tectonic controls on epicontinental black-shale deposition: Devonian-Mississippian examples from North America, in: J. Schieber, W. Zimmerle and P. Sethi (eds.), Shales and Mudstones, I. E. Schweizerbart'sche Verlagsbuchhandlung, Stuttgart, pp. 109–128.

Ettensohn, F.R., 2004, Modeling the nature and development of major Paleozoic clastic wedges in the Appalachian Basin, USA, J. Geodyn., 37, 657–681.

Ettensohn, F.R. and C.E. Brett, 1998, Tectonic components in third-order Silurian cycles: examples from the Appalachian Basin and global implications, in: E. Landing and M.E. Johnson (eds.), Silurian cycles, linkages of dynamic stratigraphy with atmospheric, oceanic, and tectonic changes, New York State Museum Bull., 491, 145–162.

Ettensohn, F.R. and C.E. Brett, 2002, Stratigraphic evidence from the Appalachian Basin for continuation of the Taconian Orogeny into Early Silurian time, in: C.E. Mitchell and R. Jacobi (eds.), Taconic convergence orogen, foreland basin, and craton, Phys. Chem. Earth, 27, 279–288.

Ettensohn, F.R. and D.R. Chesnut Jr., 1989, Nature and probable origin of the Mississippian-Pennsylvanian unconformity in the eastern United States, in: J. Yugan and L. Chun (eds.), Compte Rendu, Onzième Congr. Int. Stratigraphie Géologie Carbonifère, Vol. 4, Nanjing University Press, Nanjing, pp. 145–159.

Ettensohn, F.R. and J.C. Pashin, 1993, Mississippian stratigraphy of the Black Warrior Basin and adjacent parts of the Appalachian Basin: evidence for flexural interaction between two foreland basins, in: J.C. Pashin (ed.), New Perspectives on the Mississippian of Alabama, 30th Annual Field Trip, Tuscaloosa, Alabama Geol. Soc., pp. 29–40.

Ettensohn, F.R. and J.C. Pashin, 1997, Development of multiple unconformities during the Devonian-Carboniferous transition on parts of Laurussia, in: J. Malecka (ed.), Proc. Int. Congr. Carboniferous Permian, Warszawa, Prace Panstowowego Inst. Geol., 157 (1) 77–86.

Ettensohn, F.R., J.C. Hohman, M.A. Kulp and N. Rast, 2002, Evidence and implications of possible far-field responses to Taconian orogeny: Middle-Late Ordovician Lexington Platform and Sebree Trough, east-central United States, Southeastern Geol., 41, 1–36.

Eyers, J., 1994, The Aptian to Albian sediments of Southeast England: the application of sequence stratigraphy, 14th Int. Sedim. Congr., Recife (PE, Brazil), Abstracts E-8.

Fairbridge, R.W. and J. Bourgeois, 1978, The Encyclopedia of Sedimentology. Dowden, Hutchinson and Ross, Inc., Stroudsburg; Encyclopedia of Earth Sciences, Vol. 6, 901 pp.

Feijó, F.J., 1994a, Bacia de Pernambuco-Paraíba, Bol. Geociências Petrobrás, 8, 143–147.

Feijó, F.J. and R.G. Souza, 1994, Bacia do Acre, Bol. Geociências Petrobrás, 8, 9–16.

Fenner, J.M., A. Thies and M.E. Weber, 1998, Paleoceanographic development in an epicontinental basin (Lower Saxony Basin) during the Albian as indicated by macrofossils, benthic foraminifera and radiolaria, 15th Int. Sedim. Congr., Alicante (Spain), Abstracts, 324.

Fetter, A.M.F., W.R. Van Schmus, T.J.S. Santos, J.A. Nogueira Neto and M.A. Arthaud, 2000, U-Pb and Sm-Nd geochronological constraints on the crustal evolution and basement architecture of Ceará State, NW Borborema province, NE Brazil: implications for the existence of the Paleoproterozoic supercontinent "Atlantica", Rev. Brasileira Geociências, 30, 102–106.

Figari, E., L. Chiaramonte, R. Salvay and M.S. Cid de la Paz, 2000, Análisis estratigráfico de las secuencias cretácicas de *sag* em el Flanco Norte de la Cuenca del Golfo de San Jorge, província del Chubut, Argentina, 2° Congr. Latinoamericano Sedim., 8ª Reunión Argentina Sedim., Mar del Plata (Argentina), Resumenes, 75–76.

Fischer, A.G., 1981, Climatic oscillations in the biosphere, in: M.H. Nitecki (ed.), Biotic Crises in Ecological and Evolutionary Time, Academic Press, New York, pp. 100–131.

Fischer, A.G., 1984, The two Phanerozoic supercycles, in: W.A. Berggren and J.A. Van Couvering (eds.), Catastrophes and Earth History, Princeton University Press, Princeton, pp. 129–150.

Fitch, F.J. and J.A. Miller, 1965, Major cycles in the history of the earth, Nature, 206, 1023–1027.

Fomichev, V.A. and N.K. Volkova, 1989, Lithology of the Yamal and Gydan hydrocarbon provinces, in: Lithology of sequences in West Siberia by geophysical data, Yumen, pp. 51–59 (in Russian).

Franchi, M.R., F.E. Nullo, E.G. Sepulveda and M.A. Uliana, 1984, Las sedimentitas terciarias, 9° Congr. Geol. Argentino, Vol. 1 (9), Buenos Aires, pp. 215–266.

Franke, D. and K. Hinz, 1999, Endbericht über die Arbeiten der BGR zum Projekt Lapseis: Untersuchungen zur Neotektonik in der Laptev See, Ostsiberischen See und auf dem angrenzenden NE-Siberischen Festland mit seismologischen Breitbanddaten, Projekt Nr. 03G0529A, BGR, Hannover, 122 pp.

Franke, D., K. Hinz and O. Oncken, 2001, The Laptev Sea Rift, Marine and Petroleum Geol., 18, 1083–1127.

Franzese, J. and L. Spalletti, 2000, Extensión continental y formación de cuencas marginales mesozóicas em los Andes meridionales (30°LS-40°LS), 2° Congr. Latinoamericano Sedim., 8ª Reunión Argentina Sedim., Mar del Plata (Argentina), Resumenes, 76–77.

Frederiksen, K.S., L.B. Clemmensen and H.S. Lawaetz, 1998, Sequential architecture and cyclicity in Permian desert deposits, Brodick Beds, Arran, Scotland, J. Geol. Soc. London, 155, 677–683.

Friedman, G.M. and K.T. Johnson, 1966, The Devonian Catskill deltaic complex of New York, type example of a "tectonic delta complex", in: M.L. Shirley (ed.), Deltas in Their Geologic Framework, Houston Geol. Soc., Houston, pp. 172–188.

Friedman, G.M. and J.E. Sanders, 1978, Principles of Sedimentology, John Wiley & Sons, New York, 792 pp.

Frolov, V.I., Sh.A. Syundyukov and V.Ye. Bakin, 1987, Catagenesis of organic substance in the deep horizons of the central part of the Vilyuiskaya syneclise, DAN, Akad. Nauk SSSR, 297, 442–444 (in Russian).

Fryberger, S.G., A.M. Al-Sari and T.J. Clisham, 1983, Eolian dune, interdune, sand sheet and siliciclastic sabkha sediments on an offshore prograding sand sea, Dhahran area, Saudi Arabia, Am. Assoc. Petroleum Geol. Bull., 63, 280–312.

Fuenzalida, R. and C. Covacevich, 1988, Volcanismo e bioestratigrafia del Jurásico superior y Cretácico inferior em la Cordillera Patagónica, región de Magallanes, Chile, 5° Congr. Geol. Chileno, Santiago, Dept. Geol. Geofísica Univ. Chile, tomo III, 159–183.

Fujita, K. and D.B. Cook, 1990, The Arctic continental margin of eastern Siberia, in: A. Grantz, L. Johnson and J.F. Sweeny (eds.), The Geology of North America, Vol. I: The Arctic Ocean Region, Geol. Soc. Am., pp. 257–288.

Gaiduk, V.K., 1988, The Vilyuiskaya Middle Paleozoic rift system, Yakutsk, YaF SO RAN, 128 pp. (in Russian).

Garfunkel, Z. and D. Derin, 1984, Permian-early Mesozoic tectonism and continental margin formation in Israel and its implication for the history of the Eastern Mediterranean, in: J.F. Dixon and A.H.F. Robertson (eds.), The Geologic Evolution of the Eastern Mediterranean, Geol. Soc. London Spec. Publ. 17, pp. 187–201.

Gates, A.E., C. Simpson and L. Glover III, 1986, Appalachian Carboniferous dextral strike-slip faults: an example from Brookneal, Virginia, Tectonics, 5, 119–133.

Geiser, P. and T. Engelder, 1983, The distribution of layer parallel shortening fabrics in the Appalachian foreland basin of New York and Pennsylvania: evidence for two non-coaxial phases of the Alleghanian orogeny, in: R.D. Hatcher Jr., H. Williams and I. Zietz (eds.), Contributions to the tectonics and geophysics of mountain chains, Geol. Soc. Am., Memoir 158, pp. 161–175.

Geological Map, 1969, Geology of the USSR, Vol. XXXII, Primorye Region – Part I. Geological Description, Nedra, Moscow, 696 pp. (in Russian).

Geological Map, 1996, Geological map of the Priamorye and adjacent areas, scale 1:2,500,000, Obyasnitelnaya zapiska, Sankt-Peterburg – Blagoveschenk – Kharbin, 135 pp. (in Russian).

Ghignone, J.I., 1972, Ensaio de paleogeologia do Nordeste e as sequências sedimentares, 26° Congr. Brasileiro Geol., Belém (PA), Anais, 3, 21–28.

Glennie, K.W., 1972, Permian Rotliegendes of North-Western Europe, interpreted in the light of modern desert sedimentation studies, Am. Assoc. Petroleum Geol. Bull., 56, 1048–1071.

Gliese, J. and H. Hager, 1978, On brown coals resources in the Lower Rhine Embayment (West Germany), Geol. Mijnbouw, 57, 517–525.

Góes, A.M and F.J. Feijó, 1994, Bacia do Parnaíba, Bol. Geociências Petrobrás, 8, 57–68.

Golozubov, V.V., 2004, Tectonics of the North-Western Pacific Ocean: Jurassic and Early Cretaceous assemblages, Vladivostok, Avtoreferat dissertatsii doktora geol.-mineral, Nauka, 47 pp.

Golozubov, V.V. and A.I. Khanchuk, 1995, Taukhe and Zhuravlyovka terranes (Southern Sikhote-Alin) – a fragment of the Early Cretaceous Asia margin, Tikhookeaskaya Geol., 14 (2) 13–25 (in Russian).

Golozubov, V.V. and U. Li Dong 1997, Formation dynamics of the Cretaceous Partizansk-Sukhodol epicontinental basin, Tikhookeanskaya Geol., 16 (6) 46–57 (in Russian).

Golozubov, V.V., V.S. Markevich and E.V. Bugdaeva, 1999, Early Cretaceous changes in vegetation and environments of East Asia, Palaeogeogr., Palaeoclim., Palaeoecol., 153, 139–146.

Golozubov, V.V., A.I. Khanchuk, I.V. Kemkin, et al., 1992, Taukhe and Zhuravlyovka terranes (Southern Sikhote-Alin), Project 321 "Raspad Gondvany i accretsiya Azii", DVO RAN, Vladivostok, 82 pp. (in Russian) [Preprint].

Golubev, V.M., 1992a, Principles of the common cosmodynamics of the geosphere and biosphere, Otechestvennaya Geol., 9, 79–89 (in Russian).

Golubev, V.M., 1992b, On the fundamentals of the Earth's evolutionary theory: geosphere, biosphere, 29th Int. Geol. Congr., Kyoto (Japan), Abstracts, 1, pp. 47, 67.

Golubev, V.M., 1993a, The planetary jointing and gravirotational mechanism of earthquakes, Joint Conf. Seismology East Asia, Tottori (Japan), Proc., 78–81.
Golubev, V.M., 1993b, Cosmodynamic and geodynamic mechanism of biological and social processes, Joint Conf. Seismology East Asia, Tottori (Japan), Proc., 74–77.
Golubev, V.M., 1994a, Structure and origin of the planetary jointing of the Earth's crust, Izvestiya VUZov, Geol. Razvedka, 2, 3–11 (in Russian).
Golubev, V.M., 1994b, Geology of the bottom, geodynamics and oil potential of the Bering sea region, SPb: Nedra, 125 pp. (in Russian).
Golubev, V.M., 1994c, Tectonics and oil-and-gas potential of the Bering Sea bottom, Proc. Int. Conf. Arctic margins, Anchorage (USA), 1992, pp. 381–386.
Golubev, V.M., 1994d, Geodynamics of the Phanerozoic sedimentation, 14th Int. Sedim. Congr., Recife (PE, Brazil), Abstracts, E-10.
Golubev, V.M., 1994e, The structure and formation history of the Bering Sea sedimentary basin, 14th Int. Sedim. Congr., Recife (PE, Brazil), Abstracts, G-38.
Golubev, V.M., 1994f, A nature of short-period cyclically of earthquakes and volcanic eruptions, Int. Volcanol. Congr., Ankara (Turkey), Abstracts, 37.
Golubev, V.M., 1995, Global extinctions and the biological evolution, Palaeobotanist, 44, 1–8.
Golubev, V.M., 1996a, Gravirotational geodynamics and oceanization of the Earth, 30th Int. Geol. Congr., Beijing (China), Abstracts, 1, 4.
Golubev, V.M., 1996b, Introduction into a universal theory of the Earth, 30th Int. Geol. Congr., Beijing (China), Abstracts, 1, 6.
Golubev, V.M., 1996c, Bering Sea model of evolution of the East-Asian transition zone, 30th Int. Geol. Congr., Beijing (China), Abstracts, 1, 206.
Golubev, V.M., 2000a, Theses of the system geotectonics, in: Obschie voprosy tektoniki. Tektonika Rossii, Materialy 33 Tekton, soveschaniya, Moskva, GEOS, pp. 132–137 (in Russian).
Golubev, V.M., 2000b, Geodynamics and system geotectonics, 31st Int. Geol. Congr., Rio de Janeiro (Brazil), Abstracts, 1, 7.
Golubev, V.M., 2000c, Astenospheric subduction and oceanization, 31st Int. Geol. Congr., Rio de Janeiro (Brazil), Abstracts, 1, 9.
Golubev, V.M., 2000d, Geonomy and ecology, 31st Int. Geol. Congr., Rio de Janeiro (Brazil), Abstracts, 1, 28.
Gonzalez, M.G., 1999, Los paleosuelos de la Formación Laguna Palacios (Cretácico Superior) de Patagonia y la Formación Asencio (Cretácico

Superior-Terciario Inferior) de Uruguay, Bol. 5° Simp. Cretáceo Brasil, 1° Simp. Cretácico América del Sur, Serra Negra (SP, Brasil), pp. 65–70.
Goodman, W.M. and C.E. Brett, 1994, Roles of eustasy and tectonics in development of Silurian stratigraphic architecture of the Appalachian foreland basin, in: J.M. Dennison and F.R. Ettensohn (eds.), Tectonic and eustatic controls on sedimentary cycles, SEPM (Soc. Sedim. Geol.), Concepts Sedim. Paleont., 4, 147–169.
Goodmann, P.T., 1988, Stratigraphy and basin dynamics, upper Tonoloway and Kaiser Formation (Upper Silurian-Lower Devonian), central Pennsylvania: an episodic perspective, Northeastern Geol., 10, 231–240.
Goodwin, A.M., 1981, Archaean Plates and greenstone belts, in: Kroner, A. (ed.), Precambrian Plate Tectonics, Elsevier, Amsterdam, pp. 105–135.
Goodwin, P.W.A. and E.J. Anderson, 1985, Punctuated aggradational cycles: a general hypothesis of episodic stratigraphic accumulations, J. Geol., 93, 515–533.
Goroshko, M.V., L.P. Karsakov and V.A. Guryanov, 1999, Stratigraphy of the Lower Riphean, in: Geology and Mineral resources of the Primurye, Magellan, Khabarovsk, 38–40 (in Russian).
Grabau, A.W., 1932, Principles of Stratigraphy, 3rd ed., A.G. Seiler, New York, 1185 pp.
Grabau, A.W., 1940, The rythm of the ages, Henri Vetch, Peking, 561 pp.
Gradstein, F.M. and J.G. Ogg, 2004, Geologic time scale 2004 – why, how, and what next! Lethaia, 37, 175–181.
Grahn, Y. and F. Paris, 1992, Age and correlation of the Trombetas Group, Amazon Basin, Brazil. Rev. Micropaléontologie, 35, 197–209.
Gramberg, I.S. (ed.), 1988, Barents Sea shelf plate, in: PGO "Sevmorgeologia" Trans., Vol. 196. Nedra, Leningrad, 262 pp. (in Russian).
Gramberg, I.S. and Yu.Ye. Pogrebitsky (eds.), 1984, Geological Structure and Economic Minerals of the USSR, Vol. 9 – Seas of the Soviet Arctic, Nedra, Leningrad, 280 pp. (in Russian).
Gramberg, I.S. and V.I. Ushakov (eds.), 2000, Severnaya Zemlya, Geological Structure and Mineragenesis, VNIIOkeangeologia, St. Petersburg, 187 pp. (in Russian).
Gramberg, I.S., V.L. Ivanov and Yu.Ye. Pogrebitsky (eds.), 2004, Geology and Mineral Resources of Russia (6 vols.), Vol. 5: Arctic and Far East Seas, Book 1: Arctic Seas, VSEGEI Press, St. Petersburg, 468 pp. (in Russian).
Gramberg, I.S., Ju.N. Kulakov, Yu.Ye. Pogrebitsky and D.S. Sorokov, 1983, Arctic oil and gas superbasin, Proc. 11th World Petroleum Congr., London, pp. 93–99.

Grantz, A., S.D. May and P.E. Hart, 1990, Geology of the continental margin of Alaska, in: A. Grantz, L. Johnson and J.F. Sweeny (eds.), The Geology of North America, Vol. I: The Arctic Ocean Region, Geol. Soc. Am., pp. 257–288.

Grantz, A., S.D. May and J.F. Sweeny (eds.), 1990, The Arctic Ocean region, in: A. Grantz, L. Johnson and J.F. Sweeny (eds.), The Geology of North America, Vol. I, Geol. Soc. Am., 644 pp.

Grantz, A., M.I. Holmes and B.A. Kososki, 1975, Geologic framework of the Alaskan continental terrace in the Chukchi and Beaufort seas, in: C.J. Yorath, E.R. Parker and D.J. Glass (eds.), Canada's continental margins and offshore petroleum exploration. Canadian Soc. Petroleum Geol., Memoir 4, pp. 669–700.

Grantz, A., D.L. Clark, R.L. Phillips and S.P. Srivastava, 1998, Phanerozoic stratigraphy of the Northwind Ridge, magnetic anomalies in the Canada basin, and the geometry and timing of rifting in the Amerasia Basin, Arctic Ocean. Geol. Soc. Am. Bull., 110, 801–820.

Greb, S.F., C.F. Eble and D.R. Chesnut Jr., 2002, Comparison of the Eastern and Western Kentucky coal fields (Pennsylvanian), USA – why are coal distribution patterns and sulfur contents so different in these coal fields? Int. J. Coal Geol., 50, 89–118.

Greb, S.F., D.R. Chesnut Jr., and C.F. Eble, in press, Temporal changes in coal-bearing depostional sequences (Lower and Middle Pennsylvanian) of the Central Appalachian Basin, U.S.A, in: J.C. Pashin and R. Gastaldo (eds.), Coal-Bearing Strata: Sequence Stratigraphy, Paleolcimate, and Tectonics, Am. Assoc. Petroleum Geol. Energy Minerals Publ.

Green, D., R.N. Crockett and M.T. Jones, 1980, Tectonic control of Karoo sedimentation in mid-eastern Botswana, Transactions Geol. soc. South Africa, 83, 213–219.

Grotzinger, J.P., 1986a, Cyclicity and palaeoenvironmental dynamics, Rocknest Platform, Northwest Canada, Geol. Soc. Am. Bull., 97, 1208–1231.

Grotzinger, J.P., 1986b, Upward shallowing platformal cycles: a response to 2.2 billion years of low amplitude, high frequency (Milankovich Band) sea level oscillations, Paleoceanography, 1, 403–416.

Guiraud, M. and J.C. Plaziat, 1993, Seismites in the fluviatile Bima Sandstones: identification of paleoseisms and discussion of their magnitude in a Cretaceous synsedimentary strike-slip basin (Upper Benue, Nigeria), Tectonophysics, 225, 493–522.

Gurnis, M., 1990, Ridge spreading, subduction, and sea level fluctuations, Science, 255, 970–972.

Gurnis, M., 1991, Depressed continental hypsometry behind oceanic trenches, Eos, 72, 271.

Gurnis, M., 1992, Rapid continental subsidence following the initiation and evolution of subduction, Science, 255, 1556–1558.
Gurnis, M., 1993, Phanerozoic marine inundation of continents driven by dynamic topography above subducting slabs, Nature, 364, 589–593.
Gusev, G.S., A.P. Petrov, G.S. Fradkin, et al., 1985, The Structure and Evolution of the Earth's Crust in Yakutia, Moscow, Nauka, 248 pp. (in Russian).
Gvirtzman, G. and T. Weissbrod, 1984, The Hercynian geanticline of the Helez and the late Paleozoic history of the Levant, in: J.E. Dixon and A.H.F. Robertson (eds.), The Geologic Evolution of the Eastern Mediterranean, Geol. Soc. London Spec. Publ. 17, pp. 177–186.
Ham, W.E. and J.L. Wilson, 1967, Paleozoic epeirogeny and orogeny in the central United States, Am. J. Sci., 265, 332–407.
Hamam, K.A. and I.M. Nasrulla, 1989, Drilling operations and evaluation of Matbakh-2, a deep formation in Qatar offshore, in: Country reports and case studies presented at the seminar on deep formations in the Arab countries: hydrocarbon potential and exploration techniques, OAPEC, pp. 229–241.
Hambrey, M.J., 1985, The Late Ordovician-Early Silurian glacial period, Palaeogeogr., Palaeoclim., Palaeoecol., 51, 273–289.
Hancox, P.J. and Rubidge, B.S., 1997, The role of fossils in interpreting the development of the Karoo Basin, Paleont. Africana, 33, 41–54.
Hancox, P.J., R.M.H. Smith, B.S. Rubidge and J. Neveling, 2002, From deserts to glaciers: a 100 million year transect of the terrestrial sedimentological sequence of the main Karoo Basin, 16th Int. Sedim. Conference, Johannesburg (South Africa), Excursion Guide, 144 pp. + annexes.
Hand, B.M., J.M. Wessel and M.O. Hayes, 1969, Antidunes in the Mount Toby conglomerate (Triassic), Massachusetts, J. Sedim. Petrol., 39, 1310–1316.
Haq, B., 1991, Sequence Stratigraphy, Sea-Level Change, and Significance for Deep Sea, Int. Assoc. Sedim. Spec. Publ. 12, pp. 3–39.
Haq, B.U., J. Hardenbol and P.R. Vail, 1987, Chronology of fluctuating sea levels since the Triassic, Science, 235, 1156–1167.
Haq, B.U., J. Hardenbol and P.R. Vail, 1988, Mesozoic and Cenozoic chronostratigraphy and eustatic cycles, in: C.K. Wilgus, B.S. Hastings, C.G.S.C. Kendall, H.W. Posamentier, O.A. Ross and J.C. Van Wagoner (eds.), Sea Level Changes: An Integrated Approach, Soc. Econ. Paleont. Min., Spec. Publ. 42, pp. 71–108.
Hardie, L.A. and H.P. Eugster, 1971, The depositional environment of marine evaporites: a case for shallow, clastic accumulation, Sedimentology, 16, 187–220.
Hargreaves, R.B., J. Rehacek and P.R. Hooper, 1997, Palaeomagnetism of the Karoo igneous rocks in South Africa, South African J. Geol., 100, 195–212.

Harland, W.B., A.V. Cox, P.G. Llewellyn, C.A. Pickton and R. Walters, 1982, A geologic time scale, Cambridge University Press, Cambridge, 131 pp. + wall chart.

Harris, P.M., S.H. Frost, G.A. Seiglie and N. Schneiderman, 1984, Regional unconformities and depositional cycles, Cretaceous of the Arabian Peninsula, in: J.S. Schlee (ed.), Interregional unconformities and hydrocarbon accumulation, Am. Assoc. Petroleum Geol., Memoir 36, pp. 67–79.

Harvey, A.M., 1984, Debris flows and fluvial deposits in Spanish Quaternary alluvial fans: implications for fan morphology, in: E.H. Koster and R.J. Steel (eds.), Sedimentology of gravels and conglomerates, Canadian Soc. Petroleum Geol., Memoir 10, pp. 123–132.

Hasui, Y. and M.A.F. Oliveira, 1984 Província Mantiqueira: setor central, in: F.F.M. Almeida and Y. Hasui (coord.), O Pré-Cambriano do Brasil, Edit, Edgard Blücher Ltda, São Paulo, pp. 308–344.

Hasui, Y., F.A.M. Abreu and R.N.N. Villas, 1984a Província Parnaíba, in: F.F.M. Almeida and Y. Hasui (coord.), O Pré-Cambriano do Brasil, Edit, Edgard Blücher Ltda, São Paulo, pp. 36–45.

Hasui, Y., J.B.S. Costa and F.A.M. Abreu, 1984b, Província Tocantins: setor setentrional, in: F.F.M. Almeida and Y. Hasui (coord.), O Pré-Cambriano do Brasil, Edit, Edgard Blücher Ltda, São Paulo, pp. 187–204.

Hays, J.D. and W.C. Pittman III, 1973, Lithospheric plate motion, sea-level changes, and climate and ecological consequences, Nature, 246, 18–22.

Heckel, P.H., 1986, Sea-level curve for Pennsylvanian eustatic marine transgressive-regressive depositional cycles along mid-continent outcrop belt, North America, Geology, 14, 330–334.

Hempton, M.R., 1987, Constraints on Arabian Plate motion and extensional history of the Red Sea, Tectonics, 6, 687–705.

Herak, M., 1986, A new concept of geotectonics of the Dinarides, Zagreb, Acta Geol., 16 (1) 1–42.

Hernández-Pacheco, E., 1915, Geologia y paleontologia del Mioceno de Palencia, Trab. Comis. Invest. Paleont. Prehist., Memoir 5, 295 pp.

Hernández-Pacheco, E., 1930, Sobre la extensión del Neógeno em el norte de la altiplanicie de Castilla la Vieja, Bol. Real Soc. Esp. Hist. Nat., 30, 396–398.

Heward, A.P., 1978, Alluvial fan sequence and megasequence models with examples from the Westphalian D-Stephanian B coalfields, northern Spain, in: A.D. Miall (ed.), Fluvial Sedimentology, Mem. Can. Soc. Petroleum Geol., 5, 669–702.

Hibbard, J., 2000, Docking Carolina: Mid-Paleozoic accretion in the southern Appalachians, Geology, 28, 127–130.

Hibbard, J.P., E.F. Stoddard, D.T. Secor and A.J. Dennis, 2002, The Carolina Zone: Overview of Neoproterozoic to Early Paleozoic peri-Gondwana terranes along the eastern flank of the southern Appalachians, Earth Sci. Rev., 57, 299–339.
Hinz, K., M. Bolck and M. Kos'ko, 1997, Seismic stratigraphic report. Cruise report, marine seismic measurements, and geoscientific studies of the shelf and slope of the Laptev sea and East Siberian Sea, Arctic (1997), BGR, Hannover, part II, pp. 57–61.
Horton, W., J. Aleinikoff and W. Burton, 1995, Mesoproterozoic and Neoproterozoic terranes in the eastern Piedmont of Virginia: implications of coordinated field studies and U-Pb geochronology, Geol. Soc. Am. Abstracts Progr., 27, 397.
Horváth, F., 1993, Towards a mechanical model for the formation of the Pannonian basin, Tectonophysics, 226, 333–357.
Horváth, F., 1995, Phases of compression during the evolution of the Pannonian Basin and its bearing on hydrocarbon exploration, Mar. Petroleum Geol., 12, 837–844.
Horváth, F. and S. Cloetingh, 1996, Stress-induced late stage subsidence anomalies in the Pannonian basin, Tectonophysics, 266, 287–300.
Hughes-Clarke, N.W., 1988, Stratigraphy and rock unit nomenclature in the oil producing area of the interior Oman, J. Petrol. Geol., 11, 5–60.
Hunter, R.E., 1977a, Basic types of stratification in small aeolian dunes, Sedimentology, 24, 361–387.
Hunter, R.E., 1977b, Terminology of cross-stratified sedimentary layers and climbing-ripple structures, J. Sedim. Petrol., 47, 697–706.
Hunter, R.E., 1981, Stratification styles in eolian sandstones: some Pennsylvanian to Jurassic examples from the western interior USA, in: F.G. Ethridge and R.M. Flores (eds.), Recent and Ancient Nonmarine Depositional Environments: Models for Exploration, Soc. Econ. Paleont. Min., Spec. Publ. 31, pp. 315–329.
Husseini, M.I., 1988, The Arabian Infracambrian extensional system, Tectonophysics, 148, 93–103.
Ingersoll, R.V. and C.J. Busby, 1995, Tectonics of sedimentary basins, in: C.J. Busby and R.V. Ingersoll (eds.), Tectonics of Sedimentary Basins, Blackwell Science, Oxford, pp. 1–51 [Chapter 1].
Inman, D.L., G.C. Ewing and J.B. Corliss, 1966 Coastal sand dunes of Guerrero Negro, Baja California, Mexico, Geol. Soc. Am. Bull., 77, 787–802.
Irwin, M.L., 1965, General theory of epeiric clear water sedimentation, Am. Assoc. Petroleum Geol. Bull., 49, 445–459.
Issawi, B., 2002, Egypt during the Phanerozoic, 6th Int. Conf. Geol. Arab World, Egypt, 2, 401–450.

Ivanova, N.M., 1997, Prospective Paleozoic reefs in the southern part of the Barents Sea shelf, Petroleum Geosci., 3, 153–160 (in Russian).

Ivanova, N.M. and S.B. Secretov, 1989 Sedimentary cover succession of the central Laptev Sea (by CDP data), in: Geological Structure Near Polar Prospective for Oil and Gas Areas of the World Ocean, NPO "Sevmorgeo", Leningrad, pp. 77–91 (in Russian).

Jachariah, J.K., Y.G. Bhaskar Rao, R. Srinivasan and K. Gopalan, 1999, Rb-Sr-Nd isotopic systematics of uranium mineralised stromatolotic dolomite from Proterozoic Cuddapah Supergroup, South India and constraints on age and provenance, Chem. Geol., 162, 49–64.

Jackson, N.J. and C.R. Ramsey, 1980, Time-space relationships of Upper Precambrian volcanic and sedimentary units in the central Arabian Shield, J. Geol. Soc. London, 137, 617–628.

Jacobs, P., 1994, Eocene sequence stratigraphic siliciclastic sedimentation model in the southern Bight, North Sea, 14th Int. Sedim. Congr., Recife (PE, Brazil), Abstracts, E-11–E-12.

Jacobs, P., M. DeBatist, B. Van Bavinchove and E. Sevens, 1998, Depositional systems and facies within a sequence stratigraphical framework: the Belgian Basin during Ypresian times (Lower Eocene, Southern Bight, North Sea), 15th Int. Sedim. Congr., Alicante (Spain), Abstracts, 444–445.

Jamičić, D., 1995, The role of sinistral strike-slip faults in the formation of the structural fabric of the Slavonian Mts. (Eastern Croatia), Geol. Croatica, 48, 155–160.

Jamieson, R.A. and C. Beaumont, 1988, Orogeny and metamorphism: a model for deformation and pressure–temperature–time paths with applications to the central and southern Appalachians, Tectonics, 7, 417–445.

Jarrar, G.H., A. Baumann and H. Wachendorf, 1991, The Saramuj conglomerates: evolution of a Pan-African molasses sequence from southwest Jordan, Neues Jahrb. Geol. Paläont. Monatsh., 6, 335–356.

Johnson, D.D. and C. Beaumont, 1995, Preliminary results from a planiform kinematic model of orogen evolution, surface processes and the development of clastic foreland basin stratigraphy, in: S.L. Dorobek and G.M. Ross (eds.), Stratigraphic Evolution of Foreland Basins, SEPM (Soc. Sedim. Geol.) Spec. Publ. 52, pp. 3–24.

Johnson, J.G., 1971, Timing and coordination of orogenic, epeirogenic, and eustatic events, Geol. Soc. Am. Bull., 82, 3263–3298.

Johnson, M.R., C.J. Van Vuuren, W.F. Hegenberger, R. Key and U. Shoko, 1996, Stratigraphy of the Karoo Supergroup in southern Africa: an overview, J. African Earth Sci., 23, 3–15.

Jones, R.W. and A. Racey, 1994, Cenozoic stratigraphy of the Arabian Peninsula and Gulf, in: M.D. Simmons (ed.), Micropalentology and hydrocarbon exploration in the Middle East, Chapman and Hall, London, pp. 273–303.

Jordan, T.E., 1981, Thrust loads and foreland basin evolution, Cretaceous, western United States, Am. Assoc. Petroleum Geol. Bull., 65, 2506–2520.

Juhász, E., L.Ó. Kovács, P. Müller, Á. Tóth-Makk, L. Phillips and M. Lantos, 1997, Climatically driven sedimentary cycles in the Late Miocene sediments of the Pannonian Basin, Hungary, Tectonophysics, 282, 257–276.

Kaila, K.L. and H.C. Tewari, 1985, Structural trends in the Cuddapah Basin from deep seismic soundings (DSS) and their tectonic implications, Tectonophysics, 115, 69–86.

Kaila, K.L., H.C. Tewari, K. Roy Chowdhury, V.K. Rao, A.R. Sridhar and D.M. Mall, 1987, Crustal structure of the Proterozoic Cuddapah Basin of India from deep seismic soundings and gravity data, in: S. Asano and W.D. Mooney (eds.), Seismic Studies of the Continental Lithosphere, Tectonophysics, 140, 1–12.

Kaila, K.L. and 10 co-authors, 1979, Crustal structure along the Kavali-Udipi profile in the Indian peninsular shield from deep seismic sounding, J. Geol. Soc. India, 20, 307–333.

Kaplan, M.E., 1965, On the diagenesis and epigenesis of the South Primorye Triassic deposits, Doklady Akad. Naul SSSR, 163 (4) 976–979 (in Russian).

Kaplan, M.E., 1966, Concretions of the South Primorye Triassic deposits, Litol. i poleznye iskopaemye, 5, 114–118 (in Russian).

Karner, G.D. and A.B. Watts, 1983, Gravity anomalies and flexure of the lithosphere at mountain ranges, J. Geophys. Res., 88 (B12) 10449–10477.

Karsakov, L.P. and V.A. Guryanov, 1990, The scheme of division and correlation of Lower Precambrian complexes, in: Stratigraphy of the Precambrian and Phanerozoic in the Transbaikal Region and Southern Far East, Far East Interreg. Conf., Khabarovsk, Abstracts 4, 3–6 (in Russian).

Karsakov, L.P., Yu.F. Malyshev and N.P. Romanovsky, 1977, The problems of deep structure of the southeastern part of the Aldan shield, in: Zoning of Geophysical Fields and Deep Structure of the Far East, Vladivostok, Akad. Nauk SSSR, 80–92 (in Russian).

Karsakov, L.P., V.A. Guryanov and M.V. Goroshko, 2002, Stratigraphy of the Riphean lower horizons (southeastern part of the Siberian platform), in: Stratigraphy, Geol. Correlation, 10, 47–61 (in Russian).

Kashirtsev, V.A., 1994, Natural bitumen and bituminous rocks in the southeastern part of the Siberian platform, Novosibirsk, DSc thesis, 35 pp. (in Russian).

Kaufman, A.J., 1998, Neoproterozoic chemostratigraphy: key events in Earth history ordered by detailed intra- and interbasinal correlation, 40° Congr. Brasileiro Geol., Belo Horizonte (MG), Anais (res.), 2.

Kazmin, V., A. Shifferaw and T. Balcha, 1978, The Ethiopian basement, stratigraphy and possible manner of evolution, Geol. Rundschau, 67, 531–546.

Kegel, W., 1957, Das Parnaíba-Becken, Geol. Rundschau, 45, 522–540.

Kemp, J., C. Pettaton and J.Y. Calvez, 1982, Cycles in the chelagenic evolution of the Precambrian shield in part of southwestern Saudi Arabia, Arabian Deputy Mininstry Mineral Resources Prof. Paper 1, 27–42.

Khan, A., 1989, Stratigraphy and hydrocarbon potential of Permo-Triassic sequence of rocks in the State of Kuwait, Organization Arab Petroleum Producing Countries (OAPEC), pp. 3–29.

Khanchuk, A.I., 2000, Paleogeodynamic analysis of the Russian Far East ore mineral deposits, in: Ore mineral deposits of the continental margins, Vypusk 1, Dalnauka, Vladivostok, pp. 5–31.

Khanchuk, A.I. and I.V. Kemkin, 2003, Geodynamic evolution of the Sea of Japan region during the Mesozoic, Vestnik, DVO, 6, 94–108 (in Russian).

Khanchuk, A.I., V.V. Ratkin, M.D. Ryazantseva, et al., 1995, Geology and minerals of the Primorye. Dalnauka, Vladivostok, 82 pp. (in Russian).

Kim. B.I. and N.M. Ivanova, 2000, About seismic complexes distinguished on the Laptev Sea continental slope and adjacent part of the Eurasia Basin, in: Geological and Geophysical Characteristics of the Arctic Region Lithosphere, 3rd ed., VNIIOkeangeologia, St. Petersburg, pp. 82–92 (in Russian).

Kim, B.I. and D.S. Yashin, 1999, Peculiarities of the Laptev sea shelf structure and the southern centroclinal closure of the Eurasia Basin in connection with hydrocarbon potential, in: Trans., 4th Int. Conf. "Development of the Russian Arctic Shelves", RAO-99, St. Petersburg, pp. 148–152 (in Russian).

Kimura, T., 1987, Recent knowledge of Jurassic and Early Cretaceous floras in Japan and phytogeography of this time in East Asia, Bull. Tokyo Gakugey Univ., 4 (39) 87–115.

King, L.C., 1967, The Morphology of the Earth, 2nd ed., Oliver and Boyd, Edinburgh, London, pp. 241–309 [Chapter IX: The geomorphology of Africa].

King, W., 1872, The Kadapah and Karnul formations in the Madras Presidency, Mem. Geol. Survey India, 8 (1) 346.

Kingston, D.R., C.P. Dishroon and P.A. Williams, 1983, Global basin classification system, Am. Assoc. Petroleum Geol. Bull., 67, 2175–2193.

Klein, G.deV., 1989, Origin of coal-bearing cyclothems of North America, Geology, 17, 152–155.

Klemme, H.D., 1975, Giant oil fields related to their geologic setting: a possible guide to exploration, Bull. Can. Petroleum Geol., 23, 30–66.
Klitzsch, E., 1990, Paleozoic, in: R. Said (ed.), Geology of Egypt, Balkema, Rotterdam, pp. 393–406.
Kochurek, G., 1981, Significance of interdune deposits and bounding surfaces in aeolian dune sands, Sedimentology, 28, 753–780.
Kochurek, G., 1988, Late Palaeozoic and Mesozoic eolian deposits of the Western Interior of the United States, Sedim. Geol., 56, 413.
Kochurek, G. and R.H. Dott Jr., 1981, Distinction and uses of stratification types in the interpretation of aeolian sand, J. Sedim. Petrol., 51, 579–595.
Kochurek, G. and G. Fielder, 1982, Adhesion structures, J. Sedim. Petrol., 52, 1229–1241.
Kochurek, G. and J. Nielsen, 1986, Conditions favourable for the formation of warm-climate aeolian sand sheets, Sedimentology, 33, 795–816.
Komarnitsky, V.M. and E.V. Shipilov, 1991, New data on the magmatism of the Barents Sea, Acad. Sci. Proc., 320 (#5) 1203–1206.
Kominz, M.A. and G.C. Bond, 1991, Unusually large subsidence and sea-level events during Middle Paleozoic time: new evidence supporting mantle convection models for supercontinent assembly, Geology, 19, 56–60.
Korzh, M.V., 1959, Petrography of the South Primorye Triassic deposits and paleogeography of their time formation, Akad. Nauk SSSR, Moscow, 83 pp. (in Russian).
Kosheleva, V.A. and D.C. Yashin, 1999, Bottomset beds of the Russian arctic seas, VNIIOkeangeologia, St. Petersburg, 287 pp. (in Russian).
Kos'ko, M.K., 1977, Facies and structural zonation of the Ordovician to Devonian on Anjou Islands, in: Tectonics of the Arctic. Fold Basement of Shelf Sedimentary Basins, NIIGA, Leningrad, pp. 56–87 (in Russian).
Kos'ko, M.K., 1993, Geology of the New Siberian Islands: constraints on paleogeodynamic reconstructions, in: L.P. Zonenshain Memorial Conference on Plate Tectonics, Moscow, 1993, Inst. Oceanology RAN, GGOMAR, p. 85 (in Russian).
Kos'ko, M.K. and G.V. Trufanov, 2002, Middle Cretaceous to Eopleistocene sequences on the New Siberian Islands: an approach to interpret offshore seismic, Marine Petroleum Geol., 19, 901–919.
Kos'ko, M.K. and V.I. Ushakov (eds.), 2003, Wrangel Island: geology, minergenesis, geoecology, Trans. VNIIOkeangeologia, Vol. 200, 137 pp. (in Russian).
Kos'ko, M.K, B.G. Lopatin and V.G. Ganelin, 1990, Major geological features of the islands of East Siberian and Chukchi seas and the northern coast of Chukotka, Marine Geol., 93, 349–367.

Kos'ko, M.K., Yu.Ya. Zamansky, A.E. Langhinen and N.M. Ivanova, 2002a, Canada Basin: Central Arctic elevated area boundary in Northwind Ridge area (Amerasia Basin, Arctic Ocean), in: Geological-Geophysical Features of the Lithosphere of the Arctic Region, Vol. 4, VNIIOkeangeologia, St. Petersburg, pp. 114–130 (in Russian).

Kos'ko, M.K., V.V. Verba, Ye.A. Korago, Ye.Ye. Musatov and O.I. Suprunenko, 2002b, Basement of the Eurasian Arctic shelf: fragmental structure, in: D.A. Dodin and V.S. Surkov (eds.), The Russian Arctic: Geological History, Mineragenesis, Environmental Geology, VNIIOKeangeologia, St. Petersburg, pp. 114–130 (in Russian).

Kos'ko, M.K. and 8 co-authors, 2000, Tectonic basement of the Eurasian Arctic shelf: age and some aspects of evolution, Polarforschung, 68 (1998) 275–281.

Kosygin, Yu.A., I.P. Karsakov and Yu.F. Malyshev, 1984, Deep structure of Precambrian blocks, in: Tectonic Nature of Geophysical Fields in the Far East, Nauka, Moscow, pp. 108–132 (in Russian).

Kováč, M. and N. Hudáčková, 1997, Changes of paleoenvironment as a result of interaction of tectonic events with sea level changes in the northeastern margin of the Vienna Basin, Zentralbl. Geol. Paläont., Teil I, 5/6, 457–469.

Kováč, M., A. Nagymarosy, N. Oszczypko, L. Csontos, A. Slaczka, M. Marunteanu, L. Matenco and E. Márton, 1998, Palinspastic reconstruction of the Carpathian-Pannonian region during the Miocene, in: M. Rakús (ed.), Geodynamic Development of the Western Carpathians, Bratislava, Ministry Slov. Monograph, pp. 189–217.

Kováč, M., A. Nagymarosy, K. Holcová, N. Hudáčková and A. Zlinská, 2001, Paleogeography, paleoecology and eustacy: Miocene 3rd order cycles of relative sea-level changes in the Western Carpathian-North Pannonian basins, Acta Geol. Hung., 44, 1–45.

Kováč, M., A. Andreyeva-Grigorovich, R. Brzobohaty, L. Fodor, M. Harzhauser, N. Oszczypko, D. Pavelić, F. Rögl, B. Saftić, L'. Sliva and Z. Stránik, 2003, Karpatian paleogeography, tectonics and eustatic changes, in: R. Brzobohatý, I. Cicha, M. Kováčc and F. Rögl (eds.), The Karpathian. A Lower Miocene Stage of the Central Paratethys, Brno, Masaryk University, pp. 49–72.

Kováč, M., I. Baráth, I. Harzhauser M. Hlavatý and N. Hudáčková, 2004, Miocene depositional systems and sequence stratigraphy of the Vienna Basin, Cour. Forsch.-Inst. Senckenberg, 246, 187–212.

Kovach, K., 1979, Introduction to Upper Carboniferous stratigraphy and depositional environments of eastern Ohio, in: F.R. Ettensohn and G.R. Dever (eds.), Carboniferous geology from the Appalachian Basin to the Illinois Basin through eastern Ohio and Kentucky, 9th Int. Congr. Carboniferous Geol. Stratigraphy, Lexington, Univ. Kentucky, Field Trip No. 4, pp. 19–29.

Kovačić, M., 2004, Sedimentology of the Upper Miocene deposits from the southwest part of the Pannonian Basin, unpublished PhD thesis, University of Zagreb, Zagreb, 203 pp. (manuscript; in Croatian).
Krasny, L.I., 1981, Geology of the Yakutskaya ASSR, Nedra, Moscow, 300 pp.
Krylov, I.N., 1976, Approaches to the classification of stromatolites, in: M.R. Walter (ed.), Stromatolites, Elsevier, Amsterdam, pp. 31–43.
Kukal, Z., 1970, Geology of Recent Sediments, Acad. Publ. House Czechoslovak Acad. Sci., Prague, 490 pp.
Lababidi, M.M. and A.N. Hamdan, 1985, Preliminary lithostratigraphic correlation study in OAPEC member countries, Organization Arab Petroleum Producing Countries (OAPEC), pp. 1–87.
Lapina, N.N., N.A. Znachko-Yavorsky and N.N. Kulikiv, 1968, Polar type of lithogenesis, in: Genesis and Classification of Sedimentary Rocks, Moscow, pp. 212–217 (in Russian).
Layer, P., L.M. Parphenov, A.A. Surnin and V.P. Timopheev, 1993, First ^{40}Ar–^{39}Ar dating of magmatic and metamorphic rock of Verkhoyansk-Kolyma Mesozoides, Proc. RAS, 329, 621–624 (in Russian).
Lee, D.W. and K.H. Paik, 1990, Evolution of strike-slip fault-controlled Cretaceous Yongdong Basin, South Korea: signs of strike-slip tectonics during infilling, J. Geol. Soc. Korea, 26, 257–276.
Lee, D.W., 1999, Strike-slip fault tectonics and basin formations during Cretaceous in Korean Peninsula, Island Arc, 8, 218–231.
Leeder, M.R., 1982, Sedimentology, Process and Product, George Allen and Unwin, London, 344 pp.
Leeder, M.R., 1999, Sedimentology and Sedimentary Basins, Blackwell, Oxford, 592 pp.
Lemon, K. and M. Tucker, 2002, The onset of the Gondwana glaciation recorded in the strata of equatorial Europe, 16th Int. Sedim. Congr., Johannesburg (South Africa), Abstracts, 220.
Lien, T. and R. Walker, 2000, Meandering turbidite channels in the Ross Formation, Shannon Basin, Ireland, 20th IAS Regional Meeting Sedim., Dublin (Ireland), Abstracts, 40.
Lima Filho, M.F., J.M. Mabesoone and M.S.S. Viana, 1999, Late Mesozoic history of sedimentary basins in NE Brazilian Borborema Province before the final separation of South America and Africa, 1: tectonic-sedimentary evolution, Bol. 5° Simp. Cretáceo Brasil, 1° Simp. Cretácico América del Sur, Serra Negra (SP), pp. 605–612.
Limeres, M., G. Hinterwimmer, S.B. Ibáñez and E. Rodríguez, 2000, Modelo de facies de un complejo estuarino en una sucesión transgressiva: Formación Springhill, área La Tehuelche, Cuenca Austral, Argentina, 2° Congr.

Latinoamericano Sedim., 8ª Reunión Argentina Sedim., Mar del Plata (Argentina), Resumenes, 100.
Linn, F.M., D.A. Textoris and J.M. Dennison, 1990, Syn-depositional tectonic influence on lithofacies of the Helderberg Group (Silurian-Devonian) of the central appalachians, in: Fall Program, November 1990, Appalachian Basin Industrial Associates, Morgantowen, West Virginia University, 17, 190–213.
Loczy, L. and E.A. Ladeira, 1976, Geologia estrutural e introdução à geotectônica, Edit, Edgard Blücher Ltda, São Paulo, 528 pp.
Long, D.G.F., 1978, Proterozoic stream deposits: some problems of recognition and interpretation of ancient sandy fluvial systems, in: Miall, A.D. (ed.), Fluvial Sedimentology, Canadian Soc. Petroleum Geol. Mem., 5, 313–342.
Lopatin, B.G. (ed.), 1999, The State Geological Map of the Russian Federation, Scale 1:1,000,000 (New Series), Sheet S-53-55 – Novosibirsk Islands, Explanatory Note, VSEGEL, St. Petersburg, 208 pp. (in Russian).
Mabesoone, J.M., 1959, Tertiary and Quaternary sedimentation in a part of the Duero Basin (Palencia, Spain), Leidse Geol. Mededelingen, 24, 31–180.
Mabesoone, J.M., 1975, Desenvolvimento paleoclimático do Nordeste brasileiro, Bol. Núcleo Nordeste Soc. Brasileira Geol., 5; Atas 7º Simp. Geol. Nordeste, Fortaleza (CE), 75–93.
Mabesoone, J.M., 1977, Paleozoic-Mesozoic deposits of the Piauí-Maranhão syneclise (Brazil): geological history of a sedimentary basin, Sedim. Geol., 19, 7–38.
Mabesoone, J.M., 1988, Controle global de sedimentação: integração dos dados. Recife, Univ. Fed. Pernambuco, Dept. Geol., sér. B – Estudos Pesquisas, 9, 61–77.
Mabesoone, J.M., 1994, Sedimentary Basins of Northeast Brazil, Recife, Fed. Univ. Pernambuco, Geol. Dept., Spec. Publ. 2, 308 pp.
Mabesoone, J.M., 2000a, Ciclicidade e relevo, Rev. Brasileira Geomorfologia, 1, 68–72.
Mabesoone, J.M., 2000b, Abertura do Oceano Atlântico Sul: como, onde, quando? Bol. Núcleo Nordeste Soc. Brasileira Geol., 16; 18º Simp. Geol. Nordeste, Recife (PE); resumos, 62.
Mabesoone, J.M., 2001, História geológica da Província Borborema (NE Brasil), Bol. Núcleo Nordeste Soc. Brasileira Geol., 17; 19º Simp. Geol. Nordeste, Natal (RN); resumos, 306–307.
Mabesoone, J.M., 2002, Neoproterozoic Bambuí limestone basin in South America, 16th Int. Sedim. Congr., Johannesburg (South Africa), Abstracts, 231.
Mabesoone, J.M., 2003a, Periodicity of sedimentation: with examples from NE Brazilian Borborema Province, 3rd Latinamerican Congr. Sedim., Belém (PA, Brazil), Abstracts, 28–29.

Mabesoone, J.M., 2003b, Relief-correlated sediments: differences between mobile belts and platform areas, 3rd Latinamerican Congr. Sedim., Belém (PA, Brazil), Abstracts, 29–30.
Mabesoone, J.M. and V.H. Neumann, 2002, Cyclic development of NE Brazilian sedimentary basins since Paleoproterozoic, 16th Int. Sedim. Congr., Johannesburg (South Africa), Abstracts, 232.
Mabesoone, J.M. and J.L. Rolim, 1981, A propósito das supostas glaciações paleozóicas no Nordeste brasileiro, Bol. Núcleo Nordeste Soc. Brasileira Geol., 9; Atas 11° Simp. Geol. Nordeste, Recife (PE), 290–294.
Mabesoone, J.M., P.J. Duarte, R. van Grieken, A. Delgado and E.M.P. Freire, 1985, Parnaíba basin shales, Anais Acad. Brasileira Ciências, 57, 481–495.
Mabesoone, J.M., M.S.S. Viana and M.F. Lima Filho, 1999, Late Mesozoic History of Sedimentary Basins in NE Brazilian Borborema Province Before the Final Separation of South America and Africa, Vol. 3: Paleogeography, Bol. 5° Simp. Cretáceo Brasil, 1° Simp, Cretácico América del Sur, Serra Negra (SP), pp. 621–626.
Magnavita L.S., 1996, Sobre a implantação da fase sin-rifte em riftes continentais, 39° Congr. Brasileiro Geol., Salvador (BA), Anais, 5, 335–338.
Makarenko, G.F., 1983, Trapps in the Structure of the Continents, Nauka, Moscow, 208 pp. (in Russian).
Makar'eva, E.I., 1997, Characteristics of the Franz Jozef Land magmatic formations by geological survey data, in: Actual Problems of the Barents-Kara Shelf and Adjacent Zone Regional Geology, VNIIOkeangeologia, St. Petersburg, Abstracts, 41 (in Russian).
Makeev, V.M., V.K. Davydov and V.I. Ustritsky, 1991, Find of the Middle Carboniferous deposits with tropical fossils on De long Islands, in: Paleozoic Stratigraphy and Paleontology of the Arctic, Leningrad, PGO "Sevmorgeologia", pp. 167–170 (in Russian).
Malich, N.S. (ed.), 2002, East Siberia, in: V.P. Orlov, (ed.-in-chief), Geology and Mineral Resources of Russia, Vol. 3, VSEGEI Press, St. Petersburg, 396 pp. (in Russian).
Malinovsky, A.L., A.N. Philippov, V.V. Golozubov, et al., 2002, Lower Cretaceous deposits of the Kema river (Eastern Sikhote-Alin): a sedimentary filling of a back-arc basin, Tikhookeanskaya Geol., 21 (1) 52–66 (in Russian).
Malyshev, Yu.F., 1977, Geophysical Investigations of the Precambrian on the Aldan Shield, Nedra, Moscow, 127 pp. (in Russian).
Marini, O.J., R.A. Fuck, M.A. Dardenne and J.C.M. Danni, 1984, Província Tocantins: setores central e sudeste, in: Almeida, F.F.M. and Hasui, Y. (coord.), O Pré-Cambriano do Brasil, Edt, Edgard Blücer Ltda, São Paulo, pp. 205–264.

Markevich, P.V. and Y.D. Zakharov (eds.), 2004, Triassic and Jurassic of the Sikhote-Alin, Book I: Terrigenous Assemblage, Dalnauka, Vladivostok (in Russian).

Markevich, P.V., V.P. Konovalov, A.I. Malinovsky and A.N. Philippov, 2000, Lower Cretaceous Deposits of the Sikhote-Alin, Dalnauka, Vladivostok, 278 pp. (in Russian).

Martín Serrano, A., 1991, La definición y el encajamiento de la red fluvial actual sobre el Macizo Hespérico em el marco de sua geodinámica alpina, Rev. Soc. Geol. España, 4, 337–351.

Martinioni, D., 1997, Cretaceous-Paleogene surface stratigraphy of the Austral Basin in the southernmost Andes. New evidences from the central Tierra del Fuego, Argentina, GAEA heidelbergensis, 3; 18th IAS Regional European Meeting Sedim., Heidelberg (Germany), Abstracts, 231–232.

Martins-Neto, M.A., A.C. Pedrosa-Soares and S.A.A. Lima, 2001a, Tectono-sedimentary evolution of sedimentary basins from late Paleoproterozoic to late Neoproterozoic in the São Francisco craton and Araçuai fold belt, eastern Brazil, Sedim. Geol., 141/142, 343–370.

Márton, E., D. Pavelić, B. Tomljenović, R. Avanić, J. Pamić and P. Márton, 2002, In the wake of a counterclockwise rotating Adriatic microplate: Neogene paleomagnetic results from northern Croatia, Int. J. Earth Sci., 91, 514–523.

Mascarenhas, J.F., A.J. Pedreira, A. Misi, A.C. Motta and J.H.S. Sá, 1984, Província São Francisco, in: F.F.M. Almeida and Y. Hasui (coord.), O Pré-Cambriano do Brasil, Edit, Edgard Blücher Ltda, São Paulo, pp. 46–122.

Matheos, S.D., N.G. Cabaleri, C. Armella, S.A. Valencio and J.R. Franzese, 1994, Depositional and diagenetic features of Lower Cretaceous limestones in the south of the Neuquén Basin, Argentina, 14th Int. Sedim. Congr., Recife (PE, Brazil), Abstracts, B-9.

Matos, R.M.D., 1992, The Northeast Brazilian rift system, Tectonics, 11, 766–791.

McCann, T., 1998, The tectonosedimentary evolution of the Carboniferous foreland basin of NE Germany, 15th Int. Sedim. Congr., Alicante (Spain), Abstracts, 547–548.

McGee, W.J., 1897, Sheet flood erosion, Geol. Soc. Am. Bull., 8, 87–112.

McKee, E.D., 1966, Structures of dunes at White Sand National Monument, New Mexico (and a comparison with structures of dunes from other selected areas), Sedimentology, 7, 1–69.

McKenzie, D.P., 1978, Active tectonics of the Alpine-Himalayan Belt: the Aegean sea and surrounding regions, Geophys. J. Royal Astronom. Soc., 55, 217–254.

McKerrow, W.S., 1979, Ordovician and Silurian changes in sea level, J. Geol. Soc. London, 16, 17–145.
McLane, M., 1995, Sedimentology, Oxford University Press, Oxford, 423 pp.
Mediavilla, R., C.J. Dabrio, A. Martín Serrano, J.I. Santisteban, 1996, Lacustrine Neogene systems of the Duero Basin: evolution and controls, in: P.F. Friend and C.J. Dabrio (eds.), Tertiary basins of Spain. The stratigraphical record of crustal kinematics, Cambridge University Press, Cambridge; World Regional Geol., 6, 228–236.
Mediavilla, R., J.I. Santistebam and C.J. Dabrio, 1998, Tectonic evolution during the Alpine Cycle. Sedimentation and erosion related to capture of endorheic fluvial systems, 15th Int. Sedim. Congr., Alicante (Spain); Excursion A6, 129–148.
Mello, Z.F., 1977, Considerações tectono-petrológicas em sequências molássicas no Nordeste Oriental brasileiro, Recife, Univ. Fed. Pernambuco; Curso Pós-grad. Geociências, MSc dissertation, 254 pp.
Merriam, E.D., 1966, Symposium on cyclic sedimentation, State Geol. Survey Kansas, Bull. 169 (1, 2), 636 pp.
Mesner, J.C. and L.C.P. Wooldridge, 1964, Maranhão Paleozoic basin and Cretaceous coastal basins, North Brazil, Am. Assoc. Petroleum Geol. Bull., 48, 1475–1512.
Miall, A.D., 1978, Fluvial sedimentology, Can. Soc. Petroleum Geol., Memoir 5, 859 pp.
Miall, A.D., 1984, Principles of Sedimentary Basin Analysis, Springer-Verlag, New York, 490 pp. [cf. Chapter 8: Regional and global stratigraphic cycles, pp. 329–365].
Miall, A.D., 1996, The Geology of Fluvial Deposits: Sedimentary Facies, Basin Analysis and Petroleum Geology, Springer-Verlag, Berlin, 582 pp.
Miall, A.D., 1997, The Geology of Stratigraphic Sequences, Springer-Verlag, Berlin, 433 pp.
Middleton, G.V., 2003, Encyclopedia of Sediments and Sedimentary Rocks, Kluwer, Dordrecht, 821 pp.
Milani, E.J., 2002, Evolução tectono-sedimentar das sinéclises paleozóicas do Brasil, 41° Congr. Brasileiro Geol., João Pessoa (PB), Anais, 274 (res.).
Milici, R.C. and W. de Witt Jr., 1988, The Appalachian Basin, in: L.L. Sloss (ed.), Sedimentary Cover – North American Craton, U.S. Geol. Soc. Am., The Geology of North America, D-2, pp. 427–469.
Mironyuk, Ye.P. (ed.), 1986, Geological Map of the USSR, Sheet 0-52 (53)-Tommot, Scale 1:1,000,000 (New Series), Explanatory Note, Leningrad, 127 pp.

Miroshnikov, L.D., 1971, Review of the South Primorye history of the geological development: geology of the South Primorye Mesozoic-Cenozoic depressions, Leningrad, Trudy VNIGRI Vypusk, 302, 172–176 (in Russian).

Mitrovica, J.X., C. Beaumont and G.T. Jarvis, 1989, Tilting of continental interiors by the dynamical effects of subduction, Tectonics, 8, 1079–1094.

Miura, K., 1972, Possibilidades petrolíferas da Bacia do Acre, 26° Congr. Brasileiro Geol., Belém (PA), Anais, 3, 15–20.

Modie, B.N., 2002a, Glacial records in Botswana, 16th Int. Sedim. Congr., Johannesburg (South Africa), Abstracts, 261.

Modie, B.N., 2002b, Sedimentary basins of Botswana, 16th Int. Sedim. Congr., Johannesburg (South Africa), Abstracts, 262.

Moore, J.M., 1979, Tectonics of the transcurrent fault system, Saudi Arabia, J. Geol. Soc. London, 136, 535–561.

Moresi, L. and Gurnis, M., 1996, Constraints on lateral strength of slabs from three-dimensional dynamic flow models, Earth Planet. Sci. Lett., 138, 15–28.

Muir, I.D. and B.R. Rust, 1982, Sedimentology of a Lower Devonian coastal alluvial fan complex: the Snowblind Bery Formation of Cornwallis Island, Northwest Territories, Canada, Bull. Can. Petroleum Geol., 30, 245–263.

Muniz, G.C.B., 1979, Moluscos devonianos da Formação Inajá no Estado de Pernambuco, Anais Aced. Brasileira Ciências, 51, 651–675.

Murris, R.J., 1980, Middle East: stratigraphy, evolution and oil habitat, Am. Assoc. Petroleum Geol. Bull., 64, 597–618.

Murthy, Y.G.K., 1979, Salient features of the revised geological map of the Cuddapah Basin, Inst. Indian Peninsular Geol., Hyderabad; 3rd Workshop on Status, Problems and Programmes in Indian Peninsular Shield, pp. 1–21.

Murthy, Y.G.K., 1981, The Cuddapah Basin – a review of basin development and basement framework relations, Inst. Indian Peninsular Geol., Hyderabad; 4th Workshop on Status Problem in the Cuddapah Basin, pp. 51–72.

Nagaraja Rao, B.K. and G. Ramlingaswamy, 1976, Some new thoughts on the stratigraphy of Cuddapah Supergroup, Seminar on Kaladgi-Badami, Bhima and Cuddapah Supergroup, Mysore; Anstracts, 17–20.

Nagaraja Rao, B.K., S.T. Rajurkar, G. Ramlingaswamy and B. Ravindra Babu, 1987, Stratigraphy, structure and evolution of Cuddapah Basin, in: B.P. Radhakrishna (ed.), Purana Basins of Peninsular India, Bangalore, Geol. Soc. India, Memoir 6, pp. 33–86.

Nance, R.D., T.R. Worsley and J.B. Moody, 1986, Post-Archean biochemical cycles and long-term episodicity in tectonic processes, Geology, 14, 514–518.

Narain, H., 1987, Geophysical constraints on the evolution of Purana basins of India with special reference to Cuddapah, Godavari and Vindhyan basins, in: B.P. Radhakrishna (ed.), Purana Basins of Peninsular India, Bangalore, Bull. Geol. Soc. India, Memoir 6, pp. 5–32.

Netto, A.S.T., 1978, A implantação da fase rift na Bacia do Recôncavo, 30° Congr. Brasileiro Geol., Recife (PE), Anais, 1, 506–517.

Neves, S.P., 2001, Correlation between the Borborema province (NE Brazil) and neighbouring cratons and Pan-African belts: implications for the evolution of Western Gondwana, 8° Simp. Nac. Estud. Tectônicos, 2nd Int. Symp. Tectonics Brazilian Geol. Soc., Recife (PE), Anais, extended abstracts, 85–87.

Neves, S.P., 2003, Proterozoic history of the Borborema province (NE Brazil): correlations with neighbouring cratons and Pan-African belts and implications for the evolution of western Gondwana, Tectonics, 34, 14 pp.

Nickel, E. and A. Schäfer, 1997, A sequence stratigraphic approach to the Tertiary of the Lower Rhine embayment (NW-Germany) and adjacent areas, GAEA heidelbergensis, 3; 18th IAS Regional Eurpean Meeting Sedim., Heidelberg (Germany), Abstracts, 254–255.

Ojeda, H.A. and A. Fugita, 1974, Bacia Sergipe-Alagoas: geologia regional e perspectivas petrolíferas, 28° Congr. Brasileiro Geol., Porto Alegre (RS), Anais, 1, 137–158.

Olivero, E.B., 1997, Jurassic-Paleogene stratigraphy and basin evolution of the Eastern Andes of Tierrra del Fuego, GAEA heidelbergensis, 3; 18th IAS Regional European Meeting Sedim., Heidelberg (Germany), Abstracts, 258–259.

Olivero, E.B., 1998, Mesozoic-Paleogene geology of the Marginal-Austral basin of Tierra del Fuego. IGCP Project 381 – South Atlantic Mesozoic Correlations; 3rd Ann. Conf., Comodoro Rivadavia (Argentina); Field trip no. 3, 42 pp.

Olivero, E.B. and D.R. Martinioni, 1996, Sedimentologia de la Formación Yahgán (Jurásico-Cretácico) em Tierra del Fuego, 13° Congr. Geol. Argentino, Buenos Aires; Actas, 2, 45–59.

Olivero, E.B. and D.R. Martinioni, 1998, A review of the Mesozoic-Paleogene geology of the Andes Fueginos, Isla Grande de Tierra del Fuego. IGCP Project 381 – South Atlantic Mesozoic Correlations, 3rd Ann. Conf., Comodoro Rivadavia (Argentina); Bol. Asoc. Paleont. Golfo San Jorge, 2, Abstracts, pp. 23–24.

Olivero, E.B., R.D. Acevedo and D.R. Martinioni, 1997, Geologia y estructura de Bahia Ensenada, Tierra del Fuego. Rev. Asoc. Geol. Argentina, 52, 169–179.

Paciullo, F.V.P., A. Ribeiro, R.R. Andreis and R.A.J. Trouw, 1998, Eventos tafrogênicos e orogênicos na zona de interferência entre as faixas Brasília

e Ribeira, Sul de Minas Gerais, 40° congr. Brasileiro Geol., Belo Horizonte (MG), Anais (res.), 24.
Pamić, J., 1998, North Dinaridic Late Cretaceous-Paleogene subduction-related tectonostratigraphic units of Southern Tisia, Croatia, Geol. Carpathica, 49, 341–350.
Pamić, J., M. Belak and D. Slovenec, 1992/1993, Lower Miocene trachyandesites (shoshonites) from Mt. Krndija in Slavonija (northern Croatia), Zagreb, Rad HAZU, 463, 27–47 (in Croatian).
Pamić, J., M. Belak, T.D. Bullen, M.A. Lanphere and E.H. McKee, 2000, Geochemistry and geodynamics of a late Cretaceous bimodal volcanic association from the southern part of the Pannonian Basin in Slavonija (North Croatia), Min. Petrol., 68, 271–296.
Parenago, P.P., 1952, On the gravity potential of the Galaxy, Astronom. J., 29 (3), 261–287 (in Russian).
Parfenov, I.M., 1984, Continental Margin and Island Arcs of Mesozoides in Northeast Asia, Nauka, Novosibirsk, 192 pp. (in Russian).
Parfenov, I.M. and M.I. Kuzmin (eds.), 2001, Tectonics, Geodynamics and Metallogeny in the Sakha Territory (Yakutia), MAIK "Nauka"/Interperiodika, Moscow, 571 pp. (in Russian).
Pashin, J.C., W.E. Osborne and A.K. Rindsberg, 1991, Outcrop characterization of sandstone heterogeneity in Carboniferous reservoirs, Black Warrior Basin, Alabama, Bartelsville, Bartelsville Project Office, U.S. D.O., DOE/BC/1 4448-6, 126 pp.
Patchen, D.G., K.L. Avary and R.B. Erwin, 1985a, Southern Appalachian region. Am. Assoc. Petroleum Geol., COSUNA Chart SAP.
Patchen, D.G., K.L. Avary and R.B. Erwin, 1985b, Northern Appalchian region, Am. Assoc. Petroleum Geol., COSUNA Chart NAP.
Pavelić, D., 2001, Tectonostratigraphic model for the North Croatian and North Bosnian sector of the Miocene Pannonian Basin System, Basin Res., 13, 359–376.
Pavelić, D., M. Miknić and M. Sarkotić Šlat, 1998, Early to Middle Miocene facies succession in lacustrine and marine environments on the southwestern margin of the Pannonian basin system, Geol. Carpathia, 49, 433–443.
Pavelić, D., R. Avanić and J. Zupanić, 2000, Ottnangian lacustrine sediments on Mt. Medvednica: facies, depostional environments, and tectonic controls, in: I. Vlahović and R. Biondić (eds.), 2. Hrvatski Geol. Kongr.; Zagreb, Zbornik radova, 339–343.
Pavlidis, Yu.A., 1982, Sedimentary settings in the Chukchi Sea and shelf facies and sedimentary zones, in: Problems of Shelf Geomorphology, Lithology and Lithodynamics, Nauka, Moscow, pp. 47–75 (in Russian).

Petri, S., A.M. Coimbra, G. Amaral and W.L. Poncano, 1986, Guía de nomenclatura estratigráfica, Rev. Brasiliera Geociências, 16, 376–415.

Petri, S. and V.A. Campanha, 1981, Brazilian continental Cretaceous, Earth Sci. Rev., 17, 69–85 [Special issue: The Geology of Brazil].

Pitman III, W.C., 1978, Relationship between eustasy and stratigraphic sequences of passive margins, Geol. Soc. Am. Bull., 89, 1389–1403.

Plaziat, J.C. and B.H. Purser, 1998, The tectonic significance of seismic sedimentary deformations within the syn- and post-rift deposits of the northwestern (Egyptian) Red Sea coast and Gulf of Suez, in: B.H. Purser and D.W. Bosence (eds.), Sedimentation and Tectonics of Rift Basins: Red Sea Gulf of Aden, Chapman and Hall, London, pp. 347–652.

Pomerol, C., 1978, Evolution paléogéographique et structurale du Bassin de Paris, du Précambrien à l'Actuel, em relation avec les régions avoisinantes, Geol. Mijnbouw, 57, 533–543.

Ponikarov, V., V. Kazmin, I. Mikhailov, A. Razvaliayev, V. Krasheninnikov, V. Kozlov, E. Soulidi-Kandratiyev, K. Mikhailov, V. Kulatov, V. Faradzhev and K. Mirzayev, 1967, The geology of Syria: explanatory notes on the geological map of Syria, scale 1:500,000, Part 1: Stratigraphy, Damascus, Ministry Industry.

Ponte, F.C. (coord.), et al., 1972, Análise comparativa da paleogeologia dos litorais atlânticos brasileiro e africano, Salvador, Petrobrás/SEPES/DIPER, internal report, 20 pp.

Portero, J.M., P. Olmo, J. Ramírez and I. Vargas, 1982, Síntesis del Terciario continental de la Cuenca del Duero, Madrid, Inst. Geol. Min. España; Temas Geol. Mineral., 6, 11–40.

Posamentier, H.W., M.T. Jervey and P.R. Vail, 1988, Eustati controls on clastic deposition I-Conceptual framework, in: C.K. Wilgus, B.S. Hastings, C.G.S.C. Kendall, H. Posamentier and J. Van Wagonar (eds), Sea-level changes: An Integrated Approach, Soc. Econ. Paleont. Min., Spec. Publ. 42, pp. 109–124.

Poselov, V.A., I.S. Gramberg, R.R. Murzin, V.V. Butsenko, V.D. Kaminsky and Yu.Ye. Pogrebitsky, 2002, Structure and boundaries of continental and oceanic lithosphere in the Arctic Basin, in: D.A. Dodin and V.S. Surkov (eds.), The Russian Arctic: Geological History, Minerogenesis, Environmental Geology, VNIIOkeangeologia, St. Petersburg, pp. 121–133 (in Russian).

Powell, J., 1989, Stratigraphy and Sedimentation of the Phanerozoic Rocks in Central and South Jordan, Part A: Ram and Khreim Groups, Amman, NRA, Geol. Map. Div. Bull., 11, 130 pp.

Powers, R.W., L. Ramrirez, C. Redmon and J. Elberg Jr., 1966, Geology of the Arabian Peninsula: Sedimentary Geology of Saudi Arabia, U.S. Geol. Survey Prof. Paper 560-D, 127 pp.

Pozaryski, W. and W. Brochwicz-Lewinski, 1978, On the Polish trough, Geol. Mijnbouw, 57, 545–557.

Prelogovic, E., D. Jamičić, B. Aljinović, B. Saftić and J. Velić, 1995, Structural dynamics in the southern part of the Pannonian Basin, in: I. Vlahović and R. Biondić (eds.), 1. Hrvatski Geol. Kongr.; Zagreb, Zbornik radova, 481–486 (in Croatian).

Prelogovic, E., B. Saftić, V. Kuk, J. Velić, M. Dragaš and D. Lučić, 1998, Tectonic activity in the Croatian part of the Pannonian basin, Tectonophysics, 297, 283–293.

Preobrazenskaya, E.N., V.I. Ustritsky and E.G. Bro, 1995, Paleozoic deposits of Kalguev Island (Barents Sea), Stratigraphy Geological Correlation, 3 (5) 75–85 (in Russian).

Price, R.A., 1973, Large-scale gravitational flow of supracrustal rocks, southern Canadian Rockies, in: K.A. de Jong and R. Scholten (eds.), Gravity and Tectonics, John Wiley & Sons, New York, pp. 491–502.

Pustowalow, L.W., 1940, Petrographiya osadocnich porod (Petrography of Sedimentary Rocks), Vols. I and II, Gostoptechizdat, Moscow (in Russian).

Quin, J., 2000, Controls on the environmental setting of a thick shallow marine succession, the South Munster Basin, Ireland, 20th IAS Regional Meeting Sedim., Dublin (Ireland), Abstracts, 50.

Quinlan, G.M. and C. Beaumont, 1984, Appalachian thrusting, lithospheric flexure, and the Paleozoic stratigraphy of the eastern interior of North America, Can. J. Earth Sci., 21, 973–996.

Rabassa, J., A. Coronato, G. Bujaleski and C. Roóg, 1998, Quaternary of the Andean Region of Tierra del Fuego, in: E.B. Olivero, Mesozoic-Paleogene geology of the Marginal-Austral basin of Tierra del Fuego. IGCP Project 381 – South Atlantic Mesozoic Correlations; 3rd Ann. Conf., Comodoro Rivadavia (Argentina); Field trip no. 3, 14–19.

Radhakrishna, B.P., 1987, Purana Basins of Peninsular India, Bangalore, Geol. Soc. India, Memoir 6, 518 pp.

Rajurkar, S.T., 1979, Pre-Kurnool age of the Cuddapah Orogeny, Indian Mineralogist, 18, 32–39.

Rajurkar, S.T. and G. Ramlingaswamy, 1975, Facies variation within the upper Cuddapah strata in the northern part of the Cuddapah basin, Geol. Survey India, Misc. Publ. 23 (1) 157–164.

Rajurkar, S.T. and G. Ramlingaswamy, 1978, Atmakur fault and its bearing on the evolution of Kurnool basin, Sagar, University of Saugar; Proc. Symp. Purana Formations of Peninsular India, pp. 242–245.

Reading, H.G., 1996, Sedimentary Environments: Processes, Facies and Stratigraphy, 3rd ed., Blackwell Science, Oxford, 688 pp.

Regali, M.S.P., N. Uesugui and A.S. Santos, 1974, Palinologia dos sedimentos Meso-Cenozóicos do Brasil, Bol. Técnico Petrobrás, 17, 263–301.

Reineck, H.E. and I.B. Singh, 1980, Depositional Sedimentary Environments – with Reference to Terrigenous Clastics, 2nd ed., Springer-Verlag, Berlin, 549 pp.

Rezende, W.M., 1972, Post-Paleozoic geotectonics of South America related to plate tectonics and continental drift, 26° Congr. Brasileiro Geol., Belém (PA), Anais, 3, 205–210.

Rezende, W.M. and J.O. Ferradaes, 1971, Integração geológica regional da bacia sedimentar da Foz do Amazonas, 25° Congr. Brasileiro Geol., São Paulo (SP), Anais, 3, 203–214.

Ribeiro, H.J.P.S., 1992, Depositional architecture, relative paleo-lake base level changes of the Recôncavo basin, Bahia, Brazil, and the global sea-level changes ("Vail's Curve"), 37° Congr. Bras. Geol., São Paulo (SP); Bol. res. expand., 2, 458–460.

Riccardi, A.C., M.P. Iglesias Llanos and M. Vizan, 1997, Bio- and magneto-stratigraphy of the Pliensbachian-Toarcian (Jurassic) of the Neuquén Basin, Argentina, GAEA heidelbergensis, 3; 18th IAS Regional European Meeting Sedim., Heidelberg (Germany), Abstracts, 290.

Riegel, W., 1991, Coal cyclothems and some models of their origin, in: G. Einsele, W. Ricken and A. Seilachter (eds.), Cycles and Events in Stratigraphy, Springer-Verlag, Berlin, pp. 733–750.

Rodgers, J., 1987, The Appalachian-Ouachita orogenic belt, Episodes, 10, 259–266.

Rogers, J.W.W., 1996, A history of continents in the past three billion years, J. Geol., 104, 91–107.

Rogers, J.W.W. and M. Santosh, 2004, Continents and Supercontinents, Oxford University Press, New York, 287 pp.

Rögl, F., 1996, Stratigraphic correlation of the Paratethys Oligocene and Miocene, Mitteil. Bergbaustud. Österreich, 41, 65–73.

Ronov, A.B., 1961, Some general laws in the development of epeirogenic movements of the continents, Probl. Tektonika, 1961, 118–164 (in Russian).

Ronov, A.B., A.A. Migdisov and N.V. Barskaya, 1969, Tectonic cycles and regularities in the development of sedimentary rocks and paleogeographic environments of sedimentation of the Russian Platform (An approach to a quantitative study), Sedimentology, 13, 179–212.

Rossetti, D.F., A.M. Góes, A.E.A. Santos Jr. and J.O.S. Paz, 2004, Towards the reconstruction of the Amazonian geological history back to the Pleistocene, 23rd IAS Meeting Sedim., Coimbra (Portugal), Abstracts Book, 236.

Royden, L.H., 1988, Late Cenozoic tectonics of the Pannonian Basin System, in: L.H. Royden and F. Horváth. (eds.), The Pannonian Basin. A study in Basin Evolution, Am. Assoc. Petroleum Geol., Memoir 45, pp. 27–48.

Rubidge, B.S., P.J. Hancox and O. Catuneanu, 2000, Sequence analysis of the Ecco-Beaufort contact in the southern Karoo of South Africa, South African J. Geol., 103, 81–96.

Rubin, D.M. and R.E. Hunter, 1983, Reconstructing bedform assemblages from compound cross-bedding, in: M.E. Brookfield and T.S. Ahlbrandt (eds.), Eolian Sediments and Processes, Elsevier, Amsterdam, pp. 407–427.

Ruchin, L.B., 1958, Grundzüge der Lithologie. Lehre von den Sedimentgesteinen, Akademie-Verlag, Berlin, 806 pp. [Kap. 11: Die wichtigsten Voraussetzungen für die Bildung von Sedimentgesteinen, pp. 267–330].

Runcorn, S.K., 1962, Convection currents in the earth's mantle, Nature, 195, 1248–1249.

Rust, I.C., 1981, Lower Paleozoic rocks of southern Africa, in: C.H. Holland (ed.), Lower Paleozoic Rocks of the World, Vol. 3: Lower Paleozoic Rocks of the Middle East, Eastern and Southern Africa, and Antarctica, John Wiley, Chichester (England), pp. 165–187.

Sá, J.M., J.M. Bertrand and J. Leterrier, 1997, Geocronologia U-Pb e geoquímica de ortognaisses paleo-e mesoproterozóicos da região de Taquaretinga – PE, Bol. Núcleo Nordeste Soc. Brasileira Geol., 15; 17° Simp. Geol. Nordeste, Fortaleza (CE); res. expand., pp. 108–112.

Saftić, B., 1998, Genetic stratigraphic sequence analysis in the Pontain sediments of the western part of the Sava depression, unpublished PhD thesis, University of Zagreb, Zagreb, 136 pp. (manuscript; in Croatian).

Saftić, B., J. Velić, O. Sztanó, G. Juhász and Ž. Ivković, 2003, Tertiary subsurface facies, source rocks and hydrocarbon reservoirs in the SW part of the Pannonian Basin (Northern Croatia and South-Western Hungary), Geol. Croatica, 56, 101–122.

Sagasti, C., 1998, Facies and cyclicity in Hauterivian-Barremian carbonates of the Neuquén Basin, Argentina, 15th Int. Sedim. Congr., Alicant (Spain), Abstracts, 689.

Sagasti, G., 2000, Ciclos de Milankovitch em el Cretácico inferior de la Cuenca Neuquina Surmendoncina, Argentina, 2° Congr. Latinoamericano Sedim., 3ª Reunión Argentina Sedim., Mar del Plata (Argentina); Resumenes, 158.

Salop, L.I., 1982, Geological Evolution of the Earth in the Precambrian, Nedra, Leningrad, 343 pp. (in Russian).

Saltzman, M.R., 2002, Carbon isotope ($\delta^{13}C$) stratigraphy across the Silurian-Devonian transition in North America. Evidence for a perturbation of the global carbon cycle, Palaeogeogr., Palaeoclim., Palaeoecol., 187, 83–100.

Samson, S. and D.T. Secor, 2000, New U-Pb geochronological evidence for a Silurian Magmatic event in central South Carolina, Geol. Soc. Am. Abstracts Progr. 32, A-71.

Santisteban, J.I., R. Mediavilla, A. Martín-Serrano and C.J. Dabrio, 1996, The Duero Basin: a general overview, in: P.F. Firend and C.J. Dabrio (eds.), Tertiary Basins of Spain. The Stratigraphical Record of Crustal Kinematics, Cambridge University Press, Cambridge; World Regional Geol., 6, 183–187.

Santisteban, J.I., L. Alcalá, R. Mediavilla, M.T. Alberdi, L. Luque, A. Mazo, I. Miguel J. Morales and B. Pérez, 1997, El yacimineto de Tariego de Cerrato: El inicio de la red fluvial actual em el sector central de la Cuenca del Duero, Cuad. Geol. Ibérica, 22, 431–446.

Santos, E.J. and B.B. Brito Neves, 1984, Província Borborema, in: F.F.M. Almeida and Y. Hasui (coord.), O Pré-Cambriano do Brasil, Edit, Edgard Blücher Ltda, pp. 123–186.

Sarkisyan, S.G., 1958, Mesozoic and Tertiary deposits of Pribaykalie, Zabaykalie and Far East, Izdateltsvo Akad. Nauk SSSR, 337 pp. (in Russian).

Schaller, H., 1969, Revisão estratigráfica da bacia de Sergipe-Alagoas, Bol. Técnico Petrobrás, 12, 21–86.

Schaller, H., D.N. Vasconcelos and J.C. Castro, 1971, Estratigrafia preliminar da bacia sedimentar da Foz do Rio Amazonas, 25° Congr. Bras. Geol., São Paulo (SP), Anais, 3, 189–202.

Schobbenhaus, C., 1996, As tafrogêneses superpostas Espinhaço e Santo Onofre, Estado da Bahia: revisão e novas propostas, Rev. Brasileira Geociências, 26, 265–276.

Schwarz, E., 1999, Modelo deposicional de uma plataforma mixta silicoclástica-carbonática afectada por olas y mareas: um ejemplo del Valanginiano del norte de la Província del Neuquén, Cuenca Neuquina, Argentina, Bol. 5° Simp. Cretáceo Brasil, 1° Simp. Cretácico América del Sur, Serra Negra (SP, Brasil), pp. 175–180.

Schwarzacher, W. and A.G. Fischer, 1982, Limestone-shale bedding and perturbations of the earth's orbit, in: G. Einsele and A. Seilacher (eds.), Cyclic and Event Stratification, Springer-Verlag, Berlin, pp. 72–95.

Scotese, C.R., 1998, Quicktime computer animations, PALEOMAP Project, Department of Geology, Arlington University, Texas at Arlington.

Scott, R.W., 1990, Chronostratigraphy of Cretaceous carbonate shelf, southeastern Arabia, in: A. Robertson, A. Searle and A. Ries (eds.), The Geology and Tectonics of the Oman Region, Geol. Soc. London Spec. Publ. 49, pp. 89–108.

Seilacher, A., 1969, Fault graded beds interpreted as seismites, Sedimentology, 13, 155–159.

Seilacher, A., 1984, Sedimentary structures tentatively attributed to seismic events, Marine Geol., 55, 1–12.

Selley, R.C., 1978, Ancient sedimentary environments and their subsurface diagnosis, Chapman and Hall, London, 317 pp.

Selley, R.C., 1988, Applied Sedimentology, Academic Press, London, 446 pp. [Chapter 10, pp. 403–429].

Semenov, Yu.P., 1965, About some peculiarities of bottomset beds of the East Siberian and Chukchi Seas, in: Anthropogene in the Arctic and Sub-Arctic, Nauka, Moscow, pp. 350–353 (in Russian).

Semikhatov, M.A. and S.N. Serebryakov, 1983, The Siberian hypostratotype of the Riphean, Nauka, Moscow, 204 pp. (in Russian).

Semikhatov, M.A., et al., 1991, New stratigraphic scale of Precambrian of the USSR, Izvestiya Akad. Nauk SSSR, ser. geol., 4, 3–13 (in Russian).

Semikhatov, M.A., G.V. Ovchinnikova, I.M. Gorokhov, et al., 2003, Pb-Pb isochrone age and Sr-isotope characteristic of the Upper Yudomian carbonate deposits (the Vendian of the Yudoma-Maiskiy trough, East Siberia), DAN RF, 39, 83–87 (in Russian).

Sen, S.N. and C.H. Narasima Rao, 1967, Igneous activity in Cuddapah Basin and adjacent areas and suggestion on the palaeogeography of the basin, Hyderabad (A.P.), Proc. Symp. UMP GRB and NGRI; Publ. 8, 261–285.

Seyfert, C.K. and L.A. Sirkin, 1979, Earth History and Plate Tectonics, 2nd ed., Harper and Raw, Bonn, 600 pp.

Shanmugan, G. and R.J. Moiola, 1995, Reinterpretation of depositional processes in a classic flysch sequence in the Pennsylvanian Jackfork Group, Ouachita Mountains, Am. Assoc. Petroleum Geol. Bull., 79, 672–695.

Sharief, F.A., 1983, Permian and Triassic geological history and tectonics of the Middle East, J. Petrol. Geol., 6, 95–102.

Sherwood, K.W. and 18 co-authors, 1998, Undiscovered oil and gas resources, Alaska Federal Offshore, as of January 1995, U.S. Dept. Interior, Mineral Management Service, Alaska OCS Region, Anchorage, Alaska (OCS Monograph MMS 98-0054), 382 pp. + appendices.

Sherwood, K.W. and 6 co-authors, 2002, Structure and stratigraphy of the Hanna Trough, U.S. Chukchi shelf, Alaska, in: E.L. Miller, A. Grantz and S.L. Klemperer (eds.), Tectonic Evolution of the Bering Shelf – Chukchi Sea – Arctic Margin and Adjacent Landmasses, Boulder (Co), Geol. Soc. Am. Special Paper 360, pp. 39–66.

Shiki, T. and T. Yamazaki, 1996, Tsunami-induced conglomerates in Miocene upper bathyal deposits, Chita Peninsula, Central Japan. Sedim. Geol., 104, 175–188.

Silant'ev, S.A, O.G. Bogdanovsky, L.A. Savostin and N.N. Kononkova, 1991, Magmatism of the De Long Archipelago (East arctic), Petrography and petrochemistry of effusives and associated xenoliths (Zokhov and Vil'kitsky Islands), Geochemistry, 2, 267–277 (in Russian).

Silva, O.B., 1988, Revisão estratigráfica das Bacia de Solimões, 35° Congr. Brasileiro Geol., Belém (PA), Anais, 6, 2428–2438.

Silva Filho, M.A., 1995, A Província Borborema: nova abordagem geotectônica, Bol. Núcleo Nordeste Soc. Brasileira Geol., 14 (2); Atas 16° Simp. Geol. Nordeste, Recife (PE); res. expand., 384–389.

Silva Filho, M.A., L.F.C. Bonfim and R.A. Santos, 1978a, A geossinclinal Sergipana: estratigrafia, estrutura, evolução, 30° Congr. Brasileiro Geol., Recife (PE), Anais, 6, 2464–2477.

Silva Filho, M.A., A.C. Santana and L.F.C. Bonfim, 1978b, Evolução tectono-sedimentar do Grupo Estância: suas correlações, 30° Congr. Brasileiro Geol., Recife (PE), Anais, 2, 685–699.

Simanenko, V.P., V.V. Golozubov and A.I. Malinovsky, 2004, Early Cretaceous basalts of the Kema terrane – Moneron-Samarga island arc system fragment. Tikhookeanskaya Geol., 23 (2) 30–51 (in Russian).

Simpson, E.L. and K.A. Eriksson, 1993, Thin eolianites interbedded within a fluvial and marine succession: Early Proterozoic Whitworth Formation, Mt. Isa Inlier, Australia, Sedim. Geol., 87, 39–62.

Sims, J.D., 1975, Determining earthquake recurrence intervals from deformational structures in young lacustrine sediments, Tectonophysics, 29, 141–152.

Singh, I.B., 1980, Precambrian sedimentary sequences of India: their peculiarities and comparison with modern sediments, Precambr. Res., 12, 411–436.

Sinha, A.K. and I. Zietz, 1982, Geophysical and geochemical evidence for a Hercynian Magmatic arc, Maryland to Georgia, Geology, 10, 593–596.

Sistini, G., 1984, Tectonic and sedimentary history of the NE African margin (Egypt-Lybia), in: J.E. Dixon and A.H.F. Robertson (eds.), The Geologic Evolution of the Eastern Mediterranean, Geol. Soc. London Spec. Publ. 17, pp. 161–176.

Sixsmith, P.J., S.S. Flint, H.deV. Wickens and S.D. Johnson, 2004 Anatomy and stratigraphic development of a basin floor turbidite system in the Laingsburg Formation, main Karoo Basin, South Africa. J. Sedim. Res., 74, 239–254.

Slobodin, V.Y., B.I. Kim, G.V. Stepanova and F.Y. Kovalenko, 1990, Anoui borehole sequence by new biostratigraphic data, in: Mesozoic-Cenozoic Stratigraphy and Paleontology of the Russian Arctic, pp. 43–58 (in Russian).

Sloss, L.L., 1963, Sequences in the cratonic interior of North America, Geol. Soc. Am. Bull., 74, 93–114.

Sloss, L.L., 1964, Tectonic cycles of the North American craton, Kansas Geol. Survey Bull., 169, 449–460.
Sloss, L.L., 1972, Synchrony of Phanerozoic sedimentary-tectonic cycles of the North American craton and the Russian platform, 24th Int. Geol. Congr., 6, 24–32.
Sloss, L.L., 1976, Areas and volumes of cratonic sediments, western North America and eastern Europe, Geology, 4, 272–276.
Sloss, L.L., 1984, Comparative anatomy of cratonic unconformities, in: J.S. Schee (ed.), Interregional unconformities and hydrocarbon accumulation, Am. Assoc. Petroleum Geol., Memoir 36, pp. 1–6.
Sloss, L.L. and R.C. Spped, 1974, Relationships of cratonic and continental margin tectonic episodes, in: W.R. Dickinson (ed.), Tectonics and Sedimentation, Soc. Econ. Paleont. Min., Spec. Publ. 22, pp. 98–119.
Smith, N.D., 1974, Sedimentology and bar formation in the upper Kicking Horse river, a braided outwash stream, J. Geol., 81, 205–223.
Smith, R. and R. Swart, 2002, Changing fluvial environments in response to climatic drying in a mid-Triassic rift valley: the Omingonde Formation (Karoo Supergroup) of Namibia, 16th Int. Sedim. Congr., Johannesburg (South Africa), Abstracts, 336–337.
Smith, R.A., 1984, The lithostratigraphy of the Karoo Supergroup in Botswana, Geol. Survey Botswana, Bull. 26, 239 pp.
Smosna, R., 1988, Paleogeographic reconstruction of the Lower Devonian Helderberg Group in the Appalachian Basin, in: N.J. McMillan, A.F. Embry and G.J. Glass (eds.), Devonian of the world, Proc. 2nd Int. Symp. Devonian system; Canadian Soc. Petr. Geol., Memoir 14 (1), pp. 265–275.
Soares, P.C., P.M.B. Landim and V.J. Fúlfaro, 1974, Avaliação preliminar da evolução geotectônica das bacias intracontinentais brasileiras, 28° Congr. Brasileiro Geol., Porto Alegre (RS), Anais, 4, 61–84.
Sokolov, B.S. and M.A. Fedonkin, 1989, Early stages of life's evolution on the Earth, in: Sovremennaya Paleontologiya, Vol. 2, Nedra, Moscow, pp. 118–141 (in Russian).
Spalletti, L.A., 2000, Evolución paleoambiental de las cuencas del cono sur, 2° Congr. Latinoamericano Sedim., 8ª Reunión Argentina Sedim., Mar del Plata (Argentina); Resumenes, 170.
Spalletti, L. and C. Barrio, 1998, Channel-flood plain architecture and sequences: Triassic fluvial systems of Western Argentina, 15th Int. Sedim. Congr., Alicante (Spain), Abstracts, 733–734.
Spalletti, L.A. and J.R. Franzese, 1996, Evolution of southern South American basins during the Mesozoic. IGCP Project 381 – South Atlantic Mesozoic Correlations; SAMC News, 5, 60–62.

Spalletti, L., J. Franzese, S. Matheos and E. Schwarz, 1997, Estratigrafia secuencial de la rampa Tithoniana-Berriasiana del sur de la Cuenca Neuquina, Argentina, 1° Congr. Latinoamericano Sedim., Porlamar, Isla Margarita (Venezuela); Memorias, 2, 271–276.
Spalletti, L.A., J.R. Franzese, D. MacDonald and I.G. Pérez, 1999a, Palaeogeographic evolution of southern South America during the Cretaceous, Bol. 5° Simp. Cretáceo Brasil, 1° Simp. Cretácico América del Sur, Serra Negra (SP, Brasil), pp. 87–95.
Spalletti, L., G. Veiga, Z. Gasparini, E. Schwarz, M. Fernández and S. Matheous, 1999b, La rampa marina de la transición Jurásico-Cretácico em la Cuenca Neuquina (Argentina): facies anóxicas, procesos deposicionales y herpetofauna, Bol. 5° Simp. Cretáceo Brasil, 1° Simp. Cretácico América del Sur, Serra Negra (SP, Brasil), pp. 345–348.
Spalletti, L., J. Franzese, D. MacDonald and I.G. Pérez, 2002, Mesozoic paleogeography of south-western Gondwana basins, 16th Int. Sedim. Congr., Johannesburg (South Africa), Abstracts, 342.
Stacey, J.S. and C.E. Hedge, 1983, Direct evidence for Early Proterozoic crust in the eastern Arabian Shield, DMMR Open File Report, U.S. Geol. Survey – OF-03-63.
Steel, R.J. and T.G. Gloppen, 1980, Late Caledonian (Devonian) basin formation, western Norway – Science of strike-slip tectonics during infilling, in: H.G. Reading and P.F. Ballance (eds.), Sedimentation in Oblique Strike-Slip Mobile Zones, Int. Assoc. Sedim., Spec. Publ. 4, pp. 79–103.
Steininger, F.F., C. Müller and F. Rögl, 1988, Correlation of Central Paratethys, Eastern Paratethys, and Mediterranean Neogene stages, in: L.H. Royden and F. Horváth (eds.), The Pannonian Basin. A Study in Basin Evolution, Am. Assoc. Petroleum Geol., Memoir 45, pp. 79–87.
Stille, H., 1924, Grundfragen der vergleichenden Tektonik. Gebr. Borntraeger, Berlin, 433 pp.
Stoeser, D.B. and V. Camp, 1984, Pan-African microplate accretion of the Arabian Shield, U.S. Geol. Survey Mission, DGMR, TR4-1-R-670.
Stolbov, N.M., 1997, Discontinuity of the basite magmatism on Franz Jozef Land Archipelago, in: Actual Geological Problems of Outer Continental Shelf Boundary, VNIIOkeangeologia, St. Petersburg, Abstracts, 64–65 (in Russian).
Stolbov, N.M., 2000, Specific features of the Franz Josef Land Archipelago magmatism as reflection of its peculiarities in geodynamics, in: Geological and Geophysical Characteristics of the Arctic Region Lithosphere, fasc. 3, VNIIOkeangeologia, St. Petersburg, pp. 137–144 (in Russian).
Stollhofen, H. and I.G. Stainstreet, 1997, A record of synsedimentary tectonism during early South Atlantic Rift evolution: Carboniferous-Permian to Juras-

sic Karoo, Namibia, GAEA heidelbergensis, 3; 18th IAS Regional European Meeting Sedim., Heidelberg (Germany), Abstracts, 322.

Stow, D.A.V., J.-C. Faugeres, A. Viana and E. Gauthier, 1998, Fossil contourites: a critical review, Sedim. Geol., 115, 3–39.

Strogen, P. and C.J. O'Byrne, 2000, Field guide to the Mid-Cambrian Bray Group, Bray Head, County Wicklow, Ireland, in: J.R. Graham and A. Ryan (eds.), Filed Trip Guidebook, Vol. 5, 20th IAS Regional Meeting Sedim., Dublin (Ireland), pp. 113–128.

Suess, E., 1883–1908, Das Antlitz der Erde (3 vols.), F. Tempsky, Wien.

Suprunenko, O.I. and E.G. Bro, 1994, Possible objectives for oil fileds prospecting on the Barents-Kara shelf, in: Investors Assoc. Development Northwest Russia Prospect for oil and gas development-progress in Scient, Res. Exploration Legal Regulation, Seminar St. Petersburg (Russia), Oil Data Inc., Houston, pp. 89–105.

Surnin, A.A., A.V. Okrugin and A. I. Zaitsev, 1998, Deep xenolite in the East Yakutia basalts, Domestic Geol., 6, 44–48 (in Russian).

Sutton, J., 1963, Long-term cycles in the evolution of the continents, Nature, 198, 731–735.

Szatmari, P., J.C.J. Conceição, M.C. Lana, E.J. Milani and A.P. Lobo, 1984, Mecanismo tectônico do rifteamento sulatlântico, 33° Congr. Brasileiro Geol., Rio de Janeiro (RJ), Anais, 4, 1589–1601.

Szatmari, P., J.B.L. Françolin, O. Zanotto and S. Wolff, 1987, Evolução tectônica da margem equatorial brasileira, Rev. Brasileira Geociências, 17, 180–188.

Tankard, A.J., 1986, Depositional response to foreland deformation in the Carboniferous of eastern Kentucky, Am. Assoc. Petroleum Geol. Bull., 70, 853–868.

Tari, G., F. Horváth and J. Rumpler, 1992, Styles of extension in the Pannonian Basin, Tectonophysics, 208, 203–219.

Thomas, W.T., 1977, Evolution of Appalachian-Ouachita salients and recesses from reentrants and promontories on the Continental margin, Am. J. Sci., 277, 1233–1278.

Thomaz Filho, A., U.G. Cordani and O. Marino, 1974, Idades K/Ar de rochas basálticas da Bacia Amazônica e sua significação tectônica regional, 28° Congr. Brasileiro Geol., Porto Alegre (RS), Anais, 6, 273–278.

Thurston, D.K. and L.A. Theis, 1987, Geologic Report for the Chukchi Planning Area, Alaska. Regional Geology, Petroleum Geology, and Environmental Geology, U.S. Dept. Interior MMS, Alaska OSC region, Anchorage (Alaska), 193 pp.

Tolson, R.B., 1987, Structure and stratigraphy of the Hope Basin, southern Chukchi Sea, Alaska, in: Circum-Pacific Council for Energy and Mineral Resources, Earth Science Series, 6; Chapter 4, pp. 59–71.

Tomljenović, B. and L. Csontos, 2001, Neogene-Quaternary structures in the border zone between Alps, Dinarides and Pannonian Basin (Hrvatsko zagorje and Karlovac Basin, Croatia), Int. J. Earth Sci., 90, 560–578.
Tucker, M., 1998, Mixed carbonate-clastic high frequency sequences from the mid-Carboniferous of northern England, 15th Int. Sedim. Congr., Alicante (Spain), Abstracts, 774.
Turbay, C.V.G., P. Cesero, E.A.M. Koutsoukos and M.A. Rodrigues, 2004, Upper Albian carbonate microfacies from the Sergipe Basin, Northeastern Brazil, 23rd IAS Meeting Sedim., Coimbra (Portugal), Abstracts Book, 270.
Turcotte, D.L. and E.R. Oxburgh, 1973, Midplate tectonics, Nature, 244, 337–339.
Uesugui, N., R. Dino and E. Pedrão, 1994, The Acre Basin, in: G. Beurlen, D.A. Campos and M.C. Viviers (eds.), Stratigraphic Range of Cretaceous Mega- and Microfossils of Brazil, Inst. Geociências, Univ. Fed. Rio de Janeiro, Rio de Janeiro (RJ), pp. 400–407 [Chapter 4.5].
Uliana, M.A. and K.T. Biddle, 1988, Mesozoic-Cenozoic paleogeographic and geodynamic evolution of southern South America, Rev. Brasileira Geociências, 18, 172–190.
Ulicny, D., 1999, Sequence stratigraphy of the Dakota Formation (Cenomanian), southern Utah: interplay of eustasy and tectonics in a foreland basin, Sedimentology, 46, 807–836.
Utkin, V.P., 1980, Strike-Slip Dislocations and the Technique of Their Investigation, Nedra, Moscow, 143 pp. (in Russian).
Utkin, V.P., 1989, Strike-Slip Dislocations, Magmatic Activity and Ore Formation, Nedra, Moscow, 166 pp. (in Russian).
Vail, J.R., 1984, Pan-African crustal accretion in NE Africa, J. African Earth Sci., 1, 285–294.
Vail, P.R., 1987, Seismic stratigraphy interpretation using sequence stratigraphy, Part 1: Seismic stratigraphy interpretation procedure, in: A.W. Bally (ed.), Atlas of Seismic Stratigraphy, Am. Assoc. Petroleum Geol., Studies in Geol., 27, 1–10.
Vail, P.R., R.M. Mitchum Jr. and S. Thompson III, 1977, Seismic stratigraphy and global changes of sea level, Part 4: Global cycles of relative changes of sea level, in: C.E. Peyton (ed.), Seismic Stratigraphy – Applications to Hydrocarbon Exploration, Am. Assoc. Petroleum Geol., Memoir 26, pp. 83–97.
Vail, P.R., F. Audenard, S.A. Bowman, P.N. Eisner and C. Perez-Cruz, 1991, The stratigraphic signatures of tectonics, eustasy and sedimentology – an overview, in: G. Einsele, W. Ricken and A. Seilacher (eds.), Cycles and Events in Stratigraphy, Springer-Verlag, Berlin, Chapter 6, Sequences, part 6.1, 617–659.

Vakarcs, G., P.R. Vail, G. Tari, Gy. Pogácsás, R.E. Mattick and A. Szabó, 1994, Third-order Middle Miocene-Early pliocene depositional sequences in the prograding delta complex of the Pannonian Basin, Tectonophysics, 240, 81–106.

Vakarcs, G., J. Hardenbol, V.S. Abreu, P.R. Vail, P. Várnai and G. Tari, 1998, Oligocene-Middle Miocene depositional sequences of the Central Paratethys and their correlation with regionals tages, in: P.-C. de Graciansky, J. Hardenbol, T. Jacquin and P.R. Vail (eds.), Mesozoic and Cenozoic Sequence Stratigraphy of European Basins, SEPM (Soc. Sedim. Geol.), Spec. Publ. 60, pp. 209–231.

Valentine, J.W. and E.M. Moores, 1974, Plate tectonics and the history of life in the oceans, Sci. Am., 230 (4) 80–89.

Van de Kamp, P.C., 1973, Holocene continental sedimentation in the Salton Basin, California: a reconnaissance, Geol. Soc. Am. Bull., 84, 827–848.

Van der Vlerk, I.M. and Ph.H. Kuenen, 1948, Geheimschrift der Aarde, 5th ed., W. de Haan N.V., Utrecht, 370 pp.

Van Staal, C.R., 1994, Brunswick subduction complex in the Canadian Appalachians: Record of Late Ordovician to Late Silurian collision between Laurentia and the Gander margin of Avalon, Tectonics, 13, 946–962.

Van Staal, C.R. and J A. de Roo, 1995, Mid-Paleozoic tectonic evolution of the Appalachian central mobile belt in Northern New Brunswick, Canada: Collision, extensional collapse and dextral transpression, in: J.P. Hibbard, C.R. Van Staal and P.A. Cawood (eds.), Current Perspectives in the Appalachian-Caledonian Orogen, Geol. Assoc. Can. Spec. Paper 41, 367–389.

Van Wagoner, J.C., R.M. Mitchum, H.W. Posamentier and P.R. Vail, 1987, Seismic stratigraphy using sequence stratigraphy, Part 2: Key definitions of sequence stratigraphy, in: A.W. Bally (ed.), Atlas of seismic stratigraphy, Am. Assoc. Petroleum Geol., Studies in Geol., 27, 11–14.

Vaslet, D., 1990, Upper Ordovician glacial deposits in Saudi Arabia, Episodes, 13, 147–161.

Veiga, G.D., 1998, Base level control on the development of continental sequences: the Middle Jurassic Challacó Formation (Neuquén Basin, Argentina), 15th Int. Sedim. Congr., Alicante (Spain), Abstracts, 791–792.

Veiga, G.D. and L.A. Spalletti, 2000, Sedimentology of the aeolian to fluvial transition in the Avalé Memeber of the Agrio Formation (Lower Cretaceous), northern Neuquén Province, Argentina, 2° Congr. Latinoamericano Sedim., 8ª Reunión Argentina Sedim., Mar del Plata (Argentina); Resumenes, 178–179.

Velić, J., J. Tišljar, I. Dragičević and I. Blašković, 2000, Shoreline cross-bedded biocalcarenites (Middle Miocene) in the Podvrško-Šnjegavić area on the slope of Mt. Psunj and their petroleum signification (Povžega subdepression – Eastern Croatia), Geol. Croatica, 53, 281–293.

Viana, C.F., E.G. Gama Jr., I.A. Simões, J.A. Moura, J.R. Fonseca and R.J. Alves, 1971, Revisão estratigráfica da Bacia Recôncavo/Tucano, Bol. Técnico Petrobrás, 14, 157–192.

Visser, J.N.J., 1986, Lateral lithofacies relatonships in the glacigene Dwyka Formation in the western and central parts of the Karoo Basin, Trans. Geol. Soc. South Africa, 89, 373–383.

Visser, J.N.J., 1987, The palaeogeography of part of southwestern Gondwana during the Permo-Carboniferous glaciation, Palaeogeogr., Palaeoclim., Palaeoecol., 61, 205–219.

Vol'nov, D.A. and V.A. Litinsky, 1976, Structural and tectonic zonation of the Laptev and East Siberian shelf seas, in: Shelf geology of the East Siberian seas, NIIGA, Leningrad, 8–22 (in Russian).

Vol'nov, D.A., V.N. Voitsekhovsky, O.N. Ivanov, D.S. Sorokov and D.S. Yashin, 1970, New Siberian Islands, in: Geology of the USSR, Vol. XXVI – Soviet Arctic Islands, Nedra, Moscow, pp. 324–374.

Von Bubnoff, S., 1948, Rhythmen, Zyklen und Zeitrechnung in der Geologie, Geol. Rundschau, 35, 6–22.

Vrsaljko, D., 1999, The Pannonian paleoecology and biostratigraphy of molluscs from Kostanjek-Medvednica Mt., Croatia, Geol. Croatica, 52, 9–27.

Walcoot, R.I., 1970, Isostatic response to loading of the crust in Canada, Can. J. Earth Sci., 7, 2–13.

Watts, A.B., G.D. Karner and M.S. Steckler, 1982, Lithospheric flexure and evolution of sedimentary basins, Philos. Trans. Royal Soc., A305, 249–281.

Wehmann, A., R. Brocke, G. Hertweck, U. Jansen, P. Königshof, G. Plodowski, E. Schindler, S. Schultka and V. Wilde, 2000, Intertidal deposits in Lower Devonian siliciclastic sequences of the Mosel region, Germany, 20th IAS Regional Meeting Sedim., Dublin (Ireland), Abstracts, 62.

Weissbrod, T., 1970, The stratigraphy of the Nubian sandstone in southern Israel (Timna-Eilat area), Israel Geol. Survey Report OD/2/70, 22 pp.

Wells, S.G. and A.M. Harvey, 1987, Sedimentologic and geomorphic variations in storm-generated alluvial fans, Howgill Fells, northwest England, Geol. Soc. Am. Bull., 98, 182–198.

Werner, M., I.G. Stanistreet and H. Stollhofen, 2002, Tracing the Whitehill Formation (Karoo Supergroup) into Namibia: a combined litho-, bio-, tephro-, and sequence stratigraphic approach, 16th Int. Sedim. Congr., Johannesburg (South Africa), Abstracts, 399–400.

Willett, S.D., C. Beaumont and P. Fullsack, 1993, Mechanical model for the tectonics of doubly vergent compressional orogens, Geology, 21, 371–374.

Williams, B.P.J., R.J. Sloan and L.K. Richmond, 2000, Shallow marine to fluvial-eolian interaction in tectonically-active basins, Siluro-Devonian (ORS), Dingle Peninsula, SW Ireland, in: J.R. Graham and A. Ryan (eds.), Field Trip Guidebook, Vol. 4, 20th IAS Regional Meeting Sedim., Dublin (Ireland), pp. 71–112.

Wilson, J.T., 1966, Did the Atlantic close and then re-open? Nature, 211, 676–681.

Wilson, J.T., 1968, Static or mobile earth: the current scientific revolution, Am. Philos. Soc. Proc., 112, 309–320.

Wilson, J.T., 1970, Continental drift, transcurrent and transform faulting, in: A.E. Maxwell (ed.), The Sea, Vol. 4, Part II, John Wiley & Sons, New York, pp. 623–644.

Wilson, J.T., 1974, The life cycle of ocean basins. Stages of growth and stages of decline, in: J.A. Jacobs, R.D. Russell and J.T. Wilson (eds.), Physics and Geology, 2nd ed., McGraw Hill, New York, pp. 397–470.

Windley, B.F., 1984, The Evolving Continents, 2nd ed., John Wiley & Sons, New York, 399 pp.

Wood, G.V. and M.J. Wolfe, 1969, Sabkha cycles in the Arab/Darb Formation off the Trucial coast of Arabia, Sedimentology, 12, 165–191.

Worsley, W.R., D. Nance and J.B. Moody, 1984, Global tectonics and eustasy for the past 2 billion years, Marine Geol., 58, 373–400.

Worsley, W.R., J.B. Moody and D. Nance, 1985, Proterozoic to Recent tectonic tuning of biogeochemical cycles, in: E.T. Sundquist and W.S. Broecker (eds.), The carbon cycle and atmospheric CO_2: natural variations Archean to present, Am. Geophys. Union Geophys. Mon., 32, 561–572.

Worsley, W.R., D. Nance and J.B. Moody, 1986, Tectonic cycles and the history of the earth's biogeochemical and paleoceanographic record, Paleoceanography, 1, 233–263.

Wuestefeld, H., R. Hoffmann, W. Meyer, C. Ploschenz and J. Stets, 1997, Qunatification of crustal dynamics – the Quaternary uplift of the Rhenish Massif, GAEA heidelbergensis, 3; 18th IAS Regional European Meeting Sedim., Abstracts, 368.

Yeh Lien-tsun, 1977, Metallogenic Epochs of Sedimentary Ores and Their Significance in Historical Geology, Academia Sinica, Peking, 10 pp.

Zagwijn, W.H. and J.W.C. Doppert, 1978, Upper Cenozoic of the southern North Sea Basin: palaeoclimatic and palaeogeographic evolution, Geol. Mijnbouw, 57, 577–588.

Zakharov, V.A., N.I. Kurushin and V.P. Pochialaynen, 1996, Paleogeographical criteria of the geodynamics of the North-East Asia terranes during the Mesozoic, Geol. Geophys., 37 (11) 3–22 (in Russian).

Zakharov, Y.D., 1997, Ammonoid evolution and the problem of the stage and substage division of the Lower Triassic, Mem. Géol., Lausanne, 30, 121–136.
Zakharov, Y.D. and E.I. Shkolnik, 1994, Permian-Triassic Cephalopod facies and global phosphatogenesis, Mem. Géol., Lausanne, 22, 121–182.
Zaporozhtseva, A.S., 1958, Tufogenic argillaceous strata of the Usi-Enisei area and its origin, Trans. NIIGA, 67 (7) 104–142 (in Russian).
Zeller, E.J., 1964, Cycles and psychology, Kansas Geol. Survey Bull., 169(2) 631–636.
Zhao Chujing and Li Zhitong, 1988, Geological and tectonical peculiarity development of the eastern part of Jilin and Heilongjiang provinces, China, in the Triassic. Blagoveschenk, Glubbinoe stroeniye Tikhogo okeana i ego continentalnogo obramlenya; Tezisy dokladov, Chast'l, 83–85.
Zhemchuzhnikov, Yu.A., 1958, Similarities and differences between facies, facies-cyclic, and facies-tectonic methods of the study of the Coal Measures, Izv. Acad. Sci. USSR, Geol. Ser., 1, 1–7 (English translation).
Ziegler, P.A., 1978, North-Western Europe: tectonics and basin development, Geol. Mijnbouw, 57, 589–626.
Ziegler, P.A., 1989, Evolution of Laurussia, Kluwer Academic Publishers, Dordrecht, 102 pp.
Zielinski, T., 1982, Contemporary high-energy flows, their deposits and reference to the outwash depositional model, Prace Naukoe Uniw. Slaskiego, NR 519, Geologia, 6, 98–110.
Zwart, H.J. and U.F. Dornsiepen, 1978, The tectonic framework of Central and Western Europe, Geol. Mijnbouw, 57, 627–654.

INDEX

Pages numbers followed by t and f represent tables and figures respectively.